BIOGAS TECHNOLOGY, TRANSFER AND DIFFUSION

Proceedings of the International Conference held at the National Research Centre, Cairo, Egypt

17–24 November 1984

on

Biogas Technology, Transfer and Diffusion: State of the Art

BIOGAS TECHNOLOGY, TRANSFER AND DIFFUSION

Edited by

M. M. EL-HALWAGI

National Research Centre, Dokki, Cairo, Egypt

ELSEVIER APPLIED SCIENCE PUBLISHERS
LONDON and NEW YORK

ELSEVIER APPLIED SCIENCE PUBLISHERS LTD
Crown House, Linton Road, Barking, Essex IG11 8JU, England

Sole Distributor in the USA and Canada
ELSEVIER SCIENCE PUBLISHING CO., INC.
52 Vanderbilt Avenue, New York, NY 10017, USA

WITH 156 TABLES AND 193 ILLUSTRATIONS

British Library Cataloguing in Publication Data

Biogas technology, transfer and diffusion.
1. Methane
I. El-Halwagi, M. M.
665.7'76 TP761.P14

Library of Congress Cataloging in Publication Data

International Conference of the State of the Art on Biogas Technology, Transfer and Diffusion (1984: Cairo, Egypt)
Biogas technology, transfer and diffusion.

Includes bibliographies and index.
1. Biogas—Congresses. I. El-Halwagi, M. M.
II. Title.
TP359.B48157 1984 665.7'76 86-2133

ISBN 1-85166-000-3

Printed in Great Britain by Galliard (Printers) Ltd, Great Yarmouth

PREFACE

The International Conference on the State of the Art on Biogas Technology, Transfer and Diffusion was held in Cairo, Egypt, from 17 to 24 November 1984.

The Conference was organized by the Egyptian Academy of Scientific Research and Technology (ASRT), the Egyptian National Research Centre (NRC), the Bioenergy Systems and Technology project (BST) of the US Agency for International Development (US/AID) Office of Energy, and the National Academy of Sciences (NAS). A number of international organizations and agencies co-sponsored the Conference. More than 100 participants from 40 countries attended.

The purpose of the Conference was to assess the viability of biogas technology (BGT) and propose future courses of action for exploiting BGT prospects to the fullest extent.

The Conference emphasized a balanced coverage of technical, environmental, social, economic and organizational aspects relevant to biogas systems design, operation and diffusion. It was organized to incorporate experiences that are pertinent, for the most part, to developing countries. In addition to the wide spectrum of presentations and country programs, structured and non-structured discussions among the participants were strongly encouraged in thematic sessions at round-table discussions, and through personal contacts during poster sessions and field trips. It was clear from the enthusiastic response of most participants that the Conference, in large measure, succeeded in fulfilling its mission. Although draft papers were distributed to all participants, it was felt that the results obtained were worthy of organized and refined documentation. And this is precisely what this book intends to do.

One of the important goals of the Cairo Conference was to identify a set of guiding principles for the most promising future investment opportunities relevant to BGT in developing countries. Evidence presented at the Conference seems to indicate that BGT can be applied with the expectation of success when, for instance:

- Its potential is extended beyond the renewable energy base by better demonstrating the system multiple outputs of prospective market value including sanitation, energy, fertilizer and animal feed supplements.
- It is used as an adjunct to agricultural-processing enterprises such as relatively large animal/animal product operations.
- Its application can affect management improvement in an existing activity with resultant economic payoffs. The reçuired institutional and financial support for successful applications of these technologies is more often available at the commercial or cooperative enterprise scale than in support of smaller-scale operations. Developing such infrastructure at the smaller scale requires considerable government involvement and long periods of time to mature and would therefore be prohibitively difficult and expensive.

Throughout the pre- and post-Conference periods, many persons contributed much effort, help and support. I am greatly indebted to each of them. In particular, my special thanks go to those of the staff of the Pilot Plant Laboratory of the NRC who extended their unlimited assistance under the capable leadership of Dr A. Abdel Dayem and Dr M. A. Hamad; to Dr Paul Weatherly, Mrs Betsy Amin-Arsala and other members of the BST staff; to Mr Jay Davenport, Mr A. Nasmith, Mrs Maryalice Risdon and Miss F. R. Ruskin of the US National Academy of Sciences (NAS); to the NAS biogas project panel: Dr P. Goodrich from the University of Minnesota, Dr H. Capener from Cornell University, and Dr T. Prakasam from the Metropolitan Sanitary District of Greater Chicago; to Dr D. C. Stuckey from Imperial College, London, and Dr R. Mah from the University of California, Los Angeles. Finally, I should like to express my profound appreciation to Dr M. Kamel, President of the Egyptian Academy of Scientific Research and Technology and former President of the Egyptian NRC, for all his support and encouragement.

M. M. El-Halwagi
Editor and Conference Organizer

CONTENTS

Section 3: Sociocultural Aspects of Biogas Technology

Section 4: Economic Aspects

Section 5: Institutional and Financial Infrastructure

Section 6: Regional Programs, Networks, and Aid Agencies

Section 7: Technical Aspects

SECTION ONE

INTRODUCTION AND SUMMARY

INTRODUCTION

M.M. El-Halwagi
National Research Centre, Egypt

The overall objective of the conference "State of the Art On Biogas Technology Transfer and Diffusion" was to determine whether biogas technology (BGT) is feasible and worth pursuing, particularly for rural areas of developing countries. If so, what future direction should we take, building on our vast store of past experiences that include successes as well as failures?

In order for the strength of technological innovations like BGT to be assessed, their use in the target social population should be examined. The success of these innovations, that is, their wide-spread adoption, entails a "fit" between the innovation and the user needs, expectations, and perceptions, as well as the cost-effectiveness of the technology under prevailing conditions and constraints. The extent of the achievable success will depend upon the "goodness of fit" that can be realized.

BGT is a very controversial technology. Some believe it is a viable proposition and have given unqualified praise to the technology. Others, however, have found it to be a disappointing experience; and some reputable authorities have even completely condemned it. Generally, however, the tone has been recently coloured with subdued enthusiasm. The following two recent statements reflect this controversy.

Michael Santerre and Kirk Smith from the East West Center (1982) state:

"Anaerobic digesters have a potential for providing fuel, fertilizer and a sanitary means of waste disposal in rural areas of less developed countries (LDC's). Despite these potential benefits, digesters have had a disappointingly low success rate in many LDC's. Poor economics may explain these failures in some cases but poor fits between digesters and local conditions - a lack of appropriateness - can also be a useful indicator".

Vaclav Smil in his book Biomass Energies, (Smil, 1983) states:

"The technique is proven, the everyday operation can be made to work fairly efficiently, the benefits are aplenty, and the need for fuel is beyond any doubt. But putting the package together - and keeping it so is a taller order than a casual look would indicate"

Why this controversy? What are the characteristics of BGT that lead to discrepant assessments?

First, BGT is not a simple system; it is not merely a digester. Rather it is an integrated sequence of processes involving waste feed management, digestion, and effluent and gas handling and usage as depicted in Figure 1. These aspects of BGT will be addressed in the technical session.

Second, biogas is not merely a technology. It is a system that must fit with other systems and environments. For example as shown in Figure 2, a rural biogas system must integrate with the Household and Farming systems within the Village Surroundings. What makes a successful fit between the biogas system and the user-village setting, will be addressed in the socio-cultural session.

Third, to be effective, BGT must meet a real need at an affordable cost. Thus, subsistence peasants have no perceptible need for BGT, nor do they have the financial means for its implementation.

There is no simple formula for integrating BGT into a community or a system so that it meets real perceptible needs at an affordable cost. Many coordinated organizational efforts would be required, including strong governmental support and sound national policy that sets the stage for target-specific strategies, promotional programs, rational execution plans, and financial assistance and incentives. Coordinated and cooperative interdisciplinary endeavors are necessary at the research, development, and demonstration stages. Technical and social disciplines must work together. These issues will be addressed in the special session on financial and institutional infrastructure.

In this Conference, let us examine all relevant aspects of issues of BGT. Let us make an unbiased, open-minded search for a fair and valid assessment. Let us also evaluate the future prospects of BGT.

REFERENCES

M.T. Santere and Smith, K.R., 1982. Measures of appropriateness: the resource requirements of anaerobic digestion (biogas) systems. World Development. Resource Systems Institute, East-West Center, Honolulu. 10(3): 239-261.

Smil, V., 1983. Biomass Energies, Plenum Press, New York

Figure 1 THE BIOGAS SYSTEM
A DECEIVINGLY SIMPLE SYSTEM

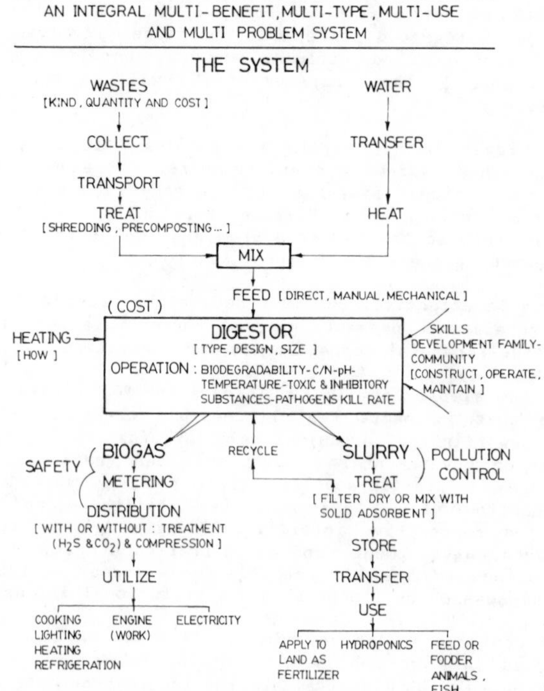

Figure 2 THE BIOGAS SUBSYSTEM
WILL INTEGRATE WITH
THE HOUSEHOLD & FARMING SYSTEMS WITHIN
THE VILLAGE & SURROUNDINGS

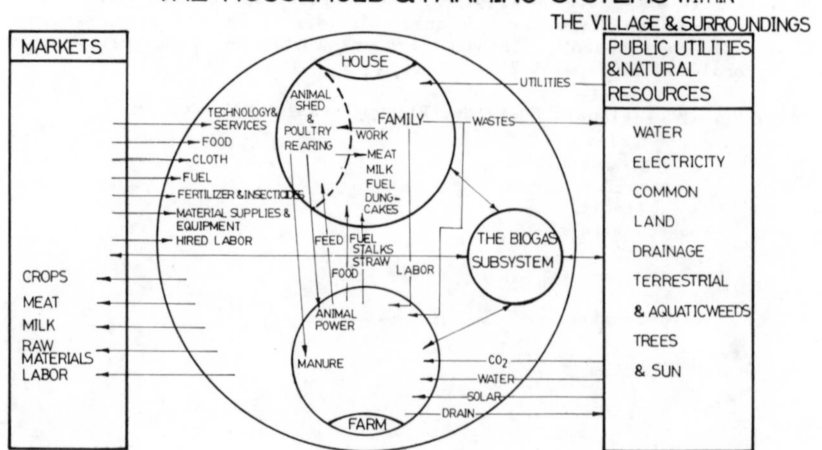

EXECUTIVE SUMMARY

Based on an extensive discussion of all aspects of biogas technology (BGT), transfer, and diffusion during this conference, the following is a summary of the state of the art of BGT and, more importantly, the areas of BGT that require further work. These areas are identified to enhance the diffusion of BGT and to enable funding agencies to identify priorities. For convenience, aspects of BGT are labeled technical, economic, social-institutional, information processing, and dissemination and technology transfer. However, all these aspects of BGT interact, and an integrated approach is needed to optimize the use of BGT in developing countries.

TECHNICAL ASPECTS OF BIOGAS TECHNOLOGY

Regarding, basic research and construction, conferees believed that balanced attention should be given to research ("know") and technology ("know-how"). The following key areas need further work:

Basic Research

Basic research is needed in the areas of microbiology, biochemistry, and advanced reactor design with emphasis on the applied aspects of these disciplines. Strong links should be forged between institutions and research groups of different parts of the world to enable advances in basic understanding to be quickly applied to practical problems.

Batch Digesters

Because of their simple design and operation, the abundance of dry agricultural residues, and the lack of available water in many developing countries, batch (dry) digesters should be investigated in considerably more depth. The production of a relatively dry organic compost obviates the need for handling a wet slurry, which is one drawback of traditional biogas plants. Anaerobic digestion of lignocellulosic materials is only partially understood at present. More research is required in the field of decomposition of solid agricultural wastes via the anaerobic routes. Areas to be investigated include optimum solids loading, seed concentration, time of operation, provision of a constant usable gas supply.

Plug Flow Digesters

More applied research is needed on this design, which has

superior kinetic performance in comparison with conventional designs, apparent lower cost per installed m^3, and ability to operate at higher total solids (TS) concentrations. Various construction materials (rubber and plastic films) should be analysed under appropriate conditions in developing country to assess their durability.

Optimization

Further efforts should be made to optimize digester design and operation based on a sound understanding of fundamental principles and digester construction methods and costs.

Standardization of Data Collection

To enable vigorous comparisons between units in different regions, concentrated effort should be made to standardize the reporting of biogas data throughout the world. Accordingly, the conferees noted the preparation of a manual in Bangkok in May 1983 and agreed to review it in order to prepare a common methodology.

Collection of Baseline Technical Data

Despite considerable information reported in the literature, more data is needed on the performance of existing field units. It is recommended that such data be collected and made freely available to interested parties.

Digestion of Wastes Other than Animal Manure

There is scant information available on full sized digesters for handling agricultural residues, aquatic plants, and industrial wastes. Residues of processing major crops and products are likely sources of feedstock for biogas systems. These include rice hulls, bagasse, coconut shell, etc. Another source is urban solid and sewage wastes. Technologies to treat these wastes and capture methane are increasingly proving cost effective and being employed in developed societies. Applicability of these systems to waste management/energy generation in developing countries offers the opportunity to handle multiple problems with integrated systems application.

Fertilizer Content of the Slurry

Economic analyses of biogas have produced uncertain results. Therefore, studies should be initiated that evaluate the net economic effect of the numerous methods available for the use of manure and residues in agriculture, for example direct application, composting and application, and digestion and application, digestion composting and application.

Heating of Digesters

Both composting and solar (passive and active) methods appear to have potential in raising digester operating temperatures without large capital investments. More information is needed on these systems with regard to optimum design and effect on overall digester performance.

Mixing in Digesters

There is little information available on the effect of mixing on detention time and digester performance. Studies should be undertaken to determine the actual detention time in typical digesters in developing countries and how this time can be improved by the judicious use of mixing.

ECONOMICS OF BIOGAS TECHNOLOGY

The economics of the various technical options of BGT in relation to other renewable energy sources and fossil fuels was of considerable interest to participants. Economics was recognized as one of the primary constraints to the use of BGT, especially for small household-sized units.

Contribution to National Energy Budgets

The economic contribution of biogas to national energy demand is likely to be small (about 1-10%). However, in rural areas it may be substantial (about 50%), and should not be dismissed lightly. BGT could considerably reduce the burden of life in these areas.

The Integrated Use of BGT

The key to the economic viability of BGT is maximizing the use of all its outputs, not just the energy content of the gas. The slurry contains nutrients and organic material, which may be used as animal or fish food and fertilizer. The gas can be used as an energy source for pumping water and powering motors, thereby promoting small-scale village industries and generating a cash income. The sanitation effect and use of the technology as a means of waste disposal and pollution abatement can also provide additional economic benefits. Thus, the potential of BGT can be realized when its place in the food-feed-fuel-sanitation system is better demonstrated.

Financial and Social Viability

The financial viability of biogas plants depends strongly on whether the gas and slurry outputs can be substituted for fuels, fertilizers, or feeds that were previously purchased with money. If so, then the resulting cash savings can be used to repay the capital and maintenance costs, and the plant will likely be viable. However, if the outputs do not generate a cash inflow, or reduce the cash outflow, then the financial viability of these plants worsens considerably. Finally, if broad social criteria are used to evaluate BGT in the form of social cost-benefit analyses (SCBA), the result will be more positive conclusions about viability than if only financial evaluations are performed.

Scale and Area of Application

Because of their small-scale economics and smaller cash flows, small household biogas units appear to be less financially viable than larger community-sized units or units treating wastes from industry or intensive animal feed lots; mostly as an adjunct to agricultural-processing enterprises.

Distortions in Markets

The economics of many developing countries are often characterized by extensive subsidies for energy and fertilizers. Such subsidies induce economic distortions that reduce the optimum use of biogas. In addition, in some rural economies, many alternative fuel sources to biogas are nonmonetized and hence are difficult to incorporate into a financial analysis.

National Policy

Because the financial viability of biogas is, at times, marginal, greater emphasis must be placed on increasing the dissemination of BGT by means of national policy changes that emphasize socioeconomic justifications i.e. "social good".

Lack of Data

There is scant information on which to evaluate the economic viability of biogas in developing countries. This lack of information is particularly acute regarding industrial and feedlot applications. Although the problem is less acute regarding household and community BGT use, most of these data come from India and are derived from theoretical design figures that use the floating cover design and cattle dung as the primary feedstock. Such a narrow data base prevents drawing conclusions about the viability of BGT under other circumstances, such as different designs and feeds in various social and environmental milieus, and in varying areas of application.

Methodology

Methods for performing economic analyses of biogas (e.g., cost-benefit ratios and SCBA) have strong limitations. In addition, it is difficult to quantify secondary benefits, such as improved public health, reduced deforestation, and improved quality of life. Furthermore, there is no commonly agreed on methodology. A consensus needs to be developed which will allow the comparison of economic data among the various applications of BGT and between BGT and other comparable energy technologies or conventional energy sources.

Prognosis

Ultimately, decisions about the economic feasibility of biogas systems must be made by individual developing countries in conjunction with the development of supportive economic policies (subsidies, financial supports etc.). Because of the high cost and limited precision of economic analyses for each individual site or specific application of biogas systems, it may be more realistic that countries should evaluate a limited number of representative sites to more closely define viable economic outcomes.

SOCIAL AND INSTITUTIONAL ASPECTS OF BIOGAS TECHNOLOGY

Current Status

Many developing countries are firmly committed to biogas development. Their programs tend to be at two different levels: (1) the experimental research and development stage, wherein the technology is undergoing testing and replication at field sites, and (2) the extension and diffusion stage, wherein large numbers of digesters are being disseminated. BGT, therefore, is an evolving technology whose importance deserves appropriate international recognition and acknowledgement.

Comparisons of Various Programs

The continued evolution of BGT requires a two-phased approach: (1) continued research and development emphasis on technical problems and (2) appropriate and necessary emphasis on the social and institutional considerations to disseminating BGT. Questions about application of BGT need to be addressed. International research funding is required in order to make comparisons between programs in different countries.

Interdisciplinary Approach to BGT

The dissemination of BGT involves two disciplines: physical and biological sciences and the social sciences. Both are essential. Creating and working successfully with biogas programs requires an effective interdisciplinary team. Organizing and sensitively managing such a team requires special attention and skill and must be recognized as essential to successful biogas development.

Institutional Infrastructure

Dissemination of BGT is occuring rapidly in China, India, Thailand, South Korea, and Taiwan, and rapid expansion is imminent in such countries like Brazil, Costa Rica, Egypt, Guatemala, Mexico, and Nepal. The crucial questions that need to be examined to expedite this expansion are: What are the appropriate forms of social organization? What types of institutional infrastructure are best suited to sponsor, support, and sustain a broad-scale program of biogas development? How can the multiinstitutional, multipurpose, and multidisciplinary features of biogas development be properly managed. Are there appropriate and useful rules for private voluntary organizations, for cooperatives, and for the private sector? Should the administration of a biogas program be placed high enough in a government administration to facilitate coordination between concerned ministries and special interest groups? Addressing these policy-related questions is becoming increasingly necessary for rapid dissemination of BGT. Special funding to pursue answers to these issues is strongly recommended.

It seems, however, that evidences tend to indicate that the required institutional and financial support for successful applications of BGT is more often available at the commercial or cooperative enterprise scales than in support of smaller-scale operations. Developing such infrastructure at the smaller scale requires considerable government involvement and long periods of time to mature and therefore is extremely difficult and expensive.

Implementation of BGT

Digesters have four main areas of use: family, community, feedlots, and industry. The implementation of biogas in each of these areas requires different patterns of management and supervision, different maintenance and operational skills, and different criteria for overall success. Careful attention should be paid to these factors in implementation of BGT.

Dissemination of BGT

There is little understanding of or empirical evidence for the factors likely to affect the dissemination of BGT. Such evidence might include the identification of people, locations, and other factors that would favour adoption, for example, the ownership of suitable inputs, adequate income, isolation from alternative energy sources, social structures that favour cooperation, credit, extension and government policy. Research should be initiated that will delineate these factors more clearly.

Case Studies

The existing processes of research, development, and dissemination of biogas systems appear to be weak and inadequately understood. Generally, scientific research has only recently been applied to small-scale plants, and a surprising lack of knowledge still exists about many biogas processes. This suggests that there are considerable lessons to be learned from detailed case histories of biogas research and implementation and from the role of research and development and implementing agencies in this and other attempts at nonagricultural rural technical change and the development of indigenous technical capacity. It is strongly recommended that empirical studies be carried out in various countries to elicit more information in these areas and that general guidelines be formulated to assist in the establishment of efficient biogas programs.

Understanding user needs

The conferees recognized that the ultimate purpose of biogas is to satisfy specific end uses, such as energy for cooking or lighting, powering motors, fertilizer requirements, and improved public health. Because other techniques, such as renewable energy sources, and low-cost sanitation, can meet equally well these needs in some situations. The specific needs of the end-user should be carefully assessed to ensure that biogas is the optimum technique to meet these needs. Especially in the rural context, more work needs to be done on assessing these end-use needs and whether they are best satisfied by BGT.

INFORMATION PROCESSING & DISSEMINATION

This area is concerned with the assimilation and transfer of all knowledge related to BGT. Hence it encompassess technical, economic, and social-institutional aspects of BGT.

Dissemination of Information

The flow of information in BGT can be analysed from two perspectives: horizontal (i.e., communication between specialists of different specific disciplines, such as technology, sociology, economy, agricultural science) and vertical (i.e., the information flow between various levels, such as political, scientific or level of the users). The impediment to information flow is the lack of consensus on methodologies in the various subdisciplines of BGT. Biogas is a multifaceted, interdisciplinary technology, that requires personnel to have a comprehensive knowledge of many other disciplines besides their own in order to communicate with other experts and personnel. One of the results of these impediments is that information on BGT that reaches policymakers is not integrated. There are few institutional arrangements available to coordinate all aspects of BGT.

Assessment

The conferees believed that the rate of implementation of biogas technology is now limited more by inadequate processing and dissemination of available information than by the technology per se.

Areas Requiring Further Work

To ameliorate the problems discussed above, the participants agreed that the following actions were needed:

- A central organization (newly formed or perhaps an appropriate existing structure) should collect information on all aspects of BGT and perhaps on competing techniques, such as power alcohol, gasifiers, single cell protein production. Transformation should be synthesized in the form of summaries that could then be disseminated as newsletters or reports in the major languages of the world. This clearing house for information, should also respond to requests for specific information in a manner appropriate to the level of the biogas user.

- Policy makers should be informed of the benefits of BGT with respect to the amelioration of problems that currently face developing countries.

- Knowledge and expertise should be transferred to appropriate nationals by means of training courses that cover all aspects of BGT. These experts then become a core of trainers in their own countries responsible for passing on their knowledge to persons at all levels of expertise including the people who build and maintain biogas units.

TECHNOLOGY TRANSFER

Unlike many technologies in the world today, the question of technology transfer from developed to developing countries is not of paramount importance. This is because the technology that is usually appropriate for use in developing countries is not highly sophisticated. However, one aspect of technology transfer that is important for all countries is technical cooperation among countries. Because much of the biogas knowledge is embodied in experts, it is imperative that this knowledge be communicated maximally. These efforts could include more exchange of experiences in training sessions, intercountry visits and consultations, conferences, seminars, and workshops.

CONCLUSION

The conference participants concluded that BGT is a viable option in developing countries for the mitigation of some of their pressing problems in such disparate areas as energy, public health, and agricultural productivity. However, BGT is not the only option available, and more work is needed to define the particular set of conditions (i.e. social, economic, and technical) under which biogas is the optimum choice for the policymaker. Areas requiring further work are suggested and funding should be made available from various multi- and bilateral aid agencies and national governments to expedite this work.

SECTION TWO
GENERAL TOPICS (THE OPENING SESSION)

SESSION OVERVIEW

Maryalice Risdon
National Academy of Sciences, USA

At the opening general session of the conference, chaired by M.M. El-Halwagi of the Egyptian National Research Centre and Paul Weatherly from the Office of Energy, Bureau for Science and Technology, U.S. Agency for International Development, five overview papers were presented to set the stage for the more focussed topics of the sessions to follow. After each presentation, the audience was invited to submit questions to the speaker.

David Stuckey from the Chemical Engineering Department, Imperial College, London, has recently completed a 2-year study for the World Bank on the status of biogas technology (BGT). His paper, "Biogas: A Global Perspective," addressed four general areas: the degree of implementation of biogas in developing countries, biogas techniques and their current problems, economic viability, and social and institutional factors. The degree of biogas technology application varies among the developing countries, reflecting the influence of such factors as government support, population density, per capita income, and the availability of noncommercial sources of fuel and fertilizer. China, India, and South Korea lead in the number of installations, and both Brazil and Nepal have growing programs.

After describing six types of digester designs as well as various techniques involved in their operation, Stuckey examined a number of specific technical problems that hinder the diffusion of BGT. One is a lack of understanding of the fundamental elements of microbiology, chemistry, and engineering involved in this multidisciplinary technology. Other problems include lack of rigorous field data and inadequate efforts to optimize existing designs.

The economic viability of biogas can be assessed by financial analysis or by social cost-benefit analysis, either of which can lead to different conclusions. Adding to the difficulty of determining economic viability is the lack of both substantive data and an agreed-on methodology.

In considering the social and institutional factors in adoption of BGT, it should be remembered that poverty effectively inhibits the demand for biogas. In addition there is a need to identify more specifically the rural user's need, taking into account inhibiting social factors, such as cultural taboos, traditional division of labor between men and women, and the educational level of the user.

Government support and adequate infrastructure to coordinate biogas activities are also factors of critical importance.

BGT has a role in meeting energy needs, but more work is needed in technical, economic,and social-institutional areas to identify favorable settings for optimum implementation and use.

An entirely different point of view was presented by Vaclav Smil from the University of Manitoba in Canada. While affirming his belief that, in theory, biogas generation is "the greatest alchemy in the world," making possible the provision of high-quality organic fertilizer and at the same time delivering convenient fuel for cooking, lighting, and electricity generation, he stated that in practice the use of biogas as a viable alternative energy source has been overrated and that in fact the current interest in the technology is a passing fad. His paper "The Realistic Potential of Biogas" stressed the numerous resource, environmental, engineering, and social constraints that prevent widespread diffusion of the technology. The steady flow of organic waste required for the effective operation of biogas digesters is often absent; climatic factors may prevent the yearround generation that is necessary for reliance on the gas as a major source of energy; the skilled workmanship needed to ensure leakproof construction of containers and ease of maintenance and repair is not always available; and the necessary commitment to daily care of the units in rural settings is often lacking, which leads to the deterioration of the units. The combined impact of these constraints has in general been underestimated.

Smil cited the Chinese experience as an illustration of the gap between theoretical expectations and practical performance. In China there is long tradition of experience in handling organic wastes, the government has given strong support, and severe rural energy shortages have provided a great incentive for biogas production. Despite these positive factors, evidence suggests that the Chinese experience is not an unqualified success. The new trend toward private enterprise is improving the economic status of the Chinese farmers enabling them to purchase fuel. In addition, beginning in 1978 farmers were again permitted to plant trees for their private use as fuel. Smil finds that these factors are causing the Chinese farmers to reject biogas as a solution to their energy needs.

M.M. El-Halwagi, head of the Pilot Plant Laboratory at the Egyptian National Research Centre, presented an analysis and grouping of ways to exploit BGT for the widest possible dissemination. His paper "Force Field Analysis of Biogas Systems and Proposed Means for Optimizing their Prospects" uses the "force field" principle of management. This principle is borrowed from the field of engineering and identifies the driving and constraining forces of a given system and then analyzes ways to maximize the driving forces and minimize the constraints. Forces that promote the adoption and diffusion of BGT and means of maximizing them are considered from both the national and individual viewpoint. Constraints and ways to minimize them are examined with respect to technical, economic, sociocultural, and organizational and infrastructural aspects.

From his analysis, El-Halwagi concludes that the key to optimum BGT propagation is the establishment of an effective organizational infrastructure. This infrastructure must include strong government support, a national coordinating agency for centralized organization of information follow-up and control, and an integrated network of centralized and decentralized agencies to plan and implement a coordinated program at all levels and to mobilize community support.

An effective infrastructure must include capabilities in the following areas: surveying energy potentials and planning; research, development, and engineering; social mobilization; marketing and technology promotion; construction and manufacturing networks with private sector participation and materials procurement capability; rural extension services and maintenance facilities; training of manpower; operational schemes of financial incentives and subsidies; and follow-up and assessment of activities.

Following this presentation, T.B.S. Prakasam recommended that conference participants examine El-Halwagi's list of advantages and constraints and arrive at a ranking of the five to ten most critical factors. This ranking would enable the conferees to present to the world the consensus of 100 BGT experts on the areas that require concerted efforts.

T.B.S. Prakasam, from the Metropolitan Sanitary District of Greater Chicago, presented a paper on "Information Needs for Diffusing Biogas Technology into the Rural Areas of Developing Countries." Information is needed to assist technologists and planners from developing countries in assessing the relevance and appropriateness of biogas systems for existing conditions in their countries. Information needs were enumerated under the following categories:

Scientific and engineering aspects

- Development and application of uniform criteria for measurement and evaluation of design and performance
- Development and demonstration of community- and industrial-scale digesters
- Methods of improving efficiency of biomethanation systems
- Optimization of integrated resource recovery systems with biomethanation as the hub
- Feasibility of application of newer designs
- Evaluation of fertilizer value of sludge derived from various stocks
- Development of cheaper and more efficient biogas utilization devices
- Determination of mass and energy flows
- Enhancement of biogas quality

Public health questions

- Does the use of biogas technology reduce public health problems?
- Is there a demonstrated improvement in water quality?
- What is the fate of toxic chemicals and pesticides?
- Is there a public health risk for owners and operators of household biomethanation systems?

Socio-economic aspects

- Development and application of uniform criteria for the evaluation of socioeconomic factors
- Evaluation of the interplay of social factors and technological factors in the performance of biomethanation systems
- Optimization of the efficiency of institutional infrastructure
- Role of women and voluntary agencies

- Development of educational materials
- Development of techniques for motivating home owners and communities to accept and maintain biomethanation systems
- Development of procedures to evaluate socioeconomic impacts of the nonquantifiable aspects of biomethanation systems
- Impact of building biogas plants on living standards

In his paper "The Integrated Digester Plant," George L. Chan from the U.S. Commonwealth of the Northern Mariana Islands described the integrated digester plant demonstration project developed during the past 5 years at the As Lito Farm in Saipan. This project is part of an integrated village development scheme that includes livestock, aquaculture, agriculture, siviculture, and industry. Chan pointed out that the Commonwealth had ample funds to develop the biogas project, which was conceived not as a community development project but as a business venture. The digesters are prefabricated, thereby avoiding possible faulty construction by the farmers. Because cement and steel must be imported to the Marianas, fiberglass proved to be the most economical material. The digester is a viable product for local small enterprises because of its simple design that facilitates quality control during manufacture, its low capital cost that can be repaid in 3 to 4 years, its easy operation, and its low repair and maintenance costs because fiberglass is corrosion and rot free, with no moving parts.

The project has demonstrated that, for an area like the Marianas that has a similar climate and natural environment with optimum conditions for photosynthesis and bacterial action, the use of a biodigester can successfully solve waste disposal and pollution problems and produce renewable fuel, fertilizer, feed, food, and various raw materials to meet the basic needs of the local population. Because of space limitations, only the abstract of Chan's paper is included. For further details, reference may be made to the author directly.

The opening session surveyed both the potential and problems of BGT. It showed promising examples of successful application, as well as a cautionary note against overoptimism about its viability for developing countries. A recurring issue in all of the presentations was the crucial importance of infrastructure as the final determinant of the successful diffusion of BGT.

BIOGAS: A GLOBAL PERSPECTIVE

David C. Stuckey
Chemical Engineering Department, Imperial College, London, U.K.

ABSTRACT

Biogas technology (BGT) can alleviate many pressing problems in developing countries, such as rural energy shortages, low agricultural productivity, and poor public health. Despite this potential and the heightened interest of many developing countries and funding agencies in the technology, BGT has not been implemented widely or rapidly, except possibly in China.

The objective of this paper is to critically assess the current status of BGT in developing countries with respect to technical, economic, and social-institutional factors and to draw some conclusions about its viability and the primary factors that inhibit its widespread application. Courses of action to enhance the viability and implementation of BGT will be suggested.

The paper examines four general aspects of BGT: (1) the degree of penetration of BGT in developing countries, (2) the technical status of various biogas digester designs, and their current problems, (3) the economic viability of biogas in the various areas of application, (i.e. domestic, community, feedlots, and industrial), and (4) the broader social and institutional factors related to BGT. Various strategies are proposed to consolidate current information and enhance the future viability of BGT.

INTRODUCTION

The process of anaerobic digestion of organic materials is commonly referred to as 'biogas' because of the biological nature of gas production. In recent years developing countries have shown considerable interest in the use and application of biogas for several reasons. First, the escalating costs of fossil fuels and the decreasing availability of renewable sources of fuel have forced many developing countries to consider the use of renewable energy technologies (RETs) for example, solar, wind, and biomass-based technologies, such as biogas, power alcohol, and gasifiers. Of these techniques, biogas has one of the lowest financial inputs per kWh of output. In addition biogas is one of the most 'mature' in terms of years of use and number of units installed, and has the potential to alleviate some of the more pressing problems in developing countries--for example deforestation and reliance on the importation of fossil fuels which leads to severe balance of payment problems. Although biogas has the potential to reduce deforestation and the importation of fossil fuels, data to show this effect have not yet been obtained.

Second, because biogas mimics natural environmental cycles, such nutrients as nitrogen, phosphorous, and potassium are conserved in the process and can be recycled back to the land in the form of a slurry. This is in contrast to the burning biomass where most of the nutrients are lost, for example, with wood stoves and gasifiers. The application of slurry reduces the need for chemical fertilizers, such as urea and superphosphate, and in addition enables humic materials to be recycled. This recycling preserves the physical properties of the soil and enables high agricultural productivities to be maintained.

Third, because the biogas process digests animal manures and nightsoil, it has the potential to considerably reduce plant, animal, and human pathogens. The cycle of reinfection is broken and considerable improvement in public health results.

Finally, because biogas is a clean-burning fuel, its domestic use can reduce the incidence of eye and lung problems that are commonly encountered with such smoke-producing fuels as firewood, agricultural residues, and coal.

Furthermore, biogas is a versatile technology and can utilise a wide variety of organic feedstocks, such as animal manures, nightsoils, agricultural residues, aquatic plants, and organic industrial wastes. Hence, in addition to being a multifaceted technology, it has potential application to many environmental and social milieus.

Despite its many benefits, biogas has a number of drawbacks, the major one being the cost of most current designs. This cost is a major impediment to BGT's diffusion in rural areas of the Third World where it may amount to a substantial fraction of a family annual income. In addition, technical problems related to maintenenance and low gas production during the winter months have occured as well as social-institutional constraints related to acceptance and diffusion of the technology.

Based on available evidence, however, the benefits of biogas in developing countries outweigh the drawbacks. It is puzzling therefore that only two developing countries, China and India, have installed large number of units, although these numbers represent only a small fraction of the potential based on biomass resources. Why has this situation arisen? Critics of biogas would answer that there are inherent problems in the technology and that BGT is not economically viable. The situation however, is complex and demands a more detailed and objective examination before firm conclusions can be drawn about the viability of this technology in developing coutries. In addition, biogas is merely one example of a technological intervention in rural areas of the third world, and past experience has shown us that such interventions are fraught with unforseen problems which all too often lead to failure.

This paper will critically assess the current status of BGT in developing countries with respect to both technical and socio-economic issues and draw some conclusions about its viability and the primary factors that inhibit its widespread application. The examination of socio-economic factors is essential to this analysis because the diffusion of technologies in developing countries has often been constrained more by social considerations than purely technical ones.

This paper is divided into five sections, the first of which examines the degree of penetration of biogas in developing countries. The second section reviews the technical status of the various designs and the current problems. The third section assesses the economic viability of biogas in the various areas of application, that is domestic, community, feedlots, and industrial. In the fourth section, the broad social and institutional factors related to biogas are discussed. Finally, based on the discussion of the previous sections, courses of action that will enhance the viability and dissemination of BGT will be suggested. The work presented draws heavily on two recent reports prepared for the World Bank by the author (Stuckey, 1983a, 1983b).

THE DISSEMINATION OF BIOGAS TECHNOLOGY IN DEVELOPING COUNTRIES

There is a large variation throughout the world in the number of digesters operating in developing countries. Three countries have installed a large number of units, and in numerical order these are: China, India, and South Korea.

China has approximately 7 million digesters, most of which are family-sized units ($6-10m^3$) used to provide gas for cooking and lighting. However, there are about 50,000 medium-($50m^3$) and large-sized (larger than $100m^3$) units of which 1,100 are generator stations with an output of 8,000 kW. This power is used mainly within communes for lighting and running machines (Stuckey, 1982). Although this number is impressive, in recent years the Chinese have admitted that a substantial fraction (30-50%) of these digesters are inoperative because of gas leaks. The main thrust of the Chinese program is toward family-sized plants, apparently for two reasons. First, in China the most common domestic animal is the pig. Most families own at least one per family member, and these are housed on the family premises. Hence there is a constant source of manure on the premises, and there is little reason to aggregate this manure to feed a community-sized plant. Second, based on current designs and construction methods, the actual cost per m^3 of digester increases with size, and therefore there is no incentive to construct larger plants to reduce the capital cost.

Three main designs are used in rural areas, and these are, in decreasing order of frequency: fixed dome (FD), floating cover (FC), and the bag type (Chan U. Sam, 1982, see later figures). Early reports in the literature (van Buren, 1979; UNEP, 1981) indicated that the primary emphasis of the Chinese biogas program was initially to rapidly disseminate the technology, use local materials to keep costs down, innovate to suit local conditions, and train people at brigade level to construct the digesters. The primary reason for installing the units appeared to be the recycling of organic fertilizers and improved sanitation.

Recent observations and information (Stuckey, 1982) reveal an important shift in thinking about biogas in China. The main reason for this change appears to be the recognition that many of the initial digesters failed (i.e., gas leak) because of poor construction. Hence the National Biogas Extension Office (NBEO) in Beijing has concentrated on building high quality digesters. As a result, the rate of dissemination of BGT slowed, and the material currently used is almost always cast concrete to prevent leaks and maintain structural strength. Furthermore, the construction of the digesters is now done primarily by well-trained technicians and although some variation in digester design exists throughout the country, there is a tendency toward a few standardized designs. This is also true for the ancillary equipment, such as burners and lights. Finally, it apppears that the Chinese are now citing energy production (E), followed by organic recycling (N) and then sanitation (S) as the most important benefits of biogas.

India has approximately 100,000 digesters; these are unevenly distributed among the various states. The primary objective of the Indian program is to provide fuel for cooking and lighting in rural domestic situations, and the main design in use is an 8-10m^3 FC digester. In recent years, in order to reduce the capital cost of units, considerable work has been done on the FD (Janata) type. Singh and Singh (1978) found that capital costs could be reduced from 50% to 70% depending on the size of the unit. Although this design is making inroads, it is still not widely accepted, probably because it requires different skills that the FC (e.g. masonry) and is a "new"technology in India. Other designs, such as the bag and plug flow, do not appear to have penetrated India at all.

In recent years, India has also become interested in the community sized unit for two reasons. First, the cost of a family-sized FC unit is high, amounting to an annual family income (i.e., $250-500). With large units there are economies of scale that can reduce the capital investment considerably. Second. The distribution of cattle among households is unequal, and cattle often have free range during the day which reduces the amount of dung available for community plants. Community plants, which can be fed a variety of organic feedstocks including agricultural residues, have the potential of providing gas to all the families in a village, thereby providing a means to reduce some of the inequities within a village.

Approximately 29,000 units have been installed in South Korea, most of these were built between 1969 and 1975 by the Government's Office of Rural Development (ORD) (Park and Park, 1981). The basic design consists of a concrete-lined cylinder with a wooden floating cover lined with plastic. However, in 1976 because of problems with cold weather operation this program was largely discontinued, and the ORD concentrated on installing village-scale units with both heating and mixing. The status of village scale implementation, in terms of numbers, is not known at this time.

The status of biogas throughout the rest of the Third World varies considerably. Of the Arab countries and Northern Africa, only Egypt has embarked on a serious effort to create a research and development organization for biogas, and at present there are only 50 units installed of either the FD or FC design. In Southern Africa, only Kenya, Tanzania, and Ethiopia have more than 100 units, and there is apparently quite a high failure rate because of inadequate maintenance and technical back up (King, 1978; Ward, 1981; Lyamchai and Mushi, 1982).

Until recently, scant information has been available about biogas in Latin America because of communication limitations. However, Umana (1982) found that there was a wide variation in efforts to establish BGT because countries in the region are net oil exporters. Brazil has by far the largest number (2,300) of biogas units and recently appears to have committed itself to a rapid expansion, the development of a strong research and development capacity, and the evolution of a comprehensive infrastructure for dissemination. Again the most common designs appear to be the FD and FC. Throughout the rest of the region, there is considerable interest in research and development but the numbers of units is low, for example in Central America (110), Mexico (150), Andean countries (120), and the South (20).

In Southeast Asia three countries have been working on biogas since the early 1970s: Thailand, the Philippines, and Taiwan. Initially the primary emphasis in Thailand was on the hygienic disposal of animal wastes. After 1973, however, the emphasis changed to energy. In 1981 there were approximately 300 digesters in the country, mostly the FC design with an average volume of $5m^3$ (FAO, 1981). Apparently, however, a substantial number of these has fallen into disuse (Ratasuk et al., 1979).

In the Philippines, Maya Farms, an agro-industrial division of Liberty Flour Mills, pioneered the use of bio-gas technology beginning in 1972. Initially, the units were installed to control the pollution caused by the disposal of swine manure. During the last decade, however, an integrated resource recovery scheme has been evolved that treats the waste from 22,000 pigs, and the gas produced is used directly as a fuel to power machinery and generate electricity (Judan, 1981). The digested slurry is separated into two fractions, liquid and solid. The liquid is used to fertilize crops and feed fish ponds, and the solids are refed to pigs, cattle, and ducks. These solids supply approximately 10-15% of the total feed requirements of the pigs and cattle and 50% of the requirements for ducks. In 1979 there were about 340 units in the Philippines; however, only 73% of these were operating (Galano et al., undated).

In Taiwan, the major biogas innovation has been the production of durable and cheap plastic made from the waste product of bauxite refining. The Red Mud Plastic (RMP) has been used to construct a plug flow bag type of digester which is primarily above ground. The unit is cheap, easy to transport and install, and has considerable potential because it can be operated at relatively high temperature under a solar greenhouse. The precise number of digesters in Taiwan is not known, but it is probably greater than 1,000.

Finally, on the Indian subcontinent, Nepal has approximately 1,200 biogas units, most of them FC although there are some FDs. In addition, there are 3 community-scale plants. The biogas produced is used primarily for cooking and lighting although in some cases it has been used to power irrigation pumps (Anon., 1981). One unique development has been the establishment of a private company, the Gobar Gas Company, to commercialize biogas, and reports indicate that it appears to have been relatively successful (Fulford, 1981). In Sri Lanka it was estimated that by the end of 1981 there would be close to 300 units (Santerre, 1981). Of the digesters installed approximately 45% were FC designs, and the remainder were FD. Most of the digesters were family sized, and the gas produced was used for cooking and lighting. Both Pakistan and Bangladesh also have biogas programs, but they are small and primarily experimental. Table 1 summarises the major biogas programs by country and the predominant designs.

Table 1: Summary of Major Biogas Programs

Country	Number	Design	Priority	Strengths	Weaknesses
China	$7X10^6$	FD[b](95)c, FC[c]bag	E,N,S[e]	Infr, R&D, Gov, Int. Ag Res	Lack of materials
India	10^5	FC(95), FD	E,N	Gov, Int, Sub	Cost, Maintenance
S. Korea	$3X10^4$	FC	E	R&D	Temp, High Inc.
Brazil	2,300	FC(60),FD(40)	E	Infr, Gov, R&D	?
Nepal	1,200	FC(75), FD(25)	E	R&D, Comm	Temp, Maintenance

Others - Taiwan, Philippines, Thailand, Sri Lanka, Mexico, Pakistan

a)data from Stuckey 1983a and 1983b.
b)FD=fixed domes
c)Approximate percentage of design.
d)FC=floating cover
e)E=energy, N=nutrient cycle, S=sanitation

TECHINICAL STATUS

The techniques or designs used to carry out the digestion of organics range from simple to extremely complex. The technique used in developing countries depends to a large extent on its level of applicability, and hence its appropriateness. There are four levels of application: domestic, communal (or village), intensive animal feedlots, and industrial. In developing countries the primary use of biogas has been domestic, with a few communal units. Feedlots and industrial areas have already been used, and only a few of these units are presently operating.

Digester Designs

The batch reactor is the simplest design option available; however, the gas production varies considerably with time, and several units must be operated simultaneously to maintain a constant gas supply. Furthermore, the labor involved in charging and emptying the units often amounts to twice the amount needed to feed and maintain a semi-continuous unit (Maramba, 1978). This fermentation can be run at normal solids contents [6-10% total solids (TS)], or at high concentrations (larger than 20%). The latter system is known as dry fermentation, and Jewell and coworkers (1981) have found that fermentation can proceed at TS concentrations up to 32%. AT 35°C gas production rates can be as high as 0.8 volume/digester volume-day. The advantage of this type of reactor is that water requirements can be reduced, which is important in arid regions, and that few materials handling problems are experienced using fibrous wastes, such as agricultural residues. In addition, operation is simple and the capital cost is low.

The FD (Chinese design - Figure 1) is by far the most numerous digester operating in developing countries and tends to be operated, at least in China, in two concurrent modes: batch and semicontinuous. Precomposted agricultural residues are normally loaded every 6 months after the digester has been partially emptied to provide fertilizer for

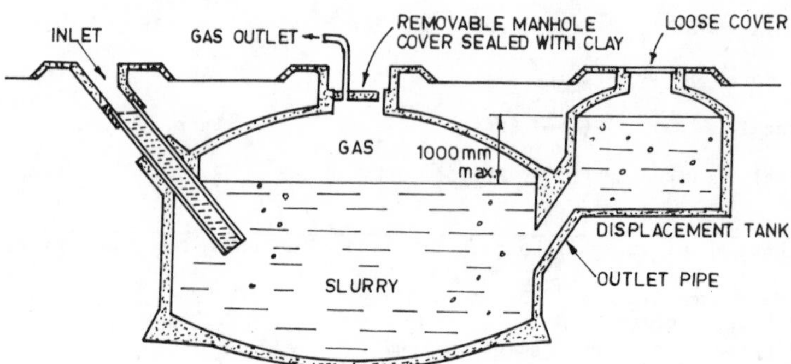

Figure 1 Fixed dome (Chinese) digester

crop planting. Between these batch loadings, swine manure and nightsoil are fed continuously to the digester, and the effluent is removed and used as top dressing. The gas produced is usually stored in the FD, and it displaces the liquid contents into the inlet and outlet chambers, often leading to gas pressures as high as 100cm of water. This high gas pressure is often a problem and has caused gas leaks in many of the early designs. Consequently, the design philosophy changed as mentioned above. In addition, the Chinese are building low-pressure FD designs where the gas produced is stored in RMP gas bags (1-1.5m^3). Although the bag is expensive, this method maintains low gas pressures and and thereby reduces gas leaks and the capital cost of construction (Stuckey, 1982). Typical gas productions are on the order of 0.1-0.2 v/v day (Chan, 1982) with detention times of 60 days at 25°C. The capital cost of these units is low, and because they are constructed below the ground, temperatures tend to remain constant and are often considerably higher than ambient temperatures in winter.

The FC (Indian, KVIC - Figure 2) design is the most popular in India and is similar to sewage sludge units in developed countries. The digester which consists of a cylindrical reactor with an H/D ratio between 2.5 and 4.1 produces gas that is trapped under a floating cover and that constitutes a volume of approximately 50% of the total daily gas production. The cover is usually constructed out of mild steel, although because of corrosion problems other materials, such as ferrocement and fiberglass, have been used. Usual detention times vary from 30 days in warm climates to 50 days in colder areas. With cattle manure at 9% TS, gas yields of between 0.2 and 0.3 v/v are achieved. These units, however, tend to be relatively expensive.

The bag digester (Figure 3) is a long cylinder (L/D=3-14) composed of either PVC, a Neoprene-coated fabric (Nylon), or RMP. The bag is placed in a trench that measures approximately half the diameter of the bag, and hence is exposed to changes in ambient temperature. The gas produced is usually stored in the reactor under the flexible membrane,

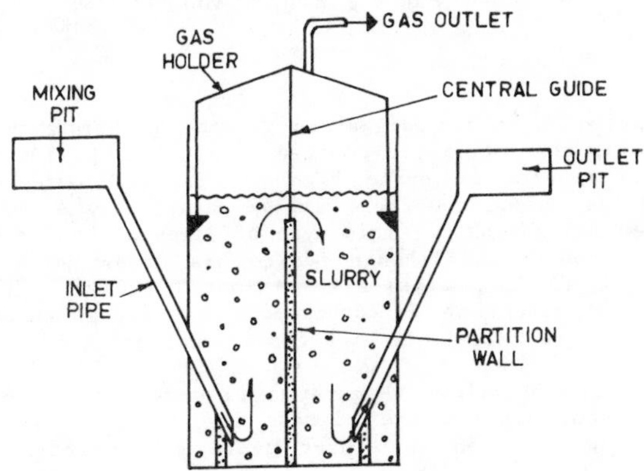

Figure 2 Floating cover (Indian) digester

although it can be stored in a separate gas bag (Park and Park, 1981). Because of their plug flow characteristics and ease of trapping solar radiation, the volumetric gas production rates of bag digesters can be quite high. The Chinese have found (Chan, 1982) that average temperature in bag digesters are 2-7° C higher than dome types. Hence volumetric gas rates can be from 50-300% higher in the bag (0.24 to 0.61 v/v day). Park and Park (1981) also found this to be true in Korea and obtained volumetric gas productions varying from 0.14 in winter (8°C) to 0.7 v/v day in summer (32°C) for swine manure.

The plug flow digester (Figure 4) is similar to the bag type but is usually placed deeper in the ground with the membrane covering only the top portion. Because of its hydraulic characteristics, the feed

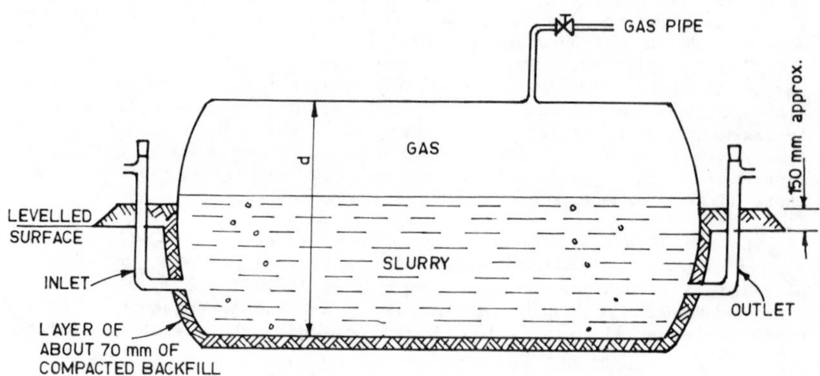

Figure 3 Bag (Taiwan) digester

concentration in these digesters can be higher than the FD and FC types. This characteristic coupled with the plug flow digester kinetic properties results in quite high gas rates. Using dairy manure at 12.9% TS, Hayes and coworkers (1979) obtained rates of 1.26 v/v at 30 days detention and 35°C. This type of digester has rarely been used in developing countries although Mexico has experimented with a similar design called the Xochicalli (Olade, 1981). This design has considerable potential in developing countries because of its low capital cost and relatively high gas production rates.

The anaerobic filter (Figure 5) is known as an immobilized growth reactor because the active biomass is maintained within the filter, either by growing on the inert media or by entrapment within the

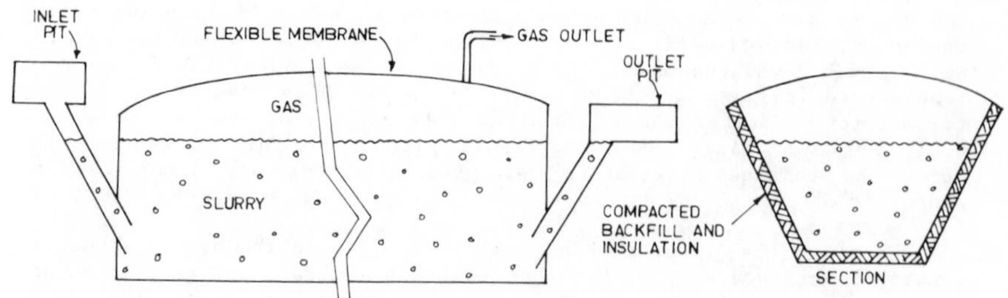

Figure 4 Plug flow digester

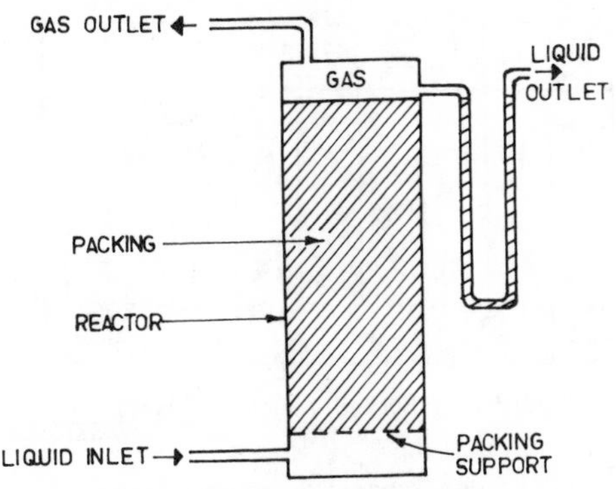

Figure 5 Anaerobic filter

media. This enables long biological solids retention times (θ_c) to be maintained at low hydraulic detention times (θ). Because of the physical configuration of the filter, only predominantly soluble wastes can be treated, although a diluted pig waste has been treated successfully with a TS content of 2.0% (Chavedej, 1980). Waste strengths from 480 mg/l up to 90,000 mg/l COD have been treated in filters, and detention times as low as 9 hours (based on void volume) are possible with COD removals of 80% (Young and McCarty, 1969). However, more usual detention times are on the order of 1-2 days (Arora and Chattopadhya, 1980), with the achievement of over 90% COD removal. Loading rates as high as 7 kg COD/m^3 are possible, and under these conditions gas production rates of 4 v/v day have been measured (Xinsheng et al., 1980). Despite its many advantages and the considerable laboratory work that has been carried out in developed countries, there are no full-sized anaerobic filter units in operation in developing countries.

The anaerobic baffled reactor (ABR - Figure 6) was recently designed by Bachmann and McCarty at Stanford University and consists of a simple rectangular tank, with dimensions similar to a septic tank, divided into 5 or 6 equal-volume compartments by means of walls from the roof and bottom of the tank. The liquid flow is alternatively upwards and downwards between the walls, and on its upward passage the waste flows through an anaerobic sludge blanket, of which there are five or six. Hence the waste is in intimate contact with the active biomass, but because of the baffled nature of the design most of the biomass is retained within the reactor even under large hydraulic shocks. With a soluble waste containing 7.1 g/l COD and a detention time of 1 day at 36°C, Bachmann and coworkers (1982) obtained 80% removal efficiencies of COD and a volumetric gas production of 2.9. Similar tests have been carried out in this author's laboratory with dilute waste (0.48 g/l COD), and similar removals were obtained at 25°C. Experimental work on the design has been carried out in China.

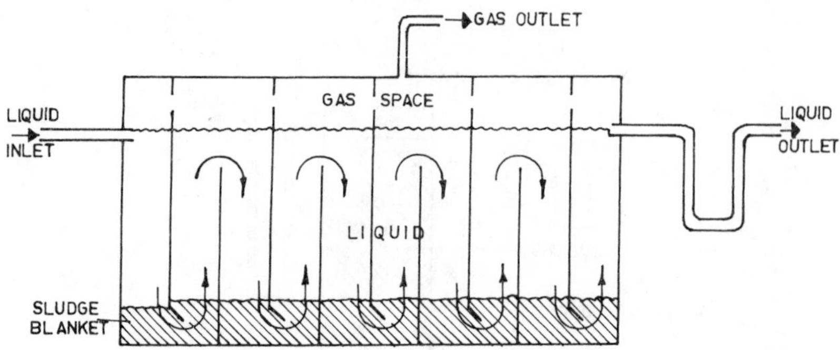

Figure 6 Anaerobic baffled reactor (ABR)

Recently a $10m^3$ unit was constructed to treat distillery wastes (Stuckey, 1982). Because of its physical configuration, this type of reactor appears to be able to treat wastes with high solids contents (e.g. sewage), and hence it may be a viable alternative in certain situations in developing countries.

Other digester designs, such as contact, UASB (upflow anaerobic sludge blanket) may be viable in particular circumstances in developing countries. Because they are relatively complex, however, they do not appear to be appropriate to the technical milieu in which they would operate.

The choice of technique depends on the feedstock(s) available, to a lesser extent on the quantity of organics to be treated, and on the socioeconomic milieu. In the domestic application of biogas the primary feedstock is animal manure (swine or cattle), and the usual design is either the FD or FC type. However, many other feedstocks, such as agricultural residues and aquatic plants, can be used. In addition, other designs, such as batch, bag, and plug flow, have the potential to produce gas with lower capital investments and higher efficiencies. In villages and intensive feedlots the most common design is the FC, although from preliminary data it appears that the plug flow reactor would be cheaper and probably more efficient. Thus, further research and development should be carried out to "unpackage" this technology and make it suitable for application in developing countries. Finally, although the industrial applications of biogas in developing countries are virtually non-existent, such designs as the anaerobic filter and ABR appear to be promising in terms of capital cost, efficiency, and appropriateness, and attempts should be made to introduce these designs. Table 2 summarises the digester designs and presents their advantages and disadvantages.

Ancillary Techniques

Reactor design, as well as such factors as heating, mixing, and feed pretreatment, have an important effect on overall digestion efficiency. In the literature it is generally agreed that the maximum rates of reaction in anaerobic digestion occur at around 35°C (mesophilic) and 55°C (thermophilic). Therefore, it has become common

Table 2. Summary of Digester Designs[a]

Option	Stage of Development	Advantages	Disadvantages	Installed Capital Cost $/m^3$	Applications
Batch (low solids)	Advanced--a number of operating units in DC[b]	1) Simple construction 2) Easy operation, with low skill requirement	1) Varying gas production with time 2) Gas storage required		1) Manures mixed with bedding 2) Animal manures 3) Aquatic plants
Batch (high residues solids)--"dry fermentation"	Pilot plant studies, some operating units in DC	1) Simple construction 2) Easy operation, with low skill requirement 3) Reduced water requirements 4) Digest agricultural residues with no pretreatment 5) Relatively high gas yields	1) Varying gas production with time 2) Gas storage required		1) Agricultural 2) Aquatic plants 3) Animal manures
Fixed dome (Chinese)	Advanced--considerable data on operation and economics. Many units throughout the World	1) Simple construction, with readily available materials--low cost 2) Relatively high pressure gas supply 3) Easy to insulate by constructing below ground	1) High structural strength required in construction 2) High quality workmanship to make gas tight 3) Low concentration feeds, hence low gas yields 4) Varying liquid levels 5) Scum control difficult	7 (simple) 13 (cast concrete)	1) Animal manures 2) Nightsoil 3) Mixtures of above with agricultural residues 4) Aquatic plants

[a]From Stuckey 1983a, 1983b.
[b]Signifies developing countries.
[c]Signifies retention time.

Table 2 (Continued)

Option	Stage of Development	Advantages	Disadvantages	Installed Capital Cost $/m^3	Applications
Floating cover (Indian)	Advanced--considerable data available on operation and and economics. Many units throughout the world	1) Gas supply constant and at stable pressure 2) Gas yields higher than fixed dome 3) Easy to control scum 4) Low structural strength of fixed dome 5) Amenable to passive solar heating	1) Higher total cost/m^3 due to floating cover 2) High heat losses due to cover 3) Short working life of gas holder	60-100	1) Animal manures 2) Nightsoil 3) Some fraction of agricultural residues 4) Aquatic plants
Bag (Taiwan)	Relatively advanced--considerable experience in Taiwan; other units in China, Latin America, Fiji, Korea	1) Low cost 2) Simple transport and installation 3) Amenable to simple passive solar heating	1) Thin membrane vulnerable to puncture 2) Low gas delivery pressure	8-30	1) Animal manures 2) Nightsoil 3) Aquatic plants 4) Some fraction of agricultural residues
Plug flow	Laboratory and pilot studies--a few units in DC	1) High solids loading possible with high efficiencies 2) High gas yields 3) Relatively simple construction 4) Relatively easy to control scum layer 5) Amenable to passive solar heating	1) Low gas delivery pressure 2) Relatively high land requirement of fixed and floating dome	50-80	1) Animal manures 2) Nightsoil 3) Aquatic plants 4) High fraction of agricultural residues

Table 2 (Continued)

Option	Stage of Development	Advantages	Disadvantages	Installed Capital Cost $/m^3$	Applications
Filter	Laboratory and pilot plant studies, relatively few full-scale units, and none in DC	1) Low θ^c possible: small reactor volumes 2) High loading rates: high gas yields 3) Low temperature operation has minimal effect on yields 4) Simple construction with readily available materials 5) High process stability	1) Feed must be low in total solids 2) Gas storage required		Soluble (2% TS) wastes ranging from low (COD 480 mg/l) to high strength (70,000 mg/l)
ABR	Laboratory studies	1) Low θ possible: small reactor volumes 2) High loading rates: high gas yields 3) Simple construction (e.g., modified septic tank) 4) High hydraulic stability 5) Capable of treating partially insoluble wastes	1) Gas storage required		Predominantly soluble wastes ranging from low (480 mg/l COD) to medium strength (2000 mg/l COD)
Contact	Laboratory studies--a few full-scale units, none in DC	1) Low θ possible: small reactor volumes 2) High loading rates: high gas yields 3) Minimal effect of temperature	1) High level of operating skills required 2) Costly		Predominantly soluble wastes
UASB	Laboratory and pilot studies--a few full-scale units, none in DC	1) Low θ possible: small reactors 2) High loading rates: High gas yields 3) Minimal effect of temperature	1) High level of operating skills required 2) Costly		Predominantly soluble wastes

practice to assume that the optimum temperature for operating digesters is 35°C (or 55°C). If the aim of the design, however, is to maximize the net energy yield (i.e. gross energy - energy used for heating and mixing), 35°C is not necessarily the optimum temperature. In cases where product gas is used to heat the incoming feed, the optimum temperature may be lower. Stevens and Schulte (1979) have shown that under some circumstances the maximum energy yield from swine wastes occurs at 25°C rather than 35°C. Also, in immobilized cell systems (e.g. filter, ABR), s sufficient mass of organisms may be present so that even at low temperatures overall gas production rates may still be quite high. Thus, the optimum temperature of operation depends on optimizing net energy yields in relation to detention time (and hence capital cost).

In developing countries, the use and complexity of heating systems depends on the kind of application. For domestic units (6-10m^3) only simple systems can be used because of cost, and such systems include insulation, composting, and solar (passive and active) methods. Insulation is an obvious method, and helps to maintain relatively constant temperatures during diurnal fluctuations. For example, with the FC, Prasad and Sathyanarayan (1979) found that 54% of the total heat loss from a digester occurred through the top of the gas cover; hence simple insulation could reduce these losses considerably. In China, digester heating is sometimes accomplished by aerobically composting agricultural residues around the digester and controlling the temperature by the judicious addition of water to the compost. Although apparently effective, this method poses the question whether the residues might not be put to better use by actually digesting them. More work needs to be done on this in order to optimize the net energy yield from a given quantity of biomass.

Passive solar methods involve the construction of a greenhouse over the top of the digester. In Guatemala this has been found to raise the temperature of an underground digester by as much as 10°C. If the digester is above ground, however, the diurnal fluctuations can inhibit the reaction rate to the exent that there is little net benefit (Mahin, 1982). Active solar methods involve the actual heating of the feed (or dilution of water). By incorporating a solar heater into the roof of a gas holder, workers at Bangalore (Reddy et al., 1979) not only heated the feed but also reduced the heat losses from the digester. Preliminary results indicate an increase of 11% in gas production using this method.

In village feedlots and industrial settings, these methods can be further refined or the actual biogas produced can be used to directly heat the digester or indirectly heat the digester by means of cooling water or exhaust gases from a biogas-powered motor. In Korea, Park and coworkers (1981) used biogas to directly heat a 157m^3 digester, and at -8°C they found that only 32% of the daily gas production was needed to maintain a temperature of 35°C. Again, optimization procedures could be used to determine the optimal fraction of gas diverted for use in engines so that the waste heat could maintain a temperature of 35°C in the digester.

Adequate mixing in digesters is extremely important for several reasons (Montheith and Stephenson, 1981):

(1) to maintain uniformity in substrate concentration, temperature, and other environmental factors;
(2) to prevent scum formation; and
(3) to prevent solids deposition.

Reports in the literature, however, sometimes present conflicting views about mixing. Hashimoto (1982) has shown that with beef cattle wastes at 55°C, continuous stirring versus stirring for 2 hours/day only resulted in small increases in methane production: 8% and 11% at 6 and 4 days retention, respectively. Other researchers (Converse et al., 1981; Mills, 1979; Smith et al., 1979) have also concluded that adequate biogas production with animal manures is achieved by intermittent mixing (e.g. for 1/2 hour every 5 hours) before and during feeding. These results seem to indicate that little mixing is required with animal manure to maintain satisfactory digester performance. However, with more heterogeneous feeds, such as aquatic plants and agricultural residues, mixing may have a greater effect on performance.

In constrast to the above findings, there is extensive documentation in the literature (Montheith and Stephenson, 1981; Smart, 1978; Verhoff et al., 1974) that actual hydraulic detention times measured by tracer studies range from a low of 18% of the theoretical value (i.e. V/Q) to an average of only 65% in one study of 11 units (Smart, 1978). Smart found no reliable, readily discernible relationships between mixing efficiencies and digester sizes, ages, types, and power input. In the only study of this kind in developing countries (El-Halwagi, 1982), the actual hydraulic detention time in an unmixed FD digester was shown to be only one-half the theoretical value.

Thus, even in completely mixed digesters, from one third to one-half of the reactor volume can be "dead" space, which will result in decreased process performance or increased reactor volumes to achieve a particular organic removal. Based on this data it appears that more tracer studies are necessary to determine actual hydraulic detention times. If these times are considerably lower than design figures, then various mixing techniques should be investigated to reduce the dead space. Such a change would have a significant effect on the cost and efficiency of digesters in developing countries.

Finally, the biodegradability of the organic substrate (1 CH_4/kg added) is one of the most important parameters affecting the viability of biogas in developing countries. For most common substrates the convertibility to methane varies between 30% and 50%. Hence the potential yield of methane could be increased by 2 to 3 times if the substrate could be rendered 100% degradable.

Three major pretreatment methods will increase biodegradability: physical-chemical, physical, and biological. Because of equipment and chemical costs, the first method is unlikely to be appropriate in developing countries. Physical pretreatment involves cutting, grinding, or shredding to increase the surface area per unit volume, and thereby increases the area open to attack by hydrolytic enzymes. Reports in the literature on this point appear to be contradictory; probable chemical composition as well as physical accessibility are important in determining biodegradability. More work is necessary to explore the potential of this method, especially given the constraints in developing countries.

The final method of pretreatment is biological, which consists of precomposting agricultural residues before digestion. Data on this method are sparse, and apparently it is practiced extensively only in China (UNEP, 1981). The method involves cutting the agricultural residue into small pieces, adding limewater and excreta, and allowing the mixture to compost in a pile. After a short time the waxy coating on the plants disintegrates, and the cellulose becomes soft and loose. In China it is claimed (UNEP, 1981) that this method increases the wettability of the substrate and prevents it from floating and forming scum layers in the digester. In addition, test results show that the biogas yield was considerably higher with precomposting. Some of the degradable organics, however, are used up in the composting process and thus there must be an optimum time in order to maximise gas production. Although this method is promising because it is easy to carry out, further work should be conducted to quantify the gas yields and to optimize the various parameters involved.

The above discussion was a general analysis of various biogas techniques. The following section raises some specific technical problems that need resolution in order to enhance the diffusion of BGT.

Specific Problems

Lack of understanding of fundamentals. Historically BGT in developing countries has evolved as an empirical technology, with most advances based on empirical observation rather than a sound understanding of the fundamentals involved. In addition, because BGT is a multidisciplinary technology that involves elements of microbiology, chemistry, and engineering and is considered a relatively mundane technology by physical scientists, it has not attracted much attention from research and development groups in developing countries. Consequently, the technology has not achieved the efficiency that is possible based on current knowledge in developed countries. Furthermore, considerable time and resources have been lost in pursuing designs and approaches that are at odds with the sound fundamentals of the technology.

For example, there is the confusion between total gas production and methane production as a measure of process performance. The latter parameter is the most important measure of performance and is related to the former by the percentage of CO_2 in the gas phase. As is well known, CO_2 is affected by such conditions as temperature, pressure, and the oxidation state of the carbon in the feedstock. Based on total gas production, however, design "innovations", such as a gas cover with a negataive pressure, have been found to increase performance! Also many workers have been concerned about decreased performance (i.e., total gas production) as a result of the high pressures (100 cm H_2O) encountered in the FD design despite the findings of Mangel and coworkers (1980) that pressures of up to 4 atm do not inhibit the rate of methane production. Finally, the addition of urine (urea) has been touted by many workers in developing countries as a solution to poor process performance in cold climates because it increases total gas production. However, because the oxidation state of carbon in urea is

-4, it can only produce CO_2 during digestion, and while the possibility exists that the digester may be nutrient limited, rarely if ever are controls run to confirm or refute this possibility.

In order for BGT to reach its full potential in developing countries there is an urgent need to increase understanding of the technology's fundamentals among research and development personnel. This can be accomplished by courses, workshops and direct links between appropriate institutes in developing and developed countries.

Lack of rigorous field data studies. Despite the use of BGT in some developing countries for many years, there is still a lack of rigorous field studies on actual process performance under a variety of operating conditions and with various feedstocks. Therefore, many biogas units are still over- or underdesigned, which leads to a waste of resources or a mismatch with user needs. An inadequate substantive data base also makes difficult the rigorous comparison of existing techniques, economic assessment of techniques, and comparisons between new and existing techniques. There is an urgent need to closely monitor a number of existing units to provide this substantive data base.

Optimization. Based on the two factors mentioned above, there have been few attempts to rationally optimize existing designs to reduce capital costs and improve performance. In one well-documented case (Subramanian et al., 1979), workers at Bangalore in India used a KVIC design and performed an optimization exercise to minimize the capital cost. They found that capital costs could be reduced by 27.5% for a small plant and by 41.2% for a large plant. Furthermore, in the large plant under identical conditions the lower cost design improved gas yields by 14% when compared to the standard KVIC design. This example shows that considerable improvements can be achieved through simple optimization techniques. Simple design change would considerably enhance the viability of BGT. This kind of exercise would also involve a careful examination of actual detention times mentioned above.

Carbon-Nitrogen ratios. The carbon to nitrogen (C/N) ratio in a substrate is important because high nitrogen levels (greater than 80 mg/l as undissociated ammonia) with low C/N ratios can cause toxicity, and low levels (high C/N ratios) can inhibit the rate of digestion. Many popular books on digestion (Meynell, 1976; National Academy of Sciences, 1977; UNEP, 1981) present tables of the C/N ratios of various feeds and state categorically that a C/N ratio of 30:1 is optimal. This ratio apparently originated with the work of Acharya (1958) and has not been disputed. The most critical factor in feedstocks, however, is the available C/N ratio, and this depends on such factors as percentage biodegradability of the substrate, the fraction of N released during digestion, and the yield of cell mass from the biodegradable fraction.

Because assessing all these factors for various feeds in a developing country context is a fairly complex undertaking, a more empirical approach is possible. Buswell and Hatfield (1936) found that ammonia concentrations of greater than 100 mg/l were necessary to prevent a retarding of the rate of gas production, and Ronnow and Gunnarson (1982) found for two pure strains of thermophilic methanogens that ammonia concentrations of greater than 120 mg/l were necessary to prevent an inhibition of the rate. Hence perhaps by monitoring soluble ammonia concentrations in the effluent it is possible to ensure that the digester will operate under optimal nitrogen conditions despite widely varying C/N ratios in the feedstock. Additional experimental work is necessary to confirm this hypothesis and delineate allowable ranges of C/N ratios with varying feedstocks.

In summary, there are a number of technical problems related to BGT in developing countries. Solutions to these problems are not simple, and they constitute interesting and complex research areas. Other problem areas not discussed here include the effect of temperature fluctuations on performance, the rate-limiting step in the degradation of lignocellulosics, and the effect of operating conditions on the fertilizer and soil conditioning properties of digested slurry as well as on the pathogens' kill rate.

ECONOMIC VIABILITY OF BGT

In addition to technical considerations of BGT there are also economic aspects that need to be examined closely because of scarce financial and material resources in most developing countries. The economic viability of biogas can be assessed using two different methods: financial and social. Finacial analysis is based on market prices, including taxes and subsidies, and usually identifies the money profit accruing to the project. In contrast, social analysis uses shadow prices that reflect the true economic worth to the society of the inputs and outputs of the project and includes secondary benefits. Because they use different methods of analysis, these methods can lead to different conclusions. In cases where a social analysis leads to a positive result and a financial one leads to a negative result (i.e. C/B* larger than 1.0), government can alleviate the financial burden on investors by providing subsidies for the capital cost, as is the case in India.

There is a dearth of substantive data on which to evaluate the economic viability of BGT, particularly in the area of industrial and feedlot applications and less so in domestic and community applications. In the latter two cases, however, most of the data come from India and are derived from theoretical design figures, using the FC design and cattle dung as the feedstock. From such a specific application, few, if any, conclusions can be drawn about the viability of biogas under other circumstances, such as various designs and feeds in various social and environmental milieus and areas of application.

In addition to lacking substantive data, existing economic evaluations lack an agreed-on methodology. Common problems include: lack of data collection on the effect of technical parameters on plant performance, evaluation of inputs, evaluation of biogas in relation to substitutable fuels and end-uses, evaluation of slurry as a fertilizer-soil conditioner and animal feed, marginal utility of output, and the evaluation of secondary benefits. A consensus on methodology needs to be developed that will allow economic data to be compared among the various applications, under varying circumstances, and enabling rigorous economic comparisons between biogas and other renewable energy technologies or with conventional energy sources.

Based on a review of the literature (Stuckey, 1983a), the financial viability of biogas plants depends strongly on whether their gas and slurry outputs can substitute for fuels, fertilizers, or feeds that were previously purchased for money. If so, then the resulting cash savings can be used to repay the capital and maintenance costs, and the plant has a good chance of financial viability. If, however, the outputs do not generate a cash inflow or reduce the cash outflow, then the financial viability of these plants becomes considerably worse. Finally, if broader social criteria are used to evaluate biogas in the form of a social cost-benefit analysis (SCBA), the result will be more positive conclusions about viability when compared to a strictly financial analysis.

Thus, in general terms, domestic biogas units will not be financially viable in countries where fuels and fertilizers are collected for free, for example, in many parts of India. However, in other countries (e.g., China) where considerable cash savings result from the installation of biogas (Stuckey, 1982), domestic units are potentially financially viable. However, because the evaluation of such secondary benefits as reduced deforestation, improved public health, and reduced reliance on imported fossil fuels presents additional problems, it is not possible at this time to draw any firm conclusions about the social viability of such domestic units.

Community-scale plants are limited by similar financial considerations as household units, although in many cases economies of scale will tend to make them a better prospect. However, based on Roy's (1980) findings the primary barriers to diffusion are social and organizational rather than technical or economic. Because the benefits from a community plant can be shared by poorer households that could not otherwise afford the investment and operating cost of domestic units, community plants also appear to be more socially viable than smaller units.

Based on cash flow considerations, industrial and commercial feedlot applications appear to be extremely promising, although more data are needed on the cash benefits of animal refeeding and integrated resource recovery schemes.

SOCIAL AND INSTITUTIONAL CONSIDERATIONS OF BGT

Despite the importance of the social and institutional implications of BGT, there have been only a few studies on the topic, and those concentrated on domestic-sized plants in rural India (Moulik, 1982; Moulik et al., 1978). However, the literature on the diffusion of technology in rural areas of developing countries (Roy et al.; Shaller, 1979) makes possible several general comments.

First, one of the most important inhibitions to the diffusion of BGT are the causes and consequences of poverty. Just as many of the needs of rural people are often poorly translated into "effective demand," so it is that technical changes, such as the introduction of biogas, are likely to be inhibited where cash is scarce and noncash sources of fuel (such as wood and crop residues) are used.

The second and perhaps most striking reason why some innovations are successful and others fail is the extent to which the new technology meets the intended user's needs (Freeman, 1974). Rural societies are increasingly recognized as complex systems, and the failure of much technical intervention appears to result from a misunderstanding by outsiders of rural problems.

At the level of the societal group, there are several social factors that tend to inhibit the spread of BGT. First, in some societies social taboos prohibit the handling of excreta, especially human wastes. Second, the assignment to either men or women of such domestic tasks as collecting fuel and cooking and the determination of who makes the major household decisions is extremely important. In developing countries women usually perform the former, and men the latter. Finally, there is a strong correlation between the educational level of the plant owner and acceptance of BGT, although this may merely reflect the superior economic status of most educated people in developing countries.

At the national level strong government interest and backing is necessary if BGT is to disseminate rapidly. This backing should be manifested in a national infrastructure that coordinates activities, provides technical assistance, and allocates resources. Unfortunately in many countries, with the exception of China, this infrastructure is often weak and leads to uneven performance. In addition, there should be strong cooperation between implementing and research and development agencies and clearly defined mechanisms whereby experience gained in the initial stages of BGT implementation can help to influence later more optimal designs and the allocation of future resources. Again, all too often these relationships are weak, and goals are rarely met, for example in India (Moulik, 1982).

Governments also have an important influence on the financial viability of BGT by means of taxes, subsidies, and pollution control laws. Kerosene, often the major monetized alternataive to biogas, is usually subsidized by governments. Thus, its price is considerably below the world import price, and biogas appears to be an expensive alternative to the consumer. Similarly, subsidies on fertilizers have a negative effect on technologies that utilize local inputs. This is not an arguement against such subsidies, but a suggestion that equal subsidies be offered to new technologies, such as biogas, which governments may consider to be desirable on a social or real resource cost basis. Promulgation and enforcement of environmental legislation also has a significant effect on the financial viability of large-scale applications, such as commercial feedlots and industry. Where a substantial cost is involved in the treatment and disposal of wastes from such facilities, installation of biogas systems can result in substantial cash savings because the generation of a valuable product, biogas, can then be used within the plant.

STRATEGIES TO ENHANCE THE VIABILITY AND DIFFUSION OF BGT

Based on this brief review of BGT throughout the world, the following areas of work should be undertaken to enhance the diffusion of BGT.

Establishment of a Global Network

A global network on BGT could promote the exchange of information and technical cooperation among developing countries (TCDC) and benefit from the knowledge in developed countries. One institute in each country would be designated as a focal point and be charged with the responsibility of being aware of all work on biogas in its country. This work would be summarized and presented at meetings of the network; other specific technical and socioeconomic papers of interest could also be presented. In conjunction with the network meeting, training courses and workshops could also be held. Because agencies such as the UN, ESCAP, USAID, and the World Bank are actively involved in supporting work in this field, it is recommended that the network cooperate fully with these agencies to prevent duplicated efforts.

Adoption of a Standardized Testing and Evaluation Methodology

Although considerable technical and economic information on biogas has been collected throughout the world, some of it is difficult to compare and evaluate because of different reporting methods and evaluation techniques. Hence there is an urgent need to agree on a common methodology to evaluate the technical and economic performance of biogas units. Recently, at a workshop in Bangkok (May 1983) the attendees prepared a draft methodology, and it is recommended that it be adopted in whole or in part.

Monitoring of Existing Field Units

Again, while considerable information exists on the laboratory performance of biogas units, there is a dearth of substantive data from the _actual_ performance of field units over a long period of time. It is recommended that a small number of units be selected in several countries that are representative of a variety of techniques and feedstocks. These should be monitored closely for 1-2 years to evaluate their performance in relation to laboratory units. This process would also identify persistent problem areas that require further research and development. This monitoring should be carried out in cooperation with the socioeconomic assessment discussed in the next section.

Socioeconomic Study of the Diffusion of Biogas Technology

There is an urgent need to more fully understand the social and economic factors that control the dissemination of BGT to ensure that the technology diffuses as rapidly as possible. This study should be carried out in selected countries with varying degrees of BGT implementation and should be performed in conjunction with the monitoring program mentioned in the previous section. Factors investigated should include financial viability, availability of capital for initial investment, technical backup and infrastructure, competition for the feedstock by other end uses, seasonal fluctuations of feedstock, and appropriateness with regard to technical skills and social customs.

DISCUSSION AND CONCLUSIONS

Based on a number of approximate quantitative analyses, BGT has considerable potential for developing countries. For example, using data from India, Lichtman (1983) concluded that biogas can essentially provide all the net usable energy currently consumed in cooking from noncommercial fuel sources in rural India. Furthermore, the analysis revealed that using the slurry from the digestion of this biomass could eliminate all fertilizer imports into India and result in considerable cash savings. Although these results are theoretically possible, Lichtman acknowledges that such widespread diffusion of BGT involves considerable capital, labour, and organizational skills. Such a large national commitment to biogas should not be undertaken unless the government has verified that it is the most (socially) cost effective investment that can be made.

Factors controlling the degree of penetration of BGT in various developing countries are extremely complex and not well understood. Strong government interest and support appear to be mandatory, for example, in China and India, if the technology is to diffuse widely. Other important factors include the population density of a country (or region) in relation to its carrying capacity; per capita income, which if too low will inhibit the purchase of units and if too high will enable the purchase of fossil fuels and the availability of noncommercial sources of fuel and fertilizer.

Regarding the technical aspects of BGT, it appears that despite its widespread use, the fundamental principles of BGT are still poorly understood. As a result there is a lack of substantive field data, and few attempts have been made to optimize existing designs. In addition, in most developing countries, there is little capacity to "unpackage" promising new technologies and adapt them to local conditions. For biogas to become more viable, it is imperative that indigenous research and development capabilities be strengthened through the transfer of knolwedge and equipment.

Assessing the economic viability of biogas is difficult, even at this stage of its development, because there is a lack of data from a wide range of conditions and an absence of an agreed on methodology. Where fuels, fertilizers, or feeds are currently purchased, however, biogas has a good chance of being financially viable.

Finally, although there is little specific information available on the social-institutional barriers to BGT diffusion, these barriers appear to be strongly related to poverty and whether biogas can meet users' needs.

In conclusion, under some defined conditions in developing countries biogas appears to be a viable technology that can alleviate some pressing problems. Nevertheless, considerably more work is needed in certain technical, economic, and social-institutional areas to more narrowly define the appropriate applications of this technology and to contribute to its more widespread diffusion. The factors that inhibit its widespread application are many and complex. At the national level these factors include lack of government interest and infrastructure, poor research and development facilities, and cheap and readily available sources of fuel and fertilizer. On the community level there are other considerations, such as initial capital cost, poor financial viability, inappropriateness, and technical problems.

REFERENCES

Acharya, C.N. 1958. Preparation of fuel gas and manure by anaerobic fermentation of organic materials. ICAR Series Bulletin No. 15, Indian Council of Agricultural Research, New Delhi, 1-58

Anonymous 1981. Progress Report, 1981. Community Biogas Irrigation System, Madhubusa, Nepal. Development and Consulting Services, Butwal, Nepal, July.

Arora, H.C., and S.N. Chattopadhya. 1980. Anaerobic contact filter process : A suitable method for the treatment of vegetable tanning effluents. Water Pollut. Control (G.B.), 501-506.

Bachman, A., V.L. Beard, and P.L. McCarty. 1982. Comparison of fixed-film reactors with a modified blanket reactor. Proceedings of the 1st International Conference on Fixed Film Biological Processes, Kings Island, Ohio, April.

Buswell, A.M., and W.D. Hatfield. 1936. Anaerobic Fermentation. Bulletin No. 32. State of Illinois Division of the State Water Survey, Urbana, Illinois.

Chan, U. Sam. 1982. State of the Art Review on the Integrated Use of Anaerobic Processes in China. Internal Report prepared for IRCWD, September.

Chavedej, S. 1980. Anaerobic Filter for Biogas Production. Thai Institute of Scientific and Technological Research, Bangkok, Thailand, January.

Converse, J.C., G.W. Evans, K.L. Robinson, W. Gibbons, and M. Gibbons. 1981. Livestock waste: A renewable resource. Proceedings of the 4th Int. Symp. 122-125.

El-Halwagi. 1982. personal communication.

Food and Agriculture Organization. 1981. The Development and Use of Biogas Technology in Rural Areas of Asia. Project Field Document No. 10, Improving Soil Fertility Through Organic Recycling. FAO/UNDP Regional Project RAS/75/004.

Freeman, C. 1974. The Economics of Industrial Innovation. Penguin Books, London.

Fulford, D.J. 1981. A commercial approach to biogas extension in Nepal. Appropriate Technology 8(2).

Galano, R.I., A.G. de los Santos, R. Alicbusan, P.E. Belarmino, and G.M. Caron. undated. Biogas Technology: Social and Economic Evaluation. De La Salle University, Manila, Philippines.

Hashimoto, A.B. 1982. Effect of mixing duration and vacuum on methane production rate from beef cattle waste. Biotech. and Bioeng. 24:9-23.

Hayes, T.D., W.J. Jewell, S. Dell'Orto, K.J. Fanfoni, A.P. Leuschane, and D.R. Sheridan. 1979. Anaerobic digestion of cattle manure. in D.A. Stafford, B.I. Wheatley, and D.E. Hughes, eds. Anaerobic Digestion. Applied Science, London, G.B.

Jewell, W.J., S. Dell'Orto, K.J. Fanfoni, S.J. Fast, D.A. Jackson, R.M. Kabrick, and E.J. Gottung. 1981. Crop residue conversion to biogas by dry fermentation. Presented at the 1981 Winter Meeting, ASAE, Chicago, Illinois, USA, December.

Judan, A.A. 1981. Biogas Works at Maya Farms: Its Socio-economic Aspects. Maya Farms Bio-energy Consultants, Liberty Flour Mills. mimeo.

King, R.P. 1978. The Development of Biogas-Slurry Plants and Systems in Kenya. Prepared for Intermediate Technology Development Group (ITDG), London. Dept. of Agricultural Engineering. Nairobi University, Nairobi, Kenya, June.

Lichtman, R. 1983. Biogas Systems in India. Volunteers In Technical Assistance (VITA), Arlington, Virginia, USA.

Lyamchi, A.A., and S.J.S. Mushi. 1982. Outline of biogas technology development in Tanzania. Paper presented at the UNESCO/UNIDO Workshop on Biogas Technology in Africa, Arusha, Tanzania, February.

Mahin, D.B. 1982. Biogas in Developing Countries. Bioenergy System Report Prepared for United States Agency for International Development.

Mangel, G., J. Villermaux, and C. Prost. 1980. Methane production under pressure by fermentation of waste materials. Europ. J. Appl. Microbiol. Biotechnol. 9:79-81.

Maramba, F.D. 1978. Biogas and Waste Recycling: The Philippine Experience. Liberty Flour Mills, Manila, Philippines.

Meynell, P.J. 1976. Methane: Planning a Digester. Prism Press.

Mills, P.J. 1979. Minimisation of energy input requirements of an anaerobic digester. Agricultural Wastes 1:57-79.

Montheith, H.D., and J.P. Stephenson. 1981. Mixing efficiencies in full-scale anaerobic digesters by tracer methods. J.W.P.C.F. 53(1):78-84.

Moulik, T.K. 1982. Biogas energy in India. Academic Book Centre, Ahmedabad, India.

Moulik, T.K., U.K. Srivastave, and D.M. Shingi. 1978. Bio-gas System in India, A Socio-Economic Evaluation. Centre for Management in Agriculture, Indian Institute of Management, Ahmedabad, India.

National Academy of Sciences. 1977. Methane Generation from Human, Animal and Agricultural Wastes. National Academy of Sciences, Washington, D.C., USA.

Latin American Energy Organisation. 1984. Strategies and Technologies for Implementing Rural Biogas Programs in Latin America. Document No. 11. Latin American Energy Organisation (OLADE), Quito, Ecuador.

Park, Y.D., and N.J. Park. 1981. A Study of Feasibility on the Village Scale Biogas Plants through a Farm Demonstration Trial During the Winter Season. Research Report of the Office of Rural Development. Suweon, Korea, 23, November.

Prasad, C.R., and S.R.C. Sathybarayan. 1979. Thermal analysis of biogas plants. Part III. Studies in Biogas Technology, Proceedings Indian Academy of Science C2:377-386.

Ratasuk, S., N. Chantramonklasri, R. Srimuni, P. Ploypatarapunyo,

S. Chavedej, S. Sailamai, and W. Sunthonsan. 1979. Final Report on Pre-Feasability Study of the Biogas Technology Application in Rural Areas of Thailand. Environment and Development Research Department, Applied Scientific Research Corporation of Thailand, Bangkok. Submitted to IDRC, Canada.

Reddy, A.K.N., C.R. Prasad, P. Rajabapaiah, and S.R.C. Sathyanarayan. 1979. Studies in biogas technology. Part IV, A Novel Biogas Plant Incorporating a Solar Water-Heater and Solar Still. Proceedings Indian Academy of Science C2:287-393.

Ronnow, P.H., and L.A.H. Gunnarson. 1982. Response of growth and methane production to limiting amounts of sulfide and ammonia in two thermophilic methanoganic bacteria. FEMS Microbiology Letters.

Roy, R., F.B. Waisanan, and E.M. Rogers. 1969. The Impact of Communication on Rural Development: An Investigation in Costa Rica and India. UNESCO, Paris and National Institute of Community Development, Hyderabad, India.

Roy, P. 1980. Family and Community Biogas Plants in Rural India and China. Alternative Technology Group, The Open University, Milton Keynes, U.K.

Santerre, M.T. 1981. An Evaluation of Biogas, Windmill and Other Rural Energy Technologies in Sri Lanka. East-West Center, Honolulu, Hawaii, USA.

Shaller, D.V. 1979. A Socio-Cultural Assessment of the Lrena Stove and its Diffusion in Highland Guatemala, mimeo (March).

Singh, R.B., and K.K. Singh. 1978. Janata Biogas Plants: Preliminary Report on Design and Cost, Planning Research and Action Division, State Planning Institute, Utter Pradesh, India.

Smart, J. 1978. An Assessment of the Mixing Performance of Several Anaerobic Digesters Using Tracer Response Techniques, Research Publication No. 72. Ministry of the Environment, Ontario, Canada.

Smith, R.J., M.E. Hein, and T.H. Greiner. 1979. Experimental methane production from animal excreta in pilot-scale and farm-size units. J. Animal Science 48(1):202-217.

Stevens, M.A., and D.D. Shulte. 1979. Low Temperature Anaerobic Digestion of Swine Manure. A.S.C.E., Environ. Eng. Div., 105, EEI, 33-42.

Stuckey, D.C. 1982. Biogas in China: A Back to the Office Report on a Study Tour of China. IRCWD, Dubendorf, Switzerland, December.

Stuckey, D.C. 1983a. Technology assessment study of biogas in developing countries. Report prepared for the World Bank as executing agency of UNDP Global Project-Testing and Demonstration of Renewable Energy Technologies. IRCWD, Dubendorf, Switzerland, pp. 1-137.

Stuckey, D.C. 1983b. The integrated use of anaerobic digestion (biogas) in development countries: A state of the art review. Interim report prepared for the World Bank as executing agency of UNDP Global Project-Research and Development in Integrated Resource Recovery. IRCWD, Dubendorf, Switzerland, pp. 1-297.

Subramanian, D.K., P. Rajabapaiah, and A.K.N. Reddy. 1979. Studies in biogas technology. Part II. Optimization of Plant Dimensions. Proceedings of the Indian Academy of Science C2:365-375.

Umana, A. 1982. State of the Art Review of Anaerobic Processes in Latin America. Internal Report prepared for IRCWD, August.

United Nations Environmental Program (UNEP). 1981. Biogas Fertilizer System. Technical Report on a Training Seminar in China. UNEP, Nairobi, Kenya.

van Buren, A., ed. 1979. A Chinese Biogas Manual - Popularising Technology in the Countryside. ITDG, July, London.

Verhoft, F.H., M.W. Tenney, and W.F. Echelberger. 1974. Mixing in anaerobic digestion. Biotech. and Bioeng. 16:757-70.

Ward, R.F. 1981. Digesters for the Third World. Presented at the 2nd Intl. Symp. on Anaerobic Digestion, Travemunde, Federal Republic of Germany, September.

Xinsheng, M., C. Ruchen, L. Nian-gua, H. Chengehun, and W. Shearer. 1980. The Xinbu system: An integrated rural production system. Development Forum 8(9).

Young, J.C., and P.L. McCarty. 1969. The anaerobic filter for waste treatment. J.W.P.C.F. 41(5):R160-73.

THE REALISTIC POTENTIAL OF BIOGAS

Vaclav Smil, University of Manitoba, Canada

ABSTRACT

Introducing biogas generation into an optimally run agroecosystem is complicated by numerous resource, environmental, engineering, and social constraints that singly, but most often in combination, dramatically reduce the practicality and reliability of biogas generation and the prospects for widespread diffusion of this otherwise desirable technology. These obstacles have been recognized previously, but in practice their combined impact always has been underestimated.

The amount and steady input of organic wastes needed for continuous high-efficiency operation of biogas digesters is often absent. Climatic factors make the year-round generation of biogas impossible or reduce production during cold months to a fraction of peak flows which precludes relying on the gas as a major source of energy, and engineering complications abound. In addition, the structures may be simple, but workmanship must be first-rate if the container is to remain leakproof under pressure for years and be easily cleaned and maintained. Operation may appear straightforward in experimental set-ups, but daily care in actual rural setting will not achieve the optimum acidities, alkalinities, C/N ratios, temperatures, liquidities and uniform mixing required for the best performance or for the survival of a digester as a biogas generator rather than as a simple fermentation pit.

Chinese experience, thus far the most extensive in the world, shows all these pitfalls. There is a gap between theoretical expectations and practical performance, even in this setting where careful handling and fermentations of organic wastes are an ancient tradition, where the construction of biogas digesters was given a considerable institutional support, and where severe rural energy shortages are a great incentive to turn to biogas. Modest expectations and measured levels of research and institutional commitment are thus in the best interests of both the promoters and the users.

INTRODUCTION

The success of new energy sources and conversion processes is not decided simply by cost, convenience, and safety. Coal is now considerably cheaper than crude oil, but the world prefers the latter. Convenience appears to be the best explanation for this preference, but in many applications electricity is the most convenient energy source. Yet with just a few exceptions, we are running away from nuclear power. Although safety concerns could explain this flight, the nuclear industry's safety record clearly surpasses that of coal mining and coal-fired power generation.

Biogas generation has so far made more difference for writers and publishers of renewable energy materials and for a growing number of professional researchers and granting agencies rather than for the farmers -- and above all for the poor peasants of the three developing continents. This nontechnical paper will explore the major reasons for this sluggish diffusion of the technology; examine the changes in China's biogas generation, which is thus far the most serious attempt at widespread adoption of the technique; and analyze biogas potential in the wider context of rural energy supply (Smil, 1983).

BENEFITS OF THE TECHNOLOGY

Nothing else makes biogas generation intellectually so appealing as its obvious link to ancient alchemies, a link made even stronger by the process' highest level of ennoblement. The most base, objectionable wastes are converted to one of the finest, cleanest fuels. Moreover, this magic transformation, carried out by legions of fermentative, acetogenic and methanogentic bacteria, appears to offer many appealing benefits in almost any rural setting, but especially in the case of traditional agroecosystems currently undergoing the transition to modern intensive farming.

Processing animal and human wastes in biogas digesters could greatly improve local hygiene and eliminate or appreciably lower the frequency of several infectious diseases -- while offering a convenient fuel whose use can lighten household labor (on a larger scale biogas can enable mechanization of some crop processing tasks and even small-scale generation of electricity), reduce the drain on other biomass energies (fuelwood, crop residues), and provide an excellent organic fertilizer.

Fermented sludge that is removed from anaerobic digesters is as beneficial as the fuel. Worldwide reports show that the increasing applications of inorganic fertilizers (some developing countries now use as much or more synthetic nitrogen than the United States or even Western Europe, e.g., Egypt and China) are paralleled by declining reliance on the planting of green manures and recycling of organic wastes. Returning organic matter to the soil is widely recognized as essential for a sustained, high-yield farming, and anaerobic fermentation of farm wastes offers a unique opportunity to satisfy this essential need while delivering convenient fuel for cooking and lighting or even for electricity generation.

In addition, the process of anaerobic fermentation can be carried on in simple units which, although not free, are not impossibly expensive for many people in developing countries. No forbiddingly high technologies here, no crippling capital commitment. Then why is not biogas technology (BGT) more successful? Perhaps the most generalized answer would be to invert Amory Lovins' characterization of soft energy technologies.

DIFFUSION CONSTRAINTS

Lovins stated that renewables are "relatively simple from the user's point of view though often technically very sophisticated." Drawing an analogy to his pocket calculator, he admitted that "I could not build one and do not understand its operation in details, its functional simplicity lets me integrate it into my life as a tool rather than a machine. I run it, it doesn't run me. Yet it uses very high technology" (Lovins, 1978).

This assembly of peoplewise to the ways of biogas generation must easily see my intended inversion. Biogas digesters may be built by the users, and determined literate peasants with a knack for experimentation can get to understand their pit's operation in great detail as it does not use any complex technologies. But the sustained successful operation of the digester is only deceptively simple. Here is a technically rather simple gadget whose efficient operation is relatively complicated from the user's point of view. Thus, it cannot be switched on or off at will; it demands regular and meticulous attention in several critical ways.

Although the now so voluminous Western biogas literature includes chemical and technical details of the fermentation process and lengthy hypothetical calculations of operating costs, there is more general evidence for the complexity of BGT. Smith and coworkers (1977) examined typical central Iowa operation where the farmer must cope with nearly 150 ha of corn, soybeans, oats, and alfalfa; 31 litters of pigs per year, and 300 heads of cattle. The authors looked at this challenge for BGT from the user's perspective and cautiously concluded: "What is scientifically achievable may not be entirely on the farm. The modern farmer is very far removed from the peasant, but he still is not a superman... Adding another complex device -- perhaps not fully understood by engineers, let alone farmers -- will further burden the farmer with maintenance and management requirements that are far removed from his primary interests: raising crops and animals. Farming in the United States has been successful, due in part to one family's ability to manage a very productive enterprise. If the equipment required to improve the productivity of a farm become too complex, the farmer will not adopt it! Perhaps BGT would be more suitable for traditional farming where there is less stress on top operational efficiency and where simpler farming system, run with much greater labor force, can adopt less complicated installations to get lower but still worthwhile returns. An example is the Chinese experience with BGT; the world's largest agroeconomy whose massive biogas programme entailed the construction of millions of small family digesters.

CHINA'S BIOGAS EXPERIENCE: A CASE STUDY

Since 1978 the Chinese have released large quantities of highly critical information about many of their pursuits, which they had praised unequivocally during the last decade of Mao's reign, including their accomplishments with biogas.

Reasons for China's Program Decline

Enough is known to list the main reasons for the decline of BGT in China. Foremost among them are the difficulties with daily operation of the digesters. According to the official statistics from the National Methane Production Leadership Group (Smil, 1984) only about one-half of all digesters built in the late 1970s could be used normally, and even among these working digesters "not too many" could be used to cook rice three times a day; still fewer could be used daily for four seasons.

Thus even in the summer of 1979 when the number of biogas digesters reached the peak of just over 7 million units, fewer than one-half and probably about one-third were working as reasonably reliable energy generators. Peasants reaction was predictable, especially as the speading failures of over-promoted digesters coincided with the first steps of introducing a more free production system in China's villages, a process that resulted in virtual abandonment of communal agriculture for contract farming, where the individual peasant family has relatively wide opportunities for income maximization.

Biogas units had been costly for Chinese peasants -- on the average nearly 100 yuan for each family digester, a considerable sum in a society where per capita rural income was not more than 150 yuan in the late 1970s and just over 300 yuan in 1983. Peasants often were left with useless pits as a result of construction defects and less of anaerobicity or with hardly more than expensive manure fermenters with low gas generation rates. Even with reasonably good care, their units were gadgets whose energy output was far from sufficient for all household cooking needs.

When the rigid system of command farming and the enforced building of digesters ceased, the total number of family units dropped dramatically, by about 40% nationwide and by nearly 50% in Sichuan, the province with four-fifths of China's digesters.

Lessons Learned

China still has approximately 4 million small biogas units. Many peasant families find them to be a good value. Although it takes work to keep them running well, cleaning of the pits is awkward, and in half of the country even the well-maintained units will not generate enough during five to seven colder months, the sludge makes the efforts worthwhile. This is the key to understanding the relatively fast acceptance of anaerobic fermentation in China, and the Chinese experience suggests the following. Family biogas generation has a much better chance for sustained success when its byproduct, the fertilizer, is at least as much needed as is the gas. Consequently, rural societies with a long tradition of waste fermentation and organic manure have a considerable advantage over all other settings. Because the proper operation of small units requires time, experience, and long-term commitment, all promotion schemes should be careful not to push the farmer. Deliberate, voluntary adoption of BGT is crucial and even with these two conditions satisfied, there will be failures and disappointments leading to the abandonment of many units. In addition, appealing as the techique is, it may not be preferable when compared to other local renewable options (e.g., in China small hydros and fuelwood lots). Ignorance of these factors leads to wasted labor effort, costly capital losses, and distrust of innovation--altogether a discouraging way to modernize rural energy supply.

BIOGAS POTENTIAL IN RURAL ENERGY SUPPLY

It must be remembered that biogas can only be marginal source of rural energy. It may make a great difference for a family or even a village, but the environmental conditions and feedstock availability will always limit its regional or nationwide importance. For example, in China even if the 4 million digesters (with average volume of $8m^3$, average daily generation rate 0.2 m^3 of biogas per m^3 digester) would operate for 7 to 8 months per year, they would produce annually only about 1.5 billion m^3 of biogas, an equivalent of just 1.25 million tons of hard coal, or less than 0.5% of China's current rural energy use.

The Chinese have published many estimates of the eventual nationwide potential for BGT. These appraisals are as high as 140 billion m^3 of biogas annually. However, generous animal and human waste production and availability estimates and a good generation rate show that no more thant 50 billion m^3 could be produced providing the fermentation process was year-round even in the colder part of the country. Taking into account the higher combustion efficiency of biogas and a total focus on biogas generation in China, that country could probably provide no more than 10% of its current rural energy consumption, which averages 13 GJ/capita a year (or about 450 kg of coal equivalent).

Simple calculations show that even maximum theoretical conversions of all collectable animal and human wastes would not cover higher consumption shares. If two-thirds of the global production of roughly two billion (dry) tons of these wastes would be fermented (yielding 0.2 m^3/kg of dry feedstock), the biogas output would be equivalent to 150 million tons crude oil, roughly 2% of the current global commercial energy consuption.

Realistically, no more than one-half of that total could be converted, an equivalent just short of 10% of today's rural energy (including all biomass sources) in Asia, Africa, and Latin America. Similarly, total Egyptian biogas capacity would be equivalent to about 0.5 million tons of crude oil. Therefore, if one-half of this total would be possible, the country could derive no more than 2-3% of its current energy use.

These ratios must be kept in mind when appraising the realistic potential of biogas generation in rural energy supplies. Once again, the Chinese experience provides an example: by the beginning of 1983, 25% of the country's peasant families were allotted an average of 0.2 ha of slope or odd land for fuelwood tree planting, and the distribution of land continues. If a harvest of 5 tons of wood/ha could be sustained from fast-growing species (a conservative assumption), then a family would obtain 16GJ of storable energy per year. In contrast, an average well-maintained digester would provide only 8GJ a year; more realistically, outputs of 20 GJ versus 6 GJ should be compared. Lower combustion efficiency of wood would shrink this difference, however, with one of the better stoves, the gap would still be about 2-fold.

In any locale with a reasonably distributed annual precipitation of 400 mm or more, the one-fifth ha of land to plant poplars or leucaenas, biogas would be a poor comparison to fuelwood. Biogas would be most successful where there are confined animals (pigs best), lack or shortage of other biomass energy possibilities, need for organic fertilizer and tradition of waste handling. After assessing the world's 150 nations, only large parts of East Asia satisfy all of these conditions.

Elsewhere there will be useful pockets for BGT--a family, a village, a special region. Thus, BGT diffusion efforts have a special challenge. The technique, although useful and worthy of much greater adoption, will probably always remain only a fringe contributor of rural energy.

REFERENCES

Lovins, A. 1978. Soft energy technologies, Annual Review of Energy 3:477-517.

Smil, V. 1983. Biomass Energies, Plenum Press, New York, pp 323-388.

Smil, V. 1984. China's Energy: Advances and Limitations, IDRC, Ottawa.

Smith, R.J., R.L. Fehr, J.A. Miranowski, and E.R. Pidgeon, 1977. Pp 341-371 in R.C. Loehr ed. Food, Fertilizer and Agricultural Theories. Ann Arbor Science Publishers, Ann Arbor, Michigan.

FORCE-FIELD ANALYSIS OF BIOGAS SYSTEMS AND PROPOSED MEANS FOR OPTIMIZING THEIR PROSPECTS

M.M. El-Halwagi
National Research Centre, Egypt

ABSTRACT

An analysis of the status of rural biogas systems has been made to identify the systems' strengths and constraints. Means of maximizing the effect of a strength factor or minimizing the influence of a constraining force are suggested. Compilation of the results tends to indicate that creating a sense of national commitment and developing an effective organizational network are two key prerequisites for successful diffusion.

INTRODUCTION

Force-field analysis is a useful management tool for devising a set of rational means to enhance the productivity of a given system. As a first step in this technique, the system is thoroughly analyzed with the aim of identifying both the driving and constraining forces. Then, to optimize the system output, ways are found for maximizing the driving forces and for minimizing the constraining ones. In the final analysis, a set of coherent actions are developed by grouping together the related individual means.

This technique was used to analyze the diffusion of biogas technology (BGT) in the rural areas of developing countries. The personal experience of the author was augmented by a number of relevant publications (References 1-6). The results were compiled into a questionnaire that included a proposed general diagnosis of the status of BGT in terms of strengths and weaknesses. Symptoms of both were listed, and the means for maximizing each driving force or minimizing every constraint were suggested.

Before generalizing its use, the questionnaire was tested by sending it to a sample of the conference participants. Responses indicated that the analysis was relevant in most cases, although sometimes a bit too general because practical experience depends on a given situation. Some participants suggested that it would be better to discuss the questionnaire during the conference because it was rather lengthy and cut across varied disciplines.

After some thought and reflection, and taking into consideration the time pressure during the conference, I decided to prepare this paper. It encompasses, in condensed form, the questionnaire's contents as well as a preliminary analysis and grouping of the possible ways and means proposed for fully exploiting the prospects of BGT in terms of realistic widespread propagation. Though the conclusions may be considered personal in nature, they are backed by expert findings and opinions conveyed through the questionnaire responses and documented contemporary literature.

A. DRIVING FORCES AND THEIR MAXIMIZATION

Strength Factor (or group of)	Means to Maximize
National	
The energy crisis and the need for alternate energy resources.	Convince decision makers about the value of BGT on a national scale.
Deforestation problems and the need to safe fuel-wood.	Induce awareness through education, telling the right people the right things, and demonstration.
Pressing need to satisfy the requirements of the rural population. Efficient utilization of BGT has positive effects on the national economy and can be readily integrated with rural development, health care schemes, and the like.	Engage top economists and environmentalists.
Individual	
Multibenefits include supplying a clean energy source, nutrient recycling, and improving rural sanitation, convenience, and leisure. BGT can be made compatible with local village conditions. Practical application of many biogas systems are within the capabilities of the ordinary farmer. Digesters can be constructed using local village resources. BGT can be intimately integrated with rural life.	Publicize, demonstrate, and conduct oriented field research and development programs. Prospects can be increased by proper social research and intensive teaching campaigns. Develop effective organizational support.
Other	
Some countries have notable success diffusing the technology, e.g., China and India.	Study and analyze the causes of success and failure, then adapt and utilize.
Much reasearch and development (R&D) has been done world-wide.	Bring the already present knowledge together and make it accessible and useable.
International and donor agencies' support for renewables, recycling, and pollution abatement.	Utilize funds and technical assistance for preliminary study and demonstration.

B. CONSTRAINING FORCES AND THEIR MINIMIZATION

Constraining Force (or group of)	Means to Minimize
Lack of optimum designs appropriate for the variety of local conditions.	Develop proper R&D and engineering capability to study, transfer, adapt, and innovate.
Weak local construction and equipment manufacturing capacities.	Training, and formation of specialized integral teams furnished with the required tools and supporting equipment. Encourage establishment of manufacturing enterprises for ready-made parts and ancillaries.
Operational problems such as pinhole leakages, corroded parts, water condensation in gas lines, scum formation, blockages, and burner problems.	Proper training of users and involvement of "women-users" participation. Establish extension and maintenance services. Regular follow-up, at least initially.
Low gas production during winter.	Apply known technical solutions including insulation, solar tent, solar-heated feed water, and compost around digester, add urine/water hyacinth, build digester under shed, and heat with biogas or engine exhaust heat.
Labor-intensive tasks associated with running the digester and handling effluent.	Apply proper and tailored design, modify animal shed, and develop tools for digester cleaning.
Excessive water requirements, especially for dry areas.	Dry fermentation, recycle water from outlet, use of urine, use less dilution (higher solids concentration).
Short supply of animal wastes.	Do not build a biogas plant, dry fermentations, use other biomass substrates, develop more efficient digesters community plants.
Lack of adequate space for siting the biogas plant.	Fixed-dome digester under shed, high efficiency digester above ground (solar-heated mini-digesters), and community plants.
Problems in utilization of digested slurry, storage and use of biogas, and hygienic aspects.	Employ known technical solutions or develop suitable ones for local conditions (i.e., information and technical capability), educate users.

Economic	
High investment cost that is prohibitive for majority of rural population.	Develop less expensive designs or construction techniques, efficient (planned) implementation schemes, community plants, cash grants, and soft loans.
Low economic returns, and little if any direct cash generation.	Demonstrate indirect benefits (e.g., better fertilizer, sanitation); compare with other non-economical investments that meet human needs and increase comfort and modernization (e.g., TV, car); enact strict pollution control and to manifest the cash gains from improved sanitation and divert commercial fuels subsidies to biogas as renewable source of energy.
Sociocultural	
Culturally deep-seated prejudices against the use of some wastes and illiteracy compounded by superstition and resistance to change.	Education programs; demonstration "seeing is believing"; get visible respected people to use the technology; social research.
BGT is skewed toward the richer strata of the rural society.	Scale incentives more toward lower-income peasant; develop dry fermenters of agricultural residues for poor farmers; make digesters cheaper and more reliable; community plants.
Lack of or superficial identification of the true needs and priorities of target groups and users thus being unable to offer a technology package that suits them best.	Thorough socio-cultural studies by professionals to make biogas a felt need (needs are seldom original, mostly they are "created" by good marketing) and advertise biogas as a "market" product.
Social differences and lack of cooperation in cases of shared facilities and community plants.	Establish proper management schemes; make community plants a service which is profitable to all involved.
Organizational/Infrastructural	
Lack of nationwide governmental policies and plans.	Convince leaders and get decision-makers strongly interested.
Lack of: efficient promotional packages, proper execution schemes, well-planned and monitored field experimentation and demonstration, adequate follow-up services and maintenance capabilities, R&D and engineering, and a centralized information and coordination base as well as proper information and communication flows.	Build full-time specialized organizational mechanisms based on thorough in-depth studies with relevance to local conditions.

ANALYSIS AND CONCLUSIONS

Examination of the force-field analysis results indicate that the means for optimizing the exploitation of BGT can be linked to the rudimentary factor of building and effective organizational infrastructure. In a recent analysis[7], China's secret of success is attributed to:

- Expanding and problem-oriented research programs;
- Strong and well-defined organizational support for training in construction, installation, and management of plants;
- Strong and effective biogas popularization campaigns;
- A sound policy for financial assistance to the beneficiaries; and, above all,
- The organizational system of rural China with its strong link between commune, brigade team, and individual households.

Propagation of BGT is a serious matter that cannot be accomplished by few scattered amateurs or even highly committed professionals. Rather, it needs national commitment, policy decision at the highest level, and an integrated approach based on an adequate institutional apparatus whose primary or even sole activity is BGT. Such apparatus should have the following components and attributes:

- Strong government support.
- National coordinating agency with centralized collection, systematic organization of information, follow-up, and control.
- An integrated network with appropriate mix of centralized and decentralized agencies capable of planning and implementing a coordinated program at all levels and of mobilizing community support for it. Success mandates clear definition of roles and responsibilities of the different component agencies.

The required institutional capacities should provide for[8-10]:

- Surveying energy potentialities and planning.
- R&D and engineering capability to identify appropriate technologies, adapt them, introduce innovations, develop prototypes and their field testing for suitability and compatibility with local conditions, develop appropriate technical solutions, and effect continuous rational optimization based on sound fundamentals and techno-economic principles.
- Social mobilization, arouse sufficient consciousness among the potential customers, identify the variety of conditions that would produce a fit between BGT and the user needs, expectations, and perceptions.
- A systematic well-managed marketing and technology promotion.

- Commercial construction and equipment manufacturing networks with strong private sector participation and materials procurement capability.
- Rural extension services and maintenance facilities.
- Training of personnel and villagers at all levels to attend to the various aspects of the institutional activities.
- Operating schemes of financial incentives, subsidies, and assistance for both adoption by the villagers and commercial production, construction, and marketing.
- Follow-up and carrying out continuous assessment of ongoing activities.

Needless to say, all the responsible entities should be strongly linked with clear definition of goals and duties. The vertical diffusion networks encompassing the central and local governments, village leaders, families, and individuals should be properly interconnected as well.

In essence, what is needed is strong national interest and logical promotional strategies, implemented by proper organizational mechanisms. This package, with its component parts and many relevant details, requires much serious thought, effort, and perseverence.

REFERENCES

1. Reserve Bank of India. 1976. Report of the Inter-Institutional Group on Financial Gobar Gas Plants by Banks. Bombay.

2. United Nations. 1980. Guidebook on Biogas Development, Energy Resources Development Series No. 21. New York.

3. M.M. El-Halwagi, A.M. Abdel Dayem, and M.A. Hamad. 1981. Profeasibility Study on Biogas Technology in Rural Areas of Egypt. Conference on Energy Conservation and Utilization, Cairo.

4. Tata Energy Research Institute. 1982. Review of the Literature of the Promotion of Biogas Systems. Bombay.

5. M.M. El-Halwagi. 1983. Biogas Technology and its Prospects in the Context of Rural Sanitation. Progress Report 2, FRCU Grant No. 82024 Cairo University.

6. V. Smil. 1983. Biomass Energies, Chapter 7. Plenum Press.

7. Tata Energy Documentation and Information Centre. March 1984. Biogas Newsletter. Bombay. Issue No. 1.

8. UNIDO. 1979. Appropriate Industrial Technology for Energy for Rural Requirements. Monograph No. 5. New York.

9. Stuckey, D.C. 1983. The Integrated Use of Anaerobic Digestion in Developing Coutries: A State-of-the Art Review. IRCWD.

10. National Academy of Sciences. 1982. Diffusion of Biomass Energy Technologies in Developing Countries. Washington, D.C.

INFORMATION NEEDS FOR DIFFUSING BIOGAS TECHNOLOGY INTO THE RURAL AREAS OF DEVELOPING COUNTRIES

Tata B.S. Prakasam*, Norman L. Brown**, and Cecil Lue-Hing*

* The Metropolitan Sanitary District of Greater Chicago
** Consultant, Washington, D.C.

ABSTRACT

Ever since the energy crisis of 1973-1974, there has been a growing interest in developing alternative non-conventional sources of energy around the world. Biogas is one such source. The fact that biogas can be obtained by the digestion of biodegradable materials under anaerobic conditions is known for a long time. Numerous studies have been conducted to optimize the production of biogas from waste organic materials. Anaerobic digestion of organic matter not only produces a clean source of energy, but also yields a soil conditioner and fertilizer, and has the potential of minimizing the transmission of some diseases. However, its adoption and acceptance in developing countries is not progressing rapidly. This paper examines some of the issues that are considered to be inhibiting the diffusion of biogas technology into the rural areas of developing countries. It will also identify areas where information is needed to assist the technologists and planners of developing countries in assessing the relevance and appropriateness of biogas digesters to existing conditions of their countries. This information will also aid in determining under what conditions the adoption of biogas technology would make a visible impact on the quality of life of the rural inhabitants. These information needs will be enumerated under scientific, engineering, public health, and socio-economic categories.

INTRODUCTION

Since the energy crisis of 1973-1974, there has been a growing interest in developing nonconventional sources of energy around the world. Biogas is one such source. Most developing countries have been affected by an increase in the prices of conventional energy sources. Consequently, these countries may find the production of biogas from waste organic materials and biomass to be attractive, provided that the biogas can be obtained dependably at an affordable cost and that the technology is accepted by the communities.

Biomethanation of organic matter produces a clean source of energy, yields a soil conditioner and fertilizer, and has the potential to minimize the transmission of some diseases. In spite of these apparent advantages, its adoption and acceptance in developing countries has not progressed rapidly. Problems in the areas of technology, sociology, economics, and management of biogas systems need to be solved to bring about a successful and rapid implementation of biogas technology (BGT) in developing countries. This paper will examine these issues and also suggest several areas of research that would contribute to the optimization of the biomethanation process and the dissemination of its technology in developing countries.

INFORMATION AND RESEARCH NEEDS IN THE ENGINEERING AND SCIENTFIC ASPECTS OF BIOMETHANATION SYSTEMS

Development of Cheap and Affordable Systems

A vast majority of the rural households in developing countries cannot afford biomethanation systems because of their cost. Research efforts that will contribute to lowering the cost of family-sized and community-sized digesters should be continued. Development and use of cheaper, durable, and locally available materials for the construction of digesters and gasholders without jeopardizing their performance, should be an ongoing matter of investigation.

Development and Demonstration of Community-Scale Digesters

Because community-scale biomethanation systems can serve the energy and fertilizer needs of the poor as well as the rich households of a community, several technologists have considered these systems as a possible solution for extending the benefits of biomethanation to all segments of the population of developing countries. However, their appropriateness in a given locale may be dictated not only by their technical feasibility, but also by the interaction of the technological and socioeconomic factors relevant to the specific locale. Work needs to be done to identify the technological factors that would minimize the adverse impact of any socioeconomic factors and to identify the socioeconomic factors that would encourage the acceptance of BGT by the potential beneficiaries.

The danger of generalizing the successes or failures of designs or systems based on a single experience should be realized by the proponents and opponents of biomethanation systems.

Methods to Improve the Performance of Biomethanation Systems

A systematic evaluation of various possibilities to improve the performance of biomethanation systems is required. Research should be directed in the following areas:

Mixing of Digesters. Unmixed digesters usually have considerable dead space and hence have less active volume for digestion. Cheap mixing methods should be devised particularly for household systems. Also, the advantage of mixing community- and industrial-scale digesters should be clearly documented by appropriate data measurements in terms of energy consumed for mixing and the resultant increase in gas production, if any. Designs should be improved so as to achieve good mixing.

Heating. Controlled studies should be performed on community- and industrial-scale digesters to demonstrate whether there will be a net benefit if a portion of the generated biogas is used to heat the digester contents. This information can be used to develop more efficient designs. In addition, the use of solar radiation as a supplemental source of energy for heating digesters should be fully explored and more efficient designs developed to effect a uniform heating of the digester contents.

Scum Control. Scum formation can be a problem in biomethanation systems. Mechanical, chemical, and microbiological scum control methods should be investigated, particularly in commercial- and industrial-scale digesters because such methods can be more easily implemented in these units than in household digesters.

Pretreatment of Feed Stocks. Cheap and efficient pretreatment techniques for enhancing the biodegradability of difficult-to-degrade feed stocks need to be developed, particularly for application in community- and industrial-scale digesters, where operational control and management are expected to be carried out by skilled personnel.

Optimization of Fertilizer and Biogas Quality

Information should be collected on the various feedstocks and the operational protocols that would yield a biogas that contains a high percentage of methane, and a digester slurry that conserves nitrogen, phosphorus, and potassium. Also useful would be the development of cheap and efficient methods for biogas clean-up to remove hydrogen sulfide and carbon dioxide, particularly in large-scale operations, if biogas needs to be stored in cylinders or used in compressers.

Evaluation of the Performance of Various Types of Digester Designs with Various Feed Stocks

Although various types of digesters, such as plug-flow, dry fermentation, two-phase digestion, and attached growth systems, have been reported for digesting various wastes in developed countries, reports of their individual applications to wastes in developing countries are relatively sparse. Although the application of principles of such systems may not be practical to family-sized digesters, it may be possible to apply them to the industrial- or community-scale for producing biogas from industrial wastes, agricultural wastes, and aquatic weeds. Laboratory and pilot studies on the optimization of loading rates, detention times, and nutrient requirements for various designs and substrates should precede any full-scale demonstration and evaluation.

Development and Application of Uniform Measurements and Procedures for Evaluating the Performance of Biomethanation Systems

Data published in the literature on biogas system performance are difficult to evaluate because of nonuniform measurements and reporting. The development of a uniform system for measurements, performance, and reporting of the data would permit the comparative evaluation of biomethanation systems operating in various parts of the world. Such a system has been developed in a recent workshop conducted in Bangkok[1]. The adoption of the suggested system by various countries is urged so that in the future a comparative evaluation of biomethanation systems operating under different environmental and socioeconomic settings is possible.

PUBLIC HEALTH ASPECTS OF BIOGAS TECHNOLOGY

It is a popular notion that biomethanation systems would decrease the incidence of disease and illness in a community by minimizing the dissemination of pathogenic organisms and helminthic ova. However, there are no data available that substantiate this belief. Reliable information is needed to answer the following questions.

- Does the introduction of biogas technology in a community reduce the incidence of helminthic and other waterborne diseases?

- Can the public health benefits of BGT accrued in a community, if any, be quantified (e.g., savings in medical expense; increased income due to more working days that otherwise would be lost due to illness; increased productivity of goods and services, etc.)?

- Is there a public health risk to the owners, operators, and other workers associated with biomethanation systems who usually come in contact with the feed stocks and digested slurry?

- What is the fate of pesticides and other toxic chemicals in the biomethanation systems?

To obtain answers to the above questions, studies with appropriate controls should be conducted to delineate any discernable differences between control and experimental populations. Some of these may be long-term studies particularly if the area is endemic with helminthic and other waterborne diseases.

SOCIOECONOMIC ISSUES OF BIOGAS TECHNOLOGY

Perhaps the most difficult and intriguing issues regarding the diffusion of BGT in developing countries are those related to socioeconomic and institutional factors. Information is lacking on what makes a biomethanation system perform successfully in one community or in the hands of particular individuals and not in others. In addition, there is no consensus on the cost-benefit analysis of biomethanation systems. Although some investigators believe that biomethanation systems are justified on the basis of their cost-benefit analysis, and hence do good to society, others believe they are not justified because of their exorbitant cost relative to their derived benefits.

For successful implementation of BGT in developing countries, it is essential to determine why apparently similar biomethanation systems perform satisfactorily in some areas and not in others. Such knowledge can be used to develop guidelines and protocols for introducing BGT to new areas or to rejuvenate biomethanation systems that are not functioning satisfactorily.

Following are some socioeconomic research areas on BGT where information is either scant or lacking.

Development and Application of Uniform and Appropriate Socioeconomic Criteria for a Comparative Evaluation of Biomethanation Systems

It is difficult to assess the performance of biomethanation systems from a societal and institutional viewpoint unless the investigators justify and make reasonably uniform the assumptions they use in their socioeconomic criteria. Hence, it is important to strive for such justification and uniformity in assumptions.

Information is required to gain an understanding of the shortcomings of BGT and the forces that are counter productive to the dissemination of BGT in various countries. This can be achieved by surveying numerous biomethanation systems that have either succeeded or failed in various environmental and socioeconomic conditions around the world. Such information can be used to identify communities that are likely to accept and benefit from biogas systems or those where BGT may not be accepted or do not play an important role in improving the living standard of the population.

Development of Procedures for Evaluating the Socioeconomic Impacts of the Nonquantifiable Aspects of Biomethanation Systems

Technologists, sociologists, and economists differ in the assumptions they use to evaluate the impact of biomethanation systems. A great deal of confusion exists on the relevance of the assumptions made or not made in the evaluation. Examples of these include the value of recreation or leisure time gained from the use of biogas, value assigned to improved public health, value of fertilizer produced, cost of feedstocks used, value of aesthetics and improvement in overall living standard, human skills required and value of employment potential, value of pollution control achieved, and value of benefits accrued because of alleviation of deforestation and soil erosion. It will be useful if future research can ascertain how a community values these intangibles. Other areas of investigation that should provide useful information include the following:

1. development of educational and motivational materials for propagating the diffusion of BGT,

2. development of techniques for encouraging the participation of women and for sustaining a key role for women in ensuring the satisfactory performance of biogas systems, and

3. evaluation of the current and potential roles that various voluntary agencies can play in the diffusion of BGT in developing countries.

PROGRAM SPECIFIC ISSUES

Before embarking on a national or regional plan for implementing biogas generation, a planning organization should gather information as recommended by the National Academy of Sciences in its 1977 report [2] and then make the decision regarding implementation. This evaluation should include the questions below. If information is lacking for one or several questions, it will be prudent to gather such information first and then proceed with the decision-making process.

• What impact will a national or regional policy of building biogas plants have on the living standard of the population?

• What is the potential of biogas for satisfying the overall energy requirement of the target area?

• Will the biogas energy, not heretofore used, make a direct impact on the standard of living?

• Will the biogas satisfy a public health need by providing a safe, practical, and economical method of handling human wastes?

• Will the fertilizer produced significantly improve agricultural productivity?

• Will the biogas production help solve a serious or potential deforestation problem by reducing the need for firewood?

• Are there sociological or cultural problems to be considered?

  (1) Are there taboos that would interfere with the handling of the raw materials or residue or the use of the gas as cooking fuel?
  (2) Are there traditions that affect whether individual or community digesters are used?

• Should individual or community digesters be built?

  (1) Are trained technicians available to operate and maintain community digesters?
  (2) Would community digesters create employment opportunities?
  (3) What would be the consequences of a "stuck" community digester in terms of energy dependence, material flows, and acceptance by the community?

REFERENCES

1. ESCAP. 1983. Report of the Workshop on Uniformity of Information Reporting on Biomethanation Systems. Bangkok.

2. National Academy of Sciences. 1977. Methane Generation from Human, Animal, and Agricultural Wastes. National Academy of Sciences, Washington, D.C.

THE INTEGRATED DIGESTER PLANT

George L. Chan
Commonwealth of the Northern Mariana Islands

ABSTRACT

This paper describes the biogas project of the U.S. Commonwealth of the Northern Mariana Islands within its Comprehensive Energy Plan aimed at utilization of the locally available energy resources. After 5 years of environmental survey and sanitation work in almost every country and territory of the South Pacific Commission Region, followed by 10 years of research and development in many of these islands and 5 years of field trials in the Northern Marianas, the author has successfully demonstrated the Integrated Digester Plant as a valid answer to the energy and food problems of the rural areas and isolated islands in the tropics. This fact has been recognized by the Resolution passed by the political leaders of the Commonwealth in support of this development concept.

The fiberglass digester adopted by the Northern Marianas meets all the important criteria for a viable commercial product that can be made by a small enterprise to serve the local market:

- simple design that facilitates quality control during manufacture;
- low capital cost that can be repaid in 3 or 4 years;
- easy operation that requires no special skill and costs very little; and
- low repair and maintenance costs because fiberglass is corrosion and rot free, and there are no moving parts.

To improve the cost-benefit ratio, this digester and a series of fiberglass tanks and long-lasting red mud plastic bags (RMPB) and ponds are integrated in a symbiotic system of livestock, aquaculture, agriculture, silviculture, and small processing industry that recycles all the wastes and residues to produce most of its fuel, feed, and fertilizer for productivity optimization. This multidisciplinary project has involved selected members of the public and private sectors in its planning and evaluation and also included an intensive public information and education program through a series of workshops (both national and regional), newsletters, posters, stickers, radio and television spots, slide shows, and booklets for the target audiences: farmers, business people, government officers, and students.

The biogas project of the Northern Marianas has addressed almost all the technical and thematic topics related to the agenda of the present meeting. What is needed now is to establish the Integrated Digester Plant on a 4 to 5 acre farm and to use it as a training center for farmers and a field station for further research and development. This is where the thrust of the Northern Marianas program is aimed, and it is hoped that other governments will establish similar projects to demonstrate this appropriate technology to their people, with or without the help of the regional and international agencies.

SECTION THREE: SOCIOCULTURAL ASPECTS OF BIOGAS TECHNOLOGY

OVERVIEW OF THE THEMATIC SESSION AND KEY RESULTS

This thematic session was designed to have three major components. An initial overview presentation by Capener and El-Hawagi highlighted the vital role played by social and cultural factors in determining biogas project success. Subsequent corroboration by specific biogas work in Rwanda and Sri Lanka provided additional examples of the effects of cultural and social patterns on all aspects of biogas system elements, including technology design. The third session component was a small group exercise to determine major constraints and facilitators that affect successful project implementation of four different scales of biogas systems.

GOAL AND DISCUSSION DESIGN

The goal of the session was to elicit from conference participants a consensus of constraining and facilitating factors in biogas implementation in order to identify areas where greatest effort could be applied to maximize the potential of anaerobic digestion. It was presumed that for a given scale of digester, constraints might tend to cluster in certain categories.

To encourage active discussion and to determine if any consensus could be established among participants, the participants were divided into eight working groups. Participants selected in advance the biogas system scale of most interest to them within the following categories:

- Small individual household system,
- Multi-family communal system,
- Public municipal system, and
- Private commercial system.

Leaders and recorders were chosen and instructed on the style of the discussion groups. Each group used a form prepared by organizers (see the Annex), which listed user needs that can be met in some degree by projects employing anaerobic digestion. The form also provided a set of illustrative constraints and facilitators.

Participants were expected to analyze--based on their own experiences with the technology at the given scale--where most constraints seemed to inhibit meeting targeted needs. Any areas of facilitation were also to be noted.

Conference organizers attempted to identify for decision makers considering biogas implementation those relevant system elements requiring the most attention during implementation.

GENERAL OBSERVATIONS

Participants met in eight smaller groups. Four groups concentrated on individual family units, two on private commercial units, and one each on multi-family communal and public municipal units.

Dynamics of these groups varied considerably. One group would not use the illustrative form. Several groups misunderstood the task but had lively discussions. Several groups worked through the process in completely different ways, but were able to present a consensus on the major constraints impeding biogas implementation.

As individuals, some participants were annoyed by the proposed activity because they could not see the value of it or they were not convinced that constraints from social/cultural factors are important. Other participants became intrigued with the process.

RESULTS

The thematic paper and the two corroborating presentations stimulated the thinking and interest of the members of the conference to a considerable degree. Several major issues raised by the speakers carried the problems of biogas development beyond technology questions.

The major point raised was that biogas as a tool of intervention has to fit the natural local system in which it is placed. The more local system activities that biogas contributes to in a positive way, the greater will be its goodness of fit with the local system. The case of the small farm household digester was used an an illustration. The point was demonstrated that when biogas contributes positively to processes like waste handling, energy, land fertilizer generation, labor, health and sanitation, and greater social prestige, the chances for its acceptance are enhanced exponentially.

Another major point was that different types of biogas installations will call for different patterns of management and operational supervision. The family household digester will use the family members as managers and laborers in a closely integrated kinship pattern of interaction. On the other hand, a multihousehold communal system will require a more formal managerial style with a clear separation between supervisor and laborers in order to establish and monitor performance since all involved do not occupy the same household. Important differences in motivation, incentives, and direct rewards for performance characterize these two patterns of administrative organization. The inappropriate management of digesters was identified as a primary reason for failure of biogas systems.

A series of other key points were identified that impact heavily on successful biogas development. These were as follows:

1. Cultural and traditional patterns were highlighted for the powerful influence they exert on the goodness of fit of biogas. Most important were the customary beliefs and attitudes about the proper place of animals in farming and household systems, housing arrangements for animals, the handling of cow dung, definitions of women's and men's work with animals, and making use of the beneficial output of a digester.

2. The influence of diffusion theory on who are likely to become the clients and users of biogas. Required cash expenditures and capabilities for instruction and maintenance or operational performance preclude the probability that BGT will be available on an extensive basis to the landless laborers and the poor-who perhaps greatly need this type of resource. The type of digesters that can best address the requirements of the needy are the multi-household communal digesters or public municipal units.

3. Private small farm household digesters were identified as of greatest concern and use by those attending the conference. This did not account for the type of communal digesters that may be required for the needy as indicated in the previous point.

4. The private commercial digesters probably have the best chance of rapid diffusion since they lend themselves to being rationalized, capitalized, and operationalized on an integrated business basis. The simplicity of the selling/buying process is not as complex as working through village-level systems to gain acceptance and adoption.

5. The evidence strongly supports the fact that the technological soundness of biogas digestion is basically proven. Of course, there is always a need for more fine tuning. The area that has not been adequately researched is the application side of the equation.

6. Conference participants responded favorably to exploring social, ecological, and cultural constraints and facilitators in biogas development.

DISCUSSION RESULTS BY TYPE AND SCALE OF DIGESTERS

The following results summarize the participants discussions during these deliberations.

Small Individual Family System

Constraints:

1. High initial investment cost.
2. Lack of appropriate information on the system among potential users.
3. Educational level of the people.
4. Traditional socio-cultural values of the people.
5. Priority of other needs that cannot be met directly by a biogas system.
6. Lack of government support.
7. Availability of cheap fuel.
8. Subsidized conventional fuel.
9. Lack of quality construction materials.
10. Low productivity of system or inappropriateness of some design.

Facilitators:

1. Availability of soft financing for system acquisition.
2. Energy contribution of the system.
3. Fertilizer/soil conditioner value.
4. Improved sanitation, health, and environmental quality.

Multi-Family Communal System

Constraints:
1. Benefit/cost calculations.
2. Diffusion of technology patterns of bureaucracy and social organizations.
3. Technological soundness and feasibility.
4. Traditions, values, beliefs, and attitudes.
5. Availability of resources and raw materials.

Facilitators:
1. Financial and institutional infrastructures.
2. Availability of resources and raw materials.
3. Integrated system level analysis.
4. Health and sanitation.

Public Municipal Systems

Constraints:
1. Financial and institutional infrastruture.
2. Traditional values, attitudes, and beliefs about making use of a product that bears a negative image as garbage or sewage.
3. Management skills, maintenance, and operation.

Facilitators:
1. System-level analysis of the multi-family benefits that can be derived.
2. Availability of resources and raw materials.
3. Health and sanitation benefits.
4. Technology soundness and feasibility.

Private Commercial Systems

Constraints:
1. Inability to prove economic viability and performance of commercial-scale biogas systems in cost/benefit calculations.
2. Inadequate or unavailable financing for commercial-scale systems.
3. Inability of initiating agencies or organizations (governments) to sponsor or implement biogas development, especially in competition with other government policies providing subsidies for energy, fertilizer, and other such elements.
4. Lack of adequate system analysis.
5. Poor technical performance of biogas systems.
6. Lack of acceptance because of traditions, values, beliefs, and attitudes.
7. Availability of resources and raw materials.

Facilitators:

1. Contributes to an effective waste management system.
2. Contributes to energy requirements.
3. Contributes to important health and sanitation needs.
4. Appropriate government policies such as pollution control and clean water subsidies.
5. The system level designs that encourage the integrated functional relationships.

6. The value of the biogas products as they ease labor and cash expenditures for alternative processes or products used previously.

THE ANNEX

Instructions for Systems Analysis Worksheet

General

1. This worksheet is not a matrix requiring participants to fill in all the boxes.
2. The worksheet is an attempt to describe visually two sets of concerns and how they might interact to either constrain or facilitate the implementation of development projects using anaerobic digestion.
3. Throughout all sessions of the conference we are seeking to identify ways in which more successful application of biogas technology can be attained. In this Thematic Session: Sociocultural Realties of Biogas Development, we hope to concentrate on social dimensions and how they can impede or assist development. We expect participants to reflect on their experiences in their countries and environments in order to draw generic conclusions about how and where social and cultural reality "participates" in the success of a project.

Specific

1. The first use of the worksheet will be a recording of your own experiences. In designing or implementing the kinds of biogas projects that you have actually handled, which major areas represented the greatest constraints, and which proved most useful? Was most of your frustration with the technology itself? Or did the social organization of the "user" community preclude proper operation and maintenance? Or were the institutional requirements, such as financing or extension, impossible to pull together?

2. The second use of the worksheet will involve small workgroups of the session. Each of these workgroups will discuss constraints and facilitations relating to a specific scale of implementation: small individual family system, multi-family system, or private commercial system (to include industrial waste).

In these exercises, we hope to elicit a consensus of each workgroup on which constraints are most serious. From this analysis we can imply the areas requiring future work, application of funds, or other actions.

SYSTEMS ANALYSIS WORKSHEET
INTERNATIONAL BIOGAS CONFERENCE
CAIRO, EGYPT, NOVEMBER 1984

	USER NEEDS						
CONSTRAINT/FACILITATION	Waste Management	Energy	Sanitation and Health	Resource Conservation	Labor (male/ female)	Market Valued	Discretionary Income
Systems Level of Analysis							
Availability of Resources and Raw Materials							
Traditions, Values, Beliefs, Attitudes							
Diffusion of Technology Through Patterns of Bureaucracy and Social Organization							
Types of Initiation and Sponsorship for Development							
Management Skills Including Operation and Maintenance							
Health and Sanitation							
Technological Soundness (Feasibility)							
Financial and Institutional Infrastructure							
Cost-Benefit Calculation							

SOCIAL, ECOLOGICAL, AND CULTURAL REALITIES OF BIOGAS DEVELOPMENT

Harold R. Capener, Cornell University
and
M. M. El-Halwagi, National Research Centre, Egypt

ABSTRACT

This paper draws from project data derived from a concluding five-year research and demonstration program concerned with the technical feasibility, economic viability, and social acceptability of biogas use in rural Egyptian villages. A series of propositions are set forth that highlight those social, ecological, and cultural considerations associated with biogas installation, and implementation in village settings. Issues examined are: the dimension of human needs for a sustainable source of energy; fitting biogas into total system functions; constraints and facilitators of diffusion; initiation, management, and administration of various types of units; and criteria for measuring overall success.

INTRODUCTION

In a book written by Seymour Serason entitled The Creation of Settings and the Future Societies (San Francisco, California: Jossey-Bass 1972) the author points out that when new projects or programs are initiated, virtually all time is spent detailing elements of the present and future to be developed or changed. Unfortunately, he writes, too little time is devoted to a critical review of the structural preconditioning of the past which the new project or program seeks to change. A similar message was the theme of a recent symposium entitled "Unanticipated Consequences of Social Structure," i.e., that the whole planning focus rests on the anticipated benefits of new development efforts, with little or no attention devoted to the realistic potentialities of the unanticipated consequences.

In this thematic session of the conference, the intention is to employ a holistic systems-level perspective. While it is not feasible to try to relate biogas to the whole society, the more important interactive connections biogas has within subsystems will be portrayed.

As a means of highlighting a series of social, ecological, and cultural phenomena considered basic to evaluating the impact of the anticipated as well as to unanticipated consequences of biogas development, the following eight propositions are presented.

PROPOSITION NO. 1

Biogas research and development activities will achieve greater measures of success to the extent that they are compatible and operate within their system-level requirements.

An illustration of this proposition is found in a small farm household system set in the Egyptian ecological environment. A primary goal in designing and installing an experimental biodigester in the farm family household could be expected to focus on an alternative method of waste management and provision of a new source of energy. When a concerted study effort was made to better understand the farm and household system dynamics within which the digester would play a part, the following insights took on significance.

First are operational factors. The size of the unit is about 5 acres. The rotation of the multicropping and mixed cropping patterns between summer and winter seasons requires careful planning. Competition for the scarce land requires allocation for animal feed versus crops for market or family consumption. The interdependence in the man-land-animal relationship is especially strong. The animals provide power for farm tasks, manure for fertilizer, and an assortment of products such as milk, meat, and hides. One of the most important of the animal products is manure, as it is a source for both fertilizer and fuel.

The importance of fertilizer in the farming system is shown by the extensive allocation of labor resources to the hauling of large quantities (half a ton or more per day) of earth from the fields to the animal shelter as an extender to the manure. This soil is hauled in bags on the backs of animals. The added material helps soak up the urine and animal waste, thus creating a greater fertilizer bulk. When the mass becomes too deep for the animals to wade in, they are tethered in a different area of the shelter. As soon as the built-up batch dries out sufficiently, it is removed to the field. Because the mixture is so heavy and bulky, it is normally composted until the land is prepared for the next season's crops, and it can be ploughed into the soil.

Estimates of the value of man and donkey labor inputs equal about 15 Egyptian pounds per month. The arduous hauling task was carried out by the son of the farmer. Outside labor was hired at an equivalent cost to work in the field, lending needed assistance to the father. The cash savings on hired labor would be sufficient to pay for the installation of the family-sized digester in about three years. With the potential reallocation of the son's labor to assist his father, and counting the value of alternate use of donkey labor, the overall cost effectiveness of the digester on labor saved alone had special attraction for the men of the household.

From the perspective of the household system, a combination of time-consuming and distasteful labor requirements were identified. Foremost was having to arise about 4:30 in the morning to tend to the animals. The first of these onerous tasks was feeding, then wading through the built-up mixture of urine, cow dung, and mud to collect sufficient fresh material for the supply of cow dung cakes, then attempting to clean the animals so that they could be milked. This operation was repeated twice each day. A series of attendant problems were also identified, such as the unsanitary quality of the milk, the great difficulty in washing one's clothes by hand, the pervasive odor of the manure from the animal shelter attached to the house, the swarms of flies, and the virtual impossibility of maintaining a reasonably clean household.

The source of energy used in the household for cooking was bottled gas, made available to urban and rural areas of Egypt at highly subsidized rates by the government. In addition, cow dung is used extensively for cooking, especially for bread baking, which is done by the women in cooperative small group settings and has sufficient redeeming social value not to give way easily to household gas or even biogas.

The design of the experimental digester included the innovative notion of installing a sloped concrete floor under the cows with a small trough to the rear. This feature enabled the daily supply of dung to be swept into the trough and with small amounts of water rinsed into the digester placed just outside the animal shelter. The solid floor thus kept the animals much cleaner.

Once installed, the biogas unit proved to possess a range of beneficial features extending beyond the primary waste management and energy goals. Something of a contest arose between the men and the women as to who were the greatest beneficiaries of the biogas plant. From the men's side, the elimination of hauling earth to the animal shelter and the built-up mixture to the field was cited first, along with the digested material's lack of strong odors and the harmful antibodies being neutralized. The fertilizing properties of the digester material proved to yield better results on the crops than the larger quantities of the earth-diluted manure. The digested effluent tended to dry and crumble into lightweight, flexible material that could be carried to the field in the morning on routine trips. Furthermore, it could be used with the compost to plow into the soil at the time of crop rotation or to be applied as fertilizer side dressing. Thus for the men, the digester provided equivalent or better fertilizer, freed up their labor resources, and allowed them to concentrate on the more pleasant and productive features of the farming enterprise.

For the women, the digester's ability to eliminate the most time-consuming and aesthetically distasteful tasks was its strongest feature. Easing the burdens of working with the animals, and particularly the waste management aspects, was highlighted. The presence of the biogas led to new multiple burner units for the kitchen, a new kitchen sink for washing dishes, an ability to heat plenty of water, a new washing machine to utilize the hot water and take the drudgery out of doing the laundry, a general cleaning up and painting which manifested a new sense of pride, and a measure of social prestige in the household. The latter was partly evidenced by the family's attaching a long plastic hose to the biogas system so the family could take a burner into the sitting room and make tea while visiting with their friends or guests.

By increasing their ability to view the farm and household system dynamics through the eyes of the farm men and women, researchers and development strategists will gain much clearer notions about how they should proceed. A new academic philosophy now rapidly gaining popularity is known as a "farming systems" approach. It advocates the necessity of looking at the system level functions from the bottom up, particularly through the eyes of the farmers and the members of their households. Gaining this type of understanding will ensure better attunement and goodness of fit of biogas technology at the farm family household level.

PROPOSITION NO. 2

Worldwide human needs for a sustainable source of energy are staggering. The raw materials for biogas-generated energy are yet to be exploited.

It is estimated that about 90% of the people in developing countries depend upon firewood for their principal source of fuel. The growing crisis in fuel wood supply spans much of Asia, Africa, and Latin America. The growing shortage of fuel wood will place increasing pressures on the ecosystem, such as deforestation, soil erosion, flooding, and loss of biomass and organic matter. In previous times the searching and gathering process for fuel wood was hard but possible. Now it is becoming more nearly impossible and increasingly detrimental to the environment (See Sandra Postel, "Protecting Forests," pp. 74-94 in State of the World 1984, by Lester R. Brown et al. New York: W. W. Norton/Worldwatch Institute, 1984.)

It has been noted that in the Indian subcontinent the increasing shortage of fuel wood will place heavier demands on the use of dung cakes for cooking. The estimate is that between 300 and 400 million tons of wet dung, which shrinks down by four-fifths when dried, is basically denied to the land and burned as fuel. The equivalent plant nutrients burned and thereby going into the atmosphere add up to more than one-third of India's chemical fertilizer use. (See Government of India, Ministry of Agriculture, Interim Report of the National Commission on Agriculture on Social Forestry, New Delhi, August 1973, p. 37.)

The economics of fuel wood scarcity cause an increase in its value. For some families, this amounts to as much as one-quarter of their earnings. (See Erik Eckholm, The Other Energy Crisis: Firewood, Worldwatch Paper 1 Washington, D.C.: Worldwatch Institute, 1975, adapted from his book Losing Ground: Environmental Stress and World Food Prospects (New York: W. W. Norton, 1976.)

The same developing nations in which fuel woods are in critically short supply are the nations where population expansion is high and where the rural peasant economies retain a strong traditional man-land-animal relationship. When people can neither find firewood nor afford kerosene, charcoal, or bottled gas, they will be forced to find something else to burn. In certain places they have resorted to gathering leaves, branches, twigs, dried weeds, and dead stalks and to stripping bark off live trees. A relatively available and sustainable source of energy must be found so that people can cook their food and partially supply their needs for warmth and light.

The existence of a close relationship of man, land, and animals in Asia, Africa, and Latin America offers the indispensable ingredients for supplying the basic raw materials for biogas as a source of energy. Land, of course, is crucial in being able to supply the agricultural wastes that can go into the digester and to supply food for the animals and/or humans whose wastes supply good digester materials. Clear advantages lie with landholders in terms of their access to and control over the resources that go into both the production and the fullest use of biogas technology (BGT).

On the other hand, the number of rural households that are completely landless or lack secure access to adequate farmland is estimated to contain more than 600 million people. This number of landless and nearly landless people is likely to increase, however, as rural populations are growing at nearly 2% a year and therefore will double in 35 years. (See Erik Eckholm, <u>The Dispossessed of the Earth: Land Reform and Sustainable Development</u>, Worldwatch Paper 30 Washington, D.C.: Worldwatch Institute, 1979). It is important to note, however, that many tenant and landless rural families own animals. The value of the animals lies not only in their edible and usable products but in their dung for fuel. Considerable concern could be expected to arise among the landless and those who do not own animals if biogas systems began to command the greater portions of heretofore free salvageable cow dung from streets and grazing areas.

The likelihood is that landless laborers and other village residents are not appropriate candidates for individual household digesters. They are more likely to become participants in some form in a communal accumulation of a combination of agricultural, animal, and human waste materials that would go into a multihousehold digester system.

By definition, raw materials would have to be available and in ample supply before either a public municipal system or a private commercial system could be seriously contemplated or constructed. In the former case, waste products from government owned and operated enterprises or municipal sewage are examples. In the latter case private dairy operations, beef cattle feed lots or desert watering stations, or poultry or swine enterprises are illustrations of sustainable sources of raw materials for biogas-supplied energy.

<u>PROPOSITION NO. 3</u>

The successful design and implementation of biogas systems will be strongly influenced by the cultural and traditional patterns of values, beliefs, and attitudes among local peoples.

Whenever a new technology is to be introduced into a cultural system, its intervention will fit more easily if its component features flow with the grain of the cultural environment. Where a new intervention crosses the traditional grain of the cultural system, its accommodation will be more problematic. It is from the roots of the past that a society's norms, values, beliefs, and attitudes are molded. Cultural definitions evolve that signify what is proper work for men and women, which kind or portions of animals are edible and which are not, what odors are considered tolerable, what is defined as being dirty or clean, the appropriateness of handling cow manure with one's hands, and whether it is more appropriate for women rather than men to handle it to make dung cakes.

The traditional grain of rural agrarian cultures can be expected to provide strong sustaining support for individual or communal household biogas development. The close man-animal-land relationships include a dependence on the use of animal dung for fuel and fertilizer and for its adhesive qualities in certain construction. The pattern of housing animals in shelters closely connected to the family dwelling for their care and protection, and the additional dependence on the animals for transportation, draft power, and their products such as milk, meat, and hides all attest to the central place animals occupy in such a system of relationships.

In the peasant rural agrarian cultural contexts where the small farm household or the communal multihousehold systems are the models, the compatibility of these systems with the cultural and traditional patterns allows such digesters to add significant benefits in the form of saving labor, easing heavy burdens, improving health and sanitary conditions, and contributing quality of life and aesthetic improvements along with an integrated system of waste management and sustainable energy.

In contrast, a biogas unit recently installed on a large dairy farm in a northeastern state in the United States followed U.S. traditional values in its design and installation. That is, there is considerable physical and psychological distance separating the barn from the family dwelling. Dairy cows are normally cared for by the men rather than by women, and cow manure is considered to be dirty and is never handled or molded with one's hands. Thus is seen the powerful influence of cultural and traditional patterns on fitting BGT into the grain and flow of existing societal values, beliefs, and attitudes.

The U.S. dairy farm model was purposely designed on a private commercial system pattern. The uses of the digester products were calculated against waste management handling, labor saving, and uses of the energy for various farmstead operations. The system design did not make any use of the methane gas in the household, partly because of the physical distance from the barn to the house, but primarily for aesthetic reasons. The woman of the household choose to use available electricity at a considerably higher cost rather than to have an animal-produced source of energy in her house.

From the above illustrations, the powerful influence of cultural traditions on appropriate biogas system designs can be seen.

PROPOSITION NO. 4

Patterns of bureaucracy and social control impose powerful constraints on the diffusion of biogas technology.

If a relatively inexpensive sustainable source of energy such as biogas could be made available to the neediest family households, it should go to those in developing countries where the greatest proportion of the population resides. These would be the households largely dependent upon collectible fuel wood, dung cakes, or other more commercial sources. A strategy of diffusion of a new technology like biogas to such a target population is called a "worst first" approach. In other words, the technology is aimed at those with the worst need first. In a social structural sense this group is the hardest to reach with any new technology because they occupy the rungs towards the bottom of the socioeconomic ladder.

The moral mandate of most development schemes is underwritten by the worst first philosophy. This is true whether the schemes come from private foundations, international davelopment agencies, or the five-year plans of governments. The rationale of public development is to render assistance to those in need, and monies can best be morally justified to assist those having the greatest need. The problem with this rationale and philosophy is that it is built on a theory that basically does not work. It is not so much that it cannot work as that it does not work. Two examples will illustrate.

In the early 1960s, the government of the United States developed a scheme wherein they declared a war on poverty. The moral edict was that for a nation with so many resources it was unconscionable for there to be so many of its people hungry, homeless, jobless, and lacking in medical care. In the March 22, 1979, issue of the Wall Street Journal, William E. Simon, a former secretary fo the Treasury, observed that despite quintupling of the federal budget over a period of 18 years, the record of accomplishment in conquering poverty was dismal. He indicated that had the money simply been given directly to the poor each person would have received about $8,000 or a family of four would have had $32,000. Instead, the greatest proportion of the war on poverty funds were absorbed by the bureaucratic system administrating the program.

The second example comes from a recent television newscast and New York Times article on the current devastating drought in Africa which affects some 24 of 51 African nations. Of around 35 million persons in and around the Sahel, it is estimated that 6 million may die because there is no way to get food delivered to them. One United Nations report calculates that despite the millions of dollars being mobilized for this most critical of all causes, only a small percentage of the resources will reach starving people. It will be contrained by a combination of physical features such as shipping ports, manpower, transportation, and distances, and also by layers of administrative procedures.

Stages in the Diffusion of Innovations

The dynamics of the diffusion process work from the stratification theory, which holds that social class differentiation is a structural property of virtually all societies. From the diffusion research conducted extensively beginning in the period 1955-1960, primarily on agricultural practices, the following findings were revealed (Everett M. Rogers, Diffusions of Innovations New York: Free Press of Glencoe, 1962).

- Five stages were associated with the process adoption: awareness, interest, evaluation, trial, and adoption.
- Five levels or propensities toward adoption were delineated:

innovators, early adopters, early majority, late majority, and general others.

The discriminating characteristics of individuals occupying each of the five levels are very significant. In brief, the characteristics of the innovators are that they have more of everything. They have more land, wealth, education, venturesomeness, position, status, and access to information. They travel more, read more, talk to more people, have a broader resource base, and can afford to take risks in trying out innovations without the danger of losing everything. They occupy prestige positions high enough in the social class structure that if they take risks that fail they are not severely put down by being laughed at or ridiculed. It is apparent that the innovators and early adopters are those in the best position to develop awareness and interest and begin to think about the merits of a new technology and to try it out or see it tried elsewhere and become convinced to adopt it.

The opposite is true of the characteristics of the lowest category, the "general others " These individuals tend to have less of everything. They lack land, wealth, education, travel, venturesomeness, position, status and access to new information. They do not talk to a wide circle of people, are locality oriented and tradition-bound in their thinking, and, perhaps most important of all, cannot afford to take two kinds of risks. The first is the economic risk of allocating very scarce resources to a biodigester that has not been proven and accepted by important others in the local situation. The second is the unreasonable notion that people in the general other class position could or would run the risk of smiles, joking, ridicule, and loss of face for presuming to bring new technology into the local system. They generally do not see themselves as the proper candidates for demonstrating or teaching their neighbors who are in an advantaged social status position around them.

One other sensitive dynamic in the demonstration and diffusion process is the "one-strike principle," using the analogy of the rules of baseball, in which a batter is allowed three strikes before being called out. In diffusion dynamics, particularly with an open public demonstration like a biogas unit, the technology interventionist or development worker will normally only have "one strike," or one attempt to achieve success or failure. The reason is that a public demonstration is very different from a laboratory experimental process, in which a failure can be counted a success since it reveals what will not work and the laboratory technician can try again and again until success is achieved. The public demonstration has all eyes fixed upon it. If it works, it is declared a success, but if it does not work it is considered a failure and public sentiment does not allow a clean slate for successive experimentation. The experimental ground becomes a kind of burned over territory, and it is much more difficult to light the sparks of interest or cooperation a second or third time.

The dynamics of the diffusion process indicate that there is no direct and easy access to the "worst first" target audiences in spite of the moral and political rhetoric of development schemes. Thus it becomes a critical requirement for researchers and development strategists to give ample attention to the anticipated as well as the unanticipated consequences of biogas projects.

PROPOSITION NO. 5.

Different types of initiation and sponsorship of biogas technology will produce a wide variety of outcomes.

The initiation and sponsorship of BGT will generally originate from one or more of the following kinds of sources:

- Private commercial sector,
- Donor governments to recipient governments,
- Private voluntary organizations, and
- International or world development agencies.

The purposes of sponsorship can range from basic research and development for technological soundness to various means of dissemination, from private commercial sales to larger scale public extension and development schemes.

Obviously the research and development sponsorship precedes the application initiatives. This was the case with the Egyptian biogas project that organized this conference. Joint funds from the United States and Egypt have been involved in the project, which has been located in the National Research Centre and directed by the project manager, Dr. M.M. El-Halwagi. The first-stage mandate for the project was to produce prototypes of technologically sound biogas digesters. The second stage was to field test the prototypes under typical village conditions and to make observations on their technological appropriateness. A final expected result is a set of observations about the suitability of the various prototypes for either their private commercial value or their rural village development applicability.

Several observations can be drawn about outcomes that might be expected from a research and development sponsored project with a mission such as that given the National Research Centre. The following key points deserve attention:

1. From the outset an interdisciplinary team was envisaged. Included were engineers, biological scientists, chemists, entomologists, agronomists, and social scientists. The project director was skillful in molding the team to the point of listening and learning from each other, especially as to how the contributions of each discipline would become dependent upon others for the whole project to succeed.
2. The core members of the R&D team planned and completed extensive travel to a series of other countries where biogas technology was an important component of their own national development plans.
3. All of the possible experiments within the laboratory were tested at the bench scale and then were moved up to the prototype level. Village conditions were simulated at the laboratory, complete with cattle and crop fertilizer demonstrations.
4. At the same time the R&D work was proceeding at the laboratory, the social science team was comissioned to carry out a general preintervention survey. The purpose was to reassess general characteristics of Egyptian rural villages and to make a selection of two villages that would be somewhat typical of predominant patterns. The villages had to be within a reasonable commuting distance of the

National Research Centre (NRC) in Cairo. This work was undertaken and completed by an Egyptian social scientist. Of the two villages chosen, one was an old village with centuries of tradition. The second was a new 20-year-old village carved out of the desert and made possible by irrigation.

5. When the prototype digesters had completed their rigorous testing at the NRC, the time had arrived to replicate their performance at the village level. By assigning highly trained and responsible members of the R&D team to go to the two villages and take complete charge and oversight, five small farm household units were constructed. Two were built in the old village and three were constructed in the new village. Cooperating farm families were nominated by village leaders, so they represented neither the richest nor the poorest families but more nearly the middle groups. The families agreed to provide the land, animals, and labor, and continuing operation and maintenance. The R&D team agreed to provide the digester components and participate in the total installation and training in their management and operation.

6. From the beginning of the installation and operation of the digesters, their solid performance became an object of interest and envy by other farm families. In particular, they desired that the NRC also provide them with a free digester. They declared they would be more than happy to receive it.

7. With the first two stages completed in what is obviously a multistage process, the logical question arises, where does the project go from here? It may very well be that a research and development organization like the NRC has carried the project as far as it reasonably can or should. It has successfully completed the research and development work. It remains for the proper development ministry to assume the larger tasks of the tranfer of what appears to be an appropriate technology from the standpoint of engineering, biology, economic feasibility, and social and cultural adaptability.

Utilizing the advantage of hindsight, one can see that it would have been better if original funds had been available for a development ministry to be incorporated in the program from stage one. Not surprisingly, however, other national program policies have gone forward to address the energy plight of the rural and low-income populations. Since Egypt has oil resources, a national program of subsidized bottled gas called butagas has been made available at a rather cheap price. This policy, which will be difficult to change, will act as a strong variable in the national energy policy equation.

The team of the NRC is in a key position to serve in an advisory capacity to a sponsored program of rural biogas dissemination. It would appear that a needed component of such a program strategy would be some form of low-cost loan or subsidy which could provide the initial cash outlay for digester installation. Careful attention in following design specifications and providing essential training, education, supervision, and monitoring services would also be prime requisites.

Private Commercial Client

A comparative model of sponsorship can be seen in a development experiment the NRC carried out with a private commercial farmer/businessman. This individual owned a poultry broiler enterprise located in the middle of a citrus orchard. Following the successful R&D work to design and install the biogas digester, its functions proved to be highly satisfactory. It utilized the poultry waste by providing energy for brooding the young chicks, grinding grains, pumping water, providing fertilizer for the orchards and fields, and facilitating a fairly integrated commercial operation.

Of special note in the sponsorship of this project was the relative simplicity of the interactions between the principal parties from the NRC and the farm system. The interesting fact is that the majority of the time and attention was paid to the R&D characteristics of the biogas unit, since in typical fashion this innovator type farmer-leader did not require all of the education, motivation, incentives, and convincing of normal public sector clients.

A review of the subject content of the biogas technology transfer interactions highlighted the following:

1. Assurances that the type of poultry waste materials were of sufficient quality and quantity to sustain the operation of a digester.

2. Assurances that the biogas digester design would operate in an ecologically sound fashion within the environmental setting.

3. Assurances that the necessary design and construction materials were available and that skilled personnel were available to either perform the construction work or supervise it.

4. Assurances that the necessary training, supervision, management, and operational skills would be provided to the local personnel to enable them take charge of the system with confidence.

5. Assurances that technical services, maintenance, monitoring, and any needed repairs would be available in the event assistance was needed.

6. Assurances that the cost-benefit calculations could show the investment to be practical, feasible, and financially sound.

7. Assurances that the full system-level capabilities of the digester would have goodness of fit and would be fully compatible in the poultry/citrus farming combination.

8. Assurances that the overall benefits of adopting the digester would outweigh the risks in terms of the technical, economic, and social consequences.

The dynamic that stands out in a striking manner in these interactive relationships is the relative simplicity of the agreements forged. Absent were the layers of protocol, institutional and organizational hierarchies, numerous rules about procedures and bureaucratic regulations, and social system constraints of customs, values, traditions, and attitudes.

The advantages of private sector commercial development of biogas technology build around self-interest, motivation, individual control, and visible incentives and returns. The private sector can make a significant contribution in adopting and adapting biogas development. Necessary advertisement, legitimation, and social and economic feasibility can all be provided by vanguard efforts from the private sector.

The basic problem of biogas diffusion to the masses, however, will remain as a major problem and challenge for the public sector. This type of initiation and sponsorship is far more complex and to date virtually outpaces our creative capacities for viable solutions.

PROPOSITION NO. 6.

The management requirements and skills will differ markedly, as will the responsibilities for operation and maintenance, within each of the four types of biogas systems.

The biological processes of anaerobic digestion will basically be the same whether in an individual household, in a communal digester serving many households, or in a public municipal or private commercial digester. For each of these systems there will be, however, significant differences in organizational environments, in managerial roles, and in the individuals involved. First, a sharp contrast can be drawn between the organizational environments of a small farm household digester and a public municipal system.

In the case of the public municipal biogas system:

1. It will be organized and rationalized on an industrial model, which creates a clear separation between management and labor.

2. The industrial model generally assumes that all managers and workers are men and that they will be paid according to their positions.

3. The industrial model emphasizes specialization in tasks and dependence upon authority.

4. The industrial model discourages employing members of the same family, as this is considered nepotism.

5. The industrial model uses fixed job responsibilities, fixed hours, and fixed pay and benefit schedules as ways of organizing and rationalizing the management and control of the system.

In the case of the small farm household biogas system:

1. It will be organized and rationalized on a family kinship model, in which management and labor are closely integrated. While some elements of the industrial model are present, the inclusion of kinship elements creates the significant difference.

2. The kinship model allows for some exchange of management functions between men and women in terms of who really makes decisions and utilizes the extended family and children as unpaid labor.

3. The kinship model emphasizes generalization in tasks and sharing of responsibilities.

4. The kinship model encourages utilization of all members of the family in the system.

5. The kinship model uses interchangeable job responsibilities, flexible hours, unpaid labor, and family shared benefits as ways of organizing, rationalizing, managing, and controlling the system.

With regard to the management skills needed for supervising the small farm household digester, the central roles performed by the women would suggest their active training and involvement in these processes. This does not raise the question of who is the boss or authority in the household; it simply recognizes that in the peasant cultural contexts where such digesters have application, women will perform key roles. They will care for the animals, handle the waste, feed the digester, utilize the gas, and be sensitive to problems that may arise. The men in such households will have more to do with installation of the digester and with utilization of the fertilizer materials coming out of it.

Much of the deliberation about the division of labor between men and women has centered around questions of crucialness by various criteria at hand. The question arises then, crucial for what ends? Men have traditionally been considered crucial because of their muscles, hard work, wage earning, decision making, and the authority structure. But the reciprocity of women's roles is also increasingly and of necessity being recognized as crucial. In a study of a Pakistani village, Ruth Dixon shows in Table 1 how the nearly 16 daylight hours of women's work daily activities were distributed.

Table 1
Time Spent on Daily Domestic Activities by Women in a Pakistani Village

Activity	Time Spent (hours)	Percentage
Care and feeding of livestock	5.50	35%
Milking and churning	1.00	7
Cooking	1.75	11
Carrying food to fields and feeding children	1.50	10
Housecleaning and making dung cakes for fuel	.75	5
Carrying water	.50	3
Child care	.50	3
Other domestic chores (e.g.,food processing, crafts)	3.00	19
Afternoon rest	1.00	7
Total waking hours	15.50	

Source: Ruth B. Dixon, Rural Women at Work (Baltimore: Johns Hopkins University Press for Resources for the Future, 1978).

Over one-third of the day is devoted to the care and feeding of the livestock. These animals are essentially under the women's management. The women's coordinating role is also clear as they engage in the processing, servicing, maintenance, and nurturing functions that keep the whole of the farm and family systems tied together and functioning.

While the planning of biogas development projects may be perceived to be more the domain of men in the rural economies, the day-to-day operation and management of the household-level digester is very likely to be carried out by women. The criticality of women's role in the small farm biogas technology must therefore be seriously considered.

The private commercial system may be organized more like the small farm household system if it is not too large. It may be a family-run operation and therefore closely parallel the kinship model. If it is a larger commercial operation, it will more likely be patterned along the lines of the public municipal system following the industrial model.

The communal multihousehold system will have a special set of management, maintenance, and operational requirements placed upon it. Important among these will be matters of dependable service, equal service in the distribution system, trouble shooting and maintenance of equipment in numerous households, and safeguarding against jealousies and factional disputes that may arise using biogas as a problematic cause. Other management and operational issues, such as dependence upon a steady source of raw materials for the digester, the likelihood of a more diverse mixture of materials, and the inherently greater problem of diverse communal and customer satisfaction, places this type of unit more on the industrial model.

On a kinship-industrial scale, the four types of units would be arrayed as follows:

Kinship Model			Industrial Model
Small farm household system	Private commercial system	Communal multihousehold system	Public municipal system

PROPOSITION NO.7.

Biogas technology makes a positive contribution towards improved health, sanitation, and aesthetics.

Biogas technology directly addresses a number of factors affecting the health and wellbeing of families and population groups. The capability of the digester to handle both human and animal waste and under proper conditions to break down the life cycles of parasitic and pathogenic materials creates a strong deterrent for disease transmission. Considering that the lack of proper sanitation is a dominant factor in the prevalence of the enteric diseases dangerous to infants and young children, the digester's interruption of the transmission cycle is a significant contribution. The fact that waste mater materials also placed inside the digester are rendered odorless and less attractive to flies, rodents, and insects is another contribution to sanitary conditions. These benefits would be expected to accrue to each of the types of biogas systems under discussion.

In the case of the small farm household digester, several additional benefits were identified. Included were the easing of some especially burdensome tasks related to waste management and reducing the amount of fuel gathering and storage and of smoke from cooking fires. More sanitary conditions in the household made for safer food preparation, processing, and storage. The availability of hot water improved the overall laundry tasks as well as general household cleaning.

The net effects of the household digester were observed to not only improve health and sanitation but also generate a new sense of status and pride in the overall household and family management. These contributions to the aesthetic emotions and sensations of the family were of notable consequence.

PROPOSITION NO. 8

It is far easier to achieve desired results on the test for soundness of biogas technology than of its social and economic feasibility.

The technical soundness of biogas technology has been proven time and again from the standpoint of its biological functioning and various engineering adaptations. When it comes to the question of its feasibility, one has to place the digester in a variety of cultural, social, environmental, and orgaizational settings, and its feasibility may fare differently in each setting. To illustrate this latter point, reference is made to three experimental situations at village levels in Egypt which met with differential success, not on their technical soundness but on their social feasibility.

The first case is the one already described where the small farm household found their biogas digester to have solid technical soundness and a highly satisfactory social feasibility. Further elaboration on this example is unnecessary.

The second case concerns a second digester placed in the same experimental village, but in this case in connection with a social service center. The center had a large covered building and a feedlot where sizable herd of young male buffalo calves were fed and prepared for market. The animals' waste was fed into the digester. The technical performance of the digester was highly satisfactory, but the social feasibility was very unsatisfactory for several reasons. First, the center operated on a fixed schedule, and when normal working hours were over everyone went home. There was no one to look after the digester or to make any effective use of its products. A subsequent system-level analysis revealed that the location of this installation would prove less than viable unless more meaningful connections could be made with nearby farmto utilize the digester effluents for fertilizer and to either make some family use of the energy or allocate it to some cottage-type industry. The adverse dynamics of the social center siting were not unanticipated. The experiment was followed through to confirm the predictions of several of the researchers.

The third case is located in a second experimental village. Three separate small farm household digesters were installed. Two of these digesters have been judged both technically sound and socially feasible. They are serving the needs of the men on the farms and the women in the households in a manner highly satisfactory to all concerned. The third household is a different story. The digester itself has proven to be technically sound but has failed on social feasibility.

The physical location of the digester proved to be the stumbling block. Two neighboring families who had been close friends for many years had their animal sheds side by side. An agreement was struck for the two families to place the digester in a spot where it could receive the waste from both animal sheds. The volume of gas to be produced was judged to be sufficient for the needs of both families.

These two families were not related to each other. They had no kinship ties or bloodline connections. Not long after a successful beginning there began to be questions and suspicions raised among the women as to whether the other family was putting its fair share of manure into the digester or whether some waste was being withheld for other uses. Further questions arose as to whether one or the other family was using more than their fair share of the biogas for cooking. As family feelings arose, words were exchanged and emotions were heightened even among the children. Finally an impasse was reached in which the families stopped operating the digester altogether to reduce the level of conflict.

From these three examples it can be seen that the tests for social feasibility of biogas will be more difficult to meet than those of technical soundness. It is also apparent that social and economic feasibility tests will have to be worked out in the public sector, where the needs of the rural populations remain to be addressed in crucial matters of energy.

TECHNICAL AND SOCIAL PROBLEMS FOR THE PROPAGATION OF BIOGAS IN RWANDA

W. E. Edelmann, ARBI, 8933 Maschwanden, Switzerland

ABSTRACT

Rwanda is a small country in central-east Africa. It has a very high population density. Deforestation for cooking purposes has led to severe problems. An extensive study has shown that the improvement of cooking stoves and the introduction of biogas production technology is of the very first order of importance for the country.

In a FAO program, some Chinese design plants have been constructed. It is shown that for geographical, demographical, and sociocultural reasons, other designs are more appropriate to the local conditions. The most suitable solution for the small farms in Rwanda seem to be small-scale, discontinuously-fed digesters for the handling of solid wastes.

INTRODUCTION

Rwanda is a small African country at the sources of the White Nile (Ruwenzori) and the Zaire. Its neighbors are Tanzania, Uganda, Zaire, and Burundi. Rwanda is one of the most densely populated countries of Africa. It is very poor and has no fossil energy resources of its own. One of its main problems is deforestation; most of its trees have been cut down for cooking purposes.

A vast study by the Swiss Federal Institute of Technology led to the conclusion that: (a) the propagation of improved stoves for more efficient use of scarce wood, and (b) the propagation of biogas technology, are of first priority for the country. Another less important possibility is the import of small hydropower technology to use the energy potential of the many small rivers and streams of the country. (Rwanda is a very hilly area about 2,000 m above sea level with rainfall throughout the year.)

The Swiss government is sponsoring the construction of a biogas installation for the agricultural school of Nyamishaba at the borders of Lake Kivu. Our group is responsible for the technology transfer. We focused on finding a solution appropriate to the school and one that could be modified for the needs of the average farmer.

THE BIOGAS POTENTIAL OF RWANDA

In Rwanda there are very few towns or large villages. Most of the population lives in small isolated farms on thousands of hills. On one farm there lives the whole extended family with children and grandparents. From the pressure of the rapidly growing population, soon more than half of the farms will have an area of less than 5,000 m^2.

The data on the animal distribution of Rwandian farms is most rudimentary. The cow is very often used as a dowry resulting in the fact that many farmers own just one cow. Nevertheless, it may be concluded that a farm in the region of Lake Kivu has an average 2.3 cows, 7.2 pigs, and a few goats and sheep. It may be estimated that more than half of the total number of animals live on farms of four animal units or more; only a few farms have more than four animal units while many farms have less than two units. Very few farms have more than 20 animal units.

The production of biogas is believed to be an important resource for the country[1-5] but biogas would only play a major role if there can be found an appropriate, applicable solution for farms with more than three animal units. In this case, in the region of Kibuye, where Nyamishaba is located, around 10 to 15,000 m^2 of biogas could be produced per day.

ACTUAL BIOGAS PRODUCTION IN RWANDA

Two organizations, the AIDR (Association Internationale pour le Developpement Rurale) and the CEAER (Centre des Etudes et d'Application de l'Energie au Rwanda) are building biogas plants in Rwanda. The FAO has also put a considerable amount of money into a program for the propagation of Chinese biogas plants in Rwanda (Programme RWA/78/004).

So far, more than 12 plants have been constructed. Most of them are not working well or are out of order. The principal reasons for this were mostly technical in nature. These included:

- construction failures,
- insufficient know-how for the start-up procedure and for maintaining the installation. For example one digester was not inoculated though there was a working plant nearby; another digester was fed daily for nearly one year though it ran sour during the very first days of function,
- the available substrate was not well suited to the process, and
- the responsible technical person left the job.

Nevertheless, there has been seen a small Chinese-type digester which worked well and produced energy and light on the farm. The costs of the plants installed are around 14,000 FRW/m^3.

SOCIOCULTURAL CONSTRAINTS FOR THE PROPAGATION OF BIOGAS IN RWANDA

An analysis of the local conditions led to several conclusions. The Rwandian people are "mountain people," i.e., they live very isolated in their farms and separate from their neighbors. They are rather conservative. There are few interactions between different families. This situation is not favorable for the construction of community biogas plants where several families bring their wastes to a single plant.

The houses are located on or about very steep slopes of the hills. Water has to be carried for a long distance in many cases. This leads to the conclusion that digestion technologies which do not need too much liquid are preferred.

The families raising animals are not accustomed to producing a liquid manure in a pit. The animals normally are standing in a shelter on a bedding of straw and plant material. A solid manure is thus produced in a form suitable for use as a fertilizer. The use of liquid fertilizer in the case of anaerobic digestion would necessitate new tools and new habits for a rather conservative people.

THE SITUATION AT THE SCHOOL OF NYAMISHABA

For this case, the number of animals equals about 11 animal units of cattle and milking cows as well as 8 units of pigs, sheep, and chickens. For bedding around 140 kg/d of straw and grass are used. There is no significant amount of easily digestible organic waste material, such as vegetable wastes. It may be estimated that under optimal conditions in the mesophilic temperature range, about 50 m^3 biogas/d could be expected (total production of energy needed for heating: 20-25 percent at Rwandian climatic conditions). At ambient temperature, the total achievable production is lower, even if the retention time of the substrate is significantly prolonged[6]. The total gas production is lowered also by increasing the total solids content of the substrate above around 8 percent TS[7]. A last factor influencing the production is the lignin content of the bedding -- highly lignified straw and very dry grass are not well suited for anaerobic breakdown. Other factors inhibiting a good fermentation seem not to be present in the substrate.

Considering these parameters, a gas production of 20 to 30 m^3/d may be obtained at ambient temperature depending on the technology used and on the time spent by the cattle outside the stable; actually, the cattle stay about four hours per day on the fields and the sheep for about six hours.

The main use of the biogas is substitution for firewood. It was estimated that actually 10 m^3/d could be used for cooking (taking into account the efficiency of the burners and the amount of food to prepare for the students). This amount may increase when the number of people increases in the near future. Other energy supply needs are: the motor of a grain mill ($4m^3/d$) and a cheese factory near the plant. This shows that the energy need equals roughly the energy production expected.

On the farm of the school, a solid manure is produced and applied. The responsible Swiss person indicated, however, that it is possible to produce and apply a liquid manure if necessary from the anaerobic digestion. This needs new organization and construction in the stables. Indeed, a lot of changes have already been made on the school of Nyamishaba. Therefore additional changes may not be advantageous.

POSSIBLE DIGESTION TECHNOLOGIES

Indian Digester (KVIC Gobar Gas Plant)

The Indian digester consists of a cylindrical, mostly vertical hole with a gas drum of metal or ferrocement on top. It is fed continuously, and the gas is produced at constant pressure. Disadvantages of this system include:

- expensive as a gasholder;
- corrosion of the gasholder if made of metal and not handled with care regularly, and cracks in the ferrocement case; and
- liquid manure necessary, requiring a water supply.and work to dilute the dry manure.

Chinese Digester (Water Pressure Digester)

The Chinese digester (in India: Janata plant) is constructed of concrete with a fixed gas dome also of concrete. It is built underground. If the gas is not used, pressure builds up, i.e., it is a variable gas pressure system. The major advantage of this technology is that it is cheaper than the Indian design. Disadvantages include:

- gas leakage through the gas dome due to cracks in the concrete and to the sometimes very high gas pressure obtained;
- gas losses through inlet and outlet tubes (the gas losses of Chinese plants are in general significantly higher than in Indian plants where gas is lost between the drum and digester wall);
- not comfortable to clean (sediment has to be brought out twice a year; the installation has to be emptied then); and
- liquid manure necessary, requiring a water supply.

The construction of water pressure digesters has been reduced significantly in contemporary China as pointed out by Chen Ru Chen[8]. He states, "The peasants become much richer. They would rather pay more to obtain reliable and convenient digesters. Peasants complain about the inconvenience in building, operating, and cleaning water pressure digesters."

Bag Digester (The Taiwanese digester)

This digester consists of a bag made of plastic material where the substrate enters on one side and leaves on the other. The gas is stored within the bag (neither high pressure, nor constant pressure). Liquid manure is necessary. The advantages are:

- It may be installed without large problems (e.g., no big construction works); and
- It is rather cheap and not too difficult to maintain.

The disadvantages include:

- The aging characteristics of cheap plastic materials are not yet known; and
- the plastic material may have to be imported.

Plug Flow Digester (as Developed by Jewell)

The plug flow digester developed at Cornell University consists of a tub dug into the ground covered by a plastic membrane for storage of the gas produced (neither high gas pressure nor constant pressure). The advantages are:

- needs less plastic material than the bag digester;
- very easy to maintain (just put the cover away);
- cheap and reliable; and
- high gas production.

The disadvantages are:

- It may be necessary to import the plastic cover material; and
- liquid manure is necessary.

The German Design (GTZ)

The German[9] design consists of a combination of a Chinese underground construction with a floating metal drum (Indian) instead of the fixed dome. Some disadvantages of the Chinese plant could thus be overcome, but the disadvantages of the metal drum still remain. There is not enough practical experience to give a final answer on this "BORDA" design.

Batch Digester (Italian design)

It consists of a pit covered with a plastic membrane.[10] It may be filled with liquid as well as with solid manure. The advantages are:

- cheap and reliable;
- easy to maintain;
- has just to be fed sporadically (e.g., after a period of two to three months at ambient temperatures) -- it does not need daily feeding; and
- suited for solid manure, i.e. no transport of water for substrate dilution and no liquid fertilizer.

The disadvantages include:

- plastic covering may have to be imported; and
- no constant gas production.

The disadvantage of varying gas production may be overcome by constructing two to three smaller units instead of a single large one. If they are refilled periodically in rotation there is always enough gas produced for the users.

In China this design is built very often. The farmers, who have applied dry composted manure, speak very positively about "dry fermentation."[8]

SOLUTION PROPOSED FOR NYAMISHABA

Considering the facts cited above, it was decided to construct three batch units for the digestion of solid manure. This solution has the advantage that it can be modified for application on the various small farms present in Rwanda. Since it is almost impossible to obtain a homogeneous slurry because of large particles occasionally present in the bedding, the water is often far from the farms, and the farmers are not accustomed to producing and applying liquid manure, the propagation of plant designs that necessitate use of liquid manure does not seem to be a viable option for Rwanda.

Figures 1-3 show the proposed design. A hexagonal form was chosen, as the pit is rather large (about 20m^3). The danger of cracks in the walls is reduced accordingly. The hexagon has the advantage that it has a large volume in relation to the length of the walls and the construction material needed. Several units may be combined without loss of area and with common walls (Figure 4).

The gasholder may be made from plastic or metal. The metal gasholder could be welded on the spot, but it has the disadvantage that it cannot be used as a storage container (as there is no slurry above which it can float). It was decided, therefore, to import a plastic membrane gasholder from Switzerland but of UV-proof, tissue-reinforced PVC. The costs for one unit are $550 US. Since there is a PVC factory in Rwanda producing plastic sheets, the gas holders may be produced within the country at a later stage with foreign know-how. As the Nyamishaba installations have an experimental character, a pit was

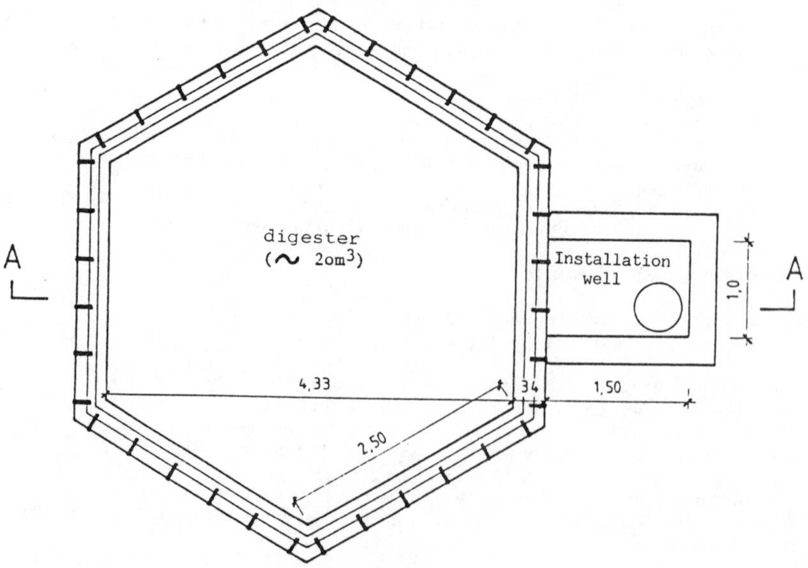

Figure 1. Proposed digester design - Plan (top view without cover)

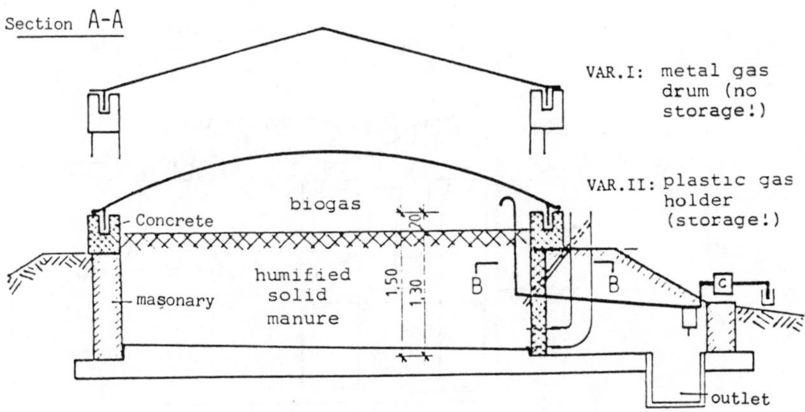

Figure 2a side view

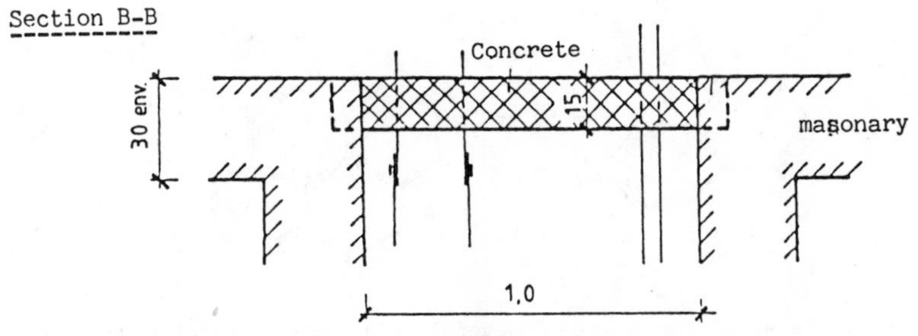

Figure 2b Detail wall

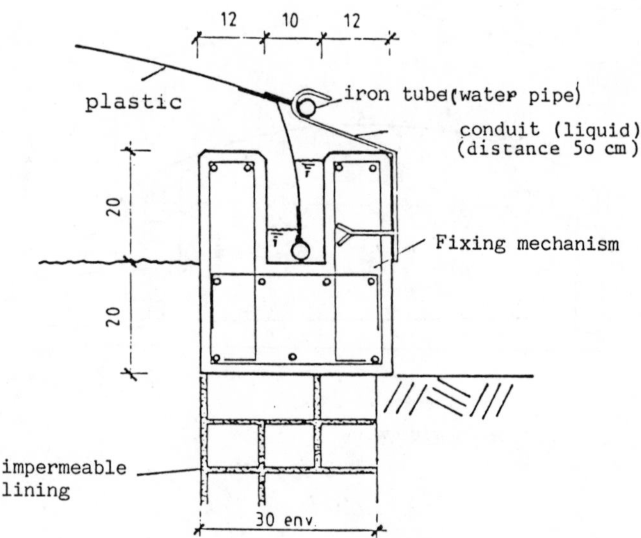

Figure 3. Detail fixation plastic membrane

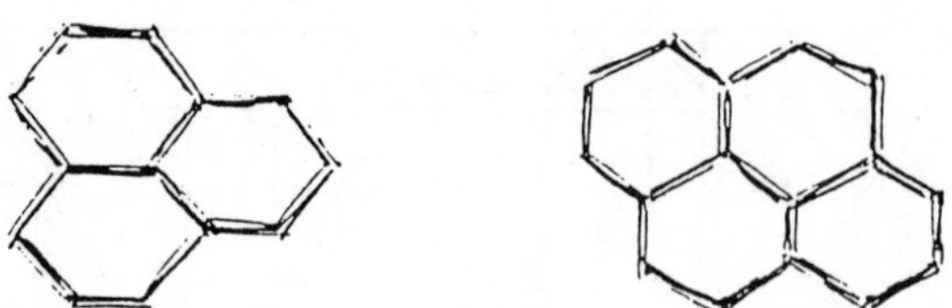

Figure 4. Arrangements of several units of hexagonal digesters

planned for easier access when measuring digestion parameters. The digesters are constructed with local materials. A crown out of cement is necessary for statical reasons at this digester size.

DIGESTION PARAMETERS

The pits will be filled with the solid manure obtained from the different stables. From pigs, a liquid manure is produced, which is stored in a small pit. On top of this pit, a latrine is built to demonstrate to the pupils that human excreta may also be used for biogas production.

The liquid of this pit is used to dilute the solid manure filled into the digester. This has the advantage that the solid manure is more or less "under water," i.e., that the walls of the digester do not need to be totally gastight, as there is some liquid around the solid manure. (The minimal amount of liquid to moisten the dry manure has to be determined. Probably the digestion will be something between a totally dry digestion and wet digestion with extremely high TS-content).

Since the local climatic conditions maintain an average temperature of at least 20°C within the digester, a digestion time of nine weeks was proposed. This means that three weeks after filling a digester, the next one has to be emptied and refilled. Such a pattern fits well with the cycle used for cleaning the stables.

Cheap solutions were proposed when possible. Figures 5 and 6 show a Chinese gas valve and the outlet for the liquid made from an old inner tube of a car wheel. A part of the liquid stays in the installation for the inoculation of the next batch while another part equalling the content of the pit at the pig stable is taken out. The solid part is taken out by forks.

FUTURE PROSPECTS

At the time of the presentation of this paper, one installation has already been built and is in the start-up stage in Nyamishaba. Two more units will be built after having obtained results from the first unit. Then a long gas line will be extended to the kitchen. Actually the gas may be used in the cheese factory nearby. The gas-pressure will be generated by a small gas compressor since it is not reasonable to put too much weight on a large membrane gas holder.

It is planned to construct small units for Rwandian farms after gaining some experience with the large units. There, the gas pressure will be generated by a weight on the plastic membrane. The membrane will show chambers similar to an air mattress which will be filled with water (Figure 7).

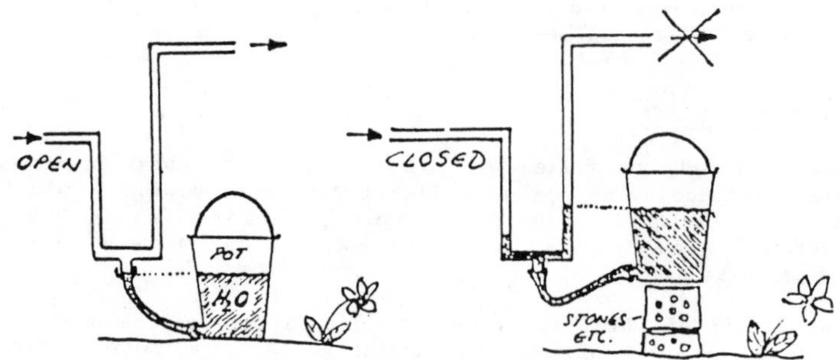

Figure 5. Simple gas valve

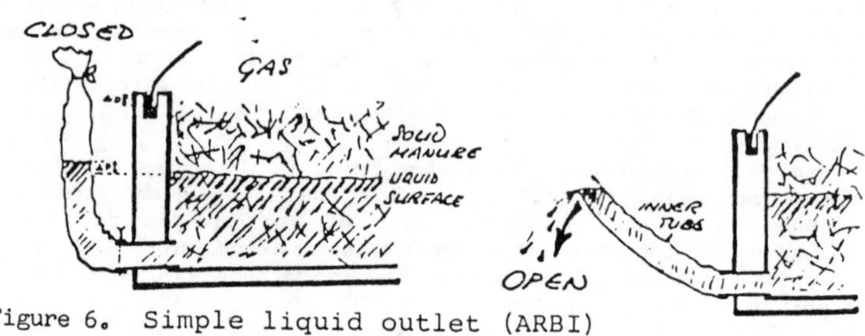

Figure 6. Simple liquid outlet (ARBI)

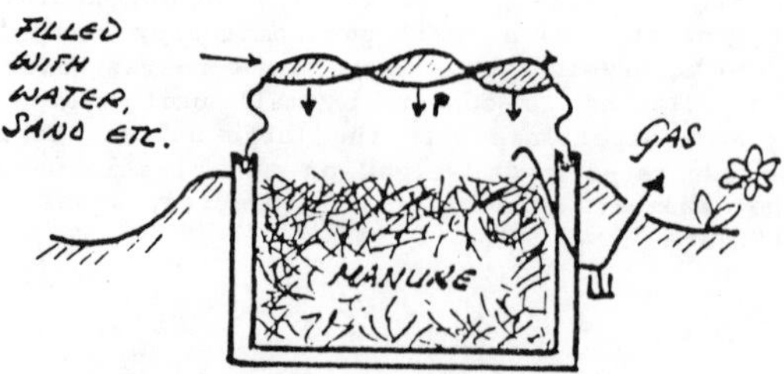

Figure 7. The ARBI design of a simple, small plant for the digestion of solid manure.

SUMMARY

A case study for the construction of biogas plants on an agricultural school in Rwanda is described. Present plants in Rwanda are not appropriate to the local situation. Therefore, it was proposed to build digesters for the fermentation of solid manure at the school. The design can be modified for installation on small farms, as soon as the design criteria for Rwandian substrates are determined with the plants currently built. Other plant designs are also described.

REFERENCES

1. Etude du secteur énergetique au Rwanda, Tome I. 1983. Syntheèse générale. Bunep/EPFL. Lausanne.

2. Rapport PAK. 1981. Tome II Eléments, analyses, résultats de l'enquête au milieu rural. 1981. DEH. Berne.

3. Rapport d'evaluation projet agricole deKibuje. 1982. DEH/Minagri. Berne/Kigali.

4. Edelmann, W., and P. Siegl. 1983. Possibilitées de la production de biogaz à l'école agro-forestière de Nyamishaba. ARBI. Maschwanden.

5. Etude du secteur énergetique au Rwanda, Tome V. Energies non conventionelles. 1983. Bunep/EPFL. Lausanne.

6. Stevens, M.A. and D.D. Schulte. 1979. Low temperature digestion of swine manure. Journ. of the Env. Eng. Div. p. 33.

7. Baserga, U. 1983. Zur Dimensionierung einer Biogasanlage. Sonnenenergie. 1/83. Zürich.

8. Chen Ru Chen. 1983. Up-to-date status of anaerobic digestion technology in China. AD 83 proceedings. p. 415-428.

9. Pluschke, P. 1984. Small and simple biogas plants for developing countries. Referate at Bioenergy 84. Götheborg.

10. Balsari, Bonfanti, and Sangiorgi. 1980. Un impianto per la produzione die biogas in piccoli e medi allevamenti. Referate at 2th. Intecsol, Verona.

11. Edelmann, W. 1983. Digestion systems for agricultural wastes. AD 83 Proceedings, Boston, USA.

TOWARDS OBTAINING OPTIMUM BENEFITS FROM BIOGAS TECHNOLOGY

M. Amaratunga
Department of Civil Engineering, University of Peradeniya, Sri Lanka

ABSTRACT

Biogas technology has not had the acceptance it deserves in many countries because of inadequate emphasis on the multifold benefits of using the technology as part of an integrated system. Attempts to popularize the technology as a means of obtaining a cooking fuel and other benefits are not given due publicity. Doubts may be raised about the economic advantages of introducing the technology as a source of fuel since it is only a small yield of the full potential of an integrated biogas system.

To appreciate fully the real potential of biogas technology as a means of rural advancement, it is necessary to consider a number of parameters including socio-cultural aspects and national policies and schemes relating to the supply of food, fuel, and other commodities. This paper describes the basic concepts of an integrated biogas system within which the technology can be evaluated for a given set of conditions.

INTRODUCTION

Various developments have been made in many countries to obtain gas that could be used as a fuel. Since the oil crisis of 1973, biogas technology has been given wide publicity as an alternative source of renewable energy, but the other direct and indirect benefits of a comprehensive integrated system have not received the attention they deserve.

When advocating the use of biogas technology, it would be relevant to describe and popularize many of the components which form a comprehensive integrated system. This paper discusses components of importance to rural societies in developing countries. Although gas is the most attractive component, it may be only a small percentage of the total potential of a comprehensive system, and a system may be economically viable even if the gas is not used.

A COMPREHENSIVE BIO-ENERGY SYSTEM

General Requirements

When advocating the use of biogas technology, the availability of the essential requirements and the means of using all products should be given careful consideration. All too often, projects are undertaken without proper feasibility evaluation and end in failure. The more essential requirements are:

- an adequate extent of land,
- a plentiful supply of water,
- the ready availability of animal and vegetable wastes, and
- adequate financial resources.

The extent to which these requirements should be met can be judged from considerations outlined below. Social traditions and expectations and state policies regarding social amenities all play an important role in the successful implementation of a biogas technology program.

Components of an Integrated System

The main components which could be considered as appendages to a biogas unit are:

- agricultural fields,
- algae ponds, and
- fish and duck ponds.

A cycle of operations can be set up, with the biogas unit forming the core of the system. The amount of benefits which could be derived from any one component will vary from time to time and will also depend on the location.

Inclusion of Other Energy Forms

Other energy forms can be integrated with this system wherever feasible. For instance, solar energy can be used for lighting, water heating, and distillation, or for heating digester slurry. Crop residues that cannot be readily fermented can be used for the production of producer gas. Wind energy, when available, can be used for water supply, electricity generation, or agitation of digester slurry or ponds.

In certain locations micro- or mini-hydropower generation becomes feasible, and its appropriate aspects can be effectively linked to a bio-energy system. The use of wind, hydropower, and producer gas will, however, require large financial resources, and benefits will be applicable more on a community scale.

Output from an Integrated System

The magnitude of the output of the products of each component will depend on many factors and will vary between wide limits. Some conservative figures are given below to illustrate the order of magnitude of economic advantages.

- Biogas: 0.03 m^3/kg of wet cattle dung.
- Fertilizer: 10% increase in crop production, allowing for land area released for ponds.
- Algae: 25,000 kg/hectare per year.
- Fish: 3,000 kg/hectare per year.
- Ducks: 3,000 kg/hectare per year.

Comprehensive data pertaining to yields of the various possible sub-division of the components are not available. It is possible, however, to get an idea of the response of a given land area to utilization for various purposes based on the assumptions made above.

Consider a 10,000m^2 (1 hectare or 2.5 acres) of land area proportioned as follows:

- Buildings for human, animal, and industrial use: 350m^2
- Roadways and pathways: 900m^2
- Biogas plant: 50m^2
- Algae ponds: 200m^2
- Fish ponds: 500m^2
- Agricultural fields 8000m^2

The following table gives an economic appraisal of the proportion of additional advantages each component of the integrated system contributes to its total potential. These have been based on the dung of five head of cattle and an assesment of 500 kg extra yield of agricultural produce.

Table 1 Economic Appraisal of Integrated System Components

Component	Unit Price (Rs)*	Value of Product (Rs)	Percentage of System
Biogas (petrol equivalent 1.62 m^3=1 liter)	13.5	2025	13%
Fish (kg)	20	3000	20
Duck (kg)	25	3750	25
Algae (kg) (40% protein)	08	4000	26
Agricultural produce (kg)	05	2500	16

* 1 US$ = Rs 26

The total capital investment for the biogas tank and the ponds is estimated to be Rs.40,000. It is estimated that, with interest rates higher than 10%, biogas and increased agricultural production alone do not make the system economically attractive. Indirect advantages have been ignored here.

LONG-TERM BENEFITS

Some Social Conditions which Influence Integrated Systems

Besides the benefits discussed, there are other benefits of a long-term nature that are not often considered in an evaluation. Among the social traditions and conditions influenced and affected by the propagation of biogas systems, those related to health and employment are perhaps the most significant. Financial constraints are often given as reasons for the slow spread of the technology.

Health Aspects. The use of gas for lighting and cooking, as a substitute for kerosene oil and firewood, provides direct benefits in the form of clean and more hygienic living conditions. Other indirect benefits not always appreciated also exist.

In most developing countries, hygienic toilet facilities are almost nonexistent in rural areas. The resulting pollution, particularly of waterways, frequently causes bowel disease epidemics. (In Sri Lanka, about 60% of patients treated in hospitals suffer from preventable bowel diseases caused by drinking contaminated water.) Popularization of the use of integrated biogas systems in which toilets are connected to the digester can help reduce the incidence of such diseases.

Anaerobic digestion provides a simple and inexpensive means of rural sewage treatment, and the added advantage of obtaining a clean fuel and a valuable fertilizer should be another influencing factor. There will undoubtedly be a reluctance to use human wastes as fertilizer, but in combination with other animal and vegetable wastes forming the bulk of the sludge, there is evidence that there will be eventual acceptance.

The risk of infection is not entirely eliminated by the use of anaerobic digestion. The kill rate for most pathogens is high, given sufficient retention times. It is possible to have a separate digestion chamber for human wastes to be given further treatment after anaerobic fermentation but before use as a fertilizer.

Benefits such as easing pressure on medical institutions and staff, savings on medical supplies and transportation, and increases in work output because of improved health are all factors which need recognition in any appraisal of the feasibility of a system. Improvement in animal and plant health through destruction of pathogens is yet another benefit.

Employment. Rural unemployment poses a serious problem in many countries with large numbers of people moving away from traditional agricultural practice and migrating toward urban areas. Manufacture of bricks, collection of sand, and production of gas-utilizing equipment, including burnt clay gas burners and lamp components, provide alternative avenues of employment. Also animal husbandry offers further opportunities for diversifying employment opportunities.

Biogas, even in an unscrubbed form, is a suitable substitute for acetylene, and oxy-biogas welding can provide a useful facility in a workshop. If biogas is used to run petrol or diesel engines, then technical requirements associated with their maintenance will provide more extensive means of employment.

Financial Aspects. The technology and its concepts would seem a high-risk type of investment. It is unlikely that private capital can be easily attracted to its development if reasonably secure high interest rates are available. (In Sri Lanka, at present reasonably secure interest rates of 30% are available.)

Frequently, subsidies are given on food and fertilizer, and sometimes on fuel as well. With expanding populations, these subsidies invariably result in an increasingly large demand on a national budget. This situation could be effectively exploited and some financial resources diverted toward integrated systems that could eventually provide some degree of self-reliance in fertilizer and fuel development at rural levels, in addition to increasing food production.

CONCLUDING REMARKS

When advocating the use of biogas technology, it is best to introduce first the concept of an integrated system so that a prospective user can appreciate the full potential of the technology. Viewed from a national long-term perspective, the system could well be viable through benefits other than the usually accepted biogas and fertilizer components.

Once a decision is made to establish an integrated system at a selected location, it would be useful to proportion land areas for various purposes and to introduce component types; for example, different types of animal husbandry or crops which would give outputs of optimum benefit.

SECTION FOUR

ECONOMIC ASPECTS

SECTION OVERVIEW

The four papers included in this section are devoted exclusively to the economic aspects of biogas systems. The first three are full-length treatises, while the fourth is a shorter communication summarizing a relevant work conducted in Nepal.

Moulik's thematic paper, "A Critique on Cost-Benefit Analysis of Biogas Programs," examines the question of cost-benefit analysis of biogas systems and sheds considerable light on a number of fairly controversial points. Firstly, he questions the assumption that peasants were noninnovative as previously assumed. Data from India indicated, in fact, that farmers readily accepted biogas technology and were quite innovative.

Next, a number of methodological points are raised including uncertainties in estimating the cost of inputs and outputs since many of these are non-monetized items. In addition, severe problems exist in estimating secondary benefits such as the multiplier effect of social benefits and improved health. He suggests that an intracountry study be carried out to more closely define these benefits.

Finally, the practice of reducing all decisions on biogas to a single figure, e.g., cost/benefit ratio, or internal rate of return is strongly questioned and a proposition is put forward that decisions should be based on a matrix that sets out these figures in relation to a number of key variables assumed in their calculation.

El-Halwagi's paper, "Assessment of the Feasibility of Rural Biogas Systems," provides methodology for analyzing the overall feasibility of rural biogas systems in terms of market analysis, technical feasibility, investment and operating costs, financial appraisal, and socio-economic cost-benefit assessment. Various techniques and guidelines are recommended to accomplish these tasks. He suggests that the marketing strategy for biogas systems should be in most part a "technology-push" type since demand for the technology is limited in many cases by lack of information and financial resources. In the socio-economic area a "system net change" approach is suggested for cost-benefit appraisal. In this approach, changes in the rural "organic-waste generating" system as a result of implementing the biogas subsystem are to be evaluated, rather than valuing inputs and outputs independently. Finally, and in view of the many uncertainties and assumptions involved in the various valuations, sensitivity analyses are strongly recommended.

Pluschke in his paper, "Analysis of Economic Factors in the Dissemination of Biogas Plants," draws heavily on the work by the German Appropriate Technology Exchange (GATE) in Africa and the Caribbean. He questions the use of sophisticated cost-benefit analyses in order to make decisions on biogas systems' implementation. He feels that a more "intuitive" approach be used on a case by case situation. This approach would include assessment of the reliability of fossil fuel supply, environmental benefits, and improvement in the quality of human life.

Lau-Wong's shorter communication, "The Economics of Biogas Systems," is an excerpt highlighting an extensive work on the economic analysis of different biogas systems in Nepal. Though it is difficult to quantify economic benefits of biogas systems, nonetheless the author makes an attempt to compare it with the afforestation scheme, since biogas in Nepal is mainly used for replacing fuelwood for cooking. Comparisons are also made for a number of income-generating schemes in which biogas can be used as a substitute for diesel fuel, and results indicated much more favorable economics.

Throughout the conference, the economic aspects were touched upon in few other papers. However, they were mostly limited to indicating that economics is a major constraining factor within the context of rural biogas systems. There are some notable exceptions. One is the Baguant and Callikan paper, "The Biogas Project in Mauritius."* Among the various facets encompassed in that paper, the economic aspects of biogas production in Mauritius are considered in simple terms. Their analysis is based on two family-sized biogas plants suitable for producing the daily cooking energy requirements for an average rural household. Under the present conditions in Mauritius, preliminary results tend to indicate unfavorable economics.

The second is Obias' paper, "A Biogas System for Developing Countries," where he demonstrates the economic viability of Maya Farms' 'biogas works'. A return on investment of over 48% is calculated for a large biogas scheme established on a hog farm of 500 sow units.

The third is Henning's summary report, "Biogas Plant in the Ivory Coast," wherein the economics of the system are worked out in fair detail. Results indicated that an optimized system (2 biogas units and 25 kW generator) can produce electricity about 30% cheaper than that produced by the national grid.

---

*This paper is not included in the book for space limitations.

A CRITIQUE ON COST-BENEFIT ANALYSIS OF BIOGAS PROGRAMS

T.K. Moulik
Indian Institute of Management
Ahmadabad, India

ABSTRACT

The principles of cost-benefit analyses are examined as they are applied to evaluate the viability of a biogas program. Particular attention is given to assessing methodological issues, limitations and parameters used in estimating the monetary values of costs and benefits, since they play a crucial role in determining results.

Important outcomes of existing cost-benefit exercises are summarized and evaluated. It is emphasized that experience should be added to laboratory data and theoretical calculations. At times, field observations and results reflecting the actual impacts of secondary benefits have been in direct conflict with academic analysis. Also, the paper corroborates the finding that the economic viability of biogas plants tends to increase with size. Another important outcome, supported herein, is that a biogas program can be a viable national investment even under questionable assumptions if societal benefits are incorporated in the analysis.

INTRODUCTION

Cost-benefit analysis is one of the most commonly used methodologies to determine the economic and financial viabilities of the various alternatives of a project. As an investment decision-making aid, it maximizes benefits by choosing the best alternatives and minimizing the incidence of unproductive and non-viable investments. There are three basic aspects of an investment appraisal -- financial, economic, and social evaluations. The financial evaluation deals with the profitability of an investment at market prices, while the economic evalution is concerned with a set of prices reflecting "efficiency" benefits to the nation. Social evaluation refers to social profitability in terms of welfare implications. The basis parameter in all three aspects of cost-benefit analysis is to work out actual flows of income and expenditure out of the project investment.

VARYING RESULTS

It is clear that the result of a cost-benefit analysis is entirely dependent on the accuracy of a stream of calculations based on projected physical quantities and prices. In other words, the assumptions of the key parameters in estimating monetary values of inputs (costs) and outputs (benefits) play a crucial role in determining the results of cost-benefit analysis. The results may drastically change in favor of or against investments depending on the monetized estimates of the key parameters.

That the results of cost-benefit analysis vary widely dependent on the assumptions of the key parameters (inputs and outputs) is clearly demonstrated in the case of its application in relation to biogas projects. In a comparative review of a number of cost-benefit studies applied to biogas projects, it was shown how the results varied in ranges of indices indicating highly profitable, or viable to unprofitable, or nonviable investments[1]. Based on this review and his own work, the present author concluded in the same paper:

"In many cost-benefit analysis exercises, often a single estimate of NPV* or IRR** is presented on the basis of which an "accept-reject" decisions has to be made. One must bear in mind that this estimate of NPV or IRR is only one value amongst all possible values of outcomes. It is obvious, therefore, that some outcomes are more likely than others, given a particular set of assumptions and estimates of variates entering the calculation of NPV or IRR."

METHODOLOGICAL ISSUES

The wide variability in results of cost-benefit analysis and the implicit risk of 'uncertainty" in reaching a particular outcome brings us to the crucial question of methodological issues and problems. In the same paper referred to earlier, the present author discussed in detail some of the critical methodological problems in relation to biogas projects. For the present, it would be worthwhile to briefly summarize some of these key problems.

At the outset, it should be emphasized that the methodological problems of a cost-benefit analysis of a biogas project essentially arise out of uncertainties or inadequacies of a stream of calculations of inputs and outputs based on physical project quantities and prices. It is therefore unwise to rely excessively on the reasoning that overestimates in one place will be balanced by underestimates in another. There are variations in the importance of some variables in influencing the ultimate results.

The uncertainties or inadequacies in input-output calculations in monetized quantitative terms are largely in relations to the following four parameters:

1. Given the widely varying decentralized operating conditions of a biogas system, there is lack of standardized data about the quality and quantity of inputs and outputs. Since it is difficult to estimate accurately the supply-response factors in such uncontrolled decentralized operations, the input-output parameters are usually estimated in cost-benefit analysis of biogas projects at the ideal

* Net Present Value

** Internal Rate of Return

experimental conditions, which most often are far away from the actual field conditions.

2. For the same reasons of decentralized variable conditions, the economic life of biogas systems irrespective of different designs cannot be easily standardized.

3. Pricing of inputs and outputs poses some insurmountable problems. In pricing there are the questions of domestic market price and price movements over a period of time (i.e., inflation) and the world price in actual and potential contribution to foreign-exchange saving and earning or import and export. The more methodologically agonizing problem of pricing is in relation to traded and non-traded inputs and outputs. It is a well-known fact that in many parts of the Third World, major inputs (animal and plant waste) including labor and outputs (biogas and digested manure) are often non-traded goods without organized stable markets and are often obtained at zero private cost. Thus the perceived opportunity cost of these inputs and outputs for a biogas plan adopter may be zero, while from the nation's point of view it could be substantial. The use of world prices as "shadow prices" to overcome this problem brings the consideration of trade efficiency and distribution, raising many other critical questions while solving some.

4. The conflict between private and social profitability or benefits raises some fundamental decision issues. What is involved in a biogas project is its total impact on the economy, which cannot be readily identified and easily priced. There are large numbers of indirect or secondary costs and benefits of biogas programs, both at private micro and at national macro levels.

IMPLICATIONS ON INVESTMENT DECISIONS

Given the limitations of cost-benefit analysis as mentioned above, what use can we make of the results of the exercises in our decision-making process? Obviously, there is a considerable scope for improving the reliability of data inputs for the cost-benefit analysis of a biogas programs. Similarly, there are some methodological improvements to overcome the existing shortcomings, such as use of "shadow prices," sensitivity analysis under varying parameters, and a probability distribution of outcomes rather than single point estimates of profitability. There are all for future action. The question still remains as to whether the results so far indicated by the exercises could provide any guidelines for investment decisions.

Let us examine here three major outcomes of the existing cost-benefit exercises. First, cost-benefit exercises on biogas projects indicated that it was not a viable or profitable investment for the private owner because the monetary benefits did not outweigh the costs incurred. If we accept this outcome, it becomes difficult to expain the popularity and expansion of private household biogas owners in countries like China and India. It is almost impossible to believe that the government subsidy, credit facilities, and propaganda alone could push the private household to such a large-scale adoption of biogas plants. Fortunately, the same cost-benefit analyses also indicated the conditions in which it could become a profitable investment for a private owner. In this connection, mention has been made in diversifying end-users of biogas in cooking, lighting, and powering machines. Similarly they were emphatic to mention that if all the indirect benefits could be quantified, benefits would outweigh costs. Herein lies the answer of the anomalous observations between the outcomes of the analyses and actual field situations.

It would be appropriate to illustrate this point with some concrete observations from the field. Whatever the rigorous laboratory data and theoretical calculations suggest about maximizing the value of recycled organic manure through aerobic or anaerobic digestions, the experiences and pragmatic Chinese farmers found a considerable agronomical advantage - an increase in crop yields on the order of 10 to 15% in using anaerobic digestion. Apart from increses in agronomical yields, Chinese peasants also discovered the advantage of integrating all the agronomical factors, including humus and health improvements, over the analytical procedure of calculating the increases in nitrogen (N), phosphorus (P) and potassium (K). In our conventional cost-benefit analysis, neither the benefits of humus nor the health benefits were included. It was perhaps too difficult to induce these benefits into simple monetary units. Nevertheless, there is enough similar evidence among the Indian peasants[2,3]. There were, for example, reported savings of $42 U.S. per year on medical expenses in a peasant family using biogas, a 50% saving of costs to replace chemical fertilizers with biomanure, a saving of 2 to 3 hours in cooking time (and associated activities like washing utensils) utilized either in leisure or gainful activities, and increasing the longevity of the utensils. Some of these are quantifiable benefits in monetary terms, even though difficult. There are still other benefits which perhaps cannot be subjected to quantification, such as convenience and comfort in cooking and the prestige value of enhancing bargaining power for dowry in marriage in India.

This is not to suggest the futility of cost-benefit analysis, but simply to demonstrate the incompleteness and vulnerabililty of the analysis favoring one conclusion or another. With the arrival of a biogas system, the private households maximized the benefits not only in monetized terms but also in non-quantifiable terms thereby making biogas viable for investment.

The second major conclusion of almost all the cost-benefit analyses of biogas projects indicated the economy of scale. This means that the economic viability of biogas plants tends to increase with an increase in size. In other words, given the available resources (e.g., cattle dung and other feedstock), there seemed to be a threshold point of biogas plant size beyond which the viability increases with the increases in size of the plant. Based on this finding it was concluded that a relatively large-size community plant would be a viable investment[1] without subsidy, ignoring the problems of organization.

There are clearly three basic factors for such an outcome. Firstly, a relatively large-size plant offers the possibility of diversifying the use of gas beyond cooking. Secondly, it maximizes the recycling of available organic matter. Lastly, it ensures a minimum required supply of gas at low temperature, i.e., in winter. It is not difficult for the private owners of biogas plants to understand these three basic factors. It is precisely with this understanding that the peasants in India and China tend to have bigger than the exactly matching plant size in relation to their cooking or other energy needs and feedstock resources.

The third most important outcome of almost all the cost-benefit analyses overwhelmingly indicated that if all the macro-level long-term societal benefits were incorporated, a biogas program became a viable national investment even under seriously questionable assumptions. Apart from the direct saving of commercial and non-commercial fuel sources for better alternative uses, a biogas program in a country benefits the nation in terms of deforestation prevention, cost reductions for sanitation and health facilities, soil improvement and increases in crop productivity, contribution to or claim on foreign exchange, generation of extra employment with relatively less administrative cost, and contribution to the distribution of personal incomes. Obviously, a special premium should be attached to such long-term societal benefits in cost-benefit analysis.

On the other hand, negative factors might include the indirect cost of depriving poorer sections of free sources of renewable fuel and the cost for diverting or reallocating construction materials for biogas systems against alternative uses. It has been strongly argued, however, that even after considering these indirect costs, the societal benefits would be large enough to make the biogas program a highly desirable and viable investment decision for the nation. In fact, the strength of the argument in favor of a biogas program often goes as far as advocating a totally interest free credit flow or even a complete subsidization of the cost as an appropriate measure for a nation. The pragmatic and appropriate decision perhaps lies somewhere in-between a position to be determined by each country independently.

REFERENCES

1. T.K. Moulik. 1981. Cost-Benefit Analysis of Indian Biogas Systems: A Case Study in Renewable Sources of Energy, Volume II, on Biogas. ESCAP. Bangkok.

2. T.K. Moulik. 1982. Biogas Energy in India. Ahmedabad, India.

3. T.K. Moulik. 1984. India's experiments with Community Biogas System. Department of Non-Conventional Energy Sources. New Delhi.

ASSESSMENT OF THE FEASIBILITY OF RURAL BIOGAS SYSTEMS

M. M. El-Halwagi
National Research Centre, Egypt

ABSTRACT

Rational and effective diffusion of biogas technology (BGT) has to be preceded by objective, well-founded feasibility assessments. These assessments should include such pertinent aspects as: market analysis; technical feasibility; investment and production costs; financial appraisal; and socio-economic cost-benefit analysis. This paper attempts to present a preliminary exposition of these facets as well as certain prospective techniques recommended for tackling them.

BGT should be treated as a marketable commodity whose demand is developed through a "technology push" market-creation type of process. Market studies must, therefore, endeavor to cover the various relevant factors pertaining to uses, users, and alternative technologies, and end with proposed realistic market strategies, promotional plans, and the needed organizational structures.

The technical part of the feasibility study should encompass an assessment of the technical feasibility for the installation and successful operation of the biogas plant. In addition, it should cover the technical details relating to site location and appropriate digester and system design and size, as well as the various factors of construction and production.

Investment costs are relatively straight-forward to estimate, and a standard format is proposed for their computation. On the other hand, production costs pose difficulties that are principally related to the lack of reliable field data and to the pricing of the non-tradable inputs. Experience-based suggestions to reasonably circumvent these problems are presented.

Financial analyses are similarly complicated by the problem of quantifying the non-tradable outputs in monetary terms based on actual market prices. Though direct benefits of biogas systems are usually taken to include only the gas and the effluent, it is proposed to add the affected waste disposal as a third and important direct and tangible benefit. The three outputs may then be priced in terms of the actual alternatives they are supplementing or displacing at the point of use. In case of the effluent use as fertilizer, however, it is recommended to value it relative to farmyard manure (FYM) as manifested by the proportional agronomical yields.

Socio-economic cost-benefit analyses are the most difficult but the most revealing. They assess the feasibility on the macro level while including the impacts of social welfare factors and environmental considerations. On account of the wide variability of social systems and the insurmountable difficulty of quantifying inputs and outputs--primary and secondary--in absolute terms, the simplistic 'system net change' approach previously proposed by the author[3] is recommended for use when possible.

Because of the many sources of inaccuracies confronting the feasibility analyst, which would preclude arrival at a specific and reliable appraisal, it is strongly recommended to resort to sensitivity analyses to the maximum possible extent. Accordingly, the analyst can come up with the relevant set of appraisals covering optimistic and pessimistic scenarios to help the evaluators and decision makers in making the proper decisions.

INTRODUCTION

Feasibility assessments are the proper tools for identifying viable investment opportunities and for making sound investment decisions.

These decisions are made at all levels of society, as well as by development assistance and funding agencies.[7] At the highest level, policy makers and planners require evidence that biogas technology (BGT) is practical and economically feasible, particularly relating to national macroeconomic issues. Funding agencies have to be provided with full-fledged feasibility studies justifying the loan part of the investment. Such studies should provide an in-depth analysis of all technical, sociocultural, and administrative details, in addition to pertinent economic data and financial analyses. At the investor level, the equity part of the investment has to be justified in terms of monetary and other benefits.

Thus, in addition to the question of technical feasibility and appropriateness, other criteria such as cost efficiency and national and social welfare have to be addressed. This is further complicated by the fact that all these factors are location and case specific and depend on the type of the biogas plant intended, whether it is a household unit, a community facility, or an agro-industrial system. Lack of proper and accurate basic data and information adds much more to the burden of the feasibility analyst.

Therefore, and at the outset, there is no universally valid assessment of the techno-economic feasibility of biogas-producing systems.[6] But there are acceptable techniques that have to be adapted and judiciously utilized. And this is exactly what this paper attempts to address even on a limited and preliminary scale.

Generally, a feasibility assessment should cover aspects relating to markets, technology, economics, and financial and social cost-benefit analyses. These, together with methods recommended for tackling them, will be discussed in general terms for the case of rural biogas systems.

MARKETS

Biogas technology should be treated from the promotional point of view as a "marketable commodity". As such, its market potential must be assessed. Again, the market potential may need to be estimated on the national or macro level, as well as on the micro level of the specific locality like a village or a hamlet where implementation will take place. The first type of estimate is usually needed in the early stages of promotional activities for probing the technology's prospects on a national scale, in order to attract the attention and commitment of policy makers and planners. Micro-level estimates are essential prior to or during the propagation and implementation stages, in order to identify the specific liable cases having the right potential and the right socio-economic environment. It should also be mentioned here that identification of market opportunities would mostly evolve from recognition of technical feasiblity rather than recognition of potential demand, i.e., it is more of a "technology push" market-creation type of process than a readily accessible "market pull" situation.

Since rural biogas systems are mostly dependent on animal wastes as substrate, their market potential is closely related to the size of the animal population subject to certain restrictions and constraints. Some aggregate studies aimed at exploring the overall prospects of BGT on the national level assess the output on the basis of the reasonably collectable fraction of the total generated animal manure. Better estimates, however, can be made on the basis of the number of animal rearing operations in the contry as indicated in the national statistics, plus the number of prospective community and household gas plants as judged from the animal ownership patterns and availability of space for the siting of digesters. In the Egyptian case[1], for instance, it was estimated that out of 4 million rural families only about 250,000 would qualify for BGT, taking into account that each of these should own at least four large animals and have enough space in the homestead for installing the digester and its accessories.

On the other hand, assessing the market potential on a local basis is highly specific and much more accurate. The number of commercial animal rearing farms as well as the eligible households can be surveyed in real terms. Prospective sites for community facilities will not depend primarily on physical factors such as animal ownership and space availability, but would strongly involve sociological considerations and as such be prone to more sources of inaccuracies.

In the specific market studies, it would also be possible to cover many factors that are usually handled in standard market analyses. These would include:

- Prospective uses of the biogas system and its outputs and their opportunity costs. These encompass: energy use for household applications, mechanical work, or generation of electric energy; manure use as fertilizer or feed; and the use of the system for waste treatment purposes.
- Potential users: Households, community, and animal rearing operations.
- The types of available biomass substrates and their opportunity costs.

- Alternative technologies (to BGT) and their competitive position.
- Realistic estimates of the BGT market share taking into account its probable penetration under the set of prevailing local conditions and available technologies.
- The proposed marketing strategies, promotional plans, and required organizational structure.

TECHNICAL ASPECTS

Technical considerations should embody:

- site location;
- system design including operating conditions, pre-and-post-treatment;
- plant size, bill of materials, and construction manpower; and the various factors of production, including: raw materials (type, quantity, and characteristics), utilities, labor, and maintenance requirements.

Assessment of the 'technical feasibility' assumes a central role in the study of the technical aspects. It undertakes to identify the technical criteria for installation and successful operation of a given biogas plant.[8] Factors that determine the installation of a technically feasible plant include: suitability of the soil conditions, adequacy of the land space for the construction of the digester and its complementary handling and treatment facilities, nearness to the point of use (particularly for the gas) and to the source of the organic substrate, distance from drinking water wells, availability of water for digester feed preparation, climatic conditions particularly the temperature, and availability of cattle. The size of the plant, on the other hand, will be determined basically by such factors as the raw materials available for digestion, the quantity of gas required, and the desired quality of the effluent, specially from the standpoint of its freedom from of pathogens.

ECONOMIC ASPECTS

Economic aspects entail estimating both investment and production costs. The group of experts that convened in the "Workshop on Uniformity of Information Reporting on Biomethanation Systems"[4] suggested the format given in Table 1 for capital cost estimation. Cost of land is missing in the table and I propose adding an additional item. As may be noted, investment costs are relatively easy to assess with a high degree of accuracy.

Production costs, on the other hand, are highly controversial. Standard items of these annual costs include:

- <u>Operation costs</u>:
  - dung, water, and other feed materials;
  - utilities (in case of heating and mechanized systems);
  - operating labor;
  - repairs and maintenance; and
  - other costs such as technical services.

- Capital charges:
  - preparation, and
  - cost of capital (interest).

In estimating these cost items, difficulties arise on account of two factors: 1. lack of reliable field data over the complete life span of the various operational models; and 2. the difficulty of reasonably accurate pricing of the non-tradable type of inputs.

The price of dung, the key input to most biogas units, should be estimated on the basis of the true replacement cost to the farmer. This would normally be in proportion to the two previous uses of dung as fertilizer (FYM) and fuel (dung cakes). It should be noted, however, that neither FYM nor cakes are equivalent to the fresh dung coming out of the animal shed and fed to the digester, since both have undergone a certain amount of handling and treatment. For instance,

TABLE 1 Capital Cost of Biomethanation System*

DIGESTER Materials	Unit	Quantity	Cost per Unit		Total Cost	
Item 1						____
Item 2						____
--						____
--						____
					Subtotal	____
Labor	**Unit**	**Quantity**	**Work hours per unit**	**Cost per Work hour**	**Total Cost**	
Excavation						____
Building						____
Plastering						____
Backfilling						____
Transportation of Materials						____
Other costs (itemize)					Subtotal	____

GAS HOLDER Materials	Unit	Quantity	Cost per Unit		Total Cost	
Item 1						____
Item 2						____
____						____
____						____
					Subtotal	____
Labor	**Unit**	**Quantity**	**Work hours Per Unit**	**Cost per Work Hour**	**Total Cost**	
Fabrication						____
Painting						____
Transportation of gas holder						____
Other costs (itemize)						____
					Subtotal	____

AUXILIARY EQUIPMENT AND ACCESSORIES	Unit	Quantity	Cost per Unit	Total cost
Equipment				
I-C engine				____
Lamps				____
Burners				____
Other (itemize)				____
Accessories				____
Pipe				
Valves				____
Flame arrester				____
Water trap				____
Other (itemize)				____
			Subtotal	____
PROCESS DESIGN AND SUPERVISION				
Labor				
Fees				____
Supplies				____
Other (itemize)				____
			Subtotal	____
			Total Capital Cost	____

* It is suggested to add a separate item for land.

cakes have been collected, formed, and dried. Therefore, the estimated dung price should be less than the price of FYM or cakes by the cost involved in the handling and processing of either.

Labor costs are also difficult to calculate in the case of household units since the normal operation of the biogas system is undertaken by the family members themselves. However, errors due to neglecting this item would be of negligible magnitude since the total time required for the different tasks associated with the operation of the unit is rather short (half an hour or so).

Repairs and maintenance costs are principally related to the metallic and plastic parts of the system. For instance, according to Moulik[5], a metallic gasholder is usually painted every second year, repaired every fifth year, and replaced after eight years. Similarly, hosepipes are replaced every third year and the pipelines are repaired during the third year. Our experience in Egypt seems to indicate that a reasonably conservative estimate for maintenance costs would be 5 percent of the investment cost.

Technical services or extension is another difficult item to estimate. Generally, this cost is borne by the government, and as such would be accounted for in the social cost-benefit analysis. Otherwise, I would suggest taking another 2-5 percent of the investment cost as the extension expense.

Depreciation depends on the estimate of the economic life of the various components of the plant. Since there is no sufficient back experience, estimates of the plant economic life vary between 15 and 40 years. The lower figure would be relatively conservative but more realistic taking into account the village setting and maintenance skills. Moreover, for financial analysis purposes and when using discount methods, the assumption of longer life would not introduce high impacts. In any case, however, replacement of the metallic and plastic parts during the economic life of the plant should be taken into account since they have much shorter life.

FINANCIAL ANALYSIS

To carry on with the financial analysis, the revenues associated with the outputs (benefits) of the system have to be quantified. Thus, the money profit accruing to the project-operating entity can be computed as the difference between revenues and production costs; both based upon actual market prices. The commercial profitability of the venture can then be appraised using conventional techniques and related performance measures. These include the static methods of the simple rate of return and payback period, or the cashflow discounting techniques such as the net present value (NPV) or the internal rate of return (IRR) methods.

The tangible outputs of rural biogas systems embody the gas, the digested slurry (effluent), and the affected waste disposal. All three benefits are difficult to value in monetary terms since they do not represent traded commodities.

The valuation of gas output is dependent upon three complex considerations[10]: the quantity and quality of gas, the mix of end-uses, and the price and type of fuel considered relevant for estimating replacement costs. On account of the seasonal variability in gas output, an average 'usable' output should be carefully estimated. Occasionally excess gas that is not utilized should not be valued. The combination of the actual end-uses and their burning efficiency, as well as that of the displaceable fuels (such as wood, charcoal, kerosene, or butane) should be taken into account when valueing the biogas product.

The benefits derived from the effluent depend on its use, e.g., as a fertilizer/soil conditioner, an animal feed, or as a feed for fish ponds.[4]. In most cases, however, the digested manure is used as a fertilizer. Estimating its value would then depend upon its quantity and quality and the location of the final point of use in relation to the digester side, as well as the cost of fertilizers supplemented or displaced. Evidently, this particular valuation is surrounded with much uncertainty. The quality of the digested product depends on the affected post-digester type of treatment, its moisture content, and other constituents e.g., (NPK, micronutrients, and humus). It appears, therefore, that the most practical and reliable way is to use the value of the net increment of agricultural output relative to FYM for quantifying the digested product value. Thus, on the average, it seems that pricing the digested manure at 1.1 times the value of FYM would be a reasonable proposition (both on dry basis).

The waste disposal benefit is accrued through the use of the biogas plant itself as a system. The value of this item can be calculated as equivalent to the cost of the supplemented or displaced alternative waste disposal scheme that would be otherwise employed in the absence of the biogas system.

SOCIAL COST-BENEFIT ANALYSIS

Quantitative social analyses are definitely the most difficult to make. Yet, they are the most revealing when it comes to assessing the true socio-economic feasibility of rural biogas systems. In a sense, social analyses measure the effect of a given biogas project on the fundamental objectives of the whole economy.[9] They are typically taken at the national level and use "shadow prices" that reflect the true economic worth (excluding the effects of taxes and subsidies, for instance) to the society of the inputs and outputs of the project. The variety of social and environmental benefits, that are sometimes called secondary or indirect benefits, are of particular relevance to social analyses. These include such factors as improving the quality of life[2] and hygiene, creating employment opportunities, saving forests, reducing the uncertainty of the commercial fuel supply, and the like.

On account of the constraints arising from the wide variability of rural social systems and the uncertainties relating to the quantification of direct and indirect benefits in absolute terms, El-Halwagi et al.[3] proposed a simplistic "system net change" approach. The focal concept in their model is to assess the changes occurring in the rural "organic-waste generating" system after implanting the biogas subsystem rather than valueing the inputs and outputs independently.

SENSITIVITY ANALYSES

As would be quite obvious from the aforementioned exposition of appraising the feasibility of rural biogas systems, sources of inaccuracies are plentiful. A very helpful and recommendable way to handle the resulting uncertainties is the sensitivity analysis technique. In this technique, the value of the key parameters are varied within the range of the anticipated uncertainty, and the selected profitability measure computed over that range to investigate how it is affected by such changes. The values of the parameters are changed one at a time and in combinations representing optimistic as well as pessimistic scenarios. Thus, the decision maker would then have a wider and more realistic view of the venture prospects.

REFERENCES

1. El-Halwagi, M. M., A. M. Abdel Dayem, and M.A. Hamad. 1981. Conference on Energy Conservation and Utilization. Cairo.

2. El-Halwagi, M. M., A. M. Abdel Dayem, and M.A. Hamad. 1982. Village demonstration of biogas technology: An Egyptian case study. Natural Resources Forum, Vol. 6, No. 4.

3. El-Halwagi, M. M., M. A. Hamad, and A. M. Abdel Dayem. 1984. Cost-Benefit Analysis of Rural Biogas Systems in Terms of Their Impacts as Agents of Socio-Economic Change. BioEnergy '84. Sweden.

4. Equity Policy Center. 1983. Report on the Workshop on Uniformity of Information Reporting on Biomethanation Systems. Held in Bangkok, Thailand, May 14-18, 1983. Equity Policy Center, Washington, D.C.

5. Moulik, T. K. 1981. Cost-Benefit Analysis of Indian Biogas System: A Case Study. Renewable Sources of Energy, Vol. II: Biogas ESCAP, Bangkok.

6. Moulik, T. K. 1982 Biogas Energy in India. Academic Book Centre. Ahmedabad.

7. National Academy of Sciences. 1982. Diffusion of Biomass Energy Technologies in Developing Countries. National Academy Press. Washington, D.C.

8. Reserve Bank of India. 1976. Report of the Inter-Institutional Group on Financing Gobar Gas Plants by Banks.

9. Squire, L. and H. G. Van der Tak. 1975. Economic Analysis of Projects. A World Bank Research Publication. The Johns Hopkins University Press, Baltimore and London.

10. Stuckey, D.C. 1983. The Integrated Use of Anaerobic Digestion in Developing Countries: A State-of-the-Art Review. IRCWD

ANALYSIS OF ECONOMIC FACTORS IN THE DISSEMINATION OF BIOGAS PLANTS - CASE STUDIES FROM AFRICA AND THE CARIBBEAN

Peter Pluschke
Project Manager of Biogas Plants, Deutsches Zentrum für Entwicklungstechnologien (German Appropriate Technology Exchange), GTZ/GATE, P.O. Box 5180, 6236, Eschborn 1, Federal Republic of Germany

ABSTRACT

In this paper, economic aspects of rural biogas plants are discussed, based on information gathered by GTZ/GATE since 1977 on small, simple plants in developing countries. For our purposes, a somewhat simple cost-benefit analysis is preferred to a sophisticated social-cost-benefit approach. Two case studies are considered in some detail: one from Burkina Faso, with one Chinese-type digester; and one BORDA/Sasse model. Another case study refers to a relatively large digester of Indian design with ferrocement gasholders, constructed in Barbados. It is obvious from the case studies that the economics of rural biogas depend not only on the output of biogas, but also on the economic valuation of the digester effluent.

GENERAL COMMENT ON THE ECONOMICS OF BIOGAS PLANTS

Due to the generally well-known facts on the economy of scale of biogas plants, it is agreed in industrialized countries that, from an economical point of view, the implementation of biogas technology must depend upon construction of bigger plants. There seems to be little hope for the dissemination of small scale, family-sized plants. Due to the generally low-in terms of spatial distribution - concentration, traditional concepts of "selling" the technology will fail for small-scale, family-size digesters, i.e., digesters with a volume of approximately 5 to 20 m^3.

Conventional cost-benefit analysis (from the point-of-view of opportunity costs) is inappropriate if we are to give an accurate picture of the potential for the dissemination of the technology. Many benefits of a small-scale, rural biogas plant are not possible to evaluate in monetary terms, for instance:

- the security of the local energy supply (in regions where supply channels generally show little reliability);
- the environmental and hygienic effects of biogas plants;
- the contribution to the quality of human life (biogas being a cleaner and more versatile fuel than the traditional fuels).

In the meantime, more sophisticated concepts for the economic evaluation of biogas plants have been applied for example, El-Halwagi (1984) and Srinivasan and Gosh (1977); but the more sophisticated these concepts are the less they reflect the decision-making process of a farmer living in a developing country. It is possible that social cost-benefit analysis will prove a powerful instrument in evaluating the potential of biogas technology on a general level and may define political strategies to promote the technology.

In GATE's Biogas Extension Service, it was decided that only simple cost-benefit analysis concepts would be applied, based on the point-of-view of cash transactions. Our experience is that application of this procedure will best help us understand the basic economic constraints for the dissemination of small scale biogas plants.

This means that we have concentrated our discussion on the question of whether a single peasant accepts the biogas plant as a helpful instrument and an efficient investment in the framework of his individual "cost-benefit concept".

Any promotion campaign or dissemination program has to consider carefully local conditions. It is the case-study type of analysis rather than the general analysis which allows us to outline dissemination activities. According to this view, some case-studies will be considered later on.

THE BAND-WIDTH OF BIOGAS PLANTS REALIZED IN THE FRAMEWORK OF THE GATE BIOGAS EXTENSION SERVICE (BES)

Since 1981, GATE has constructed in the framework of its biogas plants in six countries (Barbados, Burundi, Kenya, Nicaragua, Tanzania, and Burkina Faso) covering a broad spectrum of applications and users, Pluschke (1982).

Plants of various sizes have been constructed:

- Very simple, transportable demonstration units with digestion tanks from 0.2 m^3 upwards: the purpose of these units is to familiarize farmers with the way biogas plants work and to demonstrate how combustible gas is produced. Miniature plants of this kind can also be used for fermentation tests, GATE (1983).
- "Family-size" plants with digestion tanks of 4.5 m^3 to approximately 20 m^3 capacity: These can be used to supply a family's energy needs for cooking and lighting (and, in some cases, cooling).

  Generally speaking, family-size plants can only be installed in rural areas if livestock, which supplies the fermentation material, is confined in cowsheds etc. at least part of the time. Conditions for a biogas plant are ideal when no grazing is practiced; however, in only a few places do small-scale farmers practice this form of husbandry. It is already fairly common in the Arusha region in Tanzania and on the Caribbean islands.

- Plants in institutions such as schools, medical centers, communes etc., with digestion tanks of approximately 10 m^3 to 10 m^3 capacity; many of these plants are used for waste disposal. In schools they are also used for training purposes. Obviously, such plants have a considerable multiplier effect.
- From time to time larger biogas plants are installed on large farms and in agro-industrial enterprises. These have digestion tanks with capacities of around 100 m^3. The construction of such plants is not the principal task of the BES, however, because the program as a whole is aimed at resolving energy problems in rural areas. So the building of small, decentralized plants is clearly where the emphasis lies. But it has been found that the technology is judged much more favorably and accepted more readily if there are a few biogas plants that are technically more sophisticated and of more 'spectacular' dimensions than are the rural family units.

A DETAILED VIEW ON SOME FAMILY-SIZE BIOGAS PLANTS

To give some insight into concrete situations which are typical regarding the problems and possibilities of biogas dissemination, two case studies are to be analyzed in some detail.

The Biogas Plants in Pabre (Burkina Faso)

The first case study comes from Burkina Faso. Pabre is situated about 20 km north of Ouagadougou. It is a Catholic mission with a school for 160 pupils. About 4 ha of land are cultivated where fruits (paw-paw, oranges, bananas, etc.) and vegetables are grown. Pig-breeding is also an important agricultural activity of the station. Between 40 and 70 animals supply enough dung to run a biogas plant. Water is available for irrigation purposes and can be used for a biogas plant as well. The pattern of energy supply and use is complex:

- electricity (from the regional grid) for lighting and cooling purposes;
- butane for the kitchen of the mission staff;
- fuelwood for cooking the meals of the pupils (fuelwood is available at the station);
- diesel-fuel for water-pumping.

Biogas is seen as a substitute for the butane. After a thorough analysis, it was decided to construct a biogas plant with a digester volume of 20 m^3.

Due to the fact that in the course of the year the availability of digestible biomass fluctuates, it seemed that an adequate solution would be to construct two digesters with a volume of about 10 m^3 each. From the point of view of the BES it was interesting to compare two different constructions on this place, and so the biogas advisers decided to build one Chinese dome and one BORDA/Sasse-type plant, (Figure 1).

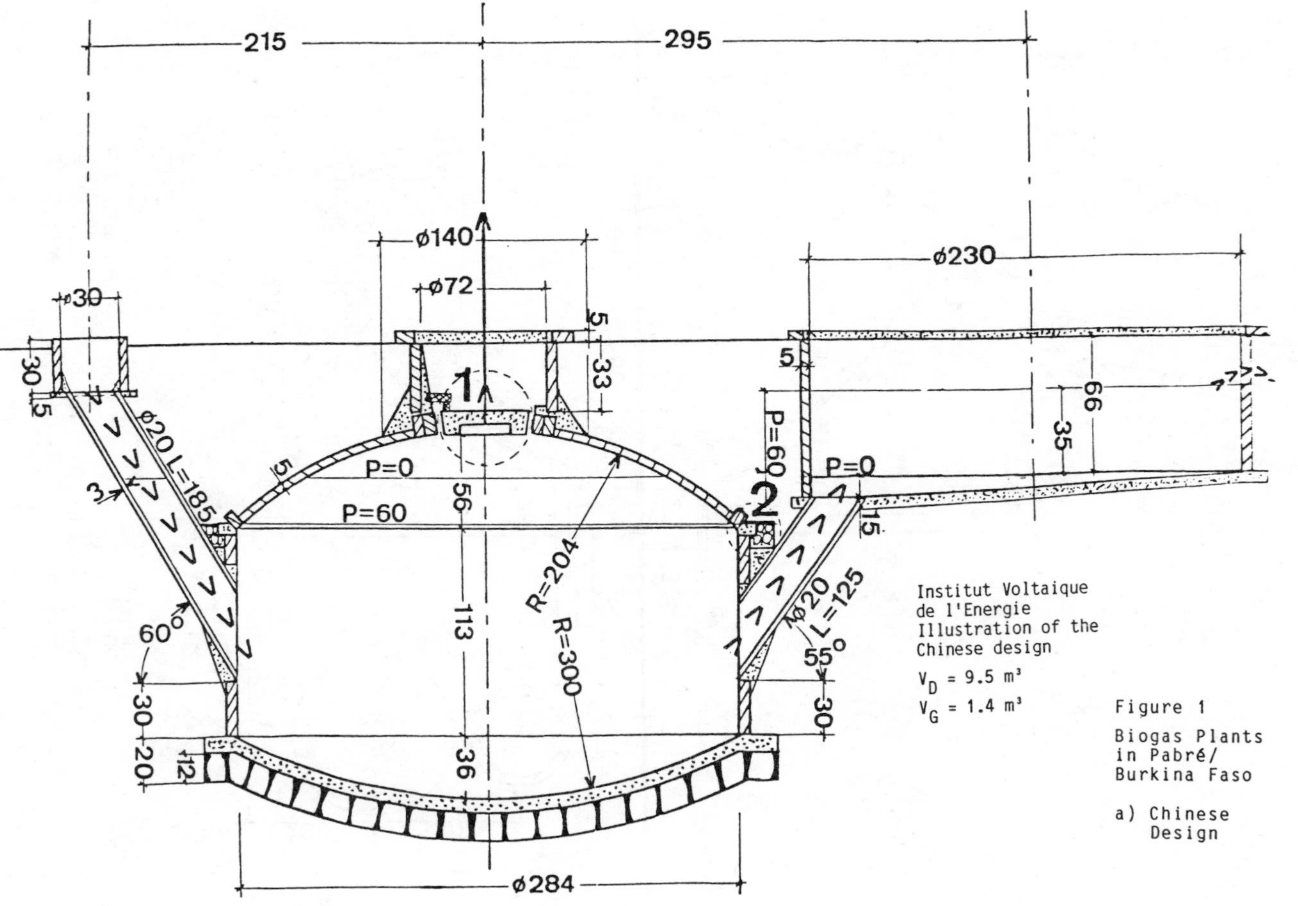

Figure 1
Biogas Plants in Pabré/ Burkina Faso

a) Chinese Design

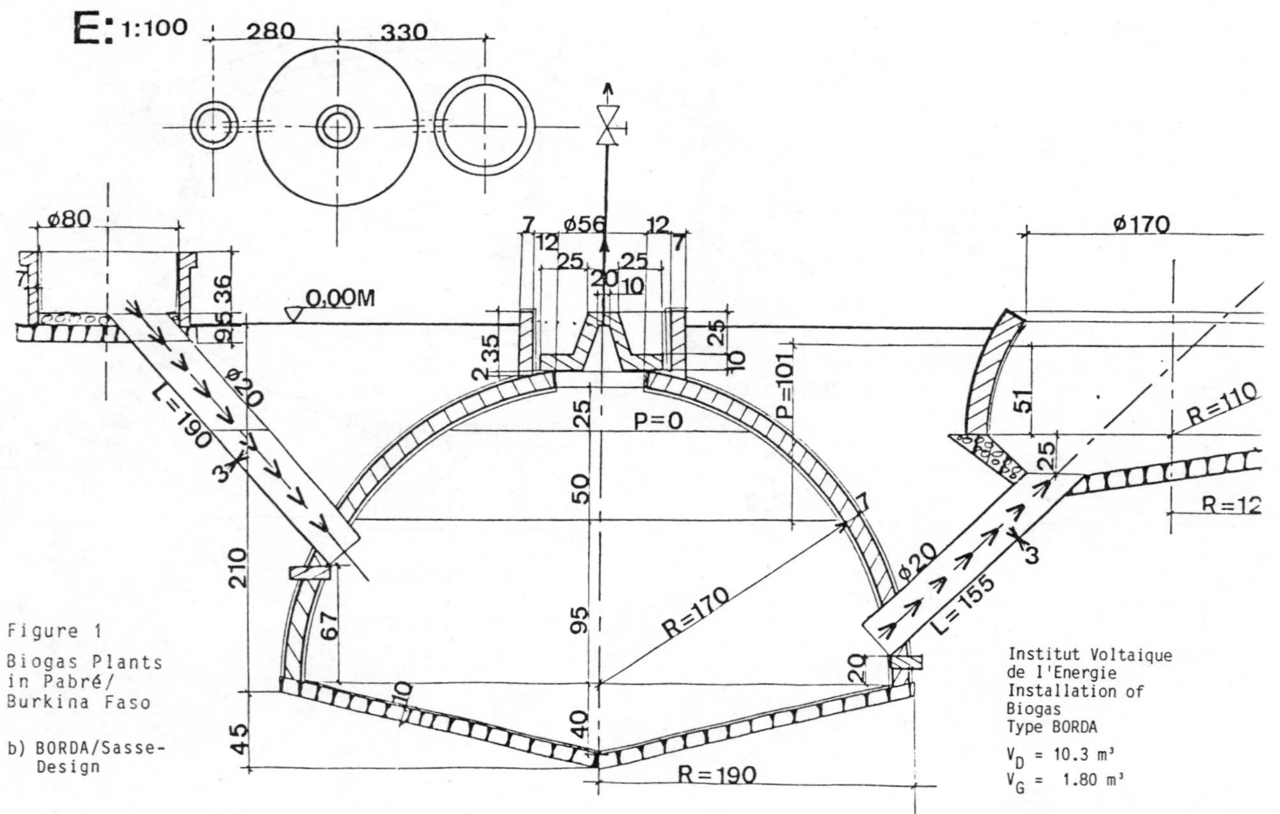

Figure 1
Biogas Plants in Pabré/ Burkina Faso

b) BORDA/Sasse-Design

Due to the spherical shell construction of the BORDA/Sasse-type plant, fewer bricks are required for the digester, than in the case of the Chinese fixed dome type. The relation is 1,251 bricks for the Chinese dome to 992 bricks for the BORDA/Sasse, i.e. about 20% less. On the other hand, about 20% more cement has been used for the construction of the BORDA/Sasse-type plant. This additional consumption is in part due to the fact that the foundation had to be reinforced because the ground caused problems.

The cost structure may be taken from Figure 2 and then compared to other biogas plants.

The overall costs for one digester are approximately 500,000 to 800,000 FCFA as compared to the yearly expenses for butane gas of about 150,000 FCFA. These are the kind of economics needed by a peasant to take the decision of constructing a biogas plant.

If we look at this situation with conventional instruments of cost-benefit analysis, we may consider a broad range of more or less plausible conditions. In Table 1 some typical results are summarized and may give a broad basis for discussion.

Table 1 Cost-Benefit Analysis of the Biogas Installations in Pabre (Burkina Faso)
two digesters: one Chinese dome (9.5 m^3)
one BORDA/Sasse type (10.3 m^3)
as shown in Figure 1

a) overall cost including all transport costs: Results 1,750,110 FCFA

	i = interest rate	lifetime (years)	running costs (FCFA/Y)	Net present value C_0=(FCFA)	internal rate of return
1.	7.5%	10	50,000	- 1,095,497	
2.	0 %	10	50,000	-798,290	
3.	7.5%	10	0	-752,293	
4.	0 %	10	0	-298,290	
5.	7.5%	20	50,000	-792,858	
6.	0 %	20	50,000	+120,350	R = 0.7%
7.	7.5%	20	0	-283,134	
8.	0 %	20	0	+1,120,135	

b) overall costs, transport excluded

	i = interest rate	lifetime (years)	running costs (FCFA/Y)	Net present value C_0=(FCFA)	internal rate of return
1.	7.5%	10	50,000	- 720,432	
2.	0 %	10	50,000	- 423,225	
3.	7.5%	10	0	- 377,228	
4.	0 %	10	0	+ 76,775	
5.	7.5%	20	50,000	- 417,793	
6.	0 %	20	50,000	+ 495,415	
7.	7.5%	20	0	+ 91,930	
8.	0 %	20	0	+ 1,495,415	r = 8.4%

i = 0% means that the biogas plant is completely financed by own capital resources

i = 7.5% means that the biogas plant is completely financed by commercial credit

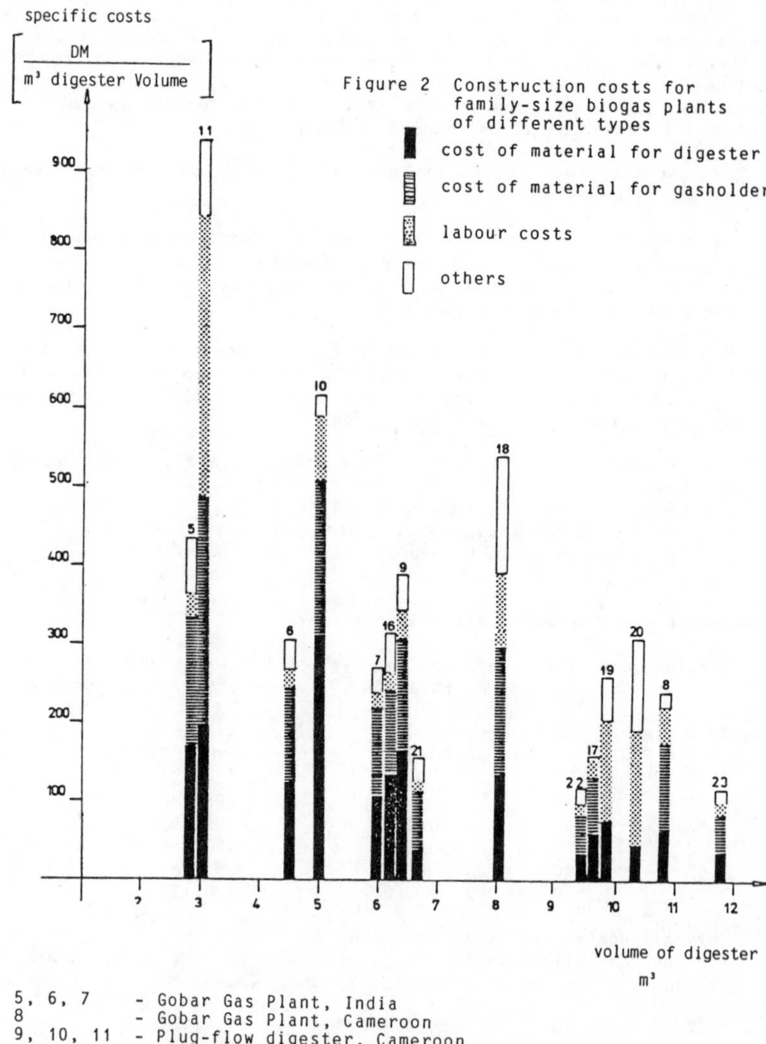

5, 6, 7 - Gobar Gas Plant, India
8 - Gobar Gas Plant, Cameroon
9, 10, 11 - Plug-flow digester, Cameroon
16, 17, 18 - BORDA/Sasse design (floating drum version), Kenya
19 - Fixed Dome Plant, Burkina Faso
20 - BORDA/Sasse design (fixed dome version), Burkina Faso
21, 22, 23 - Gobar Gas Plant (UNDARP design), India

The Hoad Dairy Farm Biogas Plant in Barbados, Pluschke (1982)

In Barbados there are 33 dairy farms with about 1,000 cows. At the 20 largest dairy farms some 900 cows produce 5,000,000 kg of biodegradable manure annually. This, if anaerobically fermented, can produce about 200,000 m^3 of biogas annually, which is equivalent to about 700,000 kWh of electricity.

The 5 x 10^6 kg of manure is only available in theory since most of the 900 cows graze in open fields and the collection of manure could be a poor economic proposition. However, dairy farmers are adopting intensive cattle rearing method and the cows are kept in sheds and fed hay, molasses and brewers' grain. This provides for easy collection of the manure.

The first dairy farm to be converted to an intensive cattle rearing method is the Hoad Dairy Farm, situated on the Northeast Coast at Morgan Lewis in St. Andrew. This model dairy farm is one of the few farms of this size (47 milk cows) which is owned and operated on a full-time basis by the owners.

The 47 dairy cows stand in a roofed shed which is divided into individual stalls for each cow. The dung and urine from the cows drop into gutters cast into the concrete floor. The gutters are sloped to one side of the shed and are cleaned once a day by manually operated sliding bars in tne gutters. The manure, which is piled into one corner of the shed after cleaning, is mixed with hay, cow bedding (bagasse), and fed leftovers. The liquid parts of this mix flow into an adjacent field and have resulted in the field becoming unusable since it is now a swamp.

The Hoad family cultivates about 18 acres of grass for feeding the cows and 4-5 harvests of grass are obtained each year. The manure spreader is used to apply wet manure onto the freshly mowed fields at intervals of three days. Ammonium sulphate is also applied.

The farm's main energy requirements are electricity for the milking machines and the milk cooler, as well as heating fuel for the gas-operated water heater used in the milking parlor. Gas is also used for heating water and cooking in the Hoad's home (a family of five). A 5 hp diesel generator is used as a standby generator in case of power failure and could be run of a biogas/diesel fuel mix. The farm and home use about 1,600 kWh of electricity and 69 kg of bottled LPG every month, which cost 350 US$.

The yearly expenditure for mineral fertilizer (ammonium sulphate) is 1,500 US$. The Hoad's major goals are to reduce energy costs for operating their milking parlor, provide a method of disposing of the run-off from their cow shed without losing fertilizer value, and to reclaim the swamped field for grass cultivation.

The biodigester, developed for the Hoad farm system, consists of two circular 60 m^3 digesters with integral floating ferrocement gas holders, a 70 m^3 storage tank, and drying pits. The digester's storage tank and drying pits are constructed by using concrete blocks, cement, and reinforcement iron. Their sites have been excavated to the water table level and the gas holders fabricated, cast, and welded on

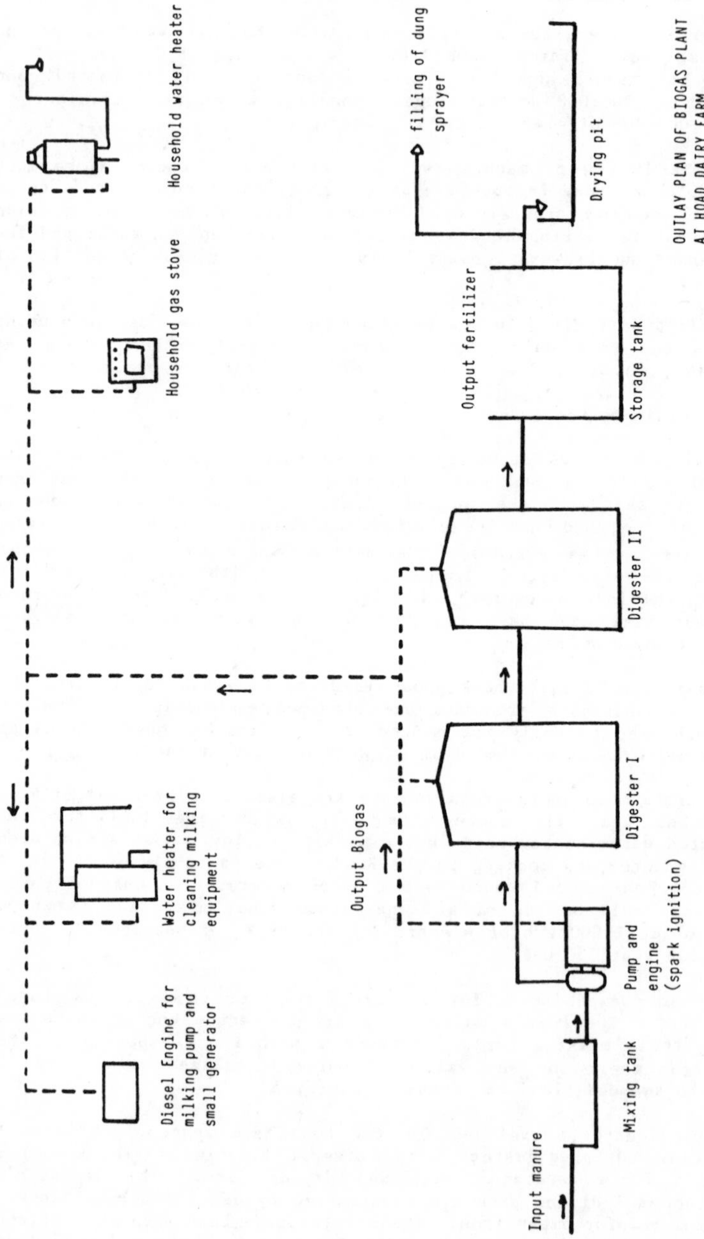

Figure 3 Flow-diagram of Biogas Plant at Hoad Dairy Farm, Barbados

site. The drying pits are covered by plastic sheet and suitable openings are made in the sides to allow air flow. Figure 3 is a schematic drawing of the biodigester system.

The biogas produced by the anaerobic fermentation of the cow manure feedstock is stored in the ferrocement gas holder, which forms the cover for the digesters and can rise to accommodate more biogas as it is produced, Frettloeh (1984). The biogas, which is pressurized by the floating gas holders, is conveyed via PVC pipes to the household gas stove and water heater, the slurry and sludge pump used to service the digester system, and the water heater and diesel engine in the milking parlor. The gas burners of the water heater and stove, the diesel engine, and digester pump engine will be converted step by step to use biogas.

The relevant economic data for the plant may be taken from he following tables:

Table 2 Basic Data for the Cost-Benefit Analysis of the Hoad Dairy Farm Biogas Plant, Barbados

a) Type of the plant
- two digesters with 60 m^3 volume each, Gobar type (Indian) ferrocement gasholders

b) Costs

- overall construction costs:	40,000 B$
- running costs:	200 B$/y

c) Benefits

- substitution of propane:	960 B$/y
- dried organic fertilizer	
o amount used on the farm:	1,700 B$/y
o sold in bags:	5,940 B$/y

The Cost-Benefit-Analysis is done as a dynamic net present value calculation and gives the following results:

SUMMARY

The two case-studies, which represent biogas plants in two very different social and economic settings, both show the biogas technology to be feasible in economic terms. The economics of the digesters in Burkina Faso are somewhat critical due to the fact that no benefits have been assumed for the application of the slurry on the fields of the mission station. Including such a contribution to the economic perspective, the prospects for the biogas technology are bright.

Even more dramatic is the contribution of the slurry to the economics of the biogas plant in Barbados. Due to the market for organic fertilizer already existing there, the biogas plant becomes a "fertilizer factory", while still serving as a methane generator.

Table 3 Net present value calculation for Hoad Dairy Farm Biogas Plant, Barbados

Year (no.)	0	1	2	3	4	5	6	7	8	9	10
Year	1984	1985	1986	1987	1988	1989	1990	1991	1992	1993	1994
Costs	(40,000)	0	0	0	0	3000.00	0	0	0	0	0
back flow	(40,000)	8400.00	8400.00	8400.00	8400.00	8400.00	8400.00	8400.00	8400.00	8400.00	8400.00
discount factor		0.900900	0.811622	0.731191	0.658730	0.593451	0.534640	0.481658	0.433926	0.390924	0.352184
present value	(40,000)	7567.57	6817.63	6142.01	5533.34	4984.99	4490.98	4045.93	3844.98	3283.77	2958.35
present value (cumulative)	(40,000)	7567.57	14385.20	20527.20	26,060.54	31,045.53	35,530.52	39,582.45	43,227.43	46,511.20	49,469.55

		Costs for Replacement included
Interest (%)	11.0	
Lifetime	10	
Benefits (B)	49,470	49,470
Costs (C)	-40,000	41,308
B/C-Ratio	1,236,738	1,197,591

Internal Rate of Return 16.40%

REFERENCES

El-Halwagi, M.M., M.A. Hamad, and A.M. Abdel Dayem. 1984. Cost-Benefit Analysis of Rural Biogas Systems in Terms of their Impacts as Agents of Socio-Economic Change. Conference Bio Energy 84. Goteborg/Sweden.

Frettloeh, G. 1984. Ferrocement Gas Holder for two 60 m^3 Digesters. International Conference on the State of Art on Biogas Technology, Transfer, and Diffusion, 1984, Cairo.

GATE. 1983. Small Transportable Biogas Transportation Units in GATE - questions-answers-information No. 4/1983. p. 19.

Pluschke, P. 1982. GATE Seconding Expert Team on Biogas. In GATE - questions-answers-information No. 1/1982. pp. 6-10.

Srinivasan, T.G. and Utpaul Ghosh. 1977. A Social Benefit-Cost Appraisal of Community Gobar-Gas Plant. In Energy Management 1(4):243-248. New Delhi.

ACKNOWLEDGEMENTS

This report is based largely on the work of my colleagues working in the field. Special thanks go to Hans Polak, Gerhard Frettloeh, Heinz-Peter Mang, Siegfried Wissing, Nico Hees, and Rainer Geppert.

THE ECONOMICS OF BIOGAS SYSTEMS - A SHORTER COMMUNICATION

Mamie M. Lau-Wong
Development and Consulting Services
Butwal, Nepal

At their present cost, domestic biogas systems for cooking and lighting are acceptable in places where fuelwood price is high, such as the capital Kathmandu and some towns in the Terai. In Butwal (a town in the Terai), a price of 0.6 NRs/kg of fuelwood is the cut-off point at which the benefit-cost ratio is almost 1. In the hills where the fuelwood price is lower, with an average of 0.36 NRs/kg, a 41 percent subsidy is required to bring the same returns.

Since subsidy comes ultimately from the society itself, it cannot be justified unless the social and economic benefits of the biogas system outweigh its cost to the society. One often-cited benefit is the conservation of forests by replacing fuelwood, which accounts for 93 percent of domestic fuel consumption. The cost of deforestation proves extremely difficult to quantify. Nevertheless, economic analysis showed that if the economic value (which may be very different from its market price) of fuelwood hits 0.39 NRs/kg, the benefit-cost ratio becomes 1 at a discount rate of 15 percent. In other words, if the economic value of fuelwood is above 0.39 NRs/kg, a subsidized domestic biogas program should be considered.

In our economic analysis, a comparison of biogas system and afforestation was made. If forest productivity can be maintained above $7m^3$/ha/year, which is a target not difficult to achieve, the afforestation scheme would cost less to society than the biogas system. However, in afforestation there is an inevitable lag between the time of investment and the first harvest, whereas for biogas the effects are immediate. Since a fuelwood shortage is imminent, the best strategy for the country is probably to implement both programs along with the development of other renewable sources of energy, such as hydroelectricity.

For income-generating activities, such as rice hulling and flour milling, a biogas system, which supplies fuel for cooking as well, is as competitive as a diesel one, provided that the digester temperature can be maintained at or above 28°C and the fuelwood price is higher than 0.3 NRs/kg. As the analysis shows, the comparison is extremely sensitive to fuelwood price and gas production, the latter being strongly dependent on temperature. Either system by itself is financially viable but sensitive to the amount of grain brought in for hulling and milling and how efficient the fuel is used. Therefore, before installing a rice and flour mill, it is essential to survey the market first, to see if potential business is enough to bring the desired returns.

As for using biogas for irrigation, the internal rate of return is high (above 50%) even without subsidy. Using diesel, the rate of return is higher than 50 percent. In fact, with only 6 percent subsidy, a biogas system would give the same net present worth as a diesel one. The subsidy may be worthwhile since utilization of biogas reduces reliance of the country on the import of diesel thereby saving foreign exchange. The economics are extremely sensitive to the actual land area irrigated. If water supply is limited, the cropping patterns should be adjusted to optimize the use of water and maximize the irrigated area.

In conclusion, the returns are high if biogas is used for income generation and irrigation. For domestic purposes such as cooking, an afforestation scheme would cost less than a biogas one. However, if latrines are attached to biogas plants for the treatment of human waste, the socio-economic value of biogas plants would definitely increase. This is actually the practice in China where waste processing is the primary objective and biogas is only a by-product. In Nepal, the idea of using gas derived from human waste is repugnant to most people. Unless this cultural barrier is overcome, a domestic gas plant using cattle dung will have very limited practicality.

SECTION FIVE

INSTITUTIONAL AND FINANCIAL INFRASTRUCTURE

INSTITUTIONAL AND FINANCIAL INFRASTRUCTURE OF BIOGAS SYSTEMS

H.D. Steingass, W.Paul Weatherly
Bioenergy Systems and Technology Project, USAID

INTRODUCTION

This paper focuses on the critical dimensions of institutional and financial infrastructures for biogas systems. The term "infrastructure" is quite broad, often referring to major physical systems as road, sewage and pipeline networks in a given location. For our discussion, infrastructure is the institutional or organizational background against which certain types of economic development can occur.

Through their experience in biogas systems and programs, conference participants are keenly aware of the need for strong institutions and financing arrangements which are a part of the development of biogas, whether it is research, extension activities or commercial development. Institutional responsibilities, particularly with a new technology or a new approach to resource utilization such as biogas, are often not defined and not clearly obvious. This discussion focuses on the questions:

- Can there be a single agency, or should it be a combination, responsible for development of systems which have applications in agriculture, energy production, sewage treatment and effluent control?
- How much regulation, if any, should be exercised over biogas implementation? How is regulatory oversight assigned?
- What is the most appropriate government intervention in the private sector to stimulate biogas, if any?

Concerning financial infrastructures, we automatically turn to the banking system. But banks are universally conservative investors, usually for good reason, and are generally reluctant to lend money for what is perceived as either unknown or unproven. What are the appropriate inducements, such as loan guarantees, purchase guarantees, etc., to encourage bank involvement in a biogas development scheme? In many developing countries, the more important financial issue is capital availability -- where will investment funds come from, especially the hard currency components, when there is much competition for limited financing?

Certainly these general institutional and financial questions have a number of answers, many of which are particular to the setting. For example, we would not expect the Chinese program to provide a complete model for biogas development in a country, even though it is the world's largest and most comprehensive. Nevertheless, it is clear that institutional arrangements are well developed in China and that they are focused on implementation and widespread use of biogas systems on at least a couple of scales; there are likely to be important lessons from this experience.

This paper briefly focuses on identifying the possible institutional/organizational arrangements and financing systems, of biogas development. Although these infrastructures are likely to exhibit strong country-specific aspects, there are relationships which can be considered general to biogas. The purpose is to examine the relevance of current infrastructures for future biogas development.

This paper is followed by a series of questions which were put forth as a stimulus for a smaller group discussion session concentrating on the issues for biogas development at three different levels. The questions are presented in the Annex, while the results of group discussions will follow separately.

INSTITUTIONAL ISSUES

Biogas is "Multi-Faceted"

To a large extent, the answers to the questions "who or what organization is responsible?" and "who or what organization finances?" differ according to the primary purpose of the biogas system. The primary applications and the institutional sectors where the application has relevance are shown in the table (next page); likely ministries with potential jurisdiction over a given sector or application are shown as well.

It is clear that several agencies could have some jurisdiction over biogas development since applications can usually serve a combination of needs. This is often the case with bioenergy systems, and is often cited as a reason for the slow penetration of bioenergy systems and approaches in both developing and developed countries alike. The number of involved groups is not infinite, however, and the table above shows some common combinations of overlapping jurisdiction, such as agriculture and energy agencies, or energy and waste management agencies. These combinations suggest the most promising areas for targetted promotion of biogas and cooperation in project development.

Historically it has been difficult to establish a new technical area of development (or regulation) without giving it some priority in terms of government policy, agency programs, funding and/or promotion. A newly created authority for biogas development may be the most effective means to achieve widespread implementation. In Jamaica, for example, a cabinet-level secretariat was established to initiate new agricultural sector investments, including some with a dual ag/energy purpose, even though there existed a strong agricultural ministry with economic development responsibilities.

Application	Sector	Agency/Ministry
Waste Management	Sewage treatment	Public Works/San. Public Health
	Agriculture & processing	Agriculture Extension Service Industrial Dev't
	Solid waste management	Public Works/San.
	Industrial sewage	Industrial Dev't Environment
Energy Production	Energy development	Elec/Gas Utility Rural Elec. Admin Public Works/San.
	Agriculture & processing	Agriculture Extension Service Elec. utility Industrial Dev't
Special Products (e.g., fertilizer)	Economic Development	Finance Rural Development Industrial Dev't

Each of the countries represented at the conference has different structures for agricultural development and extension, energy development, waste management, public health management and industry/commercial development. Further, experience differs as to the degree of inter-agency collaboration which works well in different countries. If there exists a strong interest in biogas development in, say, a powerful Ministry of Agriculture, perhaps the most effective approach is to assign responsibility for project development to it and direct other affected agencies, such as the hypothetical Rural Electrification Administration, to cooperate in terms of electrical hook-ups and purchase agreements. Some possible organizational structures for biogas development are shown in the table.

The most important institutional ingredient for the success of biogas program development is a strong agency taking a purposeful lead. It is argued by some that only a newly created agency, or at least a semi-autonomous division of an existing agency, with a mandate and resources can effectively stimulate a new development approach; existing agencies have established activities, ways of doing things and ways of thinking and are not capable of discharging a new and different mission such as biogas development on a large scale. Others argue to utilize existing strengths, avoid creation of new bureaucracies and retrain personnel and hire specialists as needed. Again, opportunities and situations which are specific to the setting in a given country will suggest the most appropriate tack to follow.

Scale of System

To some extent, the answers to institutional and financial questions break down according to the scale of biogas systems. A household or small farm digester may only involve the user, his source of information or service on the system, such as a vendor or extension service, and his means of financing or credit. The best strategy for a

Lead Agency	Secondary Agency	Remarks
Agriculture	Extension Service	Ministry funds extension svc. to instruct, promote, build
Agriculture	Rural Elec.	Have standard purchase agreements for energy
Elec. Utility	Refuse/Sanitation	Sewage & solid waste handled centrally
Public Works/ Sanitation	Elec/Gas Utility	Have standard purchase agreements for energy
New Authority	Varies	Clear relationship to secondary agencies in new agency's authorization

program whose objective is the widespread use of smaller scale biogas systems in rural settings might be to add onto an existing agricultural extension service or rural development/settlements agency, thereby gaining strength through established channels of technical advice and communication. China's massive program for small systems includes extension services for both households and farms, teams of trained digester construction workers, and some form of funding, backed up by a national research and development program (R&D).

By contrast, a large agro-industrial or municipal system may involve the user, which could be a public utility or large company, vendors, a recognized funding authority and private banks, a national ministry or specialized development agency which supports a program of incentives, and a regulatory agency which approves system designs and monitors effluents. The institutional complexity increases with the system scale, and strategies should probably relate directly to the application. For example, a careful private sector approach may achieve the greatest results for use of digestion systems in agro-processing industries. This would require fostering an aggressive investment environment (tax incentives, price guarantees) and a clear regulatory environment (e.g., effluent standards). For large municipal systems the best strategy may be to re-define the mission of an existing authority, such as a department of sanitation or public works agency, making it responsible for regional management of potential digestion feedstocks (e.g., MSW and sewage) and giving it specific contract rights in dealing with energy utilities.

The institutional infrastructures for large and small-scale systems may evolve together also. Both India and China are conducting research on larger commercial scale systems, and beginning to promote their implementation, based on their experience with household, farm and small village scale systems. It is also interesting to note that China is beginning to initiate incentives, in the interest of privatization of agriculture.

The Research Stage

Many conference presentations address engineering research questions in the design and operation of different types of digester systems. In many developing countries, biogas development is clearly in the stages of R&D and demonstration, and commercial application or use on a wider scale is not yet realized (e.g., many Latin American countries, Egypt). Institutional and financing arrangements for this level of biogas activity, though perhaps ultimately directed toward wider application, are quite different from the organizational requirements of large-scale development programs such as India, China and Nepal.

For the most part, successful biogas research needs good technical research design, a stable sponsoring institution and a stable source of funding. These factors suggest a collaboration among a country's research institutes and/or universities with agencies interested in biogas development. Agricultural development and extension agencies are often seen as the logical home for initiation of development programs since many biogas systems derive their feedstocks from livestock and crop operations, and serve agriculture by providing needed energy and waste handling systems. However, it is often the case in countries which have implementation programs as well as research to improve systems that R&D activities are not integrated with implementation. This linkage is critical for biogas development, and the organizations involved should collaborate in any research.

FINANCIAL ISSUES

In terms of financial infrastructures for biogas development, a clear break between the research, development and demonstration level on the one hand, and the commercial application level on the other seems apparent. Funding for R&D projects can, and indeed often does, come from private enterprises active in biogas as a business. Any competitive technology or equipment company must conduct research on the performance of its system(s) and promising new digestion processes and materials if it is to be active in an assumed market.

In the absence of a developed market, however, most R&D funding comes from public sources. These include government agencies and government-sponsored organizations active in the areas of agriculture, energy, waste management and public health. Additionally it includes science & technology development agencies and industrial development authorities; often these are the strongest "new idea" supporters in the government. Donor agencies, foundations, universities and non-profit development organizations can also be good sources of funding for biogas R&D and demonstration.

Some of these same public and donor organizations may well be involved in financing biogas for widespread development. The difference with R&D and demonstration funding is its designation for research purposes, and often its preferential terms: grants, forgivable loans, low- or no-interest loans, long grace and payback periods. The purpose is development of cost-effective biogas systems and knowledge about the potentially marketable applications, rather than a financial return on the loan.

Funding for biogas R&D is not easily available. The competition for preferential financial terms is strong, and financing sources are not always easily identified. The biogas developer must investigate various avenues of financing, including combinations of resources from a number of organizations. It is common, for example, for host country and donor agencies to combine resources with one another in supporting rural and economic development efforts.

The financial infrastructure for large-scale biogas projects is likely to be different. Projects typically will be examined on an individual basis in financial terms by banks, and loans would be at or near commercial rates of interest. Even for large-scale projects, however, it may be reasonable to have access to preferential financing, particularly if biogas receives priority treatment by a given government. This could result in defined programs of financing for qualified applications (thus avoiding capital shortage situations), loan guarantees and "buy-downs", purchase guarantees for energy and other co-products of biogas systems, and special investment incentives in the tax code. Similarly, special financing could be available from international banks (e.g., World Bank, ADB, etc.).

ANNEX
QUESTIONS POSED FOR GROUP DISCUSSIONS

The group was given the following questions to consider for its discussion:

1. What needs can be met by biogas technology (BGT)?
   - Is biogas a money maker?
   - What benefits does BGT bring?
   - Are we going to change anything with introducing BGT or do we maintain the status quo only more efficiently?
   - What is the context in which a BGT-institution can work towards a beneficial BGT-plant?
2. How can we evaluate the potential of BGT?
   - Is there any agreed methodology?
   - How can we assess risks and performance of a plant?
   - What happens to a BGT-plant, if development continues (e.g., when electrification comes to village)?
3. Which program or institution is needed to become active in BGT dissemination?
   - What are the models for institutions?
   - How do institutions actively function?
   - Are the benefits of BGT a result of donor activities?
   - Are there any success stories?
4. What designs are ready to promote? What technical development is needed?
   - Are the designs fitting?
   - Will we always need applied research?
   - What will be the future of basic research?
   - Shall donors coordinate research or is duplicate research still necessary?

5. What can the donors do?
   - Should they coordinate?
   - Should they demonstrate and carry technology?
   - Are there strategies for rural development and appropriate technology?

Some more statements were given for discussion:

- Is BGT significant factor in energy and fertilizer or sanitation?
- What policies would help to make BGT a significant technology?
- What is the "real" argument for a village program? How can such program be rational, accountable and justified?

SUMMARY OF GROUP DISCUSSIONS ON INSTITUTIONAL AND FINANCIAL INFRASTRUCTURE OF BIOGAS SYSTEMS

OVERVIEW

Participants were divided into more or less three equal-sized discussion groups, each handling one of the following specific areas:

1. Rural Development and Basic Human Needs.
2. National Energy and Fertilizer Needs.
3. Opportunities in Commercial Size Digesters.

Because of the relatively large group size, the varied personal experience, and thus the wide spectrum in opinion, discussions were very rich, lively, and extensive. The pointers set in the questions raised in the previous paper were used to a great advantage in guiding the discussions to conclusive ends.

In the following, the general major conclusions and recommendations based on the outcome of the three groups discussions are first outlined. Next, the highlights of each group report are presented.

GENERAL OUTCOMES

Conclusions:

- The viability of biogas sytems increases with the number of benefits they bring. Integrated systems serving objectives other than energy, such as recycling of organic matter and pollution control, are likely to be more feasible by meeting more needs.

- Assessment of the potential commercial biogas system is essential prior to any decision making. In addition to technical and financial aspects, socio-economic considerations and environmental impacts should be taken into account.

- From the institutional point of view, government support and endorsement is necessary. However, it seems that a major contribution from villagers/end users and a strong private sector role are needed for undertaking wide-scale implementation of biogas systems.

- There is no single distinct universal biogas system design available and hence particular considerations are needed for each particular country and particular case, perhaps from among the number of prospective accessible designs.

- The role of donor agencies can be very helpful in:

  - Support of R&D and demonstration plants.
  - Establishment of information and data bank, data processing, evaluations and distribution.
  - Technical assistance in the assessment of the potential uses of biomass wastes.
  - Encouraging technology transfer, exchange of information and experience among developing countries.

RECOMMENDATIONS:

- Strong need for R&D and demonstration in the developing countries.

- Better information dissemination on the different aspects of BGT and systems.

- Need for common guidelines for the evaluation of commercial biogas systems.

- Evaluation and extraction of basic data and lessons from existing biogas programs.

HIGHLIGHTS OF THE GROUP REPORT ON RURAL DEVELOPMENT AND BASIC HUMAN NEEDS

- For rural people, BGT can contribute to various basic and development needs, though the effect may not be so great. Benefits accrued from biogas systems relate to energy, organic recycling, waste management, and sanitation. Though in certain instances, the rural biogas systems can be "money-makers", the positive effects are mostly indirect such as improving the quality of life, and stopping deforestation.

- BGT is site specific and thus conclusions as to the viability of the various types of systems or conditions cannot be generalized. For instance, whether a household system would be a feasible scheme or not will largely depend on the situation.

- There is a big disparity between the cities and villages of developing countries. This gap must be narrowed, and BGT can help in this process which should be counted as an added "social" benefit.

Integration of a biogas scheme in an integrated farming system is worthwhile. Under such circumstances the biogas scheme may not only be paying by itself, but it would join into the overall improvement program.

- The governments cannot support BGT forever. The villagers must contribute the main share. Only some financial support can be given by the government. The degree of subsidy would depend (inversely) on the degree of the direct benefits BGT has for the user.

The most successful countries in BGT promotion did not receive much help from outside, but mostly support from their own governments.

• Biogas systems should not be valued by the normal (micro) economical measures since their multi-faceted contributions can be better quantified under an overall social cost-benefit analysis scheme.

HIGHLIGHTS OF THE GROUP REPORT ON NATIONAL ENERGY AND FERTILIZER NEEDS

• It is not possible to arrive to definite conclusion due to the particular nature and conditions of each country.

• Participation of biogas on the national energy plan is quite minimum, if considered from energy point of view.

• Both macro and micro levels have to be considered in planning for a biogas program as well as the society components and satisfaction of demands rather than needs.

• Consideration and quantification of the different aspects of biogas (energy fertilizer and waste management) are necessary when performing cost-benefit analysis.

• Great emphasis should be given to the formation training of the necessary technical staff to plan, design, execute, operate and maintain the biogas systems.

• Political will and endorsement of national biogas programs is high priority. This has to be on both centralized and decentralized levels.

• Insertion of biogas within the national development plans is not clear; however, interface between the different aspects of development is necessary.

• The role of donor agencies in this field seems not to be clear; however, support of R&D and training seems to be quite appropriate.

GROUP REPORT ON OPPORTUNITIES IN COMMERCIAL SIZE DIGESTERS

Background

The group tackled the issue of commercial/industrial scale biogas systems. The scale essentially includes those sizes with commercial or municipal application regardless of feedstock used (e.g., municipal or agro-livestock wastes). Based on the questions raised by AID/BST, the following points were raised with a high degree of consensus among the members:

I. Need for biogas systems

The group felt that there is a need for commercial scale biogas systems in consideration of the following realities:

a. The system makes sense economically under an integrated application in many industries/sectors.

b. The system is a useful pollution control device which is highly necessary (compulsory in some countries) in any pollutant-excreting industry.

c. There is an opportunity by which organic fertilizer generated by the systems can be tapped, which has positive implication both in the rural and national setting.

II. System evaluation/assessment

The question of how to evaluate or assess the potential of a commercial system should be based on the following issues:

a. Reduced operational expense due to energy cost savings with the displacement of conventional fuel.

b. Pollution control capability costed on the basis of business stoppage risk and/or investment requirement advantage.

c. Opportunity for land reclamation in some specific cases/countries.

d. Social forestry/resource conservation implications.

e. Opportunity for technology localization and internal expertise development.

f. Fertilizer supply implication specifically in countries where foreign exchange and commercial fertilizer availability is a problem.

III. Programs/Institutional Support

A sound multi-agency group or task force with strong private sector participation is required. It was stressed that incentives for the private sector should be established so that program planning at various levels and implementation through system installation will be assured. Establishment of operational and system standards applicable to individual countries shall be a major output of this group. When the questions of lead agency arises a suggested compromise would be to look into the objectives of a particular project/system in the order of priority and match this with the mandate of agencies concerned.

IV. System Design Considerations

The group members were highly unanimous on the idea that there is no single or standardized design that should be encouraged or recommended for adoption. Instead, the criteria for consideration as to what designs should be adopted in a particular region/country were established. The criteria are essentially premised on the strengths and limitations of a particular country/region and these major areas are identified as follows:

a. Substrate used for biogasification.

b. Skill, expertise available in the region/country for system construction, operation and maintenance.

c. Material requirement of the systems.

d. Overall socio-techno-economic advantage from the viewpoints of country concerned.

In this particular issue, the following recommendations were likewise deemed necessary:

a. Need for R&D, demonstration activities in some design aspects specifically addressed to the needs of commercial/municipal systems.

b. Both information (technical) dissemination or sharing/exchange on various designs in order to come up with an optimum or hybrid design.

c. A sort of methodology/tools for evaluating commercial design needs to be established.

V. Role of Donor Countries/Organizations

1. Establishment of a centralized databank (worldwide/regional) where vital information can be obtained. This data center can be either attached or integrated to an existing organization/network or totally new, whichever is appropriate. An essential output of this body is the production of a sort of compilation/directory containing various commercial designs.

2. Processing of information in the form of "critically evaluated" technical materials before they are distributed to member countries/organizations. This requires a qualified body/personnel whose added function would be the publication of continuous/regular publication concerning "technical updates" to be channelled to all possible countries/organizations.

3. Establishment of more demonstration plants worldwide, specifically on systems/substrates of common interests to the region and with potential replicability.

4. Technical and financial assistances on the assessment of alternatives to residue use (energy, fodder, fertilizer, chemicals, etc.) and some other areas where some changes in policy, supply, markets will or may not prove important (broadline area). It is to be emphasized that this requires "real" rather than "bureaucratic" experts from donor or developing country itself.

5. Training of both policy and technical people through specific programs and visits to operational installations.

6. Establishment of uniform methodology of system evaluation on the point of engineering, social and allied issues.

7. More conferences/workshops and training sessions emphasizing technology transfer possibilities among developing countries.

DONOR AGENCIES' RESPONSES

- Programs to be supported by the donor agencies have to be requested by the respective governments.

- There are many biogas programs in developing countries sponsored by donor agencies; however, the output of these programs is not so adequate. Donor agencies are to some extent reluctant to finance more biogas programs. The intention is towards using more thorough technological and socio-economic evaluations methods to justify project funding.

- Donor agencies are supporting training as well as demonstration projects, but it seems that less emphasis is devoted towards supporting R&D in developing countries.

SECTION SIX
REGIONAL PROGRAMS, NETWORKS, AND AID AGENCIES

SECTION OVERVIEW

A number of papers on regional programs, networks, and aid agencies are presented in this section. These encompass representations from Latin American organizations (OLADE and ICAITI), French, German and U.S. agencies or supported organizations/networks; and the U.N. System as represented by FAO. The purpose of this session was to explain the role of these various organizations relating to the area of biogas technology (BGT) and discuss possible supportive actions within a rational set of guidelines on how to proceed in the future.

The USAID indicated willingness to support prospective biogas projects, provided that there is a support on the part of government of the concerned country. The German agency GATE/GTZ is not only supporting government structures, but can as well agree to cooperate with non-government organizations. GTZ can be approached directly without necessarily going through government channels. However, GTZ will not take active part in the implementation and dissemination programs, though support with know-how and training can be provided. A sample of GATE activities is described in this section in the paper entitled "GATE Biogas Consultancy Service." An outline of GATE Biogas Extension Services (BES) is given in Pluschke's paper in the Economic Aspects Section.

The French organization GERES organized an "International Workshop on Rural Biogas Technology" in May 1984 in Marseille, France. An English summary of the workshop is presented in this section. A French group is working on biogas technology using relatively dry materials in the Sahel, Africa. GERES is actively examining the possibility of initiating an African Biogas Network.

FAO has been very responsive in supporting activities in the biogas field. These included R&D and demonstration projects, training courses and seminars, and networks and regional programs. More specifics are given in the summary paper included in this section on FAO's activities in the field of biogas.

BUN (Bioenergy Users Network), which has been strongly supported during the conceptional and inceptional stages by USAID, is intended to become a world wide network for international cooperation in the bioenergy area with the purpose of serving developing countries. A recent update on BUN is included in this section.

It was agreed that, though there are certain guiding principles that may apply in many cases regarding the future prospects and courses of action to be taken, concensus cannot, and is not even desirable to be reached in order not to put a lid on development but rather keep the possibility of supporting different viable solutions. Wide variations among various countries and situations relating to political, social, economic, and even physical conditions would preclude arriving at universally valid generalizations.

EXPERIENCES WITH RURAL BIODIGESTERS IN LATIN AMERICA

R. Caceres and B. Chiliquinga
Regional Bioenergy Program
Latin American Energy Organization
Quito, Ecuador

ABSTRACT

The present paper presents an overview of experiences and evaluations of the different biogas projects stored in the data bank of the Regional Bioenergy Program of the Latin American Energy Organization (OLADE), and discusses some important energy and rural development considerations. It also makes recommendations for planning and implementing national biogas programs.

INTRODUCTION

In 1978, the energy balance for Latin America revealed that, as the second most important source (after hydrocarbons), firewood accounted for 14.2% of the primary energy supply. In rural areas, however, biogas could substitute for some of the residential sector's 75 percent share in final biomass consumption.

In 22 Latin American countries, there are biogas projects with implementation in the field. These have varying degrees of sophistication and development. Experience began early in 1953, but since 1975, programs of a somewhat broader scope were begun. In the first stage, there was optimism that the introduction of biogas in rural areas would be rapid; however, in practice it has been slow. This makes it difficult to assure that, in the medium term, biogas could prove to be a real alternative to the scarce firewood supplies of rural areas. Nonetheless, the alarming forecasts that the firewood supplies of extensive rural zones of Latin America may well become depleted by 1995-2000 mandate the immediate consolidation of national biogas programs that could modify rural energy balances even in the long run.

BACKGROUND

OLADE carried out a project to transfer biogas technology to rural areas. This project was developed in 10 Latin American countries, (i.e., Guatemala, Honduras, Nicaragua, Jamaica, Haiti, Dominican Republic, Grenada, Guyana, Ecuador, and Bolivia) where theoretical/practical training courses and technical seminars were offered. In addition, 60 biodigesters were built, using the following technologies:

- Discontinuous (batch) models, in batteries of two cylindrical biodigesters with external gasholders of the common type;

- Continuous models, i.e., horizontal-displacement biodigesters having small external gasholders; and

- Semi-continuous models, i.e., biodigesters of the Chinese type--cylindrical and elongated, with incorporated gasholders.

As project follow-up, support was provided for exchanges among the different countries through activities such as efforts to set up a Latin American Biogas Network, with backing from OLADE and FAO, events such as the Third Latin American Seminar on Bioenergy, workshops and missions to evaluate biogas projects in countries that had already initiated biodigester construction, training courses on biodigester construction, activities to formulate national biogas programs and promotion of biogas-run equipment, of cost-reducing construction techniques, and of the use of revolving loan funds to finance biodigesters through the banks.

The following priorities have been established for this stage:

Evaluation of Biogas Potential of Latin America

Currently, the preliminary general phase is being concluded, and there are plans to continue at a more specific level in coming years. In this respect, biogas's true potential is being reevaluated, taking into account the obstacles encountered for the entry of this technology into the market.

Biogas for Rural Energy

This component specifically refers to the promotion and evaluation of biogas through (a) pilot projects for integrated rural energy systems for small and medium-sized farms, and (b) consolidation of national biogas programs.

Biogas for Agroindustrial and Urban Development

This component refers specifically to studies on the state-of-the-art for industrial-scale biodigesters, including high-efficiency models.

ASSESSMENT OF CURRENT DEVELOPMENT STATUS

In Latin America, 3950 biodigesters have been inventoried; of these, 79 are discontinuous, 2567 continuous, 996 semi-continuous, and 308 are of other types.

The degree of operability estimated for the region is around 60 percent; the discontinuous digesters work at 44 percent, the semi-continuous ones at 60 percent, and the others at 59 percent. Thus, about 40 percent of the digesters are either shut down or functioning irregularly. Of these, it is estimated that 17 to 20 percent have been completely or practically abandoned, that 40-50 percent have solvable problems, and that 35 to 43 percent are just starting up and therefore working irregularly.

In the case of digesters that have been abandoned, the main problems are tied to poor location and structural building flaws. The digesters with solvable problems basically require reactivation in terms of construction and operation.

When the biogas projects began, digester construction was highly subsidized to serve initially for demonstration. However, later evaluations showed that digesters financed for the most part by users tended to have better locations and better maintenance.

The main problems can be broken down into the following categories:

1. Technical Problems

   a. Operation and maintenance

      - Insufficient user training
      - Lack of local technical staff
      - Lack of feedstock

   b. Location

      - Biodigesters located far from feedstock sources

   c. Design and construction

      - Lack of experience in the use of construction techniques and materials

2. Institutional Problems

   a. Lack of high-level guidelines to support the promotion of biogas technology

   b. Insufficient interinstitutional coordination for the development of biogas projects

   c. Lack of national biogas programs

   d. Lack of training of agricultural extension workers in biogas technologies

3. Socioeconomic Problems

   a. High initial investment costs compared to the buying power of the small farmer

   b. Certain resistance to change in traditional uses of manure

   c. Difficulties with use of the technologies, as derived from land-extensive animal husbandry practices

4. Financial Problems

   a. Uncertain profitability

   b. Lack of soft-term credit policies

   c. Insufficient budgeting for research and demonstration projects

NATIONAL BIOGAS PROGRAMS

In monitoring the various projects in Latin America, there is a need to integrate all of the institutions involved in any one country to create the infrastructure necessary for the development of a national biogas program. For this purpose, the following requirements were developed:

- Grouping of all research, technological development, and rural development institutions;
- Development of training courses appropriate for the technologies;
- Development of facilities for procuring soft financial credits;
- Distribution of work among teams, according to specific purposes:
  - Research: universities and private firms
  - Execution/follow-up: Ministries of Agriculture and extension offices
  - Coordination: Ministries of Energy and Agriculture
  - Promotion: rural development agencies
- Coordination of national and foreign programs and projects through biogas cooperation networks; and
- Construction of biodigesters by qualified technical personnel with building permits extended by the program's board of technical supervision.

Most of the countries of the region are cooperating closely through the Latin American Biogas Cooperation Network, yielding an ongoing exchange of experiences and personnel training. At present, there are 365 specialized biogas technicians in Latin America. Countries such as Brazil, Guatemala, and Jamaica have national biogas programs whose main features are as follows:

- Interinstitutional participation,
- Revolving loan funds,

- Equipment production, and
- Participation of rural extension offices.

CONCLUSIONS

1. In the short term, it is necessary to consolidate experiences with rural biodigesters through the formulation of national biogas programs.

2. It is important to have an ongoing exchange of information through regional cooperation networks.

3. Horizontal cooperation should be tapped as a mechanism for transferring and improving technology.

4. Private organizations should become more actively involved in both equipment adaptation and marketing efforts.

APPENDICES

I. In-the-field Inventory of Biodigesters in Latin America: 1983/84

II. Latin American Institutions Involved in Biogas Field Work: 1984

APPENDIX I

IN-THE-FIELD INVENTORY OF BIODIGESTERS IN LATIN AMERICA: 1983/84

COUNTRY	TYPE	NUMBER	AVERAGE VOLUME (m^3)	OPERATING %
1. MEXICO	Cs	22	13	60
	Cm	2	40	60
	Cl	2	300	40
	Total	26	Average	58
2. GUATEMALA*	Ds	18	11	40
	Dm	5	33	40
	Dl	9	187	44
	Cs	20	13	56
	Cm	8	26	80
	Cl	3	62	60
	Ss	37	12	65
	Sm	1	25	80
	Os	4	6	80
		105		58
3. HONDURAS**	Dm	3	17	30
	Cs	1	10	60
	Cm	4	23	30
	Ss	17	12	70
		25		58
4. NICARAGUA	Dm	3	18	60
	Cm	4	19	25
	Ss	5	10	64
		12		50
5. PANAMA	Cm	5	34	54
	Ol	2	60	40
		7		50
6. EL SALVADOR	Cs	5	12	60
	Cm	1	80	60
	Om	1	50	50
		7		59
7. COSTA RICA**	Cs	22	15	60
		22		60
8. REPUBLICA DOMINICANA**	Dm	1	17	60
	Cm	1	16	50
	Cl	1	120	50
	Ss	2	12	50
	Sm	3	34	50
		8		51

APPENDIX I (cont)

IN-THE-FIELD INVENTORY OF BIODIGESTERS IN LATIN AMERICA: 1983/84

COUNTRY	TYPE	NUMBER	AVERAGE VOLUME (m^3)	OPERATING %
9. CUBA*	Dl	1	75	80
	Cs	550	6	60
	Cl	1	105	80
	Total	552	Average	60
10. HAITI	Dm	2	17	0
	Cm	3	16	0
	Ss	4	12	60
		9		27
11. JAMAICA*	Dm	2	17	50
	Cs	5	10	33
	Ss	4	12	80
	Sm	1	45	---
	Sl	2	67	50
	Om	2	17	---
		16		40
12. BARBADOS	01	1	144	80
		1		80
13. GRENADA	Ds	2	9	---
	Cm	3	21	---
	Ss	5	14	---
		9		
14. GUYANA	Dm	2	17	---
	Cs	2	10	---
	Ss	2	12	50
	Os	1	3	80
		7		19
15. VENEZUELA**	Ds	1	8	0
	Dm	1	44	80
	Cs	3	8	30
		5		34
16. COLOMBIA	Cs	11	9	60
	Cm	6	23	60
	Cl	12	185	60
	Ss	1	12	80
	Om	1	20	---
		31		60

APPENDIX I (cont)

IN-THE-FIELD INVENTORY OF BIODIGESTERS IN LATIN AMERICA: 1983/84

COUNTRY	TYPE	NUMBER	AVERAGE VOLUME (m^3)	OPERATING %
17. ECUADOR**	Ds	1	14	60
	Dm	2	17	40
	Cs	14	9	60
	Cm	4	18.5	60
	Cl	3	53	60
	Ds	3	12	53
	Dm	1	32	50
		28		58
18. PERU	Dm	3	23	80
	Cs	1	9	80
	Cm	2	25	48
	Cl	2	52	30
	Ss	28	11	59
	Sm	8	20	60
		44		60
19. BOLIVIA**	Ds	4	8	40
	Dm	3	17	27
	Cs	2	8	80
	Cm	3	16	17
	Ss	5	11	36
		17		37
20. BRAZIL*	Dm	16	17	
	Cs	1818	7.5	
	Cl	18	83	
	Ss	885	8	
	Os	291	10	
	Om	2	40	
	Ol	3	257	
		3033		60
21. URUGUAY	Cm	6	20	60
	Ss	3	8	60
		9		60
LATIN AMERICA				
TOTAL DISCONTINUOUS (batch)	Ds	26		36
	Dm	43		48
	Dl	10		48
		79		44

APPENDIX I (cont)

IN-THE-FIELD INVENTORY OF BIODIGESTERS IN LATIN AMERICA: 1983/84

COUNTRY	TYPE	NUMBER	AVERAGE VOLUME (m^3)	OPERATING %
TOTAL CONTINUOUS	Cs	2473		60
(Indian, plug-flow	Cm	52		47
Horizontal)	Cl	42		58
		2567		60
TOTAL SEMI-CONTINUOUS	Ss	978		60
(Chinese)	Sm	13		55
	Sl	2		50
		996		60
TOTAL OTHERS	Os	296		60
	Om	6		28
	Ol	6		57
		308		59
GENERAL TOTAL		3950		60

*Have a national Biogas Program
**Have only a National network
D = Discontinuous Biodigester (batch)
C = Continuous Biodigester (Indian, plug-flow Horizontal)
S = Semi-continuous (Chinese)
O = Other types
s = Small: 3 - 15m^3
m = Medium: 15 - 50m^3
l = Large: more than 50m^3
Example: Cs = Continuous small

APPENDIX II

LATIN AMERICAN INSTITUTIONS INVOLVED IN BIOGAS FIELD WORK: 1984

COUNTRY	INSTITUTION	SPECIALIZED STAFF	COOPERATION WITH REGIONAL BIOGAS NETWORKS
1. MEXICO	INSTITUTE OF ELECTRICAL RESEARCH (IIE)	4	XX
	XOCHCALLI FOUNDATION	5	XX
	MEXICAN INSTITUTE OF APPROPRIATE TECHNOLOGY (IMETA)	2	X
	NATIONAL INSTITUTE OF BIOTITIC RESOURCES	3	X
	IZTAPALAPA AUTONOMOUS METROPOLITAN UNIVERSITY	2	
	UNIVERSITY OF MICHOACAN, METALLURGICAL RESEARCH INSTITUTE	2	
	NATIONAL AUTONOMOUS UNIVERSITY OF MEXICO (UNAM), INSTITUTE OF ENGINEERING	2	
		20	
2. GUATEMALA	CENTRAL AMERICAN INSTITUTE OF INDUSTRIAL TECHNOLOGY (ICAITI)	4	XXX
	MESO-AMERICAN CENTER FOR APPROPRIATE TECHNOLOGY (CEMAT)	8	XXX
	NATIONAL TRAINING INSTITUTE (INTECAP)	2	
	ENGINEERING RESEARCH CENTER (CII/CETA)	2	X
	HIGHLANDS ASSOCIATED RESEARCH (ICADA)	1	
	PROFESSIONAL AGRONOMY OFFICE (OPINA)	3	XX

APPENDIX II (cont)

LATIN AMERICAN INSTITUTIONS
INVOLVED IN BIOGAS FILED WORK: 1984

COUNTRY	INSTITUTION	SPECIALIZED STAFF	COOPERATION WITH REGIONAL BIOGAS NETWORKS
	BIOENERG	2	X
	OTHERS	8	
		30	
3. HONDURAS	SECRETARIAT OF NATURAL RESOURCES, RURAL EXTENSION OFFICE	5	XX
	NATIONAL AGRARIAN INSTITUTE (INA)	1	
	INDUSTRIAL RESEARCH CENTER (CDI)	2	X
		8	
4. EL SALVADOR	LEMPA RIVER EXECUTIVE HYDRO-ELECTRIC COMMISSION (CEL)	2	
	CENTER FOR AGRICULTURAL/ LIVESTOCK TECHNOLOGY	1	X
		3	
5. NICARAGUA	CENTER FOR RESEARCH ON APPROPRIATE TECHNOLOGY (CITA/INRA)	5	
	NATIONAL INSTITUTE OF ENERGY (INE)	3	X
		8	
6. COSTA RICA	TECHNOLOGICAL INSTITUTE OF COSTA RICA (ITCR)	2	XX
	UNIVERSITY OF COSTA RICA (UCR)	2	
	OTHERS	3	
		7	
7. PANAMA	NATURAL RESOURCES (RENARE)	2	X
	NATIONAL ENERGY COMMISSION, INSTITUTE OF WATER RESOURCES AND ELECTRIFICATION (CONADE/IRHE)	2	
	ALTERNATIVE TECHNOLOGY GROUP (GTA)	2	X
		6	

APPENDIX II (cont)

LATIN AMERICAN INSTITUTIONS
INVOLVED IN BIOGAS FILED WORK: 1984

COUNTRY	INSTITUTION	SPECIALIZED STAFF	COOPERATION WITH REGIONAL BIOGAS NETWORKS
8. HAITI	RESEARCH CENTER, MINISTRY OF MINES AND ENERGY	5	
9. DOMINICAN REPUBLIC	LOYOLA UNIVERSITY	2	
	NATIONAL ENERGY POLICY COMMISSION	2	
	STATE SUGAR COUNCIL (CEAGANA)	2	
	AGRICULTURAL INSTITUTE OF HIGHER LEARNING (ISA)	2	X
	OTHERS	3	
		11	
10. CUBA	NATIONAL ENERGY COMMISSION	2	X
	ACADEMY OF SCIENCES	5	
	MINISTRY OF AGRICULTURE	13	
	CUBAN INSTITUTE FOR SUGARCANE BY-PRODUCTS (ICIDCA)	2	
		22	
11. JAMAICA	MINISTRY OF MINES AND ENERGY	2	X
	SCIENTIFIC RESEARCH COUNCIL	4	
	MINISTRY OF AGRICULTURE	8	
	OTHERS	2	
		16	
12. BARBADOS	MINISTRY OF AGRICULTURE	2	

APPENDIX II (cont)

LATIN AMERICAN INSTITUTIONS
INVOLVED IN BIOGAS FILED WORK: 1984

COUNTRY	INSTITUTION	SPECIALIZED STAFF	COOPERATION WITH REGIONAL BIOGAS NETWORKS
13. GRENADA	MINISTRY OF PLANNING, TRADE AND INDUSTRY	2	
	CARIBBEAN SCHOOL OF AGRICULTURE	10	
		12	
14. GUYANA	GUYANA NATIONAL ENERGY AUTHORITY (GNEA)	3	
	NATIONAL SCIENCE RESEARCH COUNCIL	2	
		5	
15. SURINAME	MINISTRY OF NATURAL RESOURCES AND ENERGY	5	
16. VENEZUELA	MINISTRY OF ENERGY AND MINES	1	X
	CENTRAL UNIVERSITY OF VENEZUELA SCHOOL OF AGRONOMY	3	X
	CENTRAL UNIVERSITY OF VENEZUELA SCHOOL OF ENGINEERING	3	X
	ANDEAN ELECTRIC POWER DEVELOPMENT COMPANY (CADAFE)	2	X
		9	
17. COLOMBIA	UNIVERSITY DEL VALLE (CIMTE)	3	X
	ARMAR COMPANY	2	XX
	OTHERS	5	
		10	
18. ECUADOR	NATIONAL INSTITUTE OF ENERGY (INE)	3	XX
	POLYTECHNIC SCHOOL OF THE COAST (ESPOL)	2	
	INSTITUTE OF APPROPRIATE TECHNOLOGY	2	

APPENDIX II (cont)

LATIN AMERICAN INSTITUTIONS
INVOLVED IN BIOGAS FILED WORK: 1984

COUNTRY	INSTITUTION	SPECIALIZED STAFF	COOPERATION WITH REGIONAL BIOGAS NETWORKS
	NATIONAL POLYTECHNIC SCHOOL	2	
	POLYTECHNIC SCHOOL OF CHIMBORAZO	3	
	OTHERS	3	
		14	
19. PERU	INSTITUTE OF INDUSTRIAL TECHNOLOGICAL RESEARCH AND TECHNICAL STANDARDS (ITINTEC)	5	XXX
	NATIONAL TECHNICAL UNIVERSITY OF CAJAMARCA	7	XX
	GLORIA COMPANY	3	
	NATIONAL TECHNICAL UNIVERSITY OF THE HIGHLANDS (UNTA)	2	
		17	
20. BOLIVIA	NATIONAL INSTITUTE OF RURAL ELECTRIFICATION (INER)	2	X
	AGROBIOLOGY PROGRAM/COCHAPAMBA (PAC)	3	X
	APPROPRIATE RURAL TECHNOLOGY PRO BOLIVIA (TRABOL)	3	X
	OTHERS	3	
		11	
21. CHILE	AGRICULTURAL DEVELOPMENT INSTITUTE	2	
	CATHOLIC UNIVERSITY OF VALPARAISO	2	
		4	

APPENDIX II (cont)

LATIN AMERICAN INSTITUTIONS
INVOLVED IN BIOGAS FILED WORK: 1984

COUNTRY	INSTITUTION	SPECIALIZED STAFF	COOPERATION WITH REGIONAL BIOGAS NETWORKS
22. BRAZIL	STATE RURAL TECHNOLOGY COMPANIES	65	XXX
	BRAZILIAN COMPANY FOR TECHNICAL ASSISTANCE TO RURAL EXTENSION (EMBRATER)	5	XXX
	BRAZILIAN COMPANY FOR AGRICULTURAL RESEARCH (EMBRAPA)	8	XX
	INSTITUTE OF TECHNOLOGICAL RESEARCH OF SAO PAULO (IPT)	4	XX
	STATE COMPANY (FAIMA)	4	XX
	SANDEPAR COMPANY	4	X
	BRAZILIAN AUXILIARY ELECTRIC LIGHT AND POWER COMPANIES (CAAEB)	2	X
	FEDERAL UNIVERSITY OF PARAIBA	3	X
	TECHNOLOGICAL CENTER-FOUNDATION OF MINAS GERAIS (CETEC)	2	
	FEDERAL UNIVERSITY OF SANTA MARIA	4	X
	CATHOLIC UNIVERSITY OF PARANA	4	X
	FEDERAL UNIVERSITY OF RIO DE JANEIRO	2	
	FEDERAL UNIVERSITY OF RIO GRANDE DO SUL	2	
	STATE ELECTRIC LIGHT AND POWER COMPANY (ELECTROBRAS)	2	
	CENTRAL DISTILLERIES JACQUES RICHER	3	
	CAUCAIA DISTILLERY	3	

APPENDIX II (cont)

LATIN AMERICAN INSTITUTIONS
INVOLVED IN BIOGAS FILED WORK: 1984

COUNTRY	INSTITUTION	SPECIALIZED STAFF	COOPERATION WITH REGIONAL BIOGAS NETWORKS
	FEDERAL UNIVERSITY OF CEARA	2	
	JACKWELL METALLURGY	4	
	FEDERAL UNIVERSITY OF PELOTAS	2	
	SAO PAULO STATE UNIVERSITY	2	
	FEEMA	2	
	NATIONAL RESEARCH COUNCIL (CNPQ)	2	
	LONDRINA STATE UNIVERSITY (UVEL)	2	
	OTHERS	15	
		148	
23. ARGENTINA	CENTER FOR PHOTOSYNTHESIS AND BIOCHEMICAL STUDIES (CEPOBI)	3	
	NATIONAL INSTITUTE FOR AGRICULTURAL TECHNOLOGY (INTA)	3	X
	OTHERS	6	
		12	
	TOTAL FOR LATIN AMERICA	365	

SOURCE: OLADE, Data Bank of the Regional Bioenergy Program.

ICAITI'S DISSEMINATION ACTIVITIES OF BIOGAS SYSTEMS

W. Ludwig Ingram, Jr.

ABSTRACT

The Central American Research Institute for Industry (ICAITI) efforts relating to the dissemination of biogas technology are outlined. These are affected through four main routes: demonstration, publications, workshops and training, and technical assistance. The main conclusions evolving from ICAITI's experience are finally highlighted.

INTRODUCTION

The Central American Research Institute for Industry (ICAITI) is a non-profit regional institution dedicated to serve the five Central American countries, namely: Costa Rica, El Salvador, Honduras, Nicaragua and Guatemala. A main function of the Institute is applied research, development and adaptation of technologies to conditions in Central America.

The on-going activities of ICAITI in the field of rural technology are a recent incorporation into the Institute's programs. Only about five or six years ago, at the most, work in this field was initiated.

At the start of the energy crisis, the Institute carried out some sporadic experimentation on anaerobic fermentation in an effort to apply alternate energy sources to ameliorate the effects of the high prices of solid fuels on the rural population, including fuelwood, which, as a result of the crisis, has greater demand, but is in short supply, thus resulting in higher prices.

The project "Fuelwood and Alternate Sources of Energy," sponsored by ROCAP, for the purpose of promoting the more efficient use of fuelwood or its substitution by an alternate energy source, allowed ICAITI to carry out during the last four years and in a systematic way work oriented towards development and dissemination of biodigesters.

The digester designed by ICAITI has as principal operational characteristics horizontal displacement and continuous operation, with daily charge and discharge and a retention time of 30 to 50 days, depending on the type of substrate and climate. The substrate that has been used up till now is cattle manure, and the digesters have been constructed with capacities varying from 5 to 70 cubic meters.

Experiments have been continued with other substrates equally abundant in the Central American rural area: hog and chicken manure, vegetable and agricultural wastes. Work is also being conducted to improve the biogas yield and reduce the construction costs of the units, with a view to facilitate a deeper penetration of the system in the rural areas and, especially, among people in the very low income bracket.

DIFFUSION AND DISSEMINATION SYSTEM OF BIOGAS DIGESTERS

ICAITI's efforts for diffusing, transferring and disseminating biogas technology are basically channeled through four main routes, namely: demonstration units, publications, workshops and courses, and technical assistance.

Demonstration Units

After setting design, construction and operational parameters for the prototype biogas plants, the dissemination stage was initiated by ICAITI with installation of demonstration systems in selected regions in the five Central American countries, with a view towards showing the largest possible number of inhabitants of the cattle raising areas the benefits derivable from biogas technology.

A total of thirteen demonstration units were built: 3 in Guatemala, 4 in Honduras, 3 in El Salvador and 3 in Costa Rica. All illustrate the use of biogas for cooking and lighting, and in four of them the use of biogas in internal combustion engines is also demonstrated. Moreover, a demonstration center of biogas technology was installed at the "Instituto Tecnico de Agricultura" in Barcena, Guatemala.

Publications

ICAITI published a series of pamphlets, bulletins and manuals for the construction and operation of biogas digesters. This material is directed to users as well as institutions and extension entities. The publications include general information on biogas, production methods, and effluent applications and utilization. They provide technical data on operation and efficiency. In a simple and illustrated form they indicate the steps to be followed for easy construction of a digester, as well as the actions to be taken for operation of the unit.

Workshops and Courses

With respect to the different demonstration units a series of workshops and practical and theoretical courses are given, in which information on methanogenic digestion is offered and the means of charging, discharging and maintaining a biogas producing digester are illustrated. Different applications of biogas are also shown.

The courses include training programs for instructors who then will be able to construct biogas digesters or give training to others. By this means the necessary chain for the massive dissemination of the technology is developed.

Technical Assistance

ICAITI provides technical assistance to parent institutions and trains their personnel. It also assists any interested person in installing a biogas digestor, and provides training covering the operation and maintenance of the unit, as well as the different applications of biogas (installation of equipment, lamps, stoves, motor conversion, etc.).

Experiences on Dissemination of Biodigestion

To date, major efforts towards disseminating biogas technology have been effected by ICAITI in the Republic of Guatemala, and most of the experiences mentioned here concern this country.

For installing prototypes and demonstration units, counterparts entities working in the rural area were selected, which could become an appropriate network for further diffusion and dissemination of this technology. For this purpose in Guatemala, the "Instituto de Ciencia y Tecnologia Agricola (ICTA)," the "Instituto Tecnico de Agricultura (ITA)," the "Direccion General de Servicios Agropecuarios (DIGESEPE)," and the "Instituto Tecnico de Capacitacion y Productividad (INTECAP)" were selected.

ICTA operates several agricultural experimental centers in different zones of the country. Some of its functions consist in generating and promoting the use of agricultural science and technology among small and medium farmers. During the investigation stage, ICTA collaborated with ICAITI in the application of the effluent as organic fertilizer and presently is giving support to the programs for disseminating the technology.

ITA is a center for technical studies which trains agricultural technicians in farming, pest control, irrigation, cattle breeding, etc. This institution cooperates with ICAITI in the research and demonstration of biogas systems, develops jointly with ICAITI research projects related to digestors, applications of biogas and biofertilizers. Professional staff and students take part in these activities.

DIGESEPE is the Division of the Ministry of Agriculture responsible for the programming, organization and dissemination of agriculture, cattle and fisheries production. It is also entrusted with providing technical services and assistance to the rural population.

INTECAP is a governmental technical institute which, in cooperation with private sector, is devoted to the development and training of human resources with a view towards improving the efficiency of the productive activities of the country.

At present ICAITI works in close cooperation with all of these institutions in dissemination efforts. DIGESEPE collaborates in training small and medium cattlemen in the biodigestion technology, and also identifies potential users and appropriate institutional extensionists dedicated to promotion, training and technical assistance.

INTECAP collaborates in the training and upgrading of labor in biogas technology and also disseminates the technology. All of these activities are implemented in close cooperation with ICAITI.

The design and operational features of ICAITI's biogas plant which up till now uses only one substrate (cattle manure), and the cattle areas selected for installing demonstration units, have determined that dissemination activities were aimed primarily towards the small and medium cattle ranches.

During a year and a half of dissemination activities, 54 biogas units were installed in Central America with the direct or indirect participation of ICAITI, that is, more than three units per month. It is known that seven other units have been constructed by persons using the construction manual prepared by ICAITI.

Most of the systems use the biogas for cooking. About 13 units use it in a combined form for lightning, while 11 units utilize the biogas for driving internal combustion engines.

Fourteen units are scheduled to be built in the next few months.

To date 62 persons have been trained in the techniques of construction and operation of biogas plants.

Most of the persons interested in these biogas systems were identified through a series of courses and seminars organized by ICAITI and the national counterparts entrusted with demonstration units.

A recent survey conducted in Guatemala in zones where demonstration units and a certain number of private ones were installed, has established a series of facts, conditions and interests which deserve to be taken into account in any dissemination program to be designed for these regions, and which could be applied to other countries with similar conditions.

First, experience has shown that there is a wide potential market for biogas systems. These systems have many applications in rural areas such as cooking, lighting, and the driving of internal combustion engines used for irrigation, grass cutting, electrical generators, tractors, etc. Through these applications, biogas systems can make a substantial contribution to lowering the prices pressure of conventional fuels on the rural population, and they can also play a significant role in the diversification of energy sources.

Field experience has also shown that a biogas system geared to the foregoing applications is technically and economically feasible. However , in some cases economic feasibility depends on certain conditions prevailing in some areas such as, for example, the type of fuel to be substituted by biogas (fuelwood that in some areas is abundant and cheap, diesel when its price is subsidized), the existence of a need for the effluent, and the required investment for installing the system. The latter is greatly influenced by the method for collecting manure. In a farm without installations for handling cattle, the construction of feeding facilities for the animals and for gravity feeding, the biodigestor can represent up to 50 per cent of the cost of installation of the system. This could result in investment costs that may exceed the annual cost for energy.

On the other hand, a series of constraining elements were identified, which must be overcome in order to achieve an effective and agile dissemination of biodigestors. In this respect it was ascertained that most small cattle farmers have a rather low educational level. They generally work in a medium where tradition is accepted as an immutable "status quo." Dissemination of knowledge and habits from generation to generation determine, or has determined, up to recent times, the insertion of the individual in a rather immutable, conformist and conservative social "habitat." Therefore, technological innovations require a considerable effort to penetrate in this social level. There is a great need for promotional programs specially designed for such inhabitants. Moreover, an effective mechanism is required for reaching them with this information.

The majority of the cattle farmers are also of low economic means. When they accept biogas as a product which can improve their standard of living at lower cost than present systems, and even if this system offers advantages over fuelwood and candles, they face the need of making an initial investment beyond their economic means. This blots out the foregoing advantages. Furthermore, owing to their rather low educational level, they have no clear idea of the meaning of return on investment.

Another aspect observed was the need for training at different levels, such as:

1. Instructors-constructors in the pertinent technology

2. Field promoters who provide orientation on the technological benefits and detect potential users

3. Users who receive training on the operation and maintenance of the biodigester and equipment

Periodic visits made by the Institute experts to the operating units have served to identify several problems regarding the systems, and the users instead of trying to solve them, just waited for the arrival of the Institute technician.

This type of assistance is very costly and the Institute cannot afford to maintain it over a long period of time. Therefore, there arises a need to implement a training program for technical instructors who will be entrusted with dissemination and formation of technical cadres in matters pertaining to construction, operation and maintenance of the biodigestion systems. This is contrary to the conventional constructors (masons) who limit themselves to effect the physical installations and do not impart training to the user on the technology to be followed.

There also arises a need for training promoters and extension workers in rural areas so that, through their institutional programs, they will be aware that biodigestion systems offer an alternative to the energy crisis in the rural area. Finally, there is a vital need to impart the appropriate training to the users.

GATE-BIOGAS CONSULTANCY SERVICE

OEKOTOP GmbH
Paul-Lincke-Ufer 41, 1000 Berlin 36, West Germany

THE IDEA

The Biogas Consultancy Service is a joint effort of the German Agency for Technical Cooperation (GTZ) and the OEKOTOP to assist Third World Countries in utilizing biomass as a renewable energy source.

Due to its unsophisticated technologies (biogas plants, alcohol distilleries, stoves etc.) a more efficient use of biomass rates high in Third World Countries strategies to reduce energy imports and environmental burden.

Heat, electricity and fuel can be supplied in a decentralized way enabling a better regional and rural development.

THE TEAM

4 Biogas Consultancy Teams have been formed, equipped with high qualification and substantial experience in various disciplines:

- agriculture and engineering
- technical crafts
- economics and social science

Most of the 12 biogas consultants have collaborated with volunteers' organizations in Third World Countries and received practical experience in grass-roots development.

THE PROGRAM

The Biogas Consultancy Service was started 1982. After 6 month advanced training in biogas technology with OEKOTOP and in several biogas training centers of other countries (PR China, India, Guatemala, Upper Volta) the biogas teams have started work in Tanzania, Kenya, Bourkina Fasso, Burundi, Nicaragua and in the Caribbean. The overriding goal of the project is to enable various countries to establish their own biogas extension programs.

The activities of the GATE Biogas-Teams are conducted in close cooperation with local or regional organizations.

FIRST EXPERIENCES

A great part of the existing biogas plants visited from the teams, does not work or work only insufficently because of the lack of regular maintenance and for technical mistakes. This fact demonstrated not only the necessity to improve the biogas-technology but also to build up a biogas-service. The biogas installations constructed by the teams range from family size types ($6m^3$) up to large scale biogas digesters for the by-products of small industries (200 m^3). This indicates the wide range of the target groups where biogas plants can be established.

OEKOTOP LTD.

OEKOTOP provides the backstopping for all Biogas Consultancy Teams, distributes relevant information on progress made, evaluates and reviews the GTZ-Biogas program.

FAO ACTIVITIES IN THE FIELD OF BIOGAS

(Extracted from a paper entitled "FAO Activities in the Field of New and Renewable Sources of Energy" prepared by the Food and Agriculture Organization (FAO) of the United Nations for the Second Session of the Intergovernmental Committee on the Development and Utilization of New and Renewable Sources of Energy, New York, 23 April-4 May 1984).

Biogas production, through anaerobic digestion of organic wastes and residues, continues to expand in many developing countries. However, to further expand biogas use, there is need for applied research to reduce digester cost and to improve efficiency. There is also need for demonstration and training in digester construction and design, and manufacture of simple, reliable biogas appliances.

FAO's biogas activities are shown in the Annex. Biogas activities initiated since the first session of the Intergovernmental Committee include: i. technical assistance projects in development in three countries, and biogas development in association with organic recycling in one country; ii) organization, in association with ESCAP, of a regional training course on biogas technology at the International Centre for Biogas Research and Development in Chengdu, China; and iii) a consultancy on the development and use of biogas in selected countries of Africa. In addition, the Biogas Network in Latin America, supported by FAO, ECLA and OLADE, continues to expand. Activities of this network include formulation of national biogas programs, preparation of a directory of institutions involved in biogas, TCDC exchange of biogas experts between countries of the region, and support to seminars and workshops on biogas.

Activities during 1984-85 will emphasize technical assistance to developing countries and promotion of a TCDC approach to regional cooperation on biogas, including: i) continued support to the Asian and the Latin American Biogas Networks; ii) a feasibility study on biogas technology development in the Caribbean; and iii) the possible establishement of a biogas network in the African region.

ANNEX
FAO BIOGAS ACTIVITIES

No.	Country	Period	Title
1.	Afganistan	1978-83	Biogas technology, research, demonstration and training
2.	Africa	1983-84	Consultancy on development and use of biogas in selected countries of Africa
3.	Asia/Pacific	1982-83	Rural energy development program: organization of two training courses on biogas technology at International Centre for Biogas Research and Development, Chengdu, China (in association with ESCAP)
4.	Bangladesh	1984	Organic recycling and biogas technology
5.	Barbados	1984-85	Small-scale biogas technology development
6.	Burundi	1982-84	Recycling organic material and biogas technology
7.	China	1982-83	Assistance to the Research and Training Centre for Biogas Development and Extension, Chengdu
8.	Honduras	1982-84	Biogas pilot and demonstration project
9.	Laos	1983-84	Small-scale bigas technology development
10.	Latin America	1982-85	Biogas network in Latin America (jointly supported by FAO/ECLA/OLADE)
11.	Latin America	1984	Feasibility of biogas technology development in the Caribbean
12.	Lesotho	1983-84	Training in small-scale biogas production
13.	Pakistan	1981-85	Coordinated national program for livestock resources and nutrition (biogas development)
14.	Singapore	1982-85	Pig-waste treatment and utilization plant (biogas development)

THE BIOENERGY USERS' NETWORK (BUN) AN EXECUTIVE SUMMARY

The Bioenergy Users' Network (BUN) is being set up to create a network of developing countries who are active in the bioenergy field and are willing to contribute information and expertise about their national bioenergy experiences.* BUN is designed to be a clearinghouse for the exchange of experiences among developing countries, a facilitator of technical sharing, and a central repository of bioenergy information. Members in the Network will be able to learn from one another's experience and to use finanical and technical responses more efficiently as a result. The focus of the Network, in support of national programs, is on national bioenergy systems development, creation of a skills bank, and promotion of a common methodology for systems evaluation and analysis.

Although BUN is a network of developing countries, other groups such as international organizations, donor agencies, consulting groups, and manufacturing firms can also participate. Each of these different groups is being offered membership status appropriate to its particular needs.

BUN's origins can be traced to a workshop held in Manila in March, 1983, which concerned "Bioenergy in National Development." Participants from seventeen developing countries decided that an Organizing Committee should be designated to consider a framework for a permanent Network. The members at the Committee were drawn from countries with an active involvement in bioenergy.

The Committee met several times during 1983-84. It concluded that a formal network mechanism was needed to enable researchers, technicians, and policymakers to interact and exchange experience and information on the development of bioenergy resources. BUN has attracted significant donor interest, with startup grants forthcoming from the U.S. Agency for International Development, the Rockefeller Brothers Fund, the Agricultural Resources Development Corporation and the International Institute for Environmental Development.

*The term "bioenergy" is a relatively new expression, coined during the past few years to reflrect the close relationship between biomass resources used for energy purposes and the resource base of agriculture. It refers to indigenous fuels derived from cellulose materials such as wood, agricultural wastes, etc. The technologies needed to process and utilize these materials as fuel supplies are well known. The needed resource management skills are not yet well developed, in many cases, although a growing body of experience exists in developing countries.

The Organizing Committee has recommended that BUN be established as a formally chartered association, but that it not seek official International Organization status. It will work with existing organizations and networks, both inside and outside the United Nations system, as well as with institutions with similar interests in d

BUN will be a member-run organization, and its activities will be geared to the priorities of members. However, the main activities of BUN are presently envisioned by the Organizing Committee as follows:

a. Program initiation: Dozens of developing countries are trying to articulate isolated activities into national bioenergy programs. BUN will provide crucial assistance in the initial stages by helping coordinate activities and identifying key elements of an appropriate program.

b. Information system on country programs. Focused on up-to-date country information including programs, institutions, resources, etc., the information system is conceived as demand-driven by country needs.

c. Skills bank of developing country expertise. With emphasis on both technical areas and organizational aspects of establishing country programs, the skills bank will be the basis for technical exchange programs.

d. Systems evaluation and analysis. The focus will be on providing common procedures and standards for evaluation and performance of bioenergy systems, not only for technologies but also for integrated systems.

e. Training programs. Based on identified skill requirements, training will incorporate the resources of developing as well as developed countries.

A basis for these activities, and for the formal launching of the Network on October 1, 1985 is being laid at the present time.

SECTION SEVEN: TECHNICAL ASPECTS

ENGINEERING DESIGNS

ENGINEERING DESIGN OF BIOGAS UNITS FOR DEVELOPING COUNTRIES

Roscoe F. Ward
Dean, School of Applied Sciences
Miami University, Oxford, Ohio, USA

ABSTRACT

While anaerobic digestion has been used for over 100 years, the approach to design and building units has been often empirical and based on myths rather than scientific principles. The purpose of this paper is to review the design concepts used in a number of developing countries and then to show how the designs can be improved by using scientific principles to obtain lower cost and more efficient units.

DESIGN CONCEPTS USED FOR FLOATING COVER INDIAN STYLE DIGESTERS

A number of guides are available that discuss the design of the floating cover digester. These include the Guidebook on Biogas Development (1980) published by the Economic and Social Commission for Asia and the Pacific (ESCAP) and Methane Generation from Human, Animal, and Agricultural Wastes by the National Academy of Sciences (1977). These are all largely based upon the KVIC publications.

The approach has been to select a gas production rate for the digester based on a feed material. For example, if 1 m^3 of gas is required per day, ESCAP has assumed that 1 kg of cattle manure mixed with 1 l of water will produce 36 l of gas in 50 days at 27°C. Thus 2 l would be added each day and held for 50 days giving a minimum volume of 2.8 m^3 for the digester which should produce 1 m^3 of gas per day.

In examining the assumption a number of issues come forward. The first is the assumption that the 50 days are required at 27°C to produce 36 l gas. This crude gas production assumption does not take into account either the feed that the cattle have been fed or the age of the manure. As shown under the section on "biodegradability", cattle manure from cows fed a high protein diet is much more digestible than from those fed crop residues.

The second is that fresh manure is also much more digestible than dried manure.

The third is, as Stuckey (1983) noted based on the kinetics of anaerobic digestion, that approximately 80% of the organics are degraded in 15 days and 90% in 30 days. Therefore, there would not seem to be a justification for a 50-day detention time if maximum gas is the objective of the system. Extra detention time means a larger digester and more cost.

The fourth issue is the assumption that 1 l of water needs to be added to each kg of manure. The addition of the water gives a solids content from 3 to 12% depending upon the age of manure. Marcheim (1983) has operated digesters at 16 to 18% solids. To keep the digester to a minimal size, one would like to have the smallest amount of water in the digester.

The inconsistencies of the approach of ESCAP become more obvious when its guide book is examined further. There does not seem to be any rationale for the detention period. For chicken manure, the assumption is that 44 l of gas will be produced per kg of feed and that the detention period should be 30 days; for pig manure, the assumption is 45 l of gas by 1 kg when digested 40 days; and for human wastes, the yield is 24 l of gas when digested for 60 days.

Some of the more recent publications have indicated that in India it may have been recognized that temperature does have an impact. The detention period for design has been changed to 30 days for south India, 40 days for central India, and 55 days for the cooler mountain areas.

The approach suggested in the National Academy of Sciences (NAS) example in determining size of the digester is to select a height-to-diameter ratio of 1:1. Then, determine the dimensions of the digester and add a gas holder which will store one half of one day's production. The second approach is to use a truncated conical (hopper) bottom tank for the digester. In the example, NAS essentially keeps the diameter the same and reduces the wall length and adds the hopper bottom. However, from an operations' point of view there appears to be no justification for consideration of the hopper bottom. The hopper bottom is an added difficulty to construction because a deeper pit would be needed.

DESIGN CONCEPTS USED FOR A CHINESE DIGESTER

Translations are available of The Chinese Biogas Manual in many languages. This manual, in its English translation (van Buren, 1979), describes the objectives of the initial development of the Chinese digester; i.e., dispose of human excreta for the prevention of disease, produce a soil conditioner (fertilizer), and provide energy which could be used for cooking.

The Chinese water pressure system ideally was designed to use a gravity feed system to collect the fresh manure and human wastes and place them in the digester. In addition, crop residues were often added to the digester to produce additional soil conditioners. Significant features include the incorporation of gas and slurry storage within the inexpensive digester vessel, which used a fixed cover and could be built in the field. While the digester met the objectives, two major disadvantages are common. The first is operational problems and the second is the labor required to remove the slurry and transport it to the field. Both of these factors are frequently overlooked.

In many areas of China, manure and night soil are applied to the fields as a soil conditioner. By making the digester larger, storage is provided to minimize the losses of ammonia nitrogen from the slurry. The slurry is removed twice a year and applied to the fields. Therefore, the primary criteria used in designing the Chinese digester in China was the quantity of soil conditioner that the farmer had the raw materials to produce and store between applications to the field.

The system was widely applied and Chen reported in 1983 that 5 to 6 million units had been built. Larger numbers have been reported but these numbers appear to be in error (Chen, 1983). Because of the successful introduction of the digester in China, other countries rushed to copy the system often without examining their own country's conditions.

One such application recently observed was in Turkey where three groups have been experimenting and conducting demonstrations using the anaerobic digester. One group had been to China, and they decided to apply the Chinese experience to Turkey. They determined that since the major cost is that of the digester, the size of the vessel should be kept to a minimum. Still, they needed to provide storage for the effluent solids so they collected the effluent from the digester in open concrete pits where it was held until it could be applied to the fields. If there were excess effluent, the pit would act like a clarification basin where the solids would settle to the bottom and the supernatant could be removed. Overlooked was the fact that the nitrogen that enters the digester is converted into the ammonium form which is volatile. Thus a portion of the nitrogen could be lost to the atmosphere.

DESIGN CONCEPTS USED FOR A BAG DIGESTER

The bag digester operates like a plug flow reactor where the materials added each day theoretically will move as a mass through the digester until the day when the theoretical detention time has been reached. Then the digested mass will flow from the digester as a unit. A portion of the effluent is reintroduced into the inlet so that it will act as seed to reinoculate the digester. The primary advantages of the bag digester are that it is purchased as a unit, it is inexpensive compared to the Indian or Chinese style digester, and can be easily installed and operated. The disadvantage is that the unit is harder to maintain at a uniform temperature.

To size a bag digester, similar approaches to that quoted for Indian digesters have been used.

The red mud used successfully to build the bag digesters in Taiwan is not universally available. In the United States, a neoprene material has been used. Other developing countries have tried locally manufactured materials without much success because of the tendency for the materials to degrade under the sunlight. Modifications of the bag digester have been observed, which include the building of a long narrow concrete trench in which a plastic cover is laced along the sidewalls. Obtaining a good seal along the walls is difficult. The plastic cover is to act only as a gas collector, and it does not allow much storage because of the rigidity of the plastic.

Length to diameter ratios of 3 to 14 are used for a bag digester. The inlet is above the digester to maintain a gas pressure of approximately 0.5 m. This maximum pressure depends upon the materials used to manufacture the bag.

If maximum gas production is desired, an approach similar to that discussed under the Indian style digester could be used. The gas production rate over a period of time would be the governing factor.

ITEMS TO CONSIDER IN EXAMINING A SYSTEM

In examining the technical and economic feasibility of anaerobic digestion systems, a number of questions need to be asked about the feedstock, the digester system, and the potential uses of the effluents. The general questions include:

- quantities and cost of the potential feedstocks;
- characteristics of the feedstock and effluents;
- cost of obtaining the feedstocks and production of the effluent products;
- cost of digester and utilization equipment;
- current and potential uses and sales of the effluents; and
- cost of alternative fuels being used.

The financial analyses of most digestion systems show that when used for only a single purpose, i.e. energy, health, or to produce a soil conditioner, they are not economical. Therefore, uses must be considered for the gaseous, liquid, and solid effluents. Even when the effluents are all used, the financial viability is positive only if the energy, soil conditioners, or feeds were previously purchased with money. In these cases the cash outflow of the user is reduced and he is able to pay for the plant.

Advocates of the system will try to argue to the counter and state what they consider are successful programs, for example in India. However, a closer examination of the Indian systems will usually show that it is only through subsidies and grants that the systems are built.

OPERATIONAL FACTORS

Six operational factors have been identified as critical to the anaerobic digestion process. Each of these factors needs to be considered when designing a system. These are:

- composition of the organic substrate;
- retention time;
- concentration of the substrate;
- organic loading rate;
- degree of mixing; and
- heating and heat balance.

Composition of the Organic Feedstock

Any organic waste that is biodegradable can be digested. As organic matter is biodegraded by bacteria, a portion of the materials produces the end product of methane and carbon dioxide, while part of the material is converted into bacterial cells. The portion converted into cells has been estimated to be from 2 to 25% for anaerobic digestion. There is always a refractory fraction that passes through a digester undegraded.

Biodegradability is usually measured in one of two ways: chemical oxygen demand (COD) or volatile solids (VS) destruction. For most applications the volatile solids are the easier to measure. It should be noted that there is no direct correlation between the two, and thus the measures of biodegradability are not comparable. Gossett et al. (1976) did note that the COD/VS ratio usually decreases during digestion so that the ratio in the effluent is less than in the influent.

Values for the biodegradability are available from the literature or can be measured in the field as VS destruction data are calculated. Some of the values from the literature are shown in Table 1.

Anaerobic digesters can use a number of materials as feedstock. These include manures, human wastes, crop residues, food processing, and other wastes, or mixtures of one or more of the residues and wastes.

Manures have been probably selected for feedstocks because they are available, have good nutrient balances, are easily slurried, and are relatively biodegradable. The range of biodegradability reported varies from 28 to 70%. This variation is partly due to what the animals have been fed. For example, Hashimoto (1981) showed that as the percentage of silage is increased over the amount of ground corn, the degradability of the manure decreases. Silage contains more lignocellulosic materials. Thus in developing countries where cattle are fed agricultural wastes, the manure would be less degradable than on cattle feedlots in the United States.

A second reason for the large variation in biodegradability of manures is that fresh manure is much more diodegradable than dried manure. The biodegradability of manure decreases as the time period between excretion and when it is fed to the digester increases.

A dairy cow that weighs about 450 kg will produce about 39 kg of manure and urine per day which will contain about 4.8 kg of dry organic solids. Cows that are larger or smaller will produce wastes in the same proportion. (See Table 2 for other animals.) The organisms in the digester require a minimum of 6 days at 35°C to reproduce and about 20 days to use efficiently the biodegradable organics in the manure.

The effects of the type of holding area and the frequency of cleaning or collecting the manure is shown in Table 3.

Jewell (1981) pointed out there is one difficulty in using pig manure where it is collected using water flush systems. In these systems, the manure arrives at the digester very diluted. As an example, in Yugoslavia at an installed site the slurry contained less than 2% solids. In the Philippines, a digester system was built on the basis of a two-to-one dilution with water but was found to be operating at four-to-one. In these cases, either a larger digester is required or some sort of concentration system is needed to keep the digester system economical.

Table 1 Biodegradability of Various Digester Feeds (Stuckey [1983] from various sources)

Substrate	Days	Temp. °C	% COD Dest.	% VS Dest.	Comments
Beef Manure					
9% Corn silage					
& 88% Corn	30.6	72.5			
91.5 Corn silage	inf.	55	52.1		
40% Corn silage		55	67.1		
53.4 Corn					
& 7% Corn silage		55	73.5		
87.6% Corn					
6- 8-week old feed		55	60.0		
Lot manure					
Cattails	120	35		59.3	batch
Cattle Manure	80			28.1	
Corn Leaves	120	35		71.8	batch
Corn Meal	90	35		84.9	batch
Corn Stalks	120	35		77.2	batch
Corn Stover	40	35	64.8		batch
	40	35	59.1		batch
<u>Chicken Litter</u>	10	60	13	20	High lignin content aged one year
Manure	15	32.5	78.1	67.8	10% TS feed
	28	31	45.3		10% TS feed
	120	35		75.6	batch
Dairy Manure	25	35	30.2	28.9	
& Straw					
Manure	120	35		57.5	batch
Manure	110	35	48.4	44.4	batch
Elephant Manure	120	35		52.5	batch
Hyacinth-Bermuda	12	35	40.3	39.2	
Kelp (sea weed)	12	35	57.7	45.1	
Kelp (treated)	120	35		62.0	batch
Meat Packing Waste 15.1	26		44.3		
Municipal Solid					
Waste and Sludge					
Municipal Solid Wastes	15	35	45.6	36.2	12.5 sewage sludge
Newsprint	90	35		88.1	batch
Peat (U.S.)	60	35		11.1	
	56	35		16.7	
<u>Pig Manure</u>	15	35	58	60	14% protein ration
	15	32.5	54.6	60.9	10% TS feed
	120	35		72.7	batch
Wheat Straw	120	35		55.4	batch
Water Hyacinth	120	35		58.8	batch
Yeast Waste		30	67.1	62.6	

Table 2 Estimated Manure and Gas Production Rates (based on 450 kg liveweight).

	Dairy Cattle	Beef Cattle	Pig	Chicken
Manure Production kg/day	39	26	23	27
Total Solids kg/day	4.8	3.4	3.3	7.9
Volatile Solids kg/day	3.9	2.7	2.7	5.8
Dig. Efficiency	35	50	55	65
COD:VS	1.05	1.12	1.19	1.28
Gas Production l/kg VS*	219	325	381	490
l/450 kg animal/day	860	870	1020	2860

NOTE: Values may vary from these values due to differences in feed ration and management practices.
*Based on theoretical gas production rate of 831 l/kg of volatile solids (VS) destroyed and assumes the CH_4:CO_2 ratio is 60:40 and the conversion of VS to COD is 1.42 (Jewell, 1981).

Table 3 Composition of Feedlot Manures.

	Environmental Lot	Concrete Lot	Dirt Lot
Water (%)	85	65	29
Total Solids (%)	15	35	71
Volatile Solids (% of total solids)	78	67	35
Biodegradable Volatile Solids (% of volatile solids)	60	31	24
Manure Production-Total Solids (kg/animal/day)	4.5	2.6	4.4*

* Includes dirt from feedlot when dry manure is collected. (Schmidt 1975)

Poultry manure is very concentrated compared with cattle or pig manure. It contains 25 to 35% solids. For 1,000 kg of live weight of chickens, Jewell (1976) estimated the methane gas production to be 2.3 m^3. In addition there is an abundance of volatile acids or ammonia in chicken manure which can under some conditions inhibit the methane production in the digester. Thus the poultry manure is either diluted or mixed with other feedstocks to avoid the possible toxic conditions in the digester.

If one is to use manure as a feedstock some of the questions that should be addressed include:

- What are the number of animals (poultry) and their size (weight)?
- Under what conditions are the animals (poultry) held?
- If the animals (poultry) are confined, what type of pens are used and how and when is the manure collected?
- What quantities of manure are collected?
- What are the existing uses and prices (if sold) of the manures?
- What are the transportation costs of the manure from the production site to the conversion facility?

Agricultural residues have been used although usually with manure. The most desirable materials for a completely mixed system are the high moisture herbaceous materials; while for dry digestion systems almost any organic material can be used. Specific materials that have been used as digester feedstocks include water hyacinths, algae, corn stover, wheat, and rice straw.

A frequent argument against the use of crop residues is that they are already used as animal feeds or for fuel. Experience in China has shown that where crop residues are used as fuel for burning, the efficiency is less than 10%. When the same crop residues are used as an anaerobic digester feedstock, small amounts of crop residues are used which produce more net energy leaving the nutrients to return to the soil. The crop residues then not burned can be used as animal feed. Thus in China, the use of the digester has allowed the animal population to increase (Stuckey, 1983).

In some parts of the world, cattle are bedded on crop residues and thus the feed for the digester contains large quantities of crop residues mixed with the manure. The bedding material is more stable, it produces less gas in the digester and leaves more solids in the effluent. For example, Jewell (1982) reported that 100 kg of manure will produce twice as much gas as 50 kg of manure and 50 kg of bedding mixed in the digester.

This stability in the agricultural residues is due to the lignocellulosic bonds. The agricultural residues have an additional disadvantage in that they may float and form a scum layer. Size reduction or chemical treatment can improve the handling and biodegradability of the residues.

Water-born plants such as water hyacinth and duckweed have been used in a limited number of cases as digester feed.

Some information that should be obtained in considering crop residues includes:

- The type of crops and areas that they are grown;
- Quantities and the form in which they are available;
- Characteristics of the residues (e.g., moisture, volatile solids, and size of particles);
- Current use and price of the residues if sold;
- Period of availability;
- Storage characteristics if known; and
- Cost of collection and transport to the conversion facility.

Human wastes represent the major problem from a health standpoint. They must either be properly used or disposed of to reduce the transmission of disease. If the wastes are proposed to be used as a feedstock for a digester, it is very likely that any community will have a number of collection systems. One may find households served by night soil collectors, while other sections are served by pit latrines, septic tanks, and/or water carriage systems. It is important if human wastes are considered as a feedstock that the sources, quantities, and characteristics of the wastes be identified.

Specific questions involve:

- Type of waste collection systems;
- Number of persons served by each system;
- Characteristics of wastes collected from each system; and
- Current use and price (if sold) of the wastes.

Retention Time

The retention time is the theoretical time that a particle or volume of liquid added to a digester would remain in the digester. It is calculated as the volume of the digester divided by the volume added per day, and it is expressed as days.

The solid retention time represents the average time that the solids remain in the system. The solid retention time can be determined by dividing the weight of volatile solids in the system by the weight per unit time of volatile solids leaving the system. The hydraulic detention time (HRT) is equal to the solid retention time in completely mixed non-recycled digester systems.

Retention time has an effect upon gas production as shown in the results of Dague (1974). There is a minimum retention time which allows the slowest growing bacteria to regenerate. Then there is a minimum retention time required to achieve a satisfactory stabilization of the solids, which Dague shows as 10 days for sewage sludge at 35°C. If the retention time is cut in half, the gas production rate will drop and the process may fail due to a condition called wash out where the bacterial cultures decrease to the point that they are no longer effective. If the retention time is greater than 10 days at 35°C., the gas production levels out and very little additional gas is produced for the additional time. Therefore, long detention times lead to low efficiency of the process.

There is a tendency to refer to gas production rates in terms of volumes of gas produced per volume of digester volume. If maximum gas is desired, the ratio may be four or more volumes/volume per day. However, if the purpose of the digester is to produce and store a soil conditioner and to have some gas, as is the case in China, then the ratio may be very low and has no meaning.

Retention time along with temperature is important from the standpoint of the destruction of pathogens. If improved health is a consideration, certain minimum values should be exceeded.

Concentration of the Feedstocks

Gas production is a function of the solid materials and their biodegradability in the digester. The more concentrated the solids, the smaller the digester and the lower the cost of the system. In sewage treatment plants, efforts are made to concentrate the solids to reduce the volume and the costs of the digester.

The literature about the Indian digesters as reported by ESCAP (1980) and elsewhere implies that an optimal solids concentration of 7 to 9% of the feed should be used. However, systems have been designed to use as little water as possible. For example, Jewell (1982) points out the advantage of dairy manure (10 to 13% solids) is that it can be added to the digester directly without dilution. If the manure has stood for a few days some water may have to be added to slurry the material for introduction into the digester. In Israel the manure is scrapped from the cattle pens with bedding which is either asphalt or concrete. Marchaim (1984) reports that he has fed this manure with only small quantities of water at 16 to 18% solids content without difficulty. Batch dry digesters have operated with solids concentrations at 60%. It is obvious that there is a conflict between the Israel experience on solids levels fed to digesters and the Indian recommendation.

The batch dry digesters with high solids concentrations appear to be an effective means of producing gas cheaply. Currently efforts are underway to manage and optimize the gas production for crop residues and urban solid wastes in landfills having very high solids.

Organic Loading Rate

The rate at which biomass is supplied to the digester is referred to as the volumetric organic loading rate and is commonly expressed in terms of grams of volatile solids per liter of digester capacity per day (gm VS/l-day). Different loading rates can be obtained by either changing the concentration of the solids in the influent or varying the flow through the digester. In practice the solids concentrations tend to be kept constant, and thus the flow rate is changed.

Degree of Mixing

Two types of systems are used in digestion. The plug flow system in which no mixing takes place and the completely mixed digester. Plug flow offers the advantage in that there is no need for mixers, but some of the effluent must be recirculated to innoculate the feed with organisms necessary to carry out the process. The advantages of the plug flow system has been addressed by Jewell (1980).

Mixing offers advantages in that the substrate is kept in contact with the microbes and temperature is kept uniformly distributed.

Mixing has three important effects upon the process as it: (1) maintains a uniformity in substrate concentration, temperature, and other environmental factors; 2) minimizes the formation of scum at the surface; and 3) prevents the deposition of solids on the bottom. The degree of mixing varies depending upon the feedstock and operating conditions. Rajabapaiah et al. (1979) reported that there was hardly any stratification inside a KVIC digester which was fed cattle manure and that the temperature profiles were within one degree throughout the digester. Hashimoto (1982) reported increases of 8% and 11% in gas production with continuous mixing, over mixing only two hours per day in digesters using cattle manure at 55°C and retention times of six and four days. Hashimoto's data would indicate that at longer detention times the effect of intermediate mixing would be minimal.

Scum formation appears to be a primary function of the feed. With the addition of large amounts of fibrous materials and fats, the formation of a scum layer is likely. If the organic materials are in the scum layer, it is likely that they will not be available as feed to the organism degrading the materials to gas. Thus the gas production rates in a digester with scum layers are reduced.

The reason for trying to avoid the accumulation of the solids at the bottom of the tank is the reduction of detention time. Detention time is a primary factor in gas production. El-Halwagi (1982) reported to Stuckey (1984) that the hydraulic detention time in an unmixed fixed dome digester was only one half the theoretical value.

Heating and Heat Balance

Digestion progresses more rapidly at a higher temperature, therefore, it is important to get the digester feed at as high a temperature as possible and to keep the heat losses to a minimum.

Insulation to reduce the heat losses from a digester is very important. In examining some of the efforts in developing countries, Prasad and Sathyanarayan (1979) reported that 54% of the total heat loss was from the cover of the Indian digesters. Thus one can see that for cold climates the Indian digester with a floating metal cover is not a viable option. The materials used to insulate the digester vary from the use of dry agricultural residues (e.g. straw, hay, and corn stalks) to commercial polyurethane materials. The example was given earlier where in Turkey the digester is built into the floor of the barn.

A number of heating techniques can be used in installation. These vary from simple solar heaters placed above the digester to heat exchangers and steam injection (bubble gun) heating.

Solar heat can be one of two types; active or passive. Active systems heat a portion of the feed during the day and it is then placed in the digester. Passive systems depend upon building a solar greenhouse that captures the radiant heat energy.

Chongging Biogas Office (1982) reported on a comparison of a membrane digester and the Chinese digester. Since the membrane digester absorbed the solar energy the temperature was higher, and it gave a larger gas production.

Ashare et al. (1977) reviewed the optimization of digesters using manure as a feed stock and Ashare et al. (1979) and West and Ashare (1980) reviewed the feasibility of using crop residues.

LOCATION OF A DIGESTER SYSTEM

A digester system needs to be located near the feed supply so the least possible amount of labor is used. As an example, in China the digester is often located near the house so that the pig pen wastes and human wastes can flow directly into the digester.

Another example of a low labor system is in Turkey where the digesters have been built under the floor in the cattle or sheep barns. This system also provides additional insulation againt cold weather and makes the digester readily accessible to the supply of manure.

Slurry Effluents

A slurry is discharged from the digester. The characteristics of the slurry depend upon the feedstock, the digester conditions, and the portion of the organic matter which is converted into gas. The effluents can all be handled by conventional liquid handling methods which would permit them to be transported to the field for use as a soil conditioner or to ponds for aquatic biomass production. In some areas the slurry has been separated using vibratory screens, settling tanks, or sludge centrifuges to separate the liquid and solid portions.

The uses of the slurry as a soil conditioner have been described by Ward (1982). The important concept to keep in mind is that the slurry should contain all the initial nutrients contained in the feedstock. The solids will have been reduced in quantity leaving a more stabilized solid which will break down very slowly. During the digestion process, the organic nitrogen will have been partly decomposed into ammonia nitrogen. The ammonia nitrogen which is produced, while it is more available to plants, is also much easier to lose through either misapplication or drying. As an example Table 4, based on Jewel (1978), shows the changes which occured when dairy manure was digested at mesophilic temperatures (32.5°C) for 20 days.

Table 4 Changes in Composition of Dairy Manure During Anaerobic Digestion (32.5° for 20 days).

Item	Raw (influent)	Slurry (effluent)
Total Solids	80.5	54.1
Reduction (%)		32.4
Volatile Solids	69.8	44.5
Reduction (%)		36.2
Total Nitrogen	3.9	3.31
Reduction (%)		15
NH_3N	1.47	1.54
Increase (%)		5

Prior et al. showed similiar changes when digesting beef cattle manure from a concrete slab at thermophilic temperatures. (See Table 5.)

Table 5 Changes in Composition of Beef Cattle Manure during Digestion (55°C for 12 days).

Item	Raw (Influent)	Slurry (Effluent)
Total Solids gm/l	9.83	5.05
Reduction (%)		49
Volatile Solids	8.57	4.82
Reduction (%)		44
Total Nitrogen	4.14	3.87
Reduction (%)		7
NH_3N	1.03	1.79
Increase (%)		70

The digester slurry often cannot be applied directly to the fields and must be stored. Some methods used to hold the materials include: (a) putting the slurry in holding ponds or tanks; (b) the sludge is placed on drying beds where the liquid containing the NH_3N either drains into the soil, is lost to the atmosphere or is contained in the sludge; or (c) where vibratory screens have been used to separate the liquid and solid fractions. Jewel (1981) showed how the effluent slurry, which is stored in open tanks or lagoons, will lose the ammonia nitrogen (NH_3N) to the atmosphere, thus reducing its fertilizer value.

Hashimoto (1978) reported on the solids recovery using a 60 mesh sieve and a sieve plus a sludge centrifuge. His data, which is shown in Table 6, shows that the recovery in the solids are lower than would be indicated in the liquid phase.

Table 6 Recovery of Solids and Nutrients by Sieves and Sieve-Centrifuge (percentage of slurry contents)

Item	Sieve	Sieve-Centrifuge
Total Solids	35	64
Volatile Solids	40	67
Suspended Solids	51	91
Total Nitrogen	19	41
Organic Nitrogen	22	60

In any economic analysis, care must be taken that the credits given to the products of the digester do not exceed the least costly alternative source.

In examining the uses of the slurry, some questions involve:

- projected characteristics of the slurry, solids, and liquid effluents;
- potential points and periods of use of the slurry;
- means and cost of storage and transport of the slurry, solids, or liquid effluents to the point of use; and
- cost of the materials that the slurry would replace (e.g., peat, fertilizer, fish, and feed.)

CONSTRUCTION MATERIALS

There is a need to build a low-cost system that will provide a gas-tight container. In China early digesters used soil cement to obtain a low-cost water displacement digester. As the gas was produced the pressure would build up and would displace liquid. Chen (1983) reported that many of these digesters soon failed under the pressure. The designers had two choices. One was to change the materials of construction, or the second was to change the gas collection and storage system to use lower pressures. In reality both changes were made. More cement was used in building the tank which could withstand the pressure. A gas storage bag system was developed to collect and store the gas at lower pressures. In northern China the bags may be contained within buildings while in southern China the bags may be outside. In Burma and other areas of the world, rubber inner tubes from auto or tractor tires have been observed being used as a gas storage device.

India's design uses an expensive floating metal gasholder. The height-to-diameter ratios recommended by KVIC are 2.5 to 4.1. Analyses of the systems show that a better ratio from a cost standpoint was to have the height equal to the diameter. However, KVIC apparently in an empirical manner had attempted to reduce the metal used in the gasholder. This meant that the tank was longer, and a deeper hole had to be dug. In areas with a high water table the tank could be floated out of the ground.

Not only is the metal gas holder expensive, but it is subject to corrosion and loses heat. To reduce the corrosion problems a water jacket ring is used around the tank into which the gas holder is installed. This means that the metal is not in contact with the digester contents but with water which is less corrosive. However, this additional ring raises the cost of the digester both in terms of materials and labor.

To reduce the heat loss through the floating cover, attempts have been made to use materials other than steel. These have included the use of ferrocement and plastics. While the ferrocement covers are relatively inexpensive and resistant to the loss of heat, they are difficult to transport and are subject to leakage. Plastic is resistant to corrosion and has lower heat losses but is about the same price as steel. In addition the raw materials often have to be imported which takes scarce foreign currency.

The fundamental problem with the Indian digesters is that they are 50% or more in cost than the Chinese of the same size.

Taiwan has developed a long plug-flow digester made of red mud plastic. This has provided an inexpensive, alternative digestion system for use by farmers. Chen (1983) reported that the red mud digesters were being used in mainland China.

SIZING OF THE DIGESTER

In sizing the digester two items must be taken into consideration: first, the type of digester that will be used, and second, the purpose or performance goals for which the process is used.

Size Based on Health Criteria

In sizing a digester, its primary purpose must be determined. If the primary purpose is for health, then reduction of the possible transmission of disease and temperature and detention time are very important criteria.

Size Based on Production of Soil Conditioner

If the purpose of the system is to produce primarily a soil conditioner than the breakdown, stabilization, and storage of the organics and nutrients will govern the system.

Size Based on Energy

If the production of energy is the most important objective, the gas production should be optimized. The primary variables that affect production of the gas are:

- biodegradability of the materials;
- concentration of the feed;
- kinetic constants; and
- detention time.

Temperature has a profound effect upon the digestion since the kinetic constants are influenced by increases in temperature. In developed countries the trend has been toward thermophilic digestion in addition to using more concentrated feeds and, in some cases, using pretreatment to increase the biodegradability.

In developing countries the digesters usually operate at ambient temperatures and therefore only the biodegradability of the feed, concentration of the feed, or detention time can be changed to optimize the process.

Therefore the critical period of the year with respect to temperature must be determined. The temperature used must be that anticipated in the digester and not the ambient air temperature. Since the anaerobic digestion process essentially stops at 10°C, the digester contents must be kept warmer than this if any gas production is to be expected.

For design, gas production rates per unit volume are often used. These rates are temperature dependent. Chen and Hashimoto (1978) adapted the Contois (1959) kinetic model to describe the mathematical relationship which would allow one to predict the volumetric methane production. The equation is:

$$V = (B_o\ S_o/HRT)\ \ (1 - K/(HRT\ u_m - 1 + K))\quad (1)$$

Where V = the volumetric methane rate in m^3/m^3 of digester
B_o is the ultimate methane yield in $m^3\ CH_4$
S_o is the influent volatile solids concentration in kg/m^3
HRT is the hydraulic retention time in days
um is the maximum specific growth rate of the microogranism in day^{-1}
K is a dimensionless kinetic parameter.

The equation states that for a given influent volatile solids concentration and a fixed hydraulic retention time, the volume of gas produced per cubic meter varies with the ultimate gas production rate of the materials, the maximum growth rate of the microorganisms, and the kinetic parameter.

Under U.S. conditions the values for B_o have been determined. Hashimoto reports them to be:

- Beef manure on grain ration concrete lot, 0.35 + or - 0.05.
- Beef manure on grain ration dirt lot, 0.25 + or - 0.05.
- Dairy cattle, 0.20 + or - 0.05.
- Pigs, 0.50 + or - 0.05.

These values indicate that because of the different feeds the ultimate gas production will vary with the animal's manure. Thus pig manure is very digestible compared to the highly lignocellulosic feed given to dairy cows.

The K value are inhibitory. Hashimoto relates the K value to S_o:

$K = 0.8 + 0.0016\ e^{0.06\ S_o}$ for cattle manure (2)

$K = 0.5 + 0.0043\ e^{0.091\ S_o}$ for pig manure

These values indicate that overloading of the digester with volatile solids (S_o), heavy metals, salts, or ammonia will be inhibited.

$u_m = 0.013\ (T) - 0.129$ can be determined empirically where T is the temperature between 20 and 60°C. This equation was developed based upon a fit of the data between these two points and could likely be extrapolated down to 15°C safely.

Since the ultimate gas yields, B_o, are hard to determine in the field, the approach of using the projected quantities of volatile solids that can be destroyed in a period of time at a temperature is used.

In the literature, one will find references to volumetric efficiencies or these can be developed from laboratory studies of the feedstocks. (See Table 7.)

Table 7 Volumetric Efficiencies (m^3 gas/m^3 of digester per day).

Type of Digester	Volumetric Efficiency	Detention time, days	Temp. °C	Feedstock	Reference
Batch	0.35	40	27	Elephant grass	Boshoff (1965)
Continuous	0.83	40	27	Elephant grass	Boshoff (1965)
	1.0	30	31	Swine manure	Maramba (1978)
	0.79	60	35	Grass	Jewell (1981)
	0.2	365	35	Grass	Jewell (1981)
Continuous Chinese	0.1-0.2	60	25	Manure/Nightsoil/ Crop residues	Chen (1982)
Indian	0.2-0.3			Cattle manure 9% solids	

For example using a batch digester, elephant grass as a feed stock at 27°C for 40 days detention, Boshoff (1965) reported a volumetric efficiency of 0.35. In a continuous digester of elephant grass the volumetric efficiency was 0.83. A designer can take the data and plot it. The effects of varying detention times can then be seen at that temperature. If the digester is to be operated at another temperature, there are relationships that can approximate the gas production at these new conditions.

The most common item that is overlooked in using these values is the effect of temperature.

DESIGN EXAMPLE

A farmer has four dairy cows that weigh 1,200 kg. The cows are confined so that all the manure can be collected for the digester. To estimate the methane gas production and design a digester for purposes of illustration, five cases will be used.

Case 1: Fresh Manure and Urine

Assuming that the cows are held in a pen where the fresh manure, urine, and small amounts of waste water can be transferred directly to the digester, Jewell (1981) reported that the cows should produce about 12.5 kg of dry solids per day and 10 kg of volatile solids. The volume of the manure and the urine would be about 90 l per day. Assuming that another 10 l of water from cleaning are added, the total volume would be 100 l per day.

The ultimate methane gas yield according to Hashimoto's (1981) review would be 0.2 m^3 per kg of volatile solids added. The equation that he reported that could be used for other times, temperatures, and solids loading to give a specific yield is:

$V = (B_oS_o$ / HRT) (1 - K /HRT u_m -1 +K) (equation No. 1)

B_o = ultimate methane gas yield per kg of volativle solids added is 0.2 for dairy cattle manure (Hashimoto).

S_o = influent volatile solids concentrations in kg/m^3.

In this case 10 kg per day times the number of days retention time gives you the S_o.

HRT (Hydraulic retention time) varies from 10 to 50 days.

The temperature coefficient u_m is equal to 0.013 (t) - 0.129 where (t) is the temperature in °C. This equation shows that the anaerobic digestion process essentially stops at 10°C.

Use design temperature of 35°C and then check the effects at 15°C, 10°C, and 55°C.

The constant K is determined by:

$K = 0.8 + 0.0016\ e^{0.06\ S_o}$

$K = 0.8 + 0.0016\ e^{0.06x100}$

$\underline{K = 1.445}$

u_m = 0.013 (T) - 0.129 at 35 degrees

u_m = 0.013 (35) - 0.129

u_m = 0.455 - 0.129

u_m = $\underline{\text{0.326 at 35 degrees}}$.

Substituting these values in the equation (1) and solving for detention times of 10, 20, 30, 40, and 50 days and then changing the temperatures to 15°C, 20°C, and 55°C results in Tables 8 and 9.

Table 8 Specific Yield - Case 1: Fresh Manure (m^3 of methane per m^3 of digester per day)

Detention Days	15°C	20°C	35°C	55°C	Digester Volume m^3
10	*	0.36	1.22	1.54	1
20	0.18	0.53	0.79	0.88	2
30	0.26	0.45	0.57	0.61	3
40	0.26	0.37	0.44	0.47	4
50	0.25	0.31	0.36	0.38	5

* Too short detention time--wash out occurs and digester would not function. Need 16 days minimum.

Table 9 Total Gas Production - Case 1: Fresh Manure (m^3 Methane per day)

Digester Size m^3	Detention Period Days	Temperature 15°C	20°C	35°C	55°C
1	10	*	0.36	1.22	1.54
2	20	0.36	1.06	1.58	1.76
3	30	0.80	1.32	1.71	1.83
4	40	1.06	1.48	1.76	1.78
5	50	1.23	1.55	1.80	1.90

* Too short detention time--wash out occurs and digester would not function. Need 16 days minimum.

If The temperature in the digester can be maintained at 35°C based on Table 9, the optimum design would be for a 2 m^3 digester and 20-day detention. The methane production would be 1.58 m^3 per day.

It would appear that if the temperature dropped to 20°C, the 2 m^3 digester still appears to be viable with 1.06 m^3 of methane per day. If the temperature would drop to 15°C for long periods of time, only marginal quantities of gas would be produced.

To examine the 20°C case more closely, Table 10 was constructed examining the addition of fresh manure to a digester operating at 20°C. It should be noted that below eight days the digester would not function because of wash out of the methane formers organisms.

Developing cost data for each size and giving the gas values would allow one to select a size for maximum return. The 15°C, 10°C, or the 35°C digester could be a plug-flow or mixed-type digester dependingupon local conditions, while for an effective 55°C digester it should be a completely mixed reactor.

Case 2: Manure and Concrete Pad Not Collected Daily

This case assumes that the cows are held in an area where there is a concrete pad. The manure dries and is scraped and collected from the pad. During the drying process, some of the volatile solids will be

Table 10 Detention Time Effect on Digester Size and Gas Yield (Case 1 20°C: Fresh Manure)

Days	Digester Size m^3	Methane Yield m^3/day
0.7	no gas - wash out	
8	0.8	0.11
9	0.9	0.22
10	1.0	0.37
11	1.1	0.46
12	1.2	0.56
13	1.3	0.65
14	1.4	0.73
15	1.5	0.80
16	1.6	0.86
17	1.7	0.95
18	1.8	0.97
19	1.9	1.01
20	2.0	1.06
21	2.1	1.09
22	2.2	1.13
23	2.3	1.16
24	2.4	1.19
25	2.5	1.22
26	2.6	1.25
27	2.7	1.27
28	2.8	1.30
29	2.9	1.32
30	3.0	1.34
..		
40	4.0	1.48
..		
50	5.0	1.55

destroyed by aerobic breakdown and thus the potential gas production will be reduced. Since the manure is dry it will be necessary to add some water to slurry the manure for feeding the digester.

Using the data of Schmidt (Table 3), the volatile solids concentration of the feedlot manure left standing is reduced by about 50%. Assuming that dairy manure behaves like feedlot manure:

- Total solids = 4 (cows) x 2.6 kg/cow = 10.4 kg/day.
- Volatile solids = (65% of 10.4) = 6.8 kg/day.

Hashimoto's data shows that the ultimate gas production of manure from cattle fed a high energy diet and whose manure has been left standing and dries out is about 70% that of fresh manure. Therefore, assume B_o = 0.14 m^3 of methane per kg of volatile solids added (i.e. 70% of 0.2), for dairy cattle manure that has fallen on a concrete pad, has dried, and then has been collected for digestion. Schmidt's data indicate that only about half of the volatile solids can be broken down, so the gas production values using B_o of 0.14 may be optimistic.

S_o = influent volatile solids x number of days
S_o = 6.8 x number of days detention period
HRT from 10 to 50 days
The temperature coefficient, u_m - 0.013 (t) - 0.129
The constant, $K = 0.8 + 0.0016 e^{0.06S_o}$

Assuming the manure is slurried so that it contains 10% solids, the following values are obtained from the equation for gas production.

Table 11 Specific Yield Case 2: Manure from Concrete Slab (m^3 of methane per m^3 digester per day)

Detention Period (days)	20°C	35°C	Digester Volume (m^3)
10	0.25	0.68	1
20	0.31	0.41	2
30	0.24	0.28	3
40	0.20	0.22	4
50	0.16	0.18	5

Table 12 Gas Production in Case 2: Manure from a Concrete Slab (m^3 methane per day)

Digester Size (m^3)	Detention Period (days)	Temperatures 20°C	35°C
1	10	0.25	0.68
2	20	0.62	0.82
3	30	0.72	0.85
4	40	0.80	0.88
5	50	0.81	0.90

If higher concentrations of solids are fed to the digester (i.e., less water is added to slurry the manure), the volume of the digester can be reduced while the total gas production should remain the same. Table 13 illustrates the effects of changing the solids concentration.

Table 13 Effect of Solids Concentration on Size -- Case 2: Manure on Concrete Slab

Solids Concentration (percent)	Size (cubic meters)
6	3.4
9	2.3
10	2.0
12	1.7
15	1.3
18	1.1

Case 3: Manure on the Ground, Partially Dried

In this case it is assumed that the cattle are held in a dirt pen where the manure is all collected after drying and is then taken for slurrying before adding to a digester. Data should be collected in the field to determine what conditions exist. However, for this case the U.S. data indicate that much dirt is collected with the manure. (See Table 3.) Schmidt found the yield to be 4.4 kg/cow per day including the dirt. Of the 4.4 kg, 70% is total solids and 35% is volatile. Thus the volatile solid production is:

4 cows x 4.4 kg manure/cow per day x 0.7 solids x 0.35 volatile = 4.3 kg/day of volatile solids

If one assumes the ultimate methane yield, B_o, in m^3 per kg of volatile solids added is 0.14 based on Hashimoto, the following will be obtained.

Table 14 Specific Yield -- Case 3: Methane from Manure from a Dirt Floor (m^3 per m^3 digester per day)

Detention Period (days)	Temperature 20°C	35°C	Digester Volume (m^3)
10	0.11	0.49	1
20	0.20	0.26	2
30	0.16	0.18	3
40	0.12	0.14	4
50	0.11	0.11	5

The digestibility as shown by Schmidt (Table 3) is only 24% or 40% that of the fresh manure, thus these values are apt to be optimisitc.

Table 15 Gas Production -- Case 3: Dirt Lot (m^3 methane per day)

Digester Size (m^3)	Detention Times (days)	Temperature 20°C	35°C
1	10	0.11	0.49
2	20	0.40	0.52
3	30	0.49	0.57
4	40	0.49	0.56
5	50	0.54	0.56

There are differences between Tables 14 and 15 because of rounding.

Cases 1, 2, and 3 show the impacts of different collection methods of the manure on gas production.

Case 4: Using Destruction of Volatile Solids

Table 1, Biodegradability of Various Digester Feeds, reports that in 110 days at 35°C, 47% of the volatile solids in cattle manure is destroyed. If you calculate the amount of gas produced in 20 days using the equation of Hashimoto, you will see there is only a potential of producing 20% additional gas after 20 days at 35°C. An empirical value is that for each kg of volatile solids destroyed, from 0.25 to

0.38 m^3 of methane will be produced. Therefore, 10 kg volatile solids x 0.47 = 4.7 kg V.S. can be destroyed. Assume that at 20 days only 80% have been destroyed. The methane gas production will be:

4.7 x 0.80 x yield per kg destroyed, or 0.9 to 1.4 m^3 methane per day.

20 days storage = 20 x liters added per day = volume of digester required, or 20 days x 100 l/day assuming fresh manure, urine, and water = 2 m^3 digester.

This agrees with the values given by the equation of Hashimoto in Case 1 at 35°C.

The difficulty in trying to use Table 1 is that it does not report the percentages of volatile solid in the feedstock. However, these values can be determined easily in the laboratory.

Case 5: Design using ESCAP (Indian) Approach

If the ESCAP basis is used for design, the following would be obtained.

Case 5A: Fresh Manure and Urine. Ninety liters of manure and urine would be obtained from the cattle to which 90 l of water would be added. Total volume to be added per day would be 180 l. The gas production would be estimated, according to ESCAP, to be 540 l of total gas or 324 l of methane using the ratio of 60% methane and 40% carbon dioxide. The digester would have to be 0.18 m^3 per day x 50 days detention = 9 m^3.

Based upon Table 3, the dry solids in the manure would be 15% of the wet manure (0.15 x 90 = 13.5 kg) of which 78% would be volatile solids (13 kg TS x 0.78 = 10.5 kg). The estimated gas production using the Hashimoto's equation for 50 days would be 1.8 m^3 of methane or at 60:40 ratio of methane to carbon dioxide 3 m^3 per day of biogas. This value is 5.5 times the ESCAP data.

If one desired to have only 270 l of methane per day at 27°C, the digester could be about 0.6 m^3 of 1/15 the size obtained using the ESCAP (Indian) approach. Since the six-day detention time is near the wash out condition you might design for 10 days, which would be for a 1 m^3 digester -- still only 1/9 the size using the ESCAP (Indian) approach.

The government of India has recommended 30 days detention time (Gunnerson 1984) for southern India which would mean that the digester would be 0.18 x 30 = 5.4 m^3 -- still much larger than needed.

Case 5B: Manure from a Concrete Pad. Using the data in Case 2, 10.4 kg of total solids should be available per day. With water added to obtain 10% solids, there would be 104 l per day. A 50-day detention time would require a 5 m^3 digester which at 27°C should yield about 830 l of methane per day rather than the 320 l ESCAP suggests.

Case 5C: Manure on Dirt. Using the data in Case 3 there would be 10 kg of solids should be available per day, which at 10% solids would require a 5m^3 digester. The gas yield should be at 27°C about 500 l of methane, rather than the 320 l KVIC suggests.

While in engineering practice it is common to use factors of safety, it is unnecessary in anaerobic digestion since it would result in larger digesters that are more expensive and would not yield appreciably larger quantities of gas.

CONSTRUCTION COSTS

The costs of building an anaerobic digestion system are very site specific. (See Table 16.) In China the initial emphasis was to use locally available materials, and labor costs were often not considered. As the economic conditions have changed, the farmers have found that their time is very valuable. Therefore, the shift has been toward using the bag (membrane) digester, which in some cases has been made of imported materials and can be easily installed and used (Chen, 1983).

In a number of countries enthusiasts, but not necessarily those with technical backgrounds, have undertaken demonstration programs and have completely neglected costs. They appear to have forgotten that while demonstrations are necessary, the system must be affordable.

Table 16 Cost of Digester Systems

Type/Location	Per Cubic Meter	Reference
Bag Digesters- China and Costa Rica	$ 10-25	Chen (1983)
Chinese Digester- China (excludes labor)	$ 20	Chen (1983)
Chinese Digester- India*	$ 100	Moulik**
Indian Digester- India*	$ 140	Moulik**
Chines Digester- Egypt (5-10 m^3)	$ 50-80	El-Halwagi**
Floating Cover-Mariana Islands	$ 700	Chan(1984)

* Per cubic meter of gas produced (not digester).
** Personal Communications (1984).

CONCLUSIONS

In examining the possibility of building an anaerobic digestion system, a number of issues must be addressed. These include:

- Availability of the feed supply;
- Use of the end products--gas and slurry;
- Materials of construction available;
- Ambient temperatures and temperatures expected within the digester; and
- Operational factors, such as:
  - composition of the organic feed,
  - retention time,
  - concentration of the feed,
  - organic loading rates,
  - degree of mixing, and
  - heat and heat balance.

When these items have been considered and included in the design process, the chances that a low cost efficient unit would result are considerably enhanced.

REFERENCES

Ashare, E., D.L. Wise, and R.L. Wentworth. 1977. Fuel Gas Production from Animal Residues. Dynatech R/D Co., Report No. 1551. U.S. Department of Energy Report No. COO-2991-10. Cambridge, Massachusetts.

Ashare, E., M.C. Buivid, and E.H. Wilson. 1979. Feasibility Study for Anaerobic Digestion of Agricultural Crop Residues. Dynatech R/D Co., Report No. 1935. Cambridge, Massachusetts.

Boshoff, W.H. 1965. Methane gas production by batch and continuous fermentation. Tropical Science, 3:155-165.

Chan, G.L. 1984. The Integrated Village. Presented at the International Conference on the State of the Art on Biogas Technology, Transfer and Diffusion, Cairo, Egypt, November 1984.

Chen, R.C. 1983. Up-To-Date Status of Anaerobic Digestion Technology in China, 1983. Third International Symposium on Anaerobic Digestion. Boston, Massachusetts.

Chongging Biogas Office. 1982. Study of Red Mud Plastic Digesters with Solar Heating. China.

Contois, D.E. 1959. Kinetics of bacterial growth: Relationship between population density and specific growth rate of continuous cultures. J. Gen. Microbiology, 21:40.

Dague, R.R. 1974. Methane Gas Production for Swine Manure. Presented at American Pork Congress, Des Moines, Iowa, March 7, 1974.

Economic and Social Commission for Asia and the Pacific (ESCAP). 1980. Guidebook on Biogas Development. United Nations Energy Resources Development Series 21.

El-Halwagi, M.M. 1982. The State of the Art of Anaerobic Digestion in the Arab Countries and West Africa. Internal Report prepared for IRCWD.

Gossett, J.M., J.B. Healy, W.F. Owen, D.C. Stuckey, L.Y. Young, and P.L. McCarty. 1976. Heat Treatment of Refuse for Increasing Anaerobic Biodegrabability. Technical Report No. 212. Civil Engineering Department, Stanford University, California.

Gunnerson, C.G., ed. 1984. Anaerobic Digestion (Biogas) in Developing Countries -- A State-of-the-Art Review. World Bank technical paper in draft.

Hashimoto, A.G., et al. 1978. Anaerobic Fermentation of Beef Cattle Manure. Report to U.S. Department of Energy.

Hashimoto, A.G., et al. 1981. Ultimate methane yield from beef cattle manure: Effect of temperature, ratio constituents, Aatibiotics, and manure age. Agric. Wastes, 3(4):241-256.

Hashimoto, A.G. 1982. Effect of mixing duration and vacuum on methane production rate from beef cattle. Biotech. and Bioeng., 24:9-23.

Jewell, W.J., et al. 1976. Bioconversion of Agricultural Wastes for Pollution Control and Energy Conservation. U.S. Department of Energy Report TID 27164. Cornell University, Ithaca, New York.

Jewell, W.J., et al. 1978. Anaerobic Fermentation of Agricultural Residues: Potential for Improvement and Implementation. Final Report. U.S. Department of Energy Report No. HGP/T2981-07. Cornell University, Ithaca, New York.

Jewell, W.J., et al. 1980. Anaerobic Fermentation of Agricultural Residues--Potential for Improvement and Implementation. U.S. Department of Energy Report No. DE-AC00Z-76-ET-20051-13. Cornell University, Ithaca, New York.

Jewell, W.J., et al. 1981. Crop Residue Conversion to Biogas by Dry Fermentation. American Society of Agricultural Engineering, Chicago meeting, Paper No. 81-3573.

Jewell, W.J., et al. 1981. Low Cost Methane Generation on Small Farms. Fuel Gas Production from Biomass, Volume II. CRC Press, Boca Raton, Florida.

Jewel, W.J., et al. 1982. The Feasibility of Biogas Production on Farms. Department of Agricultural Engineering, Cornell University, Ithaca, New York.

Marchaim, U., et al. 1982. The Israeli Anaerobic Digestion Process--NEFAH: Utilization of Agricultural Wastes to Produce Energy, Animal, and Fish Food Supplements and Industrial Products. Anaerobic Digestion 1981. Elsevier Biomedical, Amsterdam.

Maramba, F.D. 1978. Biogas and Waste Recycling: The Philippine Experience. Maya Farms Division, Liberty Flour Mills Inc., Metro Manila, Philippines.

Prasad, C.R., and S.R.C. Sathyanarayan. 1979. Studies in Biogas Technology, Part III, Thermal Analysis of Biogas Plants, Proceedings of the Indian Academy of Sciences, C2:377-386.

Prior, R.L., A.G. Hashimoto, et al. 1978. Anaerobic Fermentation of Beef Cattle Manure. Report to U.S. Department of Energy.

Rajabapaiah, P., K.V. Ramanayya, S.R. Mohan, and A.K.N. Reddy. 1979. Studies in Biogas Technology: Part I, Performance of a Conventional Biogas Plant. Proceedings of the Indian Academy of Science C2, Part 3, 357-363.

Schmidt, L.A. 1975. Feedlot wastes to useful energy--fact or fiction? American Society of Civil Engineers, Journal of the Environmental Engineering Divison, 101(EE5):787.

Stuckey, D.C. 1983. Draft report to World Bank.
van Buren, A., ed. 1979. A Chinese Biogas Manual. International Technical Publications. London.

Ward, R.F. 1983. Utilization of Biomass Digester Effluents as Soil Conditioners and Feeds. Fuel Gas Developments. CRC Press, Boca Raton, Florida.

MAIN QUESTIONS AND ANSWERS

Q: Can you quote any experience with plant stalks; for example cotton and maize.

WARD: The earliest work I recall is that done at the University of Illinois by Buswell in the 1930s. Pfeffer had done extension work from 1970 - 1980 on corn, wheat straw, and grasses. Jewel (Cornell) has done work on the pit digester using the same materials.

Q: Do you think that the inclusion of a heating system could ever be cost effectively included in a system for rural developing countries let alone operating at thermophilic conditions.

WARD: The World Bank is planning the comparison in India at a current "KVIC" thermophilic system on a large dairy farm. The complexity depends upon the degree of development.

Q: We are studying the effect of antibiotic inhibition on the Contois model and are finding a 20% reduction in methane production. Comment?

WARD: Work by Speece-Drexel shows that the organism will often adjust to specific antibiotics and eventually use them as feeds.

Q: (a) If the solids are increased in feed, the mechanism of the digester will be complex to maintain in rural India. Comment? (b) Can you suggest the type of digester using feed up to 18 to 20% which may have simpler infrastructure? (c) Would you suggest other materials instead of red mud plastic?

WARD: (a) Solids in fresh manure are from 10-13%. This is fed to digesters without difficulty. still you are recommending 9-10% in the Indian literature. (b) Thermophilic digesters in Israel operate at 18% feed (manure and bedding). (c) There are other plastic materials which can be used.

TECHNOLOGIES FOR BIOGAS PRODUCTION OF DEVELOPED COUNTRIES AND POSSIBILITIES OF TRANSFERRING THEM TO DEVELOPING COUNTRIES

W. E. Edelmann
ARBI (Arbeitsgemeinschaft Bioenergie)
Maschwanden, Switzerland

ABSTRACT

An overview on the digestion technologies of developed coountries is given. A comparison of the local conditions both of developed and developing countries shows that the premises are quite different. Thus, transfer of technical solutions does not seem to be reasonable, with the exception of some reliable and simple designs. Plastic membranes are expected to play a more important role in developing countries. Transfer of know-how is possible in some fields like the biology of anaerobic fermentations and optimizing of process parameters.

INTRODUCTION

Today biogas is produced all over the world. The aims of biogas production are basically three: (1) energy production, (2) production of an agricultural fertilizer, and (3)improvement of the hygienic conditions by reducing pollution. Depending on the local conditions, one of these three reasons can become more important than the others. This may vary among different countries; for example, in India energy production has first priority while in China hygienic reasons and the fertilizing value of the digested sludge come first. Within the same countries, the reasons for plant construction also may vary. For example, in developed countries there are many animal breeding or food processing industries that are forced by law to reduce the pollution of their organic wastewater with, for example, the help of anaerobic digestion while, on the other hand, there are farmers who want to produce energy or fertilizer in a digester.

In developing countries millions of small plants have been built. The great majority of them are variations of the Chinese water pressure digester or the Indian Gobar Gas Plant with its floating gas drum.

Because of the relatively cheap price of fossil energy in developed countries, the construction of plants started there later, i.e., at the end of the last decade. Since then thousands of plants have been built. In contrast to developing countries, a great variety of processes and plant designs have been installed. The plant sizes vary from less than 10 to several thousand m^3 per plant. From experiences with many different designs and substances and because of intense scientific research, a large know-how could be quickly established.

At this stage, an exchange of new ideas seems to be useful and necessary.

DIGESTION TECHNOLOGIES IN DEVELOPED COUNTRIES

Agricultural digesters are operated normally at mesophilic conditions. The use of thermophilic digestion of agricultural wastes is an exception, since there is greater need for process heat and because of possible problems with the sporadic presence of inhibiting substances, e.g., biotics and disinfectants, at the necessary high-loading rates. In the food processing or paper industry, where the wastewater very often is already heated to high temperatures, thermophilic digestion may be interesting. Some agricultural digesters are running successfully at psychrophylic digestion temperatures of 15-20°C. These plants have accumulation systems placed in the existing pit underneath the stable.[1] As it is forbidden by law to place manure on the snow or frozen fields, the pits must be very large (with a capacity to store the manure of three to six months). Therefore, in these cases, it is possible to convert a pit into a biogas plant without a large investment.

Most of the agricultural installations are (semi-) continuously fed with liquid substrates of a total solids content of 4 to 10%. If the substrate is very liquid, as in some types of pig farms and in industry, anaerobic filters are used. There the biomass is fixed by different methods within the digester.[2,3] The accumulation of bacteria by preventing their washout allows digestion of the substrate within a considerably shorter retention time as compared to conventional digesters. Very short retention times of less than one day have recently been obtained by using dynamic filters, where the biomass rotates on moving elements, thereby intensifying the contact with the substrate (4E AG, Dietikon, Switzerland, personal communication). For the treatment of solid wastes there are special designs varying from simple batch installations [4] to two stage designs where some liquid is circulated alternately through the substrate and through an anaerobic filter [5,6]. The breakdown of organic solids may also be carried out by an aerobic step (heat generation) and an anaerobic step (gastification of the heated substrate)[7]. It has been tried to heat up also liquid substrates by aerobic pretreatment [8].

A biogas plant consists of many components such as the digester itself, heating and mixing devices, and pumps. With many types of components on the market, there is a large variety of different combinations possible. Since a biogas plant has to be adapted to the specific needs of the user and to the local conditions, there is not nearly one plant that is identical with another. Figure 1. is a schematic of often-used process designs.

There are many possibilities for the storage of the gas. In some plants, the gas is used as it is generated. There is only a small buffer capacity of the gas volume on top of the substrate. Very often, cheap gas balloons are installed, although it is not possible to generate a gas pressure high enough for the needs of most users. In floating "wet" gas holders made out of metal (or ferrocement), the recently developed, relatively cheap dry gasholders[9], a sufficient gas pressure can be produced. If the gas is used for tractor driving, storage in pressure bottles is essential.[10] More information on existing systems can be found in literature.[11,12]

PREMISES OF BIOGAS PRODUCTION

Comparing the situations in developed and developing countries, some differences are observed. In developed countries the climate usually is temperate and the installations must be heated if driven mesophilically or thermophilically. For psychrophilic digestion, thick insulation is necessary to prevent heat losses of the substrate on its way from the animal to the installation, and within the plant itself. Generally water is abundant in developed countries; the substrates mostly are liquid since water is used for cleaning and for the transport of the organic material.

Installations in developed countries generally are capital and energy intensive. There is little labor to be done by the future user because of mechanization and automation. The installations very often are large (several hundred m^3), and cost less per unit of digester volume for construction and maintenance.[13] Therefore, in some situations expensive heat exchangers can be installed for the transfer of energy from the warm digested sludge to the cold fresh substrate, or to drive costly gas scrubbing and bottling devices. If electricity is generated, the net gas production often is maximized by further investments since in many cases the electricity can be sold to the public network.

In developing countries, the climate is usually hot. Water is often scarce and consequently there is seldom a practice in producing liquid substrates for anaerobic digestion. Devices to transport liquid substrates to the fields are generally missing. Since more people live in the agricultural sector because industrial and service sectors are less developed, farms are small and less mechanized than in developed countries. Accordingly, the cash flow of a farm is low and therefore agricultural biogas plants, which are the large majority of the plants to be built, must be very cheap, small, easy to maintain, and without expensive mechanical devices. Labor-intensive solutions are preferred compared to capital intensive ones. With the warm climate, plants do not generally need to be heated or insulated. Usually, the energy needs are low and the gas output is not optimized. Farmers in the Third World need gas for cooking and illumination, but they usually do not need biogas-generated heat for buildings in the winter. Biogas plants in developing countries therefore differ significantly from plants of developed countries. Some relevant designs are described in another paper.[14]

POSSIBILITIES OF TECHNOLOGY TRANSFER

Considering the facts cited above, it can be seen that it is difficult to transplant digesters from north to south. It is not a coincidence that during the start-up phase of biogas research, the transfer from south to north was more important than vice versa. For instance, example 4 in Figure 1 is a variation and further improvement of the Chinese water pressure digester; the buildup of pressure is used for stirring. Example 8 in Figure 1 is a further development of an Indian Gobar Gas plant. The gas drum is turned manually to break the scum layer by spokes fixed within the gas holder. In the Swiss design, there are segments of funnels instead of spokes to break and transfer the scum automatically. These developments reduce the process energy need and improve the net energy balance of biogas plants.[15]

In developed countries, there are plants that require more than half of the energy produced for plant operation. The developed countries may learn a lot from developing countries, when it is necessary to find solutions without high external energy input.

Since plants in developed countries are sophisticated and expensive, it is useless to transfer such technologies to developing countries, even if the farm is very large.[13] Therefore it is reasonable to determine the minimum requirement for biogas production. A minimal biogas plant consists of:

- A substrate tight container to digest the material; and
- A gas tight cover to collect and store the gas produced.

And:

- It must be cheap and contructed with locally available materials.

- It must be comfortable for the user and would not necessitate new habits, if possible.

Looking at Figure 1 and taking into account the points above, it should be realized that most of the solutions do not fit the presumptions. Containers out of plastic or metal are used in developed countries because there is a local container industry that looks for new markets. Energy and capital intensive solutions are not reasonable for small installations in warm climates.

The only remaining solutions suited for technology transfer are examples 13 and 15. Wooden biogas plants[16] (e.g., Figure 1, example 9) may be an alternative for countries where wood is abundant, as in Brazil or Indonesia. Filters may be erected in large industries, that anyhow depend on technology transfer.

The plug-flow digester, as described by Jewell[17], probably has many advantages compared to conventional digesters if a liquid manure is produced. In the case of solid manure,, batch digestion seems to be the optimal solution (Figure 1, example 13). Both digesters, plug flow and batch, need a plastic membrane to collect and store the biogas. Good quality plastic membranes (sometimes tissue reinforced) are probably worth being imported. Alternative materials are required

Cylindrical, overground. Gasholder out of metal, ev. plastic covered metal frame, digester out of cement or metal, ev.wood. Stirring by screw in dubblemantle tube (tube also used for heating)

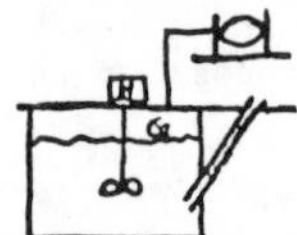

Rectangular or cubical, underground, cement. Gas storage in balloon or in wet gas holder, ev.plastic dry gas holder, stirred by paddle or propeller.

Accumulation systems mostly under building (reduction of heat losses)

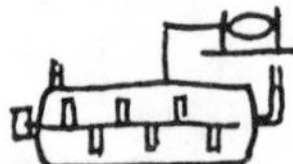

Horizontal cylinder, over- or underground, metal. Stirred by devices attached to turning axis or by blowing gas to the bottom of the digester.

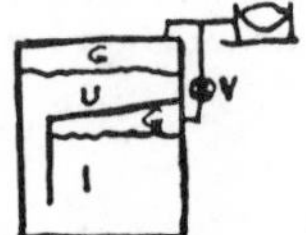

Rectangular or cubical, under- or overground, thick cement walls. Stirring by closing the valve "V": the pressure generated in the lower part "l" presses substrate into the upper part "u". When opening the valve the manure flows back rapidly and mixes the content.

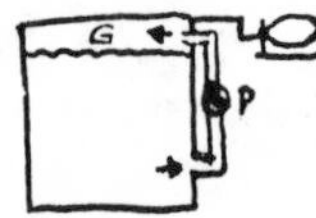

Cylindrical or cubical; metal, plastic, cement or wood. Stirring by circulating the substrate and spreading it over the scum layer.

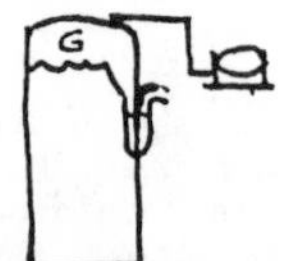

Cylindrical, fiber glass, plastic, ev. other materials. No stirring; the scum is exported by a overflow-funnel.

Sometimes additional spreading of liquid over scum or mechanical stirring.

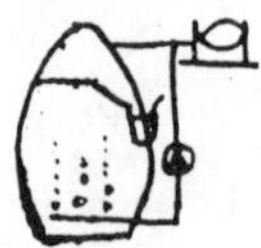

Eggshape, fiber glass. Overflow-funnel, ev. additional stirring by blowing gas into the digester.

Ev. mixing by thermoconvection (heating elements at the bottom)

Cylindrical, overground, metal or plastic (wood). Stirring by the motion of the gasholder out of metal: Attached to the gasholder are segments of funnels which transport the scum to the central outlet pipe (motion upwards) or press the scum into the liquid (motion downwards)

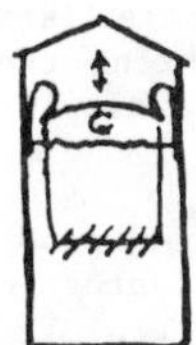

Cylindrical, overground, wood (metal, plastic, cement). Gas holder out of plastic. Gas pressure generated by mixing devices attached to the gas holder.

Ev. conventional mixing by externally driven stirrers etc.

Cylindrical, metal, overground. Gas holder drives mechanically a paddle for mixing.

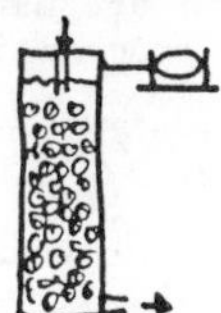

Static filter: vertical cylinder filled with filter material of a large surface. Upflow or downflow possible. Suited for thin substrates.

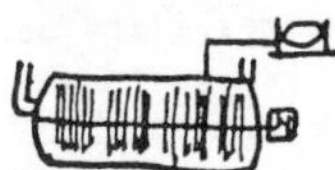

Dynamic filter: horizontal cylinder. Filter material rotates within the digester. Suited for thin substrates.

Rectangular, cubical or hexagonal, usually underground. Batch digester covered with a plastic gas holder. Liquid or solid substrates.

Two stage process: acidogenesis while sprinkling liquid over solid substrate and methanogenesis while passing through a filter.

Two stage process also in experience with liquid substrates.

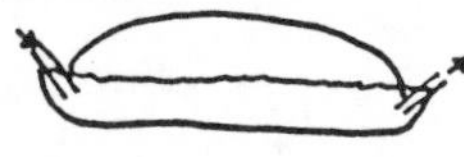

Plug flow, underground, cement or earthen walls. Covered by a plastic membrane. No stirring.

Figure 1: Schematic overview on often used digester designs in developed countries

because metal or ferrocement are too expensive and have serious disadvantages, such as corrosion or cracking. In the future, the technology of improving plastic materials (mechanical properties and ultra violet resistance) as well as optimal ways of manufacturing the gasholders must be transferred to factories and workshops in developing countries.

A comparative study on the use of plastic membranes for plant construction[18] showed that the methane permeability of synthetic sheets varies greatly among the different products (See Table 1). The permeability of the products tested varies exponentially with temperature. Criteria such as mechanical properties, longevity, resistance against destruction by chemicals, bacteria and animals, and repair possibilities seem to be more important than methane (CH_4) permeability when choosing optimal materials. For such reasons, polyethylene is not recommendable for plant construction. On the other hand, materials such as butyl rubber, polyvinyl chloride (PVC), or chloritized polyethylene have their advantages. Since there are many PVC-producing industries in developing countries, they may produce in the future improved tissue reinforced PVC close to the user.

Table 1 Methane (CH_4) Permeability of Synthetic Membranes[18]

Material	23°C cm3*/m2atm.d	35°C cm3*/m2atm.d	50°C cm3*/m2atm	Increase of permeability per 10°C
CPE	85	190	505	2.4
CPE, reinforced	70	165	415	2.4
CSR	125	290	710	2.3
CR	490	1010	2105	1.7
IIR	115	290	640	2.3
PVC	195	365	805	1.7
PVC, reinforced	190	310	615	1.3
EPR	2870	6170	8930	1.3
PE	345	760	2060	2.4

*Volume at 1 atm and 23°C. NOTE: foil thickness: 1 mm. Abbrevations: CPE: chloritized polyethylene; CSR: chlorosulfonated polyethylene; CR: chloroprenes; IIR: butyl rubber; PVC: polyvinyl chloride; EPR: polyethylene-propylene, PE: polyethylene.

CONCLUSIONS

Biogas plants must be made appropriate for local conditions. Since the conditions vary significantly between developed and developing countries, possibilities of technology transfer are very limited, with the following exceptions:

1. Some very reliable and simple solutions such as plastic-covered digesters should be adapted to the local conditions, tested, and eventually propagated in developing countries.

2. The technical collaboration of experts from developed countries with their colleagues in developing countries may be fruitful, if local conditions are known well by the foreigner. "Advice from far" does not help in most cases.

3. The biology of anaerobic digestion is the same around the world. Here it is possible to transfer much knowledge directly. However, fundamental research in this area is often limited by many factors.[19]

4. There is a wide scope of possible fruitful collaboration relating to optimizing the process parameters of a plant, such as the choice of the optimal feed mode and substrate composition and starting and optimizing the output.

When thinking of technology transfer, the most important point is that technology transfer may never be a one-way process. Both partners learn from each other to find the optimal and appropriate solution.

REFERENCES

1. Wellinger, A., and R. Kaufmann. 1982. Psychrophilic methane generation from pig manure. Proc. Biochem., 17:26-30.

2. Van den Berg, L., and K. J. Kennedy. 1983. Comparison of Advanced Anaerobic Digesters. Proc. of Int. Symp. of Anaerobic Digestion (AD) AD 83. Boston pp. 71-90.

3. Young, J.C. 1983. The Anaereobic Filter--Past, Present, and Future. Proc. AD 83. Boston pp 91-106.

4. Sangiorgi, F. 1981. Low-Cost Biogas Installations for the Italian Agriculture. In Second International Symposium on Anaerobic Digestion. Scientific Press. Fairwater, Cardiff. United Kingdom.

5. Cohen, A. 1983. Two-Phase Anaerobic Digestion of Liquid and Solid Wastes. Proc. AD 83. Boston. pp. 123-138.

6. Colleran E. et al. 1983. One- and Two-Stage Filter Digestion of Agricultural Wastes. Proc. AD 83. Boston. pp. 285-302.

7. Isman, M. 1973. A propos d'energie et plus specialment du gaz du fumier. Agriculture, No. 371.

8. Edelmann, W., J. M. Besson, and H. Engeli. 1981. On the Possibility of Aerobic Pretreatment to Heat Substrate for Anaerobic Digestion. Poster paper. AD 81. Travemunde.

9. Siegl, P. 1983. Flexible Covering for a Gas Storage Device. Weltorganization fur geistiges Eigentum. PCT. Patent WO 83/03884, 10.10.83.

10. Fankhauser, J. and A. Moser. 1983. Studie uber die Eignung von Biogas als Treibstoff fur Landwirtschaftstraktoren, Schriftenreihe FAT:18, 322.

11. Third International Symposium on Anaerobic Digestion. 1983. Proceedings. 99 Erie Street, Cambridge, Massachusetts 02139. United States.

12. Braun, R. 1982. Biogas-Methangarung organischer Abfalle. Springer Verlag, Wien und New York.

13. Edelmann, W. 1984. The Feasibility of Biogas Production on Modern Farms. Bioenergy 84 Proceedings. Goeteborg.

14. Edelmann, W. 1984. Technical and Social Problems for the Propagation of Biogas in Rwanda. International Conference on State of the ART on Biogas Technology, Transfer and Diffusion. Cairo. November 17-24.

15. Edelmann, W. 1983. Digestion Systems for Agricultural Wastes. Proc. AD 83. Boston pp. 271-284.

16. Edelmann, W. 1983. Das Biogas-Projekt am Strickhof, ETH-Bulletin No. 179, 14-16.

17. Jewell, J. J. et al. 1979. Low-Cost Methane Generation on Small Farms. 3rd Annual Symposium on Biomass Energy Systems. Golden, Colorado.

18. Weilenmann, A, and B. Kiefer. 1983. Die Verwendbarkeit von Kunststoffolien im Bereich von Biogasanlagen, Projekt Biogas, ETH-Zurich (internal report to be published).

19. Edelmann, W. 1981. Consultancy at the Department of Agricultural Microbiology. TNAU, Coimbatore, India, Report FAO, Project IND/78/020.

MAIN QUESTIONS AND ANSWERS

Q: Are the anaerobic filters useful for manure with a high content of dry matter?

Edelmann: With restrictions, the important point is that there are no large particles in it.

Q: You have mentioned that for swine manure as substrate one day retention time is possible. What is the total solids content in the substrate and the methane content of the biogas produced?

Edelmann: Approximately 1.2% and approximately 75-80%

Q: Do you recommend utilization of plastic gasholders in countries where temperatures are high?

Edelmann: Yes, some plastics.

Q: D you think that fixing bacteria in digesters could improve the production rate?

Edelmann: The rate of gas production is strongly related to the number of bacteria present.

Q: Research results are not always transferable because of different needs. Do you provide research facilities for, e.g. 22°C digestion as needed in Rwanda?

Edelmann: These measurements will be made on the full site of the plant in Rwanda. Basic data are available already (also from our own laboratories).

ENGINEERING ASPECTS OF SMALL-SCALE BIOGAS PLANTS

L. Sasse, Bremen Overseas Research and Development Agency (BORDA), Federal Republic of Germany

ABSTRACT

Small-scale gas plants are simple in construction. Yet the problems to be considered and solved are not less challenging than with highly sophisticated structures.

In this paper, an attempt is made to highlight certain design aspects that are particularly relevant to the small-size biogas plant. Though special attention is devoted to the basics of structural engineering in gas plant design, other pertinent engineering considerations are discussed. These include: sizing of the digester and the gasholder as well as some engineering facets relating to extension programs aimed at widespread diffusion of the technology.

STRUCTURAL DEMANDS

A biogas plant is a vessel containing liquid or semi-liquid matter. Any liquid creates pressure on the walls of the vessel which is in direct relation to the height of the vessel (Figure 1). To keep the pressure low, the vessel must be shallow.

The forces on a structure become less the more a statical system is indeterminate. More simply, the more a system is complicated, the lesser are the structural forces.

A sheet of paper is very unstable. If you fold it zig-zag, it is of considerable strength (Figure 2). From ancient architecture, it is known that an arc may span wide distances without reinforcement. If an arc is extended to the third dimension of a spherical structure, you have a structure of outstanding strength with a minimum of material involved (Figure 3). Here, in an Islamic country, are mosques which use this kind of structure. Perhaps we can imagine how thick an eggshell would have to be if hens laid cubes instead of eggs.

Why is a spherical shell (i.e., structure) so strong? Each direction helps carry the load. With a beam, the load can only go to the supports; the direction for the load is only one. If you have a slab supported on all four sides, the load goes roughly in two directions. If you have a shell, the load is divided into a large

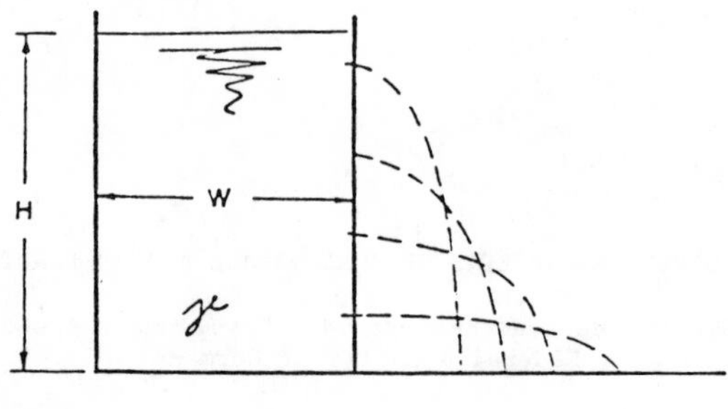

W = H . γ

Figure 1 Pressure created on the walls of the vessel by internal liquid

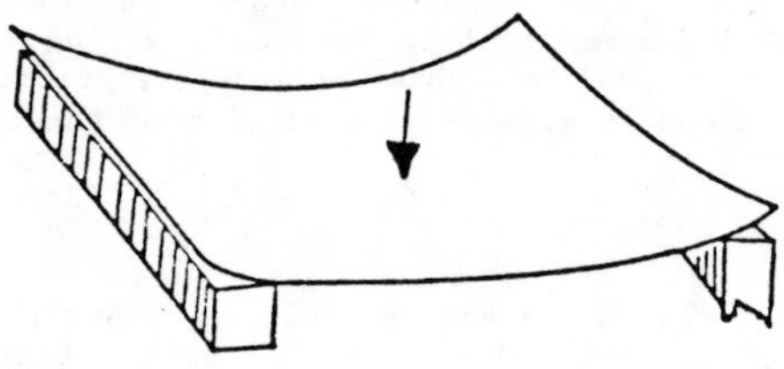

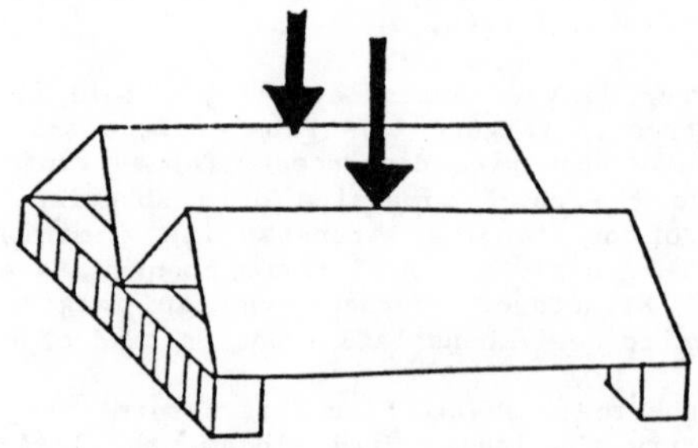

Figure 2 Folding increases the structural strength

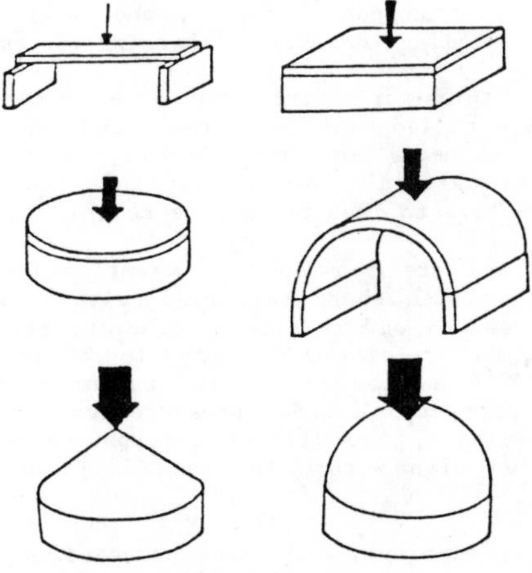

Figure 3 Spherical structures give maximum strength with minimum material

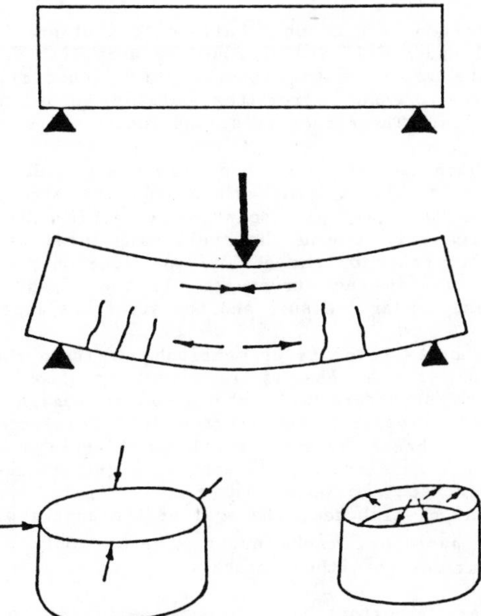

Figure 4 Illustration of compressive and tensile forces

number of radial and tangential forces, and each force is broken up into axial forces, bending forces, and shearing forces.

If shell structures are very thin, they become more fantastic. Such structures are called membranes, the most lazy of structures. If the load becomes too heavy for them, they try to escape it and give it to the neighboring material. Because of this, unbalanced loads will not cause the structure to lose too much strength.

In nature, there are thousands of examples for such structures. The problem is that such shapes are difficult to calculate. This is one of the main reasons why reinforced concrete structures are mostly straight and need much reinforcement. Any force farther on a structure creates tension. If a beam bends, the bottom part is under stress (tensile), the upper part is under pressure (compression). If a post is pressed from outside, the wall of the post is under pressure. If the post is filled with water, the wall is under stress (tension), Figure 4.

Plastic, steel, aluminum, and other homogeneous materials can take as much stress as pressure. Concrete or brick work is different. These materials have a high strength against pressure, but little ability to withstand stress, requiring reinforcement. In prestressed concrete (Figure 5), a steel bar is tight so that the concrete is continuously under pressure. If a load is put on the structure, the pressure is so high that the stress is overpowered by the pressure, and finally no or very little stress actually occurs.

This phenomenon is used when building gas plants in the ground. We build a vessel and refill its surrounding area with soil, thus creating pressure on the walls of the vessel. The liquid from inside creates stress, but the pressure from the outside keeps the stress within reasonable limits. The structure stands intact (Figure 6).

Sometimes ground soil does not create any pressure on structures because it stands alone. In such cases, the structure is in great danger. There is a simple solution to this--the spherical shell (Figure 7). The soil around the shell cannot refuse its prestressing power because it rests on the shell. Any load on the shell works as a pressure force inside the structure. If the digester is filled, the walls are already under pressure and the stress is kept under control.

There is another point to be mentioned. Within the practical range of construction, the stress is in direct relation to the amount of length that the structure will change under load. Stress makes any material longer, pressure will shorten it. Reinforced concrete under stress will not break because steel is elastic even under stress. Concrete is not that elastic. To match the difference in the length of steel, the concrete is bound to crack (Figure 8). The more the reinforcement is distributed, the smaller the cracks will be, but there is no way of avoiding cracks under stress. Only a structure under pressure is a structure without cracks.

This is very important to note especially for concrete gasholders or concrete domes in fixed dome plants. A gasholder made out of brittle material, such as concrete or brickwork, can never become tight, unless the structure is continuously and everywhere under

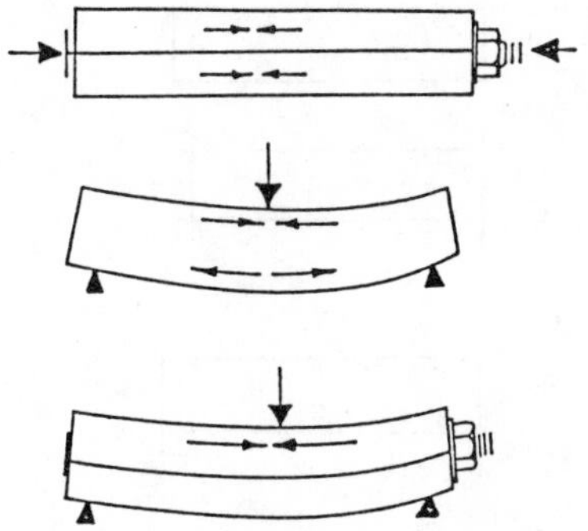

Figure 5 The concept of prestressed concrete

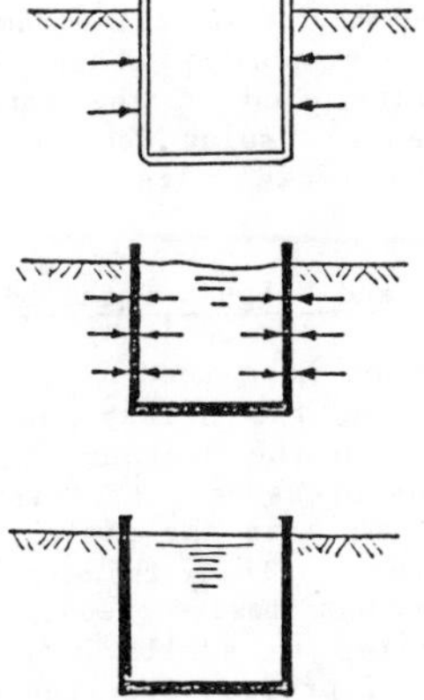

Figure 6 Intact structure - balancing inside and outside pressures

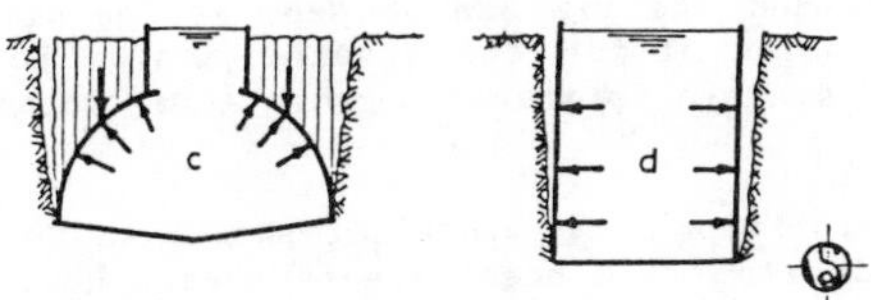

Figure 7 Spherical Shell Solution to Self-Supporting ground soil.

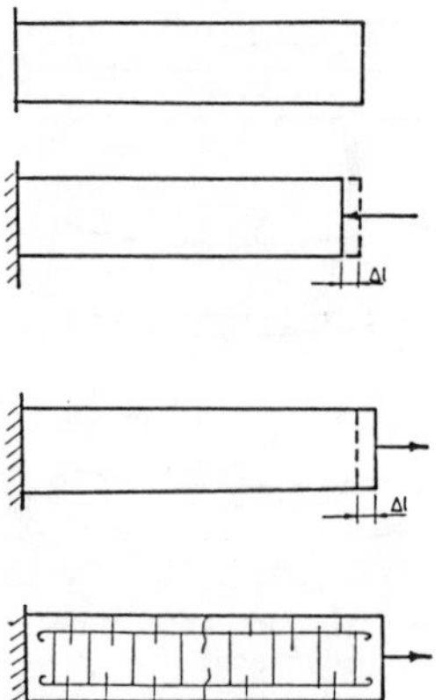

Figure 8 Only a Structure under pressure is a structure with out cracks

pressure. Therefore the dome of a fixed-dome plant should not be disturbed by corners or any abrupt changes in shape (Figure 9). Gastight paint is only to be used if the structure itself can never crack. To say it more clearly, paint does not cover cracks, with the exception of very expensive, special elastic paints.

Relation Between the Length and Height of the Bearing Structure

A rectangular beam carries more load if put in an upright position (Figure 10). This is because the stress inside the beam caused by bending is directly related to the distance between the center of the stress and the center of the pressure. The basic formula is $M = p \times z$, where p is the sum of tension, z is the distance between their centers, and M is the bending moment. This thinking model can be used to understand why a flat arc has less strength than a half round arc (Figure 11). The same applies to shells. An egg is stronger in its longer direction and weaker if pressed over the shorter dimension.

This statement becomes important when considering the bottom slab of a digester. With a flat slab, the distance between centers of tension is smaller than the thickness of the slab. If the slab is constructed like a bowl with a certain peak-height, the distance between the middle and the rim can be seen as the difference between the centers of tension inside the construction. The effect of the spherical shape adds again to its strength. As a slogan, "The rounder, the better."

Only slightly inferior to a spherical shell is the cone. The cone is more easily made from flat steel sheets than a bowl-like structure. The relationship between height and width is equally responsible for the cone's strength as it is for the spherical shell or any other structure (Figure 12). The roof of a steel gasholder is stronger the higher it becomes.

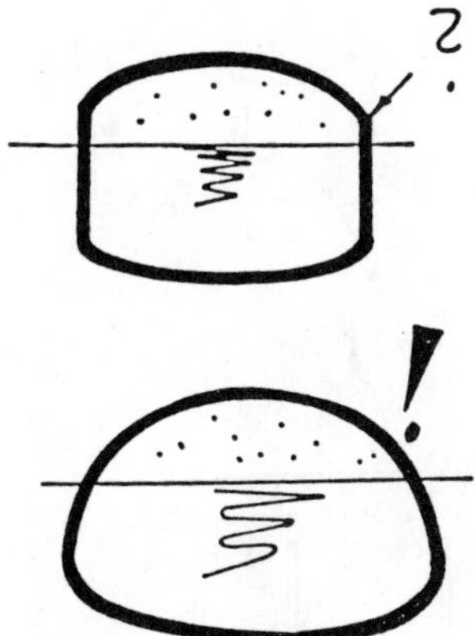

Figure 9 Avoid Corners or abrupt
Changes in shape.

Instead of adding numerous angle irons to support a flat roof of a large gasholder, it is wiser to raise its height. The rim of a cylinder, or any other shape, is the weakest point of construction because the structure cannot reject its responsibility by giving the load to the neighboring material. Therefore, any cooking vessel has a folded rim (Figure 13). Any sheet of a car's body ends with a folder. It is therefore wise to fold a rim on a gas plant's structure. In

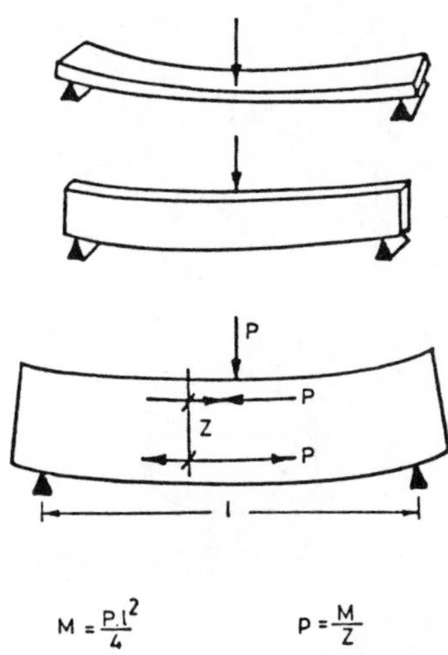

Figure 10 Bending moment depends on
its arm.

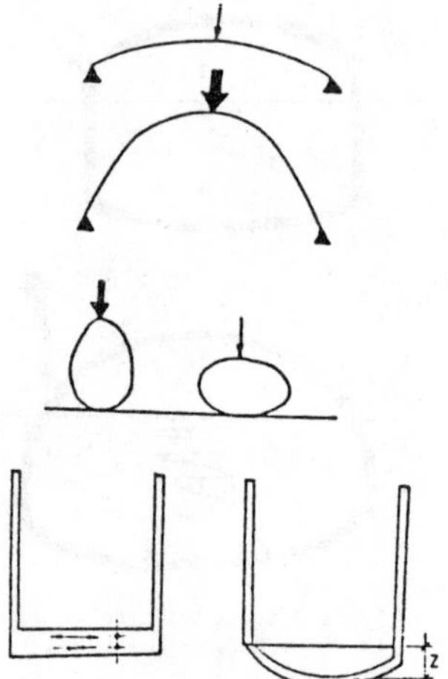

Figure 11 Illustration of the
strength of arcs and shells

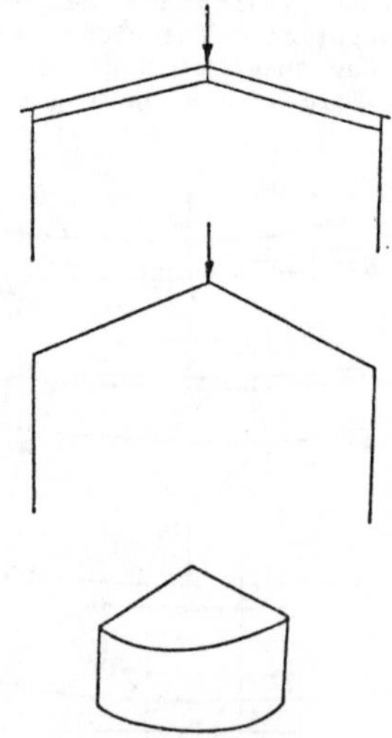

Figure 12 A Cone is stronger than
a flat structure.

brickwork, it is very simple to make the last row of the upper part of the digester thicker than the rest. In the BORDA design, a half brick is used instead of a quarter, thus gaining strength without adding much material. For the rim of a steel gasholder, such reinforcement is not needed because there is very little stress and no pressure. This stress can easily be taken by the steel mantle sheet without support.

Finally, one more formula can help in understanding a number of problems in gas plant construction. If a round vessel is filled with water, an inner force is given to the wall of the vessel (Figure 14). This force increases with the height as well as with the radius of the vessel. The relation is very direct: $p = r \times w$. Where p is the force in tangential direction, r is the radius of the cylinder, and w is the force created by the liquid. (This formula is only valid for cylinders of infinite length. In practice our cylinders are much shorter and normally fixed to the bottom slab. The connection between the slab and the wall disturbs the free movement of the cylinder. Therefore ring-tension goes toward zero and vertical bending moments occur as a result of the inner force created by the liquid. For a reinforced concrete digester this would mean that steelbars stand upright, while the undisturbed cylinder bars are laid horizontally.) For practical design, it should be remembered that in a shallow vessel the forces are generally lower than in a deeper vessel of the same volume despite the increased diameter. In BORDA designs (Figure 15), we try to consider the above-mentioned points without making construction more complicated. Great credit must be given to Chinese engineers who were first in the world to use the ancient method of building spherical shells from bricks for simple small-scale biogas plants.

Size of the Digester

The digester itself is 40-50 percent of the total cost of a floating drum plant and 65-70 percent for a fixed-dome plant. The optimum size of the digester is of great importance to the economy of a gas plant. It would thus be logical to reduce the size of the digester to its possible minimum, but it is not that simple. Biogas plants with complete temperature control and optimal conditions work well with a retention time of 22 to 30 days. This is not the case with gas plants in rural households, for example, in India. Even under rather favorable conditions, the digester temperature is only 28°C instead of 35°C. Temperature fluctuation between day and night and dry and rainy days is 2°C-5°C. Feed material is of changing quality: Animal fodder often consists of hard and dry straw, which is responsible for a high lignin content in the digestion material.

The tests BORDA undertook in cooperation with UNDARP near Poona in India proved that retention periods of 30 days are not enough to make use of the feed material. Gas production per kg of fresh buffalo dung is still increasing after 90 days of retention time (Figure 16). The digester size is in direct relation to the retention time. Taking 30 kg of cattle dung as daily feed with the same amount of water produces the following digester volumes:

30 days = 1.8 m^3, 60 days = 3.6 m^3, 90 days = 5.4 m^3,
120 days = 7.2 m^3

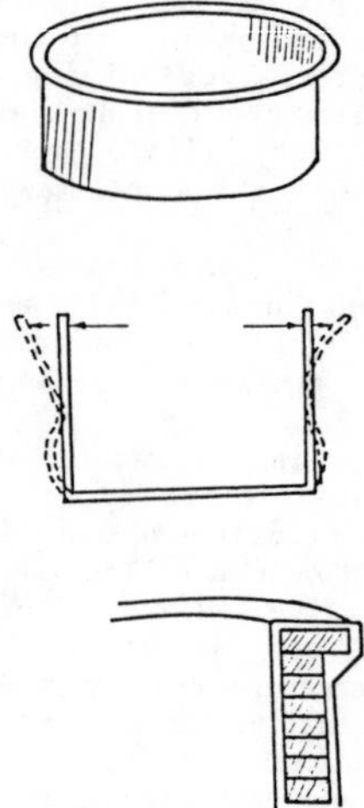

Figure 13 Fold the rim to avoid its weakness

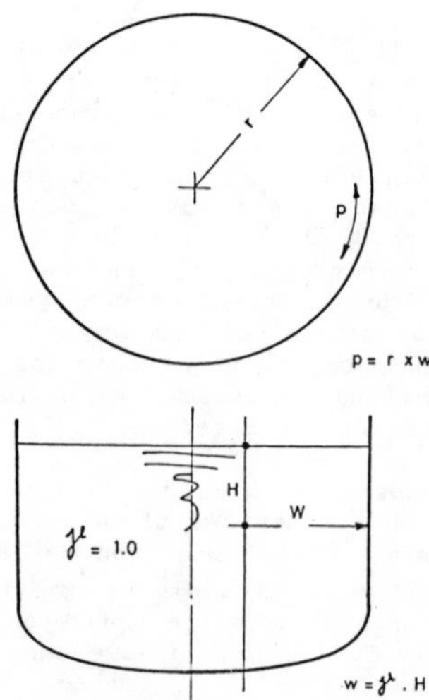

Figure 14 Inner liquid pressure forces increase with vessel height

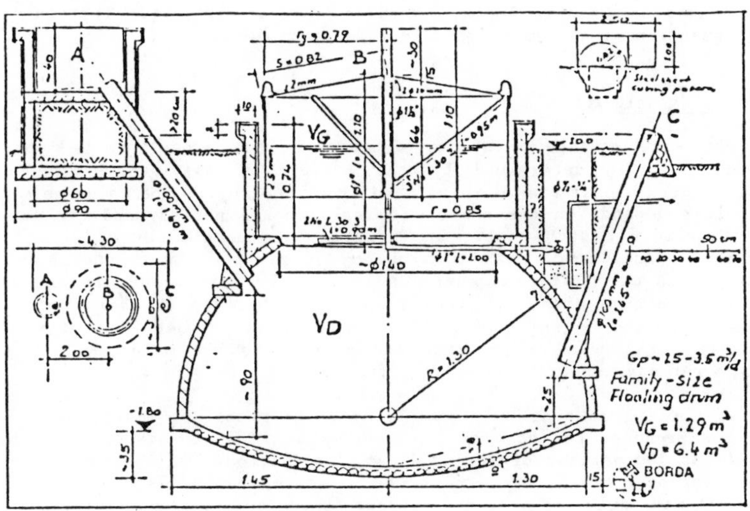

Figure 15 The BORDA design

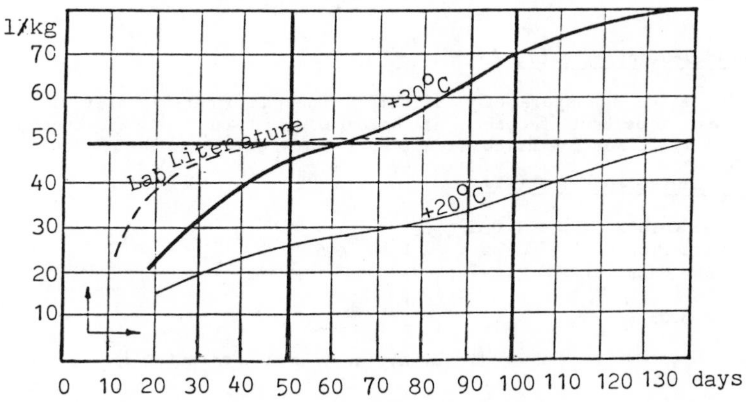

Figure 16 BORDA-UNDARP results for change in gas production with retention time and temperature for floating-drum plants

Similar tests for fixed-dome plants are under way with our partners Gram Vikas in Orissa, India. Results are not out yet. Therefore, I must restrict my statements to floating drum plants.

Size of Gasholder

The size of the gasholder depends on the amount of gas to be stored. It is determined by the pattern of gas use. For rural households, biogas is mainly used for cooking. The size of the gasholder depends on the number of meals cooked per day and the amount of gas used for each meal. In Turkey, where people eat a lot of raw vegetables and bread, less than one m^3 of gas per day is enough for a family. In many African countries one of the main dishes is beans, a cooking practice with high fuel consumption because the beans are not normally wetted before cooking. A family needs almost 2 m^3 of biogas per day. In India, where rice and pulses are the main food, 1.5 m^3 of gas per day is sufficient for a family.

If cooking is done only once a day, the gasholder must provide all the gas produced per day for this one meal. The storage capacity must be almost 100 percent of the daily gas production. If the family cooks twice per day, nearly 60 percent gas storage capacity is needed and for three times only 45 percent storage capacity is required. In the Indian example, a family would need 1.5 m^3 of gas and thus 0.9 m^3 gasholder volume at 60 percent storage capacity. For Africa, the figures would be 2.0 m^3 of gas and 1.2 m^3 storage capacity. Since beans cook slowly for a long time, perhaps 1.1 m^3 gas storage would be feasible.

For the size of the gasholder, it is important to note that gas production is not regular every day. The fluctuation might reach 25 percent below and 25 percent above average. A 25 percent buffer capacity should be included when making these calculations.

Gasholder-Digester Ratio

To decide the shape of a digester, the gasholder-digester ratio is the most important factor. It depends again on the retention time (Figure 17). For the cited examples, the following ratios are needed.

30 days Retention Time (RT):

Digester Volume (VD) = 1.8 m^3
Gas production (GP) = 30 kg x 30 l per day = 900 l = 0.9 m^3
% Storage Capacity (C) = 60% of 0.9 m^3 = 0.54 m^3 VG (gasholder volume)

VD:VG = 3.33

NOTE: 0.90 m^3 of gas per day is not sufficient for a family.

60 days RT:

VD = 3.6 m^3
GP = 30 kg x 45 l per day = 1,350 l (still short) = 1.35 m^3
C = 60% of 1.350 m^3 = 0.810 m^3
VD:VG = 4.44

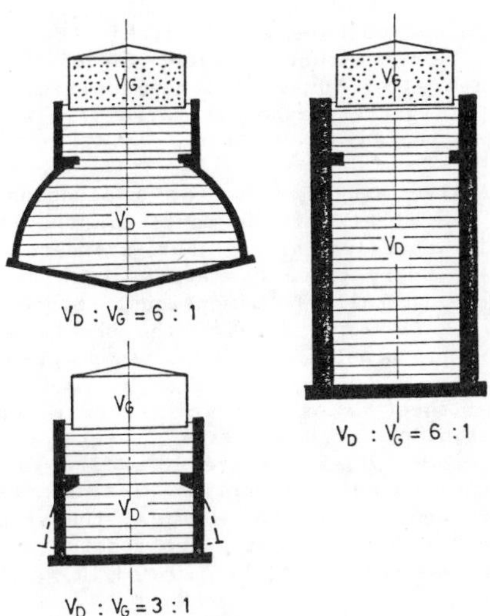

Figure 17 Gasholder/Digester Ratios

<u>90 days RT</u>:

VD = 5.4 m^3
GP = 30 kg x 60 l per day = 1,800 l (sufficient) = 1.8 m^3
C = 60% of 1.8 m^3 = 1.080 m^3
VD:VG = 5.0

<u>120 days RT</u>:

VD = 7.2 m^3
GP = 30 kg x 70 l per day = 2,100 l (even sufficient for African households) = 2.1 m^3
C = 60% of 2.100m^3 per day = 1.260 m^3
VD:VG = 5.7

These examples are of limited value in practical work. Every household has different habits and, more important, different amounts of feed material available. From the engineering viewpoint, however, it is clearly stated that for a small farmer with little digestable material available, a 30 days retention time is of no use. Further, by consequently calculating the digester-gasholder ratio, longer retention times lead to a higher ratio--about 5:1.

ENGINEERING FOR EXTENSION PROGRAMS

It is possible to design a properly sized gas plant for individual households. For extension programs, where many gas plants are constructed during a widespread diffusional campaign, the gas plant models to be introduced must be standardized. Further, a small amount of feed requires a small amount of water for mixing the slurry to get enough gas (a daily relief for women who must fetch the water). If it is decided to use a digester-gasholder ratio of 5:1, what size of a gas plant should be constructed?

The most expensive single item for a floating gas plant is the steel, from which the sheets of the mantle is the largest item. Steel sheet can come in standard sizes, but standardization is not yet achieved worldwide. Every place has access to a different market, the sheets normally come in reliable dimensions. Since we have a variety of families, we will take the size of the sheets as the base for the dimensioning of the gas plant in an extension program.

Cutting a steel sheet means more welding of a seam. It makes sense to design the gasholder so that steel sheets are used as much as possible in their given sizes (Figure 18). The gasholder consists of two main parts: the top and the mantle. The diameter of the gasholder relates to both of them. Take, for example, the standard size of steel sheets as 2,500 mm x 1,000 mm. Using two sheets for the circumference, the diameter of the gasholder will be 5.0 m/3.14 = 1.59 m.

To make better use of not properly cut sheets, we may reach to a size of 1.58 m in diameter for the mantle and 1.64 m for the top sheet. The top of the gasholder slopes and has to be a bit wider than the mantle. The size of such a gasholder would be about 1.96 m^3. This storage volume is too big for most families. If the mantle sheet is cut so that only two-thirds of the height is used, the gasholder comes to:

$$VG = (1.58^2 \times 3.14/4) \times 0.66 \text{ m} = 1.29 \text{ m}^3$$

Taking 60 percent storage capacity as a norm, this gasholder is good for a daily gas production of 1.29/0.60 = 2.16 m^3.

Using the steel sheet as a base for calculation and taking the digester-gasholder ratio of 5:1, the digester must have a volume of 5 x 1.29 = 6.45 m^3. It is now simple to design the digester accordingly. For the BORDA design, the radius of the spherical shell would come to 1.30 m and allow a net volume of 6.4 m^3.

This gas plant will produce the following amount of gas, depending on the daily feed:

30 kg per day: RT = 6,400 l : 60 l = 107 days
GP: 30 l x 67 l/kg = 2,010 l/day = 2.01 m^3

100 kg per day: RT = 6,400 l : 200 l = 32 days
GP = 30 l/kg x 100 kg = 3,000 l/day = 3.0 m^3

This gas plant will be the appropriate size for households consisting of 6 to 15 members.

Similar calculations should be made for different sizes of sheets and different cutting patterns. Then the extension program will have a few adequately sized and standardized gas plants.

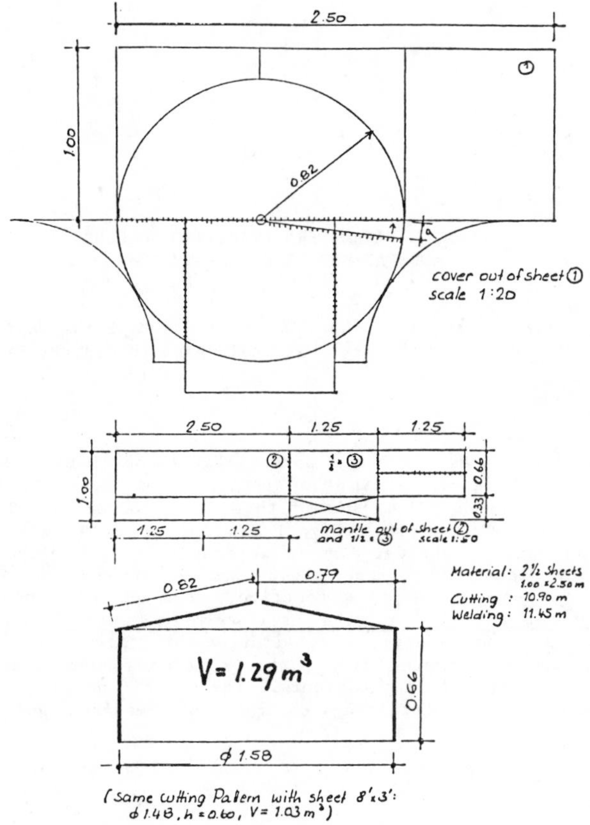

Figure 18 Steel Gasholder Cutting Pattern

CONCLUDING REMARKS

The above-mentioned points only touch a part of the problem. I have not talked about the simple method of constructing spherical shells or gigs to make the construction of gas plants easier. Nor did I mention the problems of using plastic linings in fixed-dome plants or fiber glass for gasholders.

I have only briefly summarized the basic ideas of digester design for small-scale biogas plants meant for small farmers with little feed material and little financial means.

DESIGN PARAMETERS AFFECTING SUCCESS AND FAILURE OF BIOGAS SYSTEMS

M.A. Hamad, A.M. Abdel Dayem, and M.M. El-Halwagi
Pilot Plant Laboratory, National Research Centre, Egypt

ABSTRACT

This paper stresses that a potentially successful biogas system is one that is designed to fit the specific conditions of the case for which it is planned. The truth of this basic concept is illustrated by experience gained in rural Egypt. Attention is focused on certain design parameters considered most important by the authors in their engineering development and field demonstration endeavors. These parameters encompass: gas production, storage and consumption, and their optimum matching to the user requirements; adjustment of the fermentation temperature; stabilization of gas pressure; manure handling; internal system flow patterns, scum formation and breaking; and mixing of the digester contents. The paper concludes with a number of recommendations pertaining to relevant research and development needs.

INTRODUCTION

Many biogas system designs are presently available at various degrees of sophistication[1-8]. In Egypt, as in most developing countries, attention is focused mainly on intermediate-type technologies appropriate to prevailing rural conditions. However, it was also shown that the impact of large-scale, relatively sophisticated biogas plants can be of no less importance than the small simple household units.

Village-type plausible technologies have undergone major developments in India and China. The implementation of these technologies, particularly when transferred to other countries, at times was encountered with difficulties that would even reach the stage of complete failure. Apart from socio-economic reasons, failures can be traced in many cases to the inadequate tackling of the critical design parameters relevant to the prevailing local conditions.

In the following account, the critical design parameters in the Egyptian case as manifested by the National Research Centre (NRC) field demonstration work are addressed.

KEY DESIGN PARAMETERS AFFECTING SUCCESS

Gas Production, Consumption Patterns, and Storage Capacity

Gas Production The gas production rate is a very important design parameter, which has a strong bearing on system economics. Factors affecting the gas production rate include temperature, retention time, digester type, and feed material characteristics, with temperature as the dominant one. In a previous study[10], results on a full-scale demonstration unit indicated that the gas production rate (GPR, $m^3/m^3.d$) could be correlated to temperature (T, °C) by the relation:

$$GPR = Constant .(T)^{2.7}$$

Gas Consumption Patterns The pattern of gas utilization is another important parameter. As depicted by Figure 1, the daily gas consumption for a demonstration household in Manawat village is not constant. It varies, however, according to a relatively regular sequence. Higher consumption rates are noted on Sundays (the village local market day) and Thursdays (the end part of the week). The range of variation on a given day from one week to another is also considerable, the highest value being almost double the lowest one. For this family, a biogas unit producing 1.1 m^3, which is the average value of the daily mean consumption rate, would be adequate from the

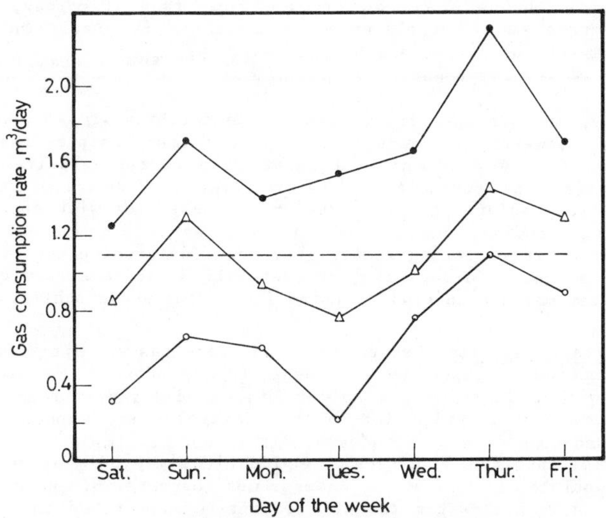

Figure 1 Variation of the family gas consumption rate over the week during the period 1st. August - 30 Sept. 1984.

- ○ Minimum consumption rate , m^3/day
- ● Maximum " " , "
- △ Mean " " , "
- ---- Average of the mean consumption , "

standpoint of the design size. Fluctuations in the consumption rate should, however, be taken care of in sizing the gas storage capacity.

In this particular case, the gasholder capacity should be increased beyond the normally recommended 33-40% of the gas production rate to minimize venting much gas on the low consumption days or using supplementary energy source on peak demand days. However, an optimum should be sought taking into account the increased initial cost of a larger gas-holder against the accrued economic returns.

The gas utilization pattern is not only important in the case of family-sized units, but can also be an equally critical parameter for large-scale plants. As a vivid example, a 50 m^3 demonstration plant serving a poultry-rearing operation in the village of Shubra Kass must meet gas needs that vary from as low as nil to as high as 100 m^3/day. This is due to the nature of the cyclic operation whereby each cycle embodies 60 days for rearing and 10 days for post-cleaning and preparing for the next cycle. During the rearing period, the temperature in the chicks house varies from 35°C at the beginning and decreases to around 22°C by the end of the cycle. Thus, gas needs vary to a great extent from the start of the cycle to its end.

By the same token, large seasonal variations take place as shown in Table 1. According to the yearly average of the biogas needs, a unit of 50 m^3/day gas production rate would normally be designed. Although the biogas plant in Shubra Kass can produce more than 70 m^3/day, it cannot meet peak demands and a supplementary gas source is used. On the other hand, large amounts of biogas are vented during the summer season as well as during periods of low gas needs.

Clearly, in this specific instance, the matching of gas production, storage requirements, and actual needs is much more complex than in the case of the family size unit. Reserve gas storage capacity together with adjusting the schedule of the digester feeding rate could only provide partial solutions. Integration of the system with another, for example, to provide energy for the irrigation requirements of an adjacent farm or generating electricity, can fully solve the problem of cyclic variations. In any case, however, the delicate economics of the biogas system mandate careful optimization of the whole scheme.

Gas Storage Capacity Design of an adequate gas-storing capacity to meet consumption patterns is dependent on the adopted digester type. For units with a floating gas holder, the needed gas storage capacity can be reasonably provided for in a relatively easy manner. In the case of fixed-dome digesters, however, it is not as simple.

With floating-drum digesters, complications may only arise when it is not possible to go deep underground because of adverse soil conditions or a high-water table, frequently encountered in Egypt. A wide and shallow digester would necessitate a more costly, large diameter gas holder. A good solution for such cases would be to design a digester with two different cross sections, the smaller being at the top where the gas holder is fitted. Examples of this design are found in Manawat[11].

The gas storage capacity of the conventional Chinese water-pressure digester[6] is very low under relatively warm conditions, similar to those prevailing in Egypt. This can be readily seen from the gas rates

Table 1. Variation of Gas Consumption Rate for Different Rearing Cycles of the Poultry Farm in Shubra Kass for One Year

Cycle No.	1	2	3	4	5
Date of the cycle	30 May- 29 July	27 August- 25 Oct.	20 Nov.- 17 Jan.	24 Jan.- 16 Mar.	15 Apr.- 5 June
Total consumption, m^3 biogas equivalent	1,680	880	3,100	4,350	2,400
Average consumption rate, m^3 biogas/d	27.5	15	52.4	71.3	47
Daily biogas average for operating cycles around the year, m^3/d			42.8		

reported under Chinese conditions--on the order of 0.15 m^3/m^3 digester-day--whereas the rate in Egypt is almost double due to the higher average temperature. A relevant study was previously published by the authors[12] concerning the gas losses and storage capacity of the Chinese-type digester. A modified outlet chamber design was proposed to allow for a higher gas storage capacity at lower gas losses. The main idea in this modification is raising the datum of the outlet chamber to minimize gas losses and increase the diameter of the effective part of the chamber thus affecting higher gas storage capacity. A complete evaluation for this new design under actual operating conditions will be published in the near future.

Fermentation Temperature

The gas production rate is greatly affected by the operating temperature. Thus, increasing the temperature level is essential to decrease the digester volume and improve the system's economics. Results of field demonstrations for the nonheated underground-built digesters indicated that the operating temperature varied from as low as 17°C during winter to a maximum of about 28°C during the hottest period of the year. As can be readily noted, the maximum attained temperature is far below the optimum for mesophilic conditions, which is around 37°C. Attention was therefore given to optimizing the operating temperature in the family-sized and large-scale digesters.

Enclosing the digester under a plastic tent to create the "greenhouse effect" is a standard practice in the NRC installed biogas systems. As seen from Figures 2-6, the greenhouse temperature is considerably higher than the ambient. Consequently, the gas productivity increased (Figure 7) by 50% on the average. It appears, however, that there is much room for improving the design of these solar passive heating systems to approach an overall optimized operation.

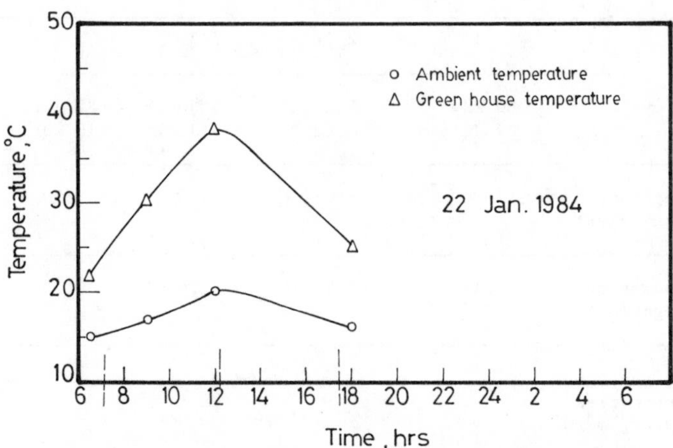

Figure 2 Green house effect (Sharkawy modified Chinese-type digester).

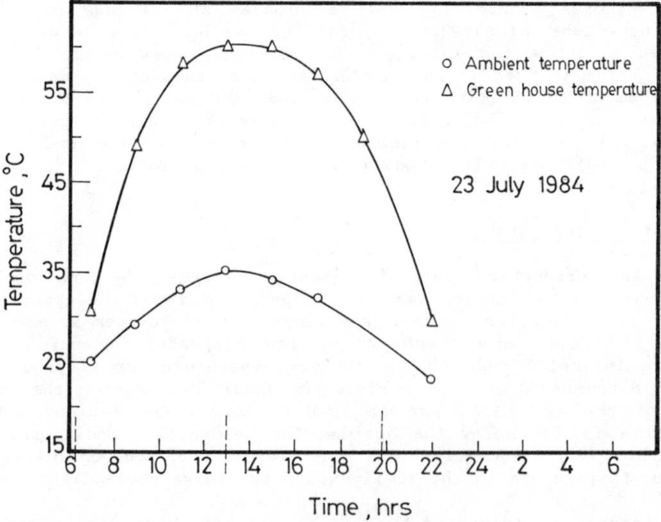

Figure 3 Green house effect (Sharkawy-modified Chinese-type digester).

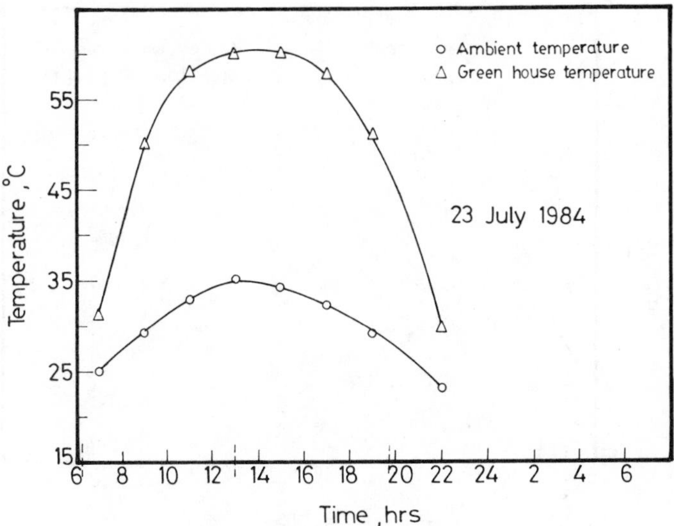

Figure 4 Green house effect (Deyab-modified Indian type digester)

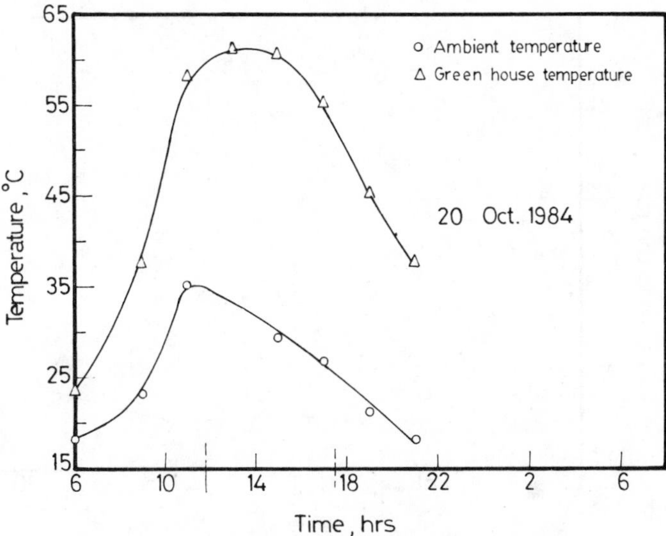

Figure 5 Green house effect (Deyab modified Indian type digester).

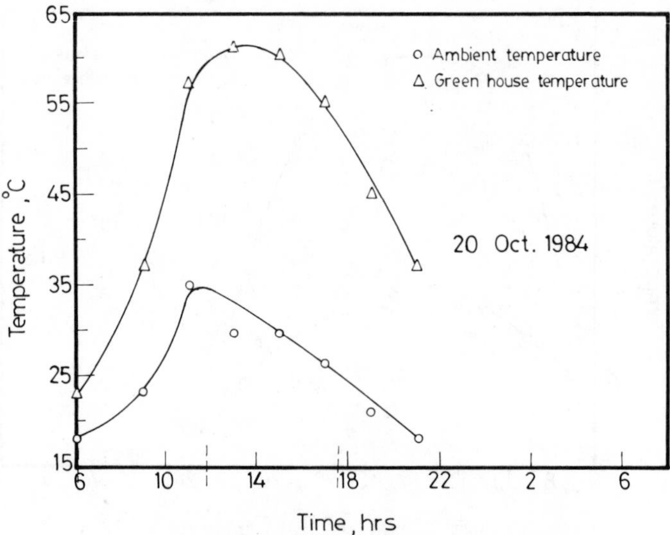

Figure 6 Green house effect (Sharkawy modified Chinese type digester).

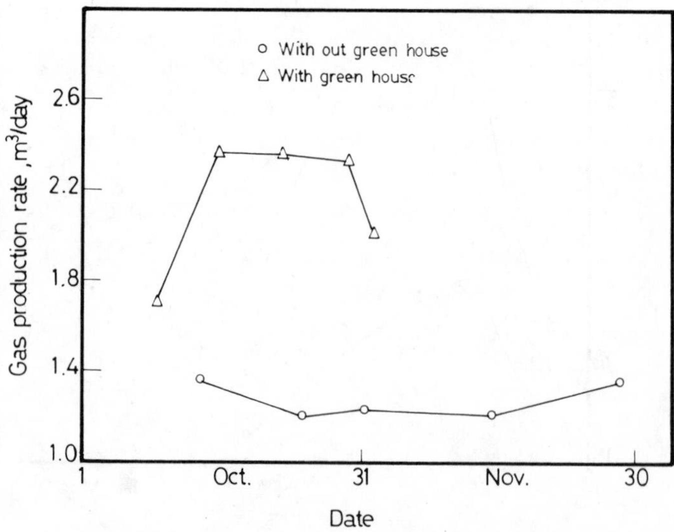

Figure 7 Effect of green house on gas production rate (9 m^3 Chinese type digester)

Direct heating (particularly for the large-scale systems) using biogas or other means such as engine exhaust heat have been tested with good results. Solar heating of feed water and digester contents is also being studied. The fundamentals of these and other heating systems (e.g., composting around the digester) are worth investigating to establish a reliable basis for the optimum design of related biogas producing schemes.

Manure-Handling System

In India, manual feeding is practiced to collect dung from the animal sheds and transfer it to the digester. Thus, urine is mostly not utilized. The Chinese practice, however, is generally based upon direct connection of the animal shed to the digester, and thus urine is used. However, in the majority of cases, the effluent from the outlet chamber is manually transported, which is a labor-intensive and dirty operation.

In Egypt, it appears that neither of the two manure effluent handling systems would be appropriate principally because of the low prices of the subsidized fuels and the relatively high labor cost. To enhance the acceptability of the biogas system in the rural areas of Egypt, the NRC engineering team has introduced a major modification to the manure handling system that is generally based on the bedding practice. The basic premise is to convert the soil floor of the animal shed into concrete with a slight slope that facilitates urine drainage into a special small canal from which it flows to the mixing chamber where it is mixed with dung that is easily swept and collected there. Small amounts of water are added when needed to prepare 8-10% solids feed; this is usually done only during summer.

The Chinese water-pressure type design was also modified to allow for self flow of the effluent by using the concept of the variable pressure of the system. Thus an amount equivalent to the feed flows from the specifically designed outlet overflow without any human intervention (Figure 8).

In the Egyptian case, the savings accrued through the modifications to the manure handling system (in effect coverting it into a sanitary disposal scheme) are much more than all other benefits combined. In fact, it is our strong belief that the proper adaptation and optimization of the manure-handling system can make the difference between success and failure of the entire rural biogas system.

Digester Internal Flow Patterns

The slurry flow pattern in anaerobic digesters is one of the important parameters. It does not only affect the gas production rate, but also affects to a great extent the pathogens and parasites kill rates. In a previous study we used the tracer technique for examining the flow patterns in Indian- and Chinese-type digesters[10]. It was found that about 50% of the volumes of both types are inactive. This has a strong bearing on the cost of the digester. In the Chinese-type digester, the tracer was found in the effluent after the second day, indicating a very high degree of bypassing and short-circuiting of the

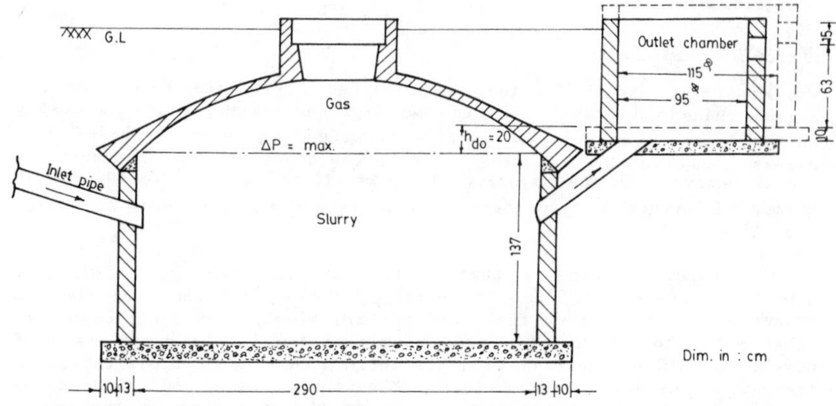

Figure 8 New modification of outlet chamber.

slurry which is very dangerous from the standpoint of pollution control. Evidently, modifications are strongly needed. The Chinese-type digester was accordingly modified by building a baffle in the digester. This can double the length of the path, decrease the dead volume, and eliminate the short circuiting and bypassing. Plug-flow digesters, e.g., the tunnel type[13], can also provide solutions for this problem. However, these designs are more suitable for large-scale digesters. A two-chamber tunnel type digester of moderate size (13 m^3) was built recently by the NRC team in the village of El-Manawat, Giza.

Scum Formation and Breaking

When dung alone is used as feeding material, no scum-related problems were noticed in both Indian- and Chinese-type digesters. This was confirmed from observations of digesters operating for more than three years. Though a relatively thick layer of scum was noticed over the slurry in the Indian-type digesters it does not affect gas production rate. This seems to be due to the continuous movement of the gasholder which affects cutting of the formed scum layer. In Chinese-type digesters, it seems that the movement of slurry up and down assists in scum breaking.

Different modifications for promoting scum breaking are given in the literature[14-16].

Mixing

Mixing is not important in small family-sized units, but it is essential in some large-scale units. For example, utilization of poultry manure mixed with bedding material, such as wheat straw or other agricultural residues, induces flotation or settling of materials which can block the digester totally, as in the case of the Shubra Kass digester. In this case, circulation of gas was found to be one of the best methods of mixing. A compressor was used for this purpose, and no problems have erupted during a relatively long period of operation (about one year). The presence of some mixing points longitudinally distributed in the NRC tunnel digester seems to operate well without affecting the plug-flow pattern of the digester. Liquid mixing is also a good method because it assists in fermenting the floating materials. It appears suitable, however, for completely mixed digesters.

Gas Pressure and Its Stabilization

In floating-drum digesters, the pressure is constant but comparatively low. It is about 10 cm water for family-sized units, but can be increased to 30-40 cm for large units. The constant pressure system is necessary for efficient operation of gas-use devices. However, attention should be devoted to the design of gas lines such that no drop in pressure can take place during periods of high gas rate utilization.

For large units, an optimum choice should be made between locating the gasholder near the digester or in the vicinity of the point of consumption, or alternatively to use a relatively sophisticated high-pressure system. In our latest design using a large unit of 320 m^3 size, it was found that locating the gasholder far away from the digester (by about 600 m) but near the consumption location is more economical.

For the Chinese water-pressure digesters, the pressure is continuously variable, which adversely affects the efficiency of gas-use devices. For this reason, a modification was introduced by the NRC engineering team to attain constant pressure by adapting the available low-cost pressure regulators used for butane gas to control the pressure at 30 cm water. Imported natural gas pressure regulators also were used, and a constant pressure of 25 cm water was obtained.

FIELDS NEEDING MORE RESEARCH AND DEVELOPMENT

The following are proposed:

- o Optimization of the means for increasing the digestion temperature in order to raise gas productivity.
- o Development of suitable low-cost effluent filters or other solid separation techniques in order to partially separate the effluent water, which can then be recycled. This would reduce the effluent volume, decrease the need for fresh water, and increase gas production.

o Development of ready-made compact digesters, including rugged plastic units.

o Development of adequate dry fermenters.

o Development of efficient biogas-use devices.

o Development of generalized and standardized design criteria for biogas systems, taking into consideration the nature of the local conditions in each country.

CONCLUSIONS AND RECOMMENDATIONS

Success and failure of biogas systems can depend very much on whether the different design parameters are adequately taken into consideration or not. Local situations should be given proper attention in every design. Satisfaction of the needs of the end-user should be the main concern. Other parameters, e.g., socioeconomic aspects, should serve and enhance this main objective. Using this basis for the transfer and adaptation of appropriate technologies would lead to potentially successful systems.

REFERENCES

1. Freeman, C. and L. Pyle. 1977. Methane Generation by Anaerobic Fermentation, an Annotated Bibliography. Intermediate Technology Publications, Ltd., London.

2. Sichuan Provincial Office of Biogas Development. 1979. Biogas Technology and Utilization. Chingdu Seminar. China.

3. Barnett, A. and S.K. Subramanian. 1978. Biogas Technology in the Third World: A Multidisciplinary Review. IDRC. Ottawa, Canada.

4. National Academy of Sciences. 1979. Microbial processes: Promising Technologies for Developing Countries. National Academy Press. Washington, D.C.

5. Hashimoto, A.G., Y.R. Chen, and V.H. Vorel. 1980. Theoretical Aspects of Methane Production: State of the Art. In Proceedings of the Conference on Livestock Waste: A Renewable Resource.

6. National Office for Biogas Development and Extension. Biogas in China. Reports at the Technical Consultation Among Developing Countries on Large-Scale Biogas Technology. People's Republic of China.

7. Jewell, W.G. et al. 1984. Anaerobic Fermentation of Agricultural Residues: Potential for Improvement and Implementation. Final Report, U.S. Department of Energy. Washington, D.C.

8. Loll, U. 1984. Biogas Plants for Animal Slurries in the Federal Republic of Germany. European Communities Seminar on Anaerobic Digestion of Sewage Sludge and Organic Agricultural Wastes. Athens, 14-15 May.

9. Hamad, M.A., and M.M. El-Halwagi. 1981. The expected role of biogas technology in rural areas of Egypt: Energy considerations. In Procedures of the Fourth International Conference on Alternative Energy Sources, Miami, Florida.

10. Hamad, M.A., A.M. Abdel Dayem, and M.M. El-Halwagi. 1981-1983. Evaluation of the performance of two rural biogas units of Indian and Chinese Design. Energy in Agriculture 1:235.

11. El-Halwagi, M.M., A.M. Abdel Dayem, and M.A. Hamad. 1982. Village demonstration of biogas technology: An Egyptian case study. Natural Resources Forum. 6:329.

12. Hamad, M.A., A.M. Abdel Dayem, and M.M. El-Halwagi. 1983. Rural biogas technology: Effect of digester pressure on gas rate and composition. J.Eng. Appl. Sciences 2:49.

13. El-Halwagi, M.M., A.M. Abdel Dayem, and M.A. Hamad. 1983. Design and construction of a new type of digester attached to an Egyptian poultry farm. In Proceedings of the Third International Symposium on Anaerobic Digestion. Boston, Massachusetts.

14. Tata Energy Documentation and Information Centre. 1984. Biogas News. No. 2, P.S. June (1984). Bombay.

15. Edelman, W.E. 1984. Technologies for Biogas Production of Developed Countries and Possibilities of Transferring Them to Developing Countries. In Proceedings of the International Conference on State of the Art on Biogas Technology, Transfer, and Diffusion. November 17-24. Cairo.

16. Edelman, W.E. 1983. Digestion Systems for Agricultural Wastes. In Proceedings of the Third International Symposium on Anaerobic Digestion. Boston, Massachusetts.

MAIN QUESTIONS AND ANSWERS

Q: Have you ever measured the temperature increase of the slurry in the digester as a result of the "greenhouse?" Heat should be transferred to the liquid rather than to the gasholder.

Hamad: The temperature was measured at one point about the centre and an increase of 2-3°C was noticed. Complete evaluation is being undertaken.

Q: How many units with plastic covers for maximizing gas production in winter months are in use?

Hamad: Biogas technology in Egypt is still in the demonstration stage. In the NRC biogas project, we have four family-size units and one large unit serving a poultry farm that are provided with plastic covers.

Q: Did you find any solution in fixed-dome gas plants to maintain a high gas-storage capacity without using this gas storage daily?

Hamad: We have modified the outlet just to enable gas storage for the 8-10 hours of the nonconsumption period. The modification is based on the relatively high gas rates attained in Egypt as compared to China. However, storage of gas from one day to another is very difficult and not recommended because of the high gas loss from water pressure digesters. Also, an additional rubber gasholder can be used in parallel if necessary.

Q: (1) How can the gas production be a parameter if it is a function of the other parameters? (2) What is the economic effect of the solar passive system on the total price? (3) In what terms do you evaluate the acceptability of the biogas systems by the farmers?

Hamad: (1) The gas production rate is the major design factor that can be assessed by evaluating a number of other parameters. (2) The impact of the greenhouse is economically promising since it pays back in about two years. In addition, accessibility to durable polyethylene films which can resist solar radiations would highly improve the economics. (3) We evaluate it in terms of the large number of farmers' requests in the demonstration villages, asking for having their own biogas systems and expressing their willingness to share in the expenses.

Q: How much did the gas production increase with the greenhouse, and did this increase correlate with the temperature rise?

Hamad: The gas production increased by about 50%, but its correlation with temperature is still being studied.

Q: What sort of cover material was used for erecting the greenhouse, and how long is its estimated lifetime?

Hamad: We use polyethylene sheets with additives to resist ultraviolent light. The lifetime of the sheets depends on the type of structure carrying it and the method of its fixation. The longest duration attained was about 15 months, and it is expected to be prolonged to two years with some modifications being undertaken. We have tried lastly to use a greenhouse attached to retaining walls, and we hope to obtain good results.

Q: With your modified Chinese model, did you have plugging in the gas outlet pipe?

Hamad: We already have two units operating for more than two years with no problems in this regard. In our new modification, to be published in the near future, the probability of the plugging of the gas outlet pipe is the same as in the conventional design or even less.

PRELIMINARY ASSESSMENT OF A FARM-SCALE DIGESTER IN RELATION TO POSSIBILITIES OF OPERATING AN INTEGRATED BIOPLEX SYSTEM

D. Bellamy, Department of Zoology, University College, Cardiff
P.C. Downing, Chediston Agri Systems, Chediston, Suffolk
B.I. Wheatley, Cardiff University Industry Centre, University College, Cardiff

ABSTRACT

A scaled-up bench-designed digester to treat slurry from a 300-sow piggery (6,000 pigs/year) has been assessed in terms of its performance over two years. Broadly speaking it functioned continuously, reducing the pollution load on the environment by 60% on the average with the other advantages of a reduction in slurry odor, easier solids disposal, and better grassland management through solids separation and irrigation of the effluent. The gas was used to provide electrical power (max 300 kW/d) sufficient to operate the farrowing house.

Preliminary trials were made on separated effluent solids as a beef feed additive and horticultural growing medium. The results indicated a potential for a considerable economic return to the operation by assembling a bioplex that recycled materials as well as energy to the farm. The major financial limitation was the low solids content of the slurry (2-4%). A preliminary assessment of the financial advantages of running a bioplex indicated that the slurry solids had to be at least 7% with biogas being used to replace fuel that costs 70% more and screen solids being sold at £200/metric ton.

INTRODUCTION

Over the past decade there has been a growing interest in the application of anaerobic digestion to treat farm wastes (Hobson et al., 1975). Much laboratory and field research has been carried out, but there have been few successful commercial applications of the new knowledge. The limiting factors are the high capital costs and the lack of management know-how in the integration of all digester products into conventional farming. Also, uncertainties arise because very few digesters have been running long enough to measure their reliability, effectiveness, and economies under real farm conditions.

This lack of detailed information on long-term digester performance in the field has prompted us to write this paper which summarizes a two-year project, from 1980 to 1982, on the integration of a full-scale digester unit into a medium-sized farm in the United Kingdom. The aims of the research were to study problems arising from the 'scale up' of a laboratory-designed digester (Hawkes et al., 1976) to treat the slurry arising from a 300-sow piggery on a 283 hectare, mixed grass/cereal farm in Suffolk. Particular attention was paid to the difficulties of using digester products on the farm. The latter aspect is considered in relation to operating a bioplex system of closed cycle farming (Bellamy and Hughes, 1980). This paper also delineates some financial and ecological aspects of agricultural anaerobic digestion and the managerial skills required for its continuous operation.

METHODS AND RESULTS

The Farm

The digester was sited at Walnut Tree Farm in the parish of Chediston, Suffolk, which is about 40 km southeast of Norwich, close to the country boundary with Norfolk. The farm consisted of a group of scattered fields (flints with boulder clay) producing cereals and grass. Animals were kept intensively under cover within several separate buildings. Pigs were on slatted floors with sluice-controlled drainage channels. During the period of study, the stock comprised 300 sows and their offspring which were grown to 100 kg. The livestock and their management have been described previously (Bellamy and Hughes, 1980). The field system consisted of heavy clay land, tiled and mole-drained, growing wheat, barley, maize, and perennial rye grass. Grass leys were ensiled for a herd of 150 dairy cows, part of the same farming enterprise.

The Digester and Ancillary Equipment

The digester was sited at a low point between the cow cubicles and the pig housing. The total tank volume was 350 m^3 with a working liquid volume of 300 m^3 (9 m diameter and 5.4 m high). Slurry was released from each pig house in turn and collected in a reception pit (capacity, about 200 m^3). The contents of the pit were mixed continuously with air. Slurry entered the pit and was fed into the tank at a daily rate which averaged about one-thirteenth of the volume of the digester contents, the pump operating for 1.5 minutes every hour. The effluent was discharged by displacement during periods when the digester contents were being mixed (10 minutes every half hour) by recirculation of the biogas. Effluent was passed through a vibrating screen (20 mesh) and the screened liquor pumped to a clay-lined lagoon dug at the highest point of the farm (capacity, 700,000 m^3). From the lagoon, the effluent was spread over adjacent grass leys with a reel-irrigator.

The untreated biogas fueled a standard Ford 1.6 liter car engine having a modified carburetor with a 20 kVA generator coupled to the crank shaft. Waste heat from the engine exhaust was used to warm water which was circulated through the digester to maintain its contents at between 34°-35^oC with a methane boiler as a back-up. About 10 m^3 of gas (about one hour of running-time for the engine) was stored in a gasometer fitted with micro-switches to operate the engine. Biogas in excess of requirements for mixing, electricity-generating and heating was vented. Further details on the digester and ancillary equipment may be obtained from Hamworthy Engineering Limited, Fleets Corner, Poole, Dorset who designed and built the unit in cooperation with the Cardiff University Industry Centre and Chediston Agri Systems.

Working of the Digester

The digester was filled with anaerobic sludge from a local sewage works in June 1980. Slurry from the pig unit was then added, starting at one-fortieth of the working volume per day, reaching one-thirteenth of the volume per day at about 160 days. The formation of gas was first recorded at about 40 days. During this start-up period, biogas production increased to a maximum at 60 days (Figure 1), then fell, to rise again reaching a second peak at 160 days. It was maintained at around this level for just over two years, when the digester was shut down for inspection (October 1982). During this time slurry was fed into the tank every day and biogas was generated continuously. The effluent solids varied over a sixfold range. Weekly means smoothed out much of this variability, and the solids content of the effluent eventually settled down at a concentration slightly lower than that in the top sampling port.

Over a six month period, selected at random, the gas output varied considerably from day to day and week to week (Figure 2). These

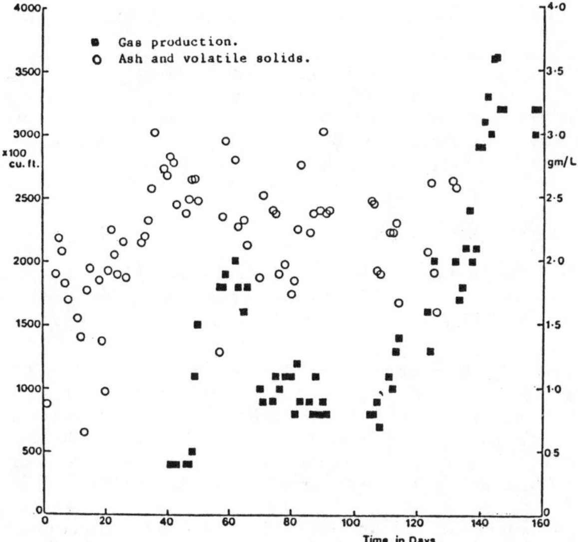

Figure 1 Changes in Gas Output and Solids Content of Effluent During Start-up.

Each Point Represents a Single Effluent Sample (Solids) on a Data-Logged Daily Gas Output.

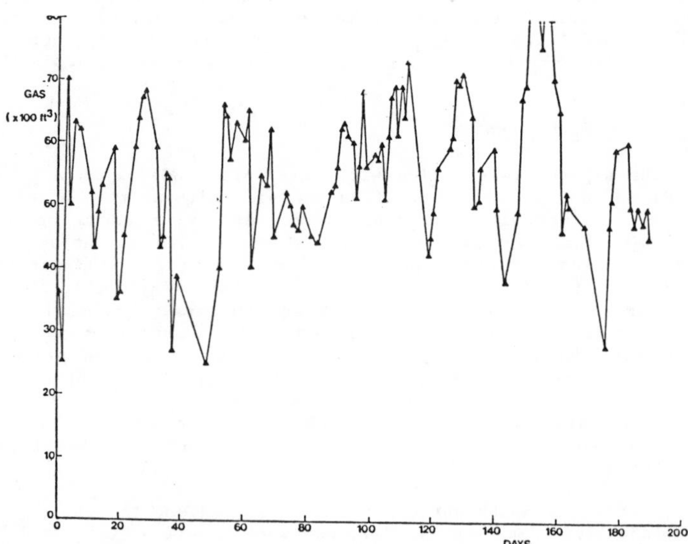

Figure 2 Daily Gas Production Over a Six-Month Period in 1981. Each Point Represents the Daily Gas Output Taken from the Data-Logged.

variations from peaks to troughs covered almost a fourfold range in gas output. A preliminary inspection of the data indicated that gas production was cyclic, with peaks and troughs occurring most frequently either every 7 or 14 days. During this time the methane content of the biogas varied little, averaging about 70% of the daily biogas output.

Chemical Analysis

Standard agricultural/water quality techniques were used throughout the work except for the element analysis in Table 2 which was carried out using electron probe microanalysis.

The tank contents were sampled from three ports positioned from top to bottom. This showed that a concentration gradient of solids became established quickly and was maintained. The concentration at the bottom of the tank (port positioned at the top of the cone) was twice that of the effluent. Solids in the top samples were slightly lower in concentration than those taken from the effluent (Figure 3).

Some average figures on long-running means are given in Table 1. Although on average the concentration of volatile solids in the effluent was between 40%-60% lower than that of the slurry (Figure 3), individual day to day samples varied widely in their solids concentration. Over the period August to November 1980, the greatest variability within the tank was found in samples from the bottom. Slurry from the reception pit also varied greatly, over a sixfold range

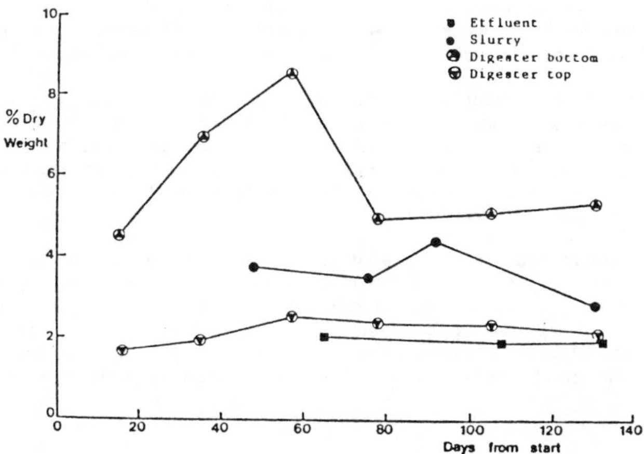

Figure 3 Changes in the Solids Content of the Digester in Relation to Slurry and Effluent During the Start-up period.

Each Point Represents the Mean of at Least Five Consecutive Daily Samples.

Table 1 Mean Values and Variations for All Data Obtained from Analysis of Slurry and Effluent (August 1980 - May 1982)

	Mean Value					Range (Max/Min)	
	Slurry	(n)	Effluent	(n)	$\frac{E}{S}$	Slurry	Effluent
*Total Solids	2.31	(159)	1.29	(147)	0.56	5.2	1.7
*Volatile Solids	1.62	(139)	0.67	(148)	0.40	4.3	2.5
*Volatile Fatty Acids	0.45	(138)	0.04	(130)	0.08	4.3	3.1
*COD	2.18	(149)	1.23	(126)	0.56	5.0	3.2
*BOD	0.79	(30)	0.23	(12)	0.29	2.1	1.4
*Total N	0.29	(35)	0.27	(12)	0.93	3.0	1.3
*Ammonia N	0.15	(135)	0.13	(123)	0.82	2.4	1.4
**Na	0.15	(8)	0.09	(8)	0.60	3.0	2.2
**K_2O	1.16	(8)	1.13	(8)	0.97	2.1	2.2
**Mg	0.25	(8)	0.25	(8)	1.00	5.1	3.1
**Cu	0.03	(8)	0.03	(8)	1.00	6.6	4.2
**P_2O_5	1.75	(8)	1.83	(8)	1.05	4.7	4.3
**Cl	0.60	(8)	0.54	(8)	0.90	3.7	2.9
pH	7.6	(5)	7.8	(5)	1.02	1.2	1.1

* g/100 ml
** g/1000

in solids concentration. Top-tank and effluent samples showed the lowest variability (Figure 4). Similar differences in short-term variability were also observed in most measurements of other substances.

After about 6 months of running, the valve draining the digester from the bottom of the cone was opened daily over a period of 10 days to remove a total of about 1,000 gallons of contents. This was just sufficient to clear the sedimented solids but had no obvious effects on gas output. The ash content of the solids removed in this way was very high (around 70% dry weight).

The proportion of ash to volatile solids varied on a day-to-day basis in all types of samples, although on a long-running means the concentration ratios of ash to volatile solids were more consistent. In general, comparisons between many samples showed that the volatile solids were directly proportional to ash. However, comparing samples of high and low solids for a given difference in volatile solids, the

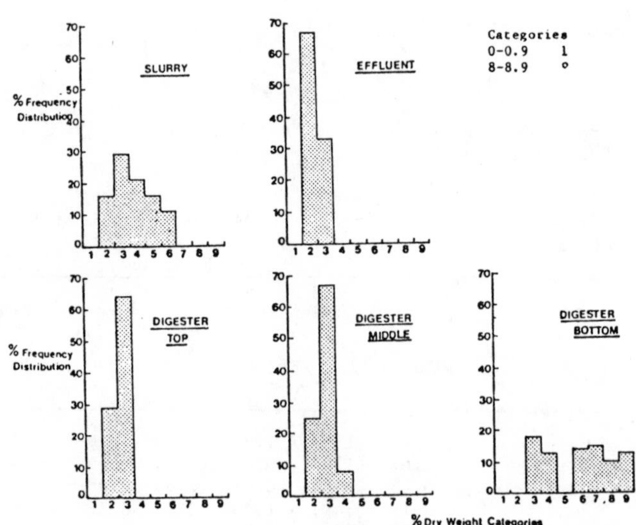

Figure 4 Variation in Total Solids of Samples of Slurry, Effluent, and Digester Contents.

Analytical Results Plotted as Frequency Distribution Histograms. Each Histogram Represents 100 Samples Taken Over two Years.

difference in ash was a smaller percentage. For example, the proportion of total solids as ash was about 40% at 1% volatile solids but dropped to only 30% at 2% solids. This relationship was less marked in samples of tank contents and slurry (Figure 5). This situation would be expected if the main variable in the samples was the volatile solids. In long-running averages the ash content of slurry did not alter these significantly in its passage through the digester. Chemical and electro-probe microanalysis indicated that a large proportion of ash was made up of calcium phosphate crystals.

After just over two years of continuous running, the digester was shut down and emptied, and the top removed for inspection. Most of the contents were sufficiently fluid to be pumped away. There was no crust at the surface. The concentrated material at the bottom of the tank was more firmly packed as a cone close to the mixer, and this was all removed by high pressure water. The heating coils had no surface deposits, and the general appearance of the inside of the tank suggested that there had been no progressive build-up of sedimented

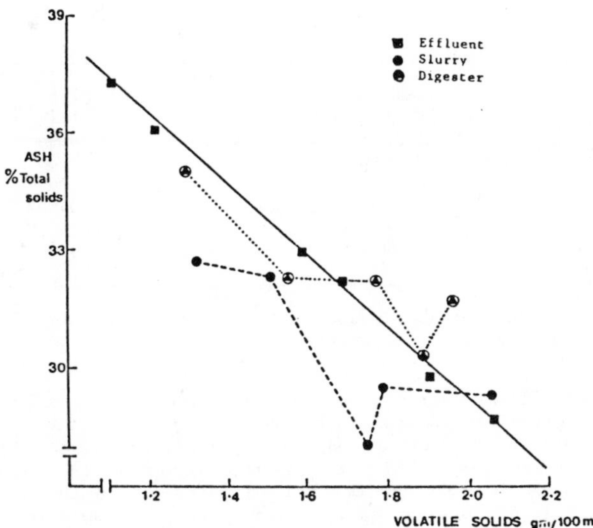

Figure 5 Relationship Between ash and Volatile Solids in Samples of Effluent Slurry and Digester Contents.

The Percentage of ash in each Sample is Plotted Against its Concentration of Volatile Solids.

solids over the two years of operation. During this time, solids amounting to about 200-300 times its volume (on a dry weight basis) passed through the digester.

Solids Separation

Passing the digester effluent through a vibrating screen removed about 25% of the solids, which consisted of a mixture of coarse vegetable fibers, pig hairs, micro-organisms, and mineral particles. This fraction contained about 20% dry matter with 10% crude protein, 15% fibre and 25% ash on a dry weight basis (Table 2). This fraction was high in calcium, phosphorus, and copper, but heavy metals were low in relation to the acceptable levels in livestock feed (Table 2). Comparative laboratory tests using stacked vibrating sieves to screen input slurry and digester effluent showed that digestion decreased the proportion of solids that could be separated through a vibrating screen by about 30%.

Table 2 Metal Content of Screen Solids

	Concentration* (percentage dry weight)
Calcium	2.4
Phosphorous	1.1
Copper	0.04
Manganese	0.03
Zinc	0.04
Arsenic	0.002
Selenium**	0
Cobalt**	0

* Mean of three samples; measured by electron probe microanalysis.
** Less than 0.5 mg/kg dry weight.

Bench centrifuges operating at the gravitational forces produced by large-scale agricultural machines removed 50% of the particulate matter of the screen liquor. These sedimented solids contained 26% of the effluent nitrogen, most of it in the form of the protein of micro-organisms. In contrast most of the nitrogen in the supernatant fraction was largely present as ammonium salts. The soluble matter in the supernatant fraction consisted mainly of the chlorides and phosphates of sodium, potassium, and calcium. By differential centrifugation of the solids sedimented in the first operation, a black syrupy "light particle" fraction was isolated, which contained a larger proportion of the screen liquor at 34% crude protein on a dry weight basis (Table 3).

Effluent Handling

No difficulties were encountered in pumping the screen liquor, but

Table 3 Composition of Solids Fractions Separated from Digester Effluent

	Fraction	
	Screen Solids[1]	Centrifuged Solids[2]
Dry Matter*	17.3	14.5
Ash**	11.4	27.0
Crude Fiber**	29.7	3.1
Lipid**	1.4	6.0
Crude Protein**	10.1	34.1

* Percentage fresh weight.
** Percentage dry weight.
[1] Mean of five samples taken over seven days.
[2] Mean of five samples of screen liquor taken over seven days.

over a period of several months a crystalline sediment, consisting mainly of phosphate salts, accumulated on the walls of the pipe taking the liquor to the lagoon. In the lagoon, the average residence time was about 90 days. The coarser particles settled to the bottom and biogas was generated from this sludge, particularly during the summer months when the surface of the lagoon was continuously disturbed by gas bubbles. The most obvious difference between the composition of the lagoon contents and that of the incoming liquor was its lower ammonia content. Eventually it is intended that the lagoon should only be used to store the effluent during the winter. From March to October the stored contents and the incoming liquor will be irrigated onto the grass leys.

The management of screen liquor by pumping had many advantages over the previous methods of traditional muck-spreading. There were no problems of churning up or compacting soil by tractors, and there was a more uniform grass cover because there were no large lumps of dense manure blocking growth of the vegetation. Also, because the screen liquor did not 'scorch' the grass, and cows were not averse to grazing almost immediately after application, the screen liquor resulted in a higher grassland productivity, which unfortunately could not be easily quantified, but was a clear advantage in the eyes of the farmer.

Solids Handling

As a Horticultural Growing Medium. The solids from the vibrating screen were used in several small-scale trials (at a 75% water content) as a peat replacement in soil-less composts. The material produced the same germination and growth as obtained with commercial peat-based products on crops such as lettuce, tomato, and cucumber. It was found that 'bag-stability' could be improved by the addition of equal volumes of dry inert wastes, such as plastic, leached fly-ash, and certain other selected low-cost materials.

As an Animal Feed Additive. After air drying, the screen solids were incorporated into the diet of beef cattle at up to 10% of total solids. Forty animals were used for a three-month trial to test its mixing properties, palatability, and possible toxicity. The starting weights of the animals, which were maintained under standard farm conditions in cubicles at Walnut Tree Farm, ranged from 200-300 kg. The animals consisted of a commercial mixture of the farm's own stock (Fresians) from its dairy unit and animals brought in from the local market (a mixture of Fresians, Fresian Hereford, and Herefords). They were fed on a diet consisting of ground barley with concentrates (10% of dry weight consisting of 3% oil, 34% protein, and 7.5% fiber with vitamins, minerals, and 'rumensin'). The barley/concentrate mixture was replaced by low levels of screen solids at 5% and 10% of dry weight respectively (Table 4).

Animals were started on a daily intake of 5 lbs. dry weight of the barley mixture, fed twice a day, and allowed free access to barley straw. This amount of food for the controls was calculated to give a growth rate slightly below maximum so that the animals would provide a sensitive indicator of any inadequacies of the experimental diets.

From the energy concentration of the food (3.0 M cal/kg DM), it was calculated that an allowance of 4.25 kg DM/day/animal would give a growth rate of about 75% of the maximum. The actual growth rates of 0.7-0.8 kg/day were as expected from this level of food intake (Figure 6).

Table 4 Composition of the Diets

	Percentage	
	Barley Plus Concentrates	Barley Plus Concentrates (10% VSS*)
Water (fresh wt.)	13.3	15.2
Protein (dry wt.)	17.9	17.1
Fat (dry wt.)	3.0	2.8
Fiber (dry wt.)	7.1	8.9
Ash (dry wt.)	10.8	12.3

* VSS - Vibrated screen solids; 10 percent ash, 25 percent fiber, and 12 percent D.C.P. on dry wt. at 35 percent moisture.

PRELIMINARY COSTINGS

Taking the mean figures for inputs and outputs over a six-month period, it is possible to draw up an economic analysis for Chediston, which is outlined in Table 5, and compare it with similar theoretical systems operating at higher slurry solids. The costs of the digester system in terms of capital have been written off over a 10-year period in the conventional manner. Returns from products have been calculated as if the gas were used to replace North Sea gas for heating water; as if the screen solids were worth £100 per dry metric ton; and as if the fertilizer value of liquor gave a 10% greater yield of arable crops when irrigated compared with raw slurry (Stewart et al. 1981). It has been assumed that the slurry management system before the digester was installed was based on a vibrating screen and storage lagoon.

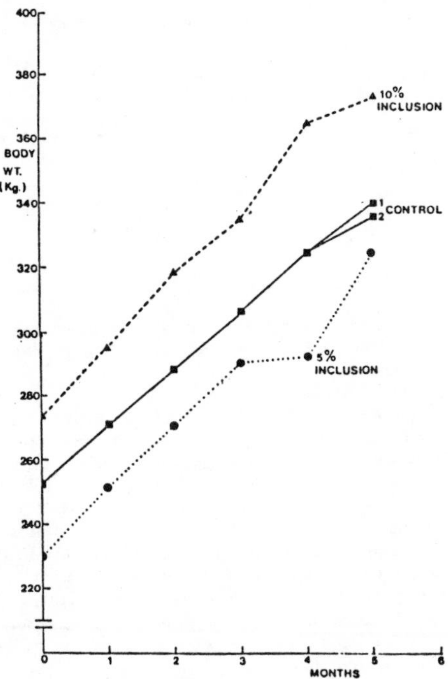

Figure 6 Growth of Barley-Beef on Diets Containing Digester Screen-Solids.

Forty Animals, Fresians and Fresian Hereford Crosses, were Tested as Described in the Text and fed at two levels of Screen-Solids as a Dietary Replacement of a Barley-Based Diet.

Compared with this conventional system of handling raw slurry, the incorporation of a digester does not require additional skills in management that cannot be quickly learned. The calculations have included the generous addition of a "half-man," and in practice this new job would be connected with, in the main, keeping the pumps running, clearing blockages in the effluent pipes and vibrating screen, and removing the screen solids. Also, it is necessary to check daily the workings of the digester and ancillary equipment from the control panel and data logging systems. The unit was fitted with alarms, which gave audible warnings of the escape of gas and failure of feed to the digester from the reception pit. In practice, the shift from a vibrating screen-pumped liquor handling system would probably not

Table 5 Annual Solids Turnover of Chediston Digester (dry metric tons/year)

	All Solids		Screen Solids		Liquor Solids	
	Total	Volatile	Total	Volatile	Total	Volatile
Slurry	210	149	49	39	161	110
Digester Effluent	117	84	29	22	88	62

require an increase in manpower, rather a re-organization of existing jobs, and this situation would introduce a very favorable economic advantage to the investment.

The Chediston system, operating with a slurry input that most frequently was around 3% solids, which is typical for British farms, using lots of washing water, has been compared with three other systems of increasing slurry concentration. The highest slurry concentration in the comparison, of 9%, is an unattainable ideal because it is very close to the concentration of undiluted feces. It can be seen that all systems put an additional cost on pig production (Table 6). On a 'pig-produced' basis, the Chediston digester operating at about 2% slurry is about as expensive as an aerobic treatment plant, taking gas as the only return. The costs of the process are sensitive to slurry solids and a practical goal is a concentration of slurry in the reception pit between 5% and 7%.

Energy returns from biogas have been calculated on the costs of town gas for heating. At Chediston, because there was a continuous demand for heating and lighting the farrowing unit, it was decided to examine the feasibility of installing a gas engine and generator to provide electrical energy from biogas.

The cost of the engine assembly amounted to a capital-plus-interest outlay of about £1,700 per annum (expected engine life about 5 years) with maintenance costs of about £1,100 per annum. The actual running time of the engine towards the end of the experiment was equivalent to 5,000 hours/year, producing electrical power worth £3,600. The generation potential from biogas of about 0.1 hr/m^3, related to the actual biogas production of the digester, gives a maximum generation time of 7,300 hours (300 kW/day), equivalent to a financial return of £5,256. Heat from the engine exhaust was more than sufficient to keep the digester at its working temperature so that all of the biogas was potentially available for electricity generation. From these figures, a crude evaluation of the economics of converting biogas to electrical energy yields a net return of £2,456/annum which is very close to the net return expected from burning biogas as a source of direct or indirect heat (Table 6).

Table 6 Economic Analysis of Anaerobic Systems at Different Solids Concentration in Relation to Total Recovery of Materials and Energy from a 300-Sow Unit

A Basis	Slurry Concentration 2.3%	5.0%	7.0%	9.0%
Slurry volume (m^3/day)	25	11.5	8.2	6.4
Digester tank volume (m^3)	325	149	107	83
Gas volume produced				
Gross (m^3/day)	201	201	201	201
Net (m^3/day)+	104	154	167	174
B Income (£/annum)				
from Gas	2255	3344	3615	3768
from Screen Solids	900	900	900	900
from Screen Liquor*	482	482	482	482
Total	5639	6728	6999	7155
C Expenditure (£)				
Capital	53704	32168	25762	21825
Operating				
Maintenance (£/annum)	1074	643	515	500
Insurance (£/annum)	537	32	258	218
Labor (£/annum)	4000	4000	4000	4000
Repayment (£/annum)	5370	3217	2576	2183
Interest (£/annum)	3222	1930	1546	1307
Total	14203	10112	8895	8211
D Net costs of treatment				
(i) (£/annum)	8568	3386	1898	968
(ii) (£/annum/pig produced)				
Gas alone	1.99	1.13	0.88	0.73
All products	1.45	0.56	0.32	0.16

Table 6 Assumptions
Size of unit, 1150 places; plant operates 360 days/year; retention time, 13 days; average air temperature, 10°C; average feed temperature, 15°C; value of biogas, £0.06/m^3 used for water or space heating; value screen solids, £100/dry metric ton; fertilizer value of slurry increased 10 percent by digestion.

+ Available for use after heating digester contents to 35°C.

* As grassland irrigation fertilizer.

DISCUSSION

The advantage of anaerobic digestion at Chediston have to be assessed against a slurry management system without digestion in which coarse solids are removed by screening, the liquid fraction being handled entirely by pumping. In this system only about 25% of the slurry volume has to be transferred to land by traditional spreading techniques. This procedure is gradually being accepted as the most satisfactory way of coping with the problems of slurry handling, particularly on heavy land, in that it reduces the volume of waste to be hauled by about 75%.

The important ecological and public health advantage of anaerobic digestion is that it reduces pollution of the environment by the organic fraction of sewage (Mosey, 1974). At Chediston this goal was achieved in that the substances causing most of its acute biological affects, when slurry drains into natural water systems, were reduced by about 60%.

Chronic pollution of the farm environment also arises because of difficulties in disposing of the more solid fraction of manure. Digestion removed about 40% of these solids, a large proportion of them being in the coarse-fibre fraction of slurry. Additionally, some of the coarse-fibre was converted into finer material that cold be pumped to the storage lagoon.

Laboratory data showed that anaerobic treatment reduced the quantity of solids to be disposed of after screening by at least 30% (Table 5). These solids contained 54% of total solids and 70% of the chemical oxygen demand, but only 6% of the biological oxygen demand (Baines, 1977).

The total volume of liquid to be disposed of was not affected by digestion, but its odor was very much reduced (Welsh et al, 1976). Apart from rendering the farm environment more acceptable to people, the deodorizing action of the digester allowed cattle to graze grass leys very soon after application of effluent. This was an important management advantage at Chediston, and would probably be a general advantage with economic gains on most dairy farms because liquid waste could be disposed of more evenly in time and space. The heavy land at Chediston was no longer overloaded with nutrients and there was a considerable reduction in organic and nutrient run-off. This has also been found to be an advantage with aerobic treatment of pig manure (Evans et al. 1975).

Three products of digestion were utilized in research and development operations on the farm. Gas was fed into a slightly modified Ford car engine to test the effects of using untreated biogas as a fuel to generate electricity. There were no special problems arising from the use of unscrubbed gas, apart from the frequent servicing and maintenance associated with a generator unit which was potentially available for continuous running.

The effluent was passed through a vibrating screen to separate the coarse solids from a liquor which could be pumped to the storage lagoon. The mesh size was determined by the precipitation problem of phosphate salts blocking the mesh as the effluent cooled. This meant that the screen had to be cleared with a wire brush each day. Agents which keep the effluent phosphates in solution are now being tested as post-digestion additives. Crystallization of phosphate also took place in the screen liquor on its way to the lagoon and, eventually, some blocked sections of the transfer pipe had to be replaced. This problem of crystallization appears to be a general one for digesters handling pig manure.

Screen solids fed to beef cattle did not cause any problems in terms of their palatability or adverse effects on the health of the animals or stockmen. Up to 10% inclusion did not affect the growth rates of any of the groups.

The main difficulties in managing the new feeding system were connected with drying and mixing the screen solids with the other dry ingredients. If the screen solids were too wet they tended to ball in the mixer. Also the relatively large area required for air-drying severely limited the quantity that could be handled in one batch. On the question of possible adverse effects occurring through pathogen transfer, samples of the screen solids and the raw material entering the digester were checked by the Ministry of Agriculture Veterinary Division under the European Economic Commission (EEC) ordinance governing recycling operations on the farm. Their main concern was to prevent the spread of *Salmonella* organisms. Although *Salmonella* was found in the slurry, it was not detected in the screen solids. This is in line with small-scale experiments (Stafford, unpublished work) which showed that anaerobic digestion kills *Salmonella* in manure. Of the metals in slurry which could be harmful if fed, copper (added to the pig food as a growth promoter) was present in the screen solids but below its toxicity limits at 10% inclusion. Other metals were present at concentrations above those for common animal feedstuffs but would not be expected to cause problems in the final diet.

In conclusion, although this short trial has opened up the possibility of coupling a beef enterprise to a digester treating pig manure, there are still some important questions that have to be answered.

What is the maximum proportion of solids that can be incorporated?

How long can animals be kept on the diet?

What is the actual nutritional value of the diet for beef cattle?

From the present work, the solids would seem to have a potential monetary value, at least that of barley, but a very large number of cattle would be needed to utilize all of the screen solids produced on the farm. For example, the Chediston digester working on the manure from a 300-sow unit would require 400 head of cattle to cope with the daily output of screen solids at 10% incorporation into their diet.

The screen solids have a better commercial potential as a peat-replacement in horticulture. No drying is necessary, the bulk may be increased by adding other inert wastes and sales of composts are by volume. The minimum bulk sales price of this type of material is around £20/m^3 which would produce an income of around £8,000/annum:

$400m^3$/year wet, 50 metric tons dry. A serious limitation in realizing this potential is the availability of year-round outlets on or close to the farm.

Bearing in mind that the whole operation at Chediston was the first application of a scaled-up pilot plant to treat farm waste, it is impossible to make a proper economic analysis of the performance of the digester from the beginning to the end of the experiment because, to cope with unexpected problems, variations in design and management were introduced from time to time. However, sufficient experience has been gained to evaluate the requirements for a successful new venture based upon the Chediston experience.

The first and foremost requirement is to manage the pig enterprise to minimize fluctuations in a slurry output from the animal units and maximize its solids content. The former will introduce an important element of predictability to digester output and the latter will minimize the cost of the digester, thus maximizing the financial return on the capital investment.

Despite the livestock units at Chediston being organized to maintain a uniform output of pigs, there were unpredictable fluctuations in the numbers and ages of growing stock. These variations, through affecting input of slurry, would have produced some of the variations seen in digester output. However, the large variations in slurry composition measured were probably more apparent than real, in that it was impossible to obtain a small-scale representative sample for chemical analysis. This was due to problems of rapid sedimentation and clumping of the raw slurry in the reception pit. There were also differences in the residence times of slurry in the various animal units where it appeared that volatile solids were reduced by pre-digestion in the warm anaerobic collecting channels. Variations in gas and effluent solids were probably also produced by intrinsic cycles in the bacterial flora within the digester, perhaps set in motion by periodicities in output.

To turn the charge of 89p per pig (produced at a 7% slurry input) into a profit would require using the gas to replace a fuel costing 70% more than North Sea gas or obtaining £200/metric ton for the screen solids. Integration of anaerobic digestion with other agricultural systems having these returns is possible, but unlikely, on conventional pig farms. The system is also very sensitive to the cost of additional labor.

Taking all of this together, future applications of anaerobic digestion to slurry treatment should involve devising a total bioplex system in order to minimize dilution of manure and link the digested effluent and gas output to other enterprises with continuous demands for high cost energy and materials input sited next to the pig unit. As much as possible of the solids in effluent should be separated by screening and the management system should be arranged to keep labor costs as low as possible. This highlights the criteria for creating a closed cycle operation which is efficient and financially sound. On some farms there would be difficulties in integrating digestion with management due to the past history of the site, i.e. the placing of its buildings and management methods which could produce inflexibility that could only be circumvented by starting a new enterprise on a 'green-field' site. Such difficulties did not exist at Chediston.

Anaerobic treatment of farm waste offers advantages over aerobic treatment. Both methods reduce polluting solids and odors of manure, but the anaerobic method is more effective in converting solids to gas. With regard to the treatment of human wastes, the aerobic method was developed to treat wastewater at low solids levels of around 2% or less, whereas the anaerobic method was devised to treat more concentrated waste streams. This difference is important in relation to applications of digestion to animal manure. It is partly a matter of the difficulty and expense in providing efficient aeration in thick suspensions and also connected with the cost of constructing large anaerobic tanks to contain big volumes of dilute wastewater. A particular disadvantage of the aerobic system is that it creates a large amount of bacterial sludge (Owens et al., 1973) which, although it sediments readily, is unstable and difficult to dispose of without creating new odor problems. In municipal sewage works, this extra sludge from aerobic treatment tanks is stabilized by anaerobic digestion as a necessary second stage.

In contrast, the anaerobic system produces a bacterial flora that does not sediment quickly and does not readily decompose to give malodorous products, with the advantage on the farm that it can be spread by pumping as a fine suspension. Ammonia nitrogen is an important fertilizer component of the effluent from anaerobic digestion, which is at about the same concentration as in the slurry feed to the digester.

Aerobic treatment may actually reduce the availability of effluent nitrogen to crops (Owens et al., 1973; Fenlon and Robinson, 1977), which is a definite financial disadvantage if the output of the unit is to be spread onto grass leys and pre-treatment of slurry is required by law. On the other hand, long-term storage of any effluent causes loss of its nitrogen.

ACKNOWLEDGMENTS

Thanks are due to the following students who helped in the field work: M. Bellamy, S. Etheridge, A. Hemp, D. Hughes, J. Joel, and C. Reynell. We would also like to thank the staff of Hamworthy Engineering Ltd., Poole, Dorset, for their ready help, advice, and cooperation at all stages of the project, and also Dr. D. Stafford, Department of Microbiology, University College, Cardiff, for allowing us to use some of his research data.

REFERENCES

Baines, S. 1977. Aerobic treatment in relation to land application of slurry. Committee of ECC, Utilization of manure by land spreading E U R 5672.

Bellamy, D. and D.E. Hughes 1979. Systems farming--the Bioplex principle. In D.A. Stafford, B.I. Wheatley, and D.E. Hughes, ed., Anaerobic Digestion, pp. 505-514.

Evans, M.R., R. Hissett, D.F. Ellam, and S. Baines. 1975. Aerobic treatment of piggery waste prior to land treatment: a case study. In Proceedings of the Third International Symposium of Livestock Wastes, 1975 at American Society of Agricultural Engineers, St. Joseph, Michigan, pp. 556-559.

Fenlon, D.R. and K. Robinson. 1977. Denitrification of stabilized pig waste. Water Res. 2:269-273.

Hawkes, D., R. Horton and D.A. Stafford. 1976. The application of anaerobic digestion to producing methane gas and fertilizer from farm wastes. Process Biochem. 11:32-36.

Hobson, P.N., A.M. Robertson and P.J. Mills. 1975. Anaerobic digestion of agricultural wastes. Agricultural Research Council Res. Rev. 1:82-85, A.R.C. London.

Mosey, F.E. 1974. Anaerobic biological treatment. Symposium on the Treatment of Wastes from the Food and Drink Industry, Newcastle Institute of Water Pollution Control.

Owens, J.D., M.T. Evans, F.E. Thacker, R. Hissett, and S. Baines. 1973. Aerobic treatment of piggery waste. Water Research 7:1745-1766.

Stewart, D.G., D.M. Badger, and M.J. Bogue. 1981. Crops and energy production. In D.E. Hughes, D.A Stafford, B.I. Wheatley, W. Baader, G. Lettings, E.J. Nyns, W. Verstraete, and R.L. Wentworth, eds., Anaerobic Digestion 1981, pp. 237-254.

Welsh, F.W., E.J.K. Schule, and H.M. Lapp. 1976. The effect of anaerobic digestion upon swine manure odors. Paper No. 76-206 at the Annual Meeting of the Canadian Society of Agricultural Engineers, Halifax, Nova Scotia, 16 pages.

AN IMPROVED PLUG-FLOW DESIGN FOR THE ANAEROBIC DIGESTION OF DAIRY CATTLE WASTE

A. Tilche, F.De Poli, L. Ercoli, and O. Tesini
Italian Commission for Nuclear and Alternative Energies (ENEA)
and
L. Cortellini, and S. Piccinini.
Research Center for Animal Production (CRPA), Italy

ABSTRACT

Of the more than 100 agricultured digesters built in Italy, starting in 1979, only a few were developed for the digestion of cow dung slurry. A great number of existing digesters treat pig wastes, and they are used for waste water treatment.

ENEA has developed a plug-flow digester suitable for dairy farms. This technology could be diffused in a very wide range of cases, not only in Italy, but wherever there is a sufficient amount of dairy cattle and a need for energy. ENEA is also developing a small size, biogas-operating congenerator to produce 5kW of electric energy and a considerable amount of hot water, suitable for digester heating.

The ENEA plug-flow digester was built in the Pellerano farm of Nonantola and began to work on June 1983. The plant was completely monitored by a data acquisition system based on a commercial personal computer, specifically studied for the monitoring of full-scale experimental anaerobic digesters. In this plant, in addition to the performance, we studied the natural movements of the unmixed slurry inside the digester, the heat exchanged between the internal exchanger and the still slurry and the heat losses from the digester concrete walls towards the soil.

The results, after nine months of work, are interesting: there is no crust formation, and there are no problems with the slurry flow. The production of biogas, with a HRT of 20 days, was about 1.7 m^3 of gas (56% of methane) per cubic meter of digester working volume.

It is now possible to prepare a mathematical model to calculate the parameters (geometrical and thermical) for plug-flow plants of different sizes, operating under different temperature conditions.

INTRODUCTION

The diffusion of biogas technology in agriculture is hampered by several factors. Most important among these are: high costs; poor reliability of the hardware; difficulties in managing the plant, and in usage of the biogas. In 1983, in order to overcome some of these bottlenecks, an experimental full-scale plug-flow plant for the digestion of dairy cattle manure was realized. The main objectives of the project were to minimize construction costs and to simplify the management of the plant. In order to reach these objectives, low cost materials and simple construction techniques were used, and all types of internal mixing have been excluded.

The plug-flow design has been chosen as the lowest cost solution among the high rate biogas plants, as demonstrated by Cornell University (Jewell et al., 1978), and by an experience parallel to ours of Landal Institute in California (Howard and De la Fuente, 1983). We have been exchanging projects, ideas and information with Landal since 1980. The plant has been financed by the Region Emilia-Romagna - Agriculture and Food Department within the regional research and demonstration program on integrative energies in animal farming (Bonazzi, et al., 1984). The plant has been designed by ENEA and built at the Pellerano farm in Nonantola by Lusetti Studi S.p.a. The experimental campaign has been carried out by ENEA and the Research Center on Animal Production of Reggio Emilia. This collaboration between the Region Emilia-Romagna and ENEA is a part of broader field of common work on energy.

METHODOLOGY

Description of the Plant

The plant treats the waste of 120 free-stabled Holstein Friesian lactating cows. Two scrapers remove daily the waste from inside the barn, which represents about 50% of all the waste produced by the cows. The waste contains a small amount of straw coming from the bedding of the cubicles. The plant is of the plug-flow type. This kind of reactor, fed with a waste of high solid content, does not need any internal mixing. A very gentle mixing is caused by the natural formation of gas bubbles and by the convective movements due to the internal heating at the bottom of the reactor. The plant is composed of: 1) the mixing and feeding tank; 2) the digester; 3) the biogas piping and storage; and 4) the heating station. Figure 1 is a diagram of the plant.

Mixing and Feeding Tank. The mixing feeding tank is a 4 m^3 concrete tank where the waste is collected by the cleaning system. A 11 kW electric Cavalmoretti chopping pump homogenized the large solid particles of the waste, cutting it into pieces 2-3 cm long before pumping the waste into the digester. The choice of the appropriate pump is very important. In the first few months of operation, this reactor was equipped with an inadequate pump that did not allow feeding the plant at a constant loading rate.

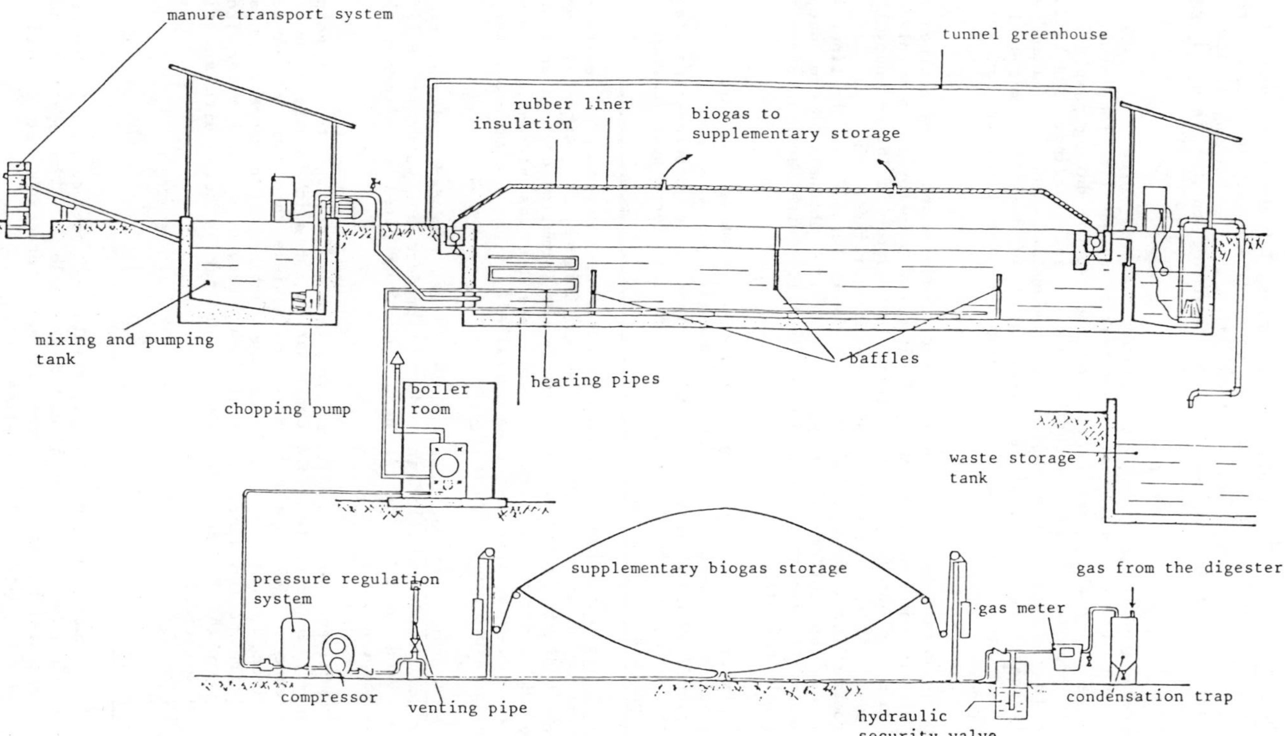

Figure 1 Schematic of the plant: longitudinal section of the reactor

Anaerobic Digester. The reactor is a rectangular shaped concrete tank 13 m long, 3 m wide and 1.75 m deep, covered by a butylic rubber liner that has a holding capacity of about 30 m^3. The working volume of the plant is between 60 and 65 m^3, depending on the internal gas pressure.

Three internal fiberglass transversal baffles force the waste to pass over or under them. Their main task is to equalize the thrust of the waste along the cross section of the reactor during pumping, in order to provide the same retention time for each loading. The baffle closer to the waste inlet bounds a chamber in which the manure is rapidly heated up to the process temperature of 36-37°C by means of a 100 m long iron, coilshaped 1 1/2" diameter pipe in which hot water circulates.

The process temperature in the rest of the plant is maintained by means of an 85 m long low-density 2" diameter polyethylene pipe placed 10 cm from the bottom of the reactor. The reactor is 160 cm under the soil level. It is neither insulated nor underdrained. An 8 cm rockwool insulation is placed over the rubber liner. The liner is hooked on the bottom of a narrow concrete channel 50 cm deep surrounding the reactor. This channel is filled with water in order to impede gas leaking.

A plastic tunnel greenhouse covers the whole reactor. In a commercial realization, this can be substituted by a simple plastic sheet placed over the insulation of the liner, which protects the plant from the rain. At the end of the reactor, the slurry overflows into a discharge tank.

Biogas Piping and Storage. Biogas, before being metered, is dewatered by means of a simple expansion chamber (Figure 2). After the gas meter, a hydraulic safety valve prevents the formation of high pressure inside the reactor. In order to increase the gas storage, a 40 m^3 capacity rubber balloon has been installed. A small compressor keeps the gas pressure around 200 mm H_2O.

Digester Heating Station. Part of the biogas produced is burned by a 25,000 kcal/h power boiler in order to produce hot water at 50-52°C that is used in the internal heat exchanging pipes to maintain the process temperature at 36-37°C.

It is always necessary to maintain the boiler over the dew-point temperature, in order to avoid condensation and acid formation due to the presence of H_2S in the biogas. To achieve this aim, a four-way valve driven by a thermostat keeps the heating water temperature at 50°C and the boiler temperature at 80°C, preventing cold water from returning to the boiler. The exhaust draught is built with an acid resistant material.

CAPITAL COSTS

Capital costs of this plant, excluding non essential devices necessary for research, are reported in Table 1. The costs do not include engineering and supplementary gas storage or gas electrical energy and water linking to the mains.

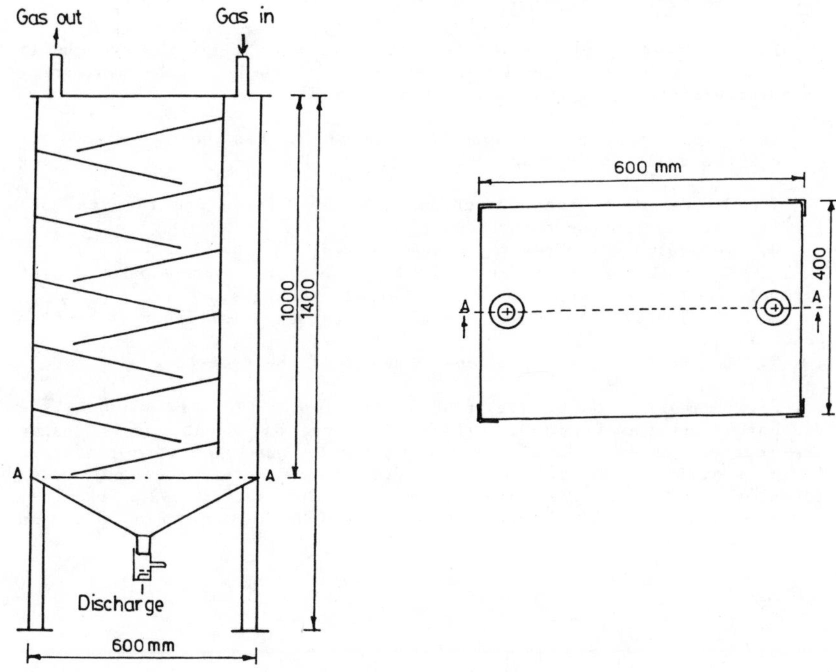

Figure 2 Schematic of the condensation trap.

Table 1. Capital Costs of the Pellerano Plug-Flow Plant

Cost component	U.S.$
Civil works*	9,600
Machinery	7,600
Labor	2,000
Total	19,200
Capital cost/m^3 vol.	295

* Labor costs included:

Note: Data to summer 1984. 1U.S.$ = 1,750 Italian lires.

TESTING AND SAMPLING PROGRAM

The plant was equipped with an automatic data logging system built by ENEA's electronic laboratory, specially designed for the monitoring of anaerobic digesters (Chiumenti et al., 1983).

This application, described in Figure 3, has been designed to execute the following operations:

1. To record biogas production, its composition and the gas used for maintaining process temperature;
2. To evaluate the reactor's heat losses;
3. To evaluate the efficiency of the heating system and the heat exchange coefficient applicable to internal exchanges;
4. To obtain information on temperature distribution in the reactor;
5. To record the electric energy needs of the digester.

Five sampling ports are placed to sample the raw manure, the discharge and the fermenting liquor at three different points inside the reactor (head, 1/3, 2/3 of the reactor). Analyses were performed twice a week. The following parameters were determined: pH, redox potential, total solids (TS), total volatile solids (VS), chemical oxygen demand (COD), total nitrogen (N-NTK)), ammonia nitrogen

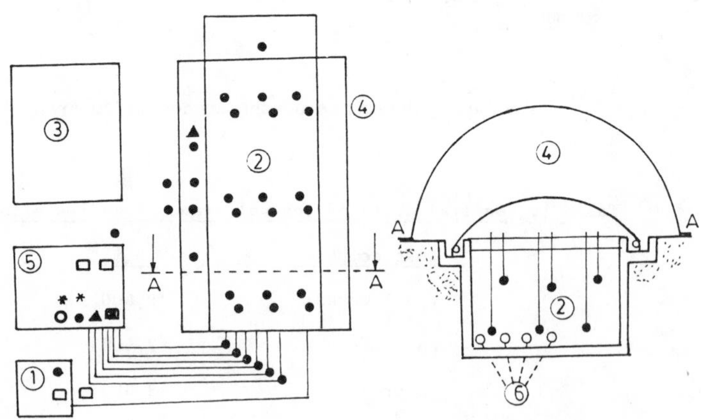

1 mixing tank
2 digester
3 biogas storage
4 tunnel greenhouse
5 heating station
6 heating coils

● temperature probes
■ watt-meters
▲ gas-metters
* hot water-meters
○ biogas analyzer
□ on/off signals

Figure 3 Schematic of the data logging system,

($N\text{-}NH_4^+$), all in accordance with the Standard Methods (APHA, 1975); volatile acids (VA) were determined using a modified steam distillation and titration method and total alkalinity by titration at pH 3.8. Gas composition (CH_4 and CO_2) was continuously monitored using a Leybold Heraeus Binos 1 IR analyzer; H_2S percentage was determined with Drager Tubes.

Over the rubber liner, three bull's eyes were installed in order to observe the surface condition of the fermenting material.

The plant operation commenced on 15 June 1983. After a period of technological tuning-up, the experimental campaign started at the beginning of October. The plant was fed once a day (except on Sunday).

On 1 June 1984, the plant was opened and emptied for a programmed inspection of the rector's internal conditions after one year of work. Fifteen days before this date, the feeding was stopped in order to follow the trend of biogas production and the volatile solids destruction in batch situation with a very long retention time. After a few minor changes, the plant at the end of August 1984 was started up again.

The biodegradable fraction of the total volatile solids was determined using the method proposed by Jewell et al., (1978).

RESULTS AND DISCUSSION

The data collected were used to design a computer model that is able to simulate--for plants of various dimensions--the thermal behavior of a plant under different loading rates. This is done to estimate biogas consumption for maintenance, to calculate the surface of the internal exchanges and the temperature of the heating water in order to keep the temperature process stable. We are already using this model for designing plug-flow plants.

Very important data is that regarding the heat exchange capacities of pipes immersed in still manure, that range from 23 to $46 W/m^2 {}^oC$ for this experience. The iron coil showed a lower heat exchange capacity compared to the polyethylene one. This means that in such condition the exchange material is relatively less important, because the heat resistance of the manure is very high. The relatively lower coefficient found for the iron pipe can be explained by the fact that the iron coil is in the first part of the reactor where the density of the manure is higher and the bubbling of biogas is lower. The overall efficiency of the heating system is between 80 and 85%. The biogas consumption for temperature maintenance goes from 25 to 65% of the biogas produced, depending on the loading rate and the temperature of the environment (Tilche et al., 1984).

As far as the need for electrical energy, it is very low, only about 18 kWh/d; 50% is used for the chopping and for the discharge pumps. The graph of Figure 4 presents biogas production, loading rate and average digestion temperature during the period from October 1983 to June 1984. The figure shows the relationship between loading rates and biogas production and between digestion temperature and biogas

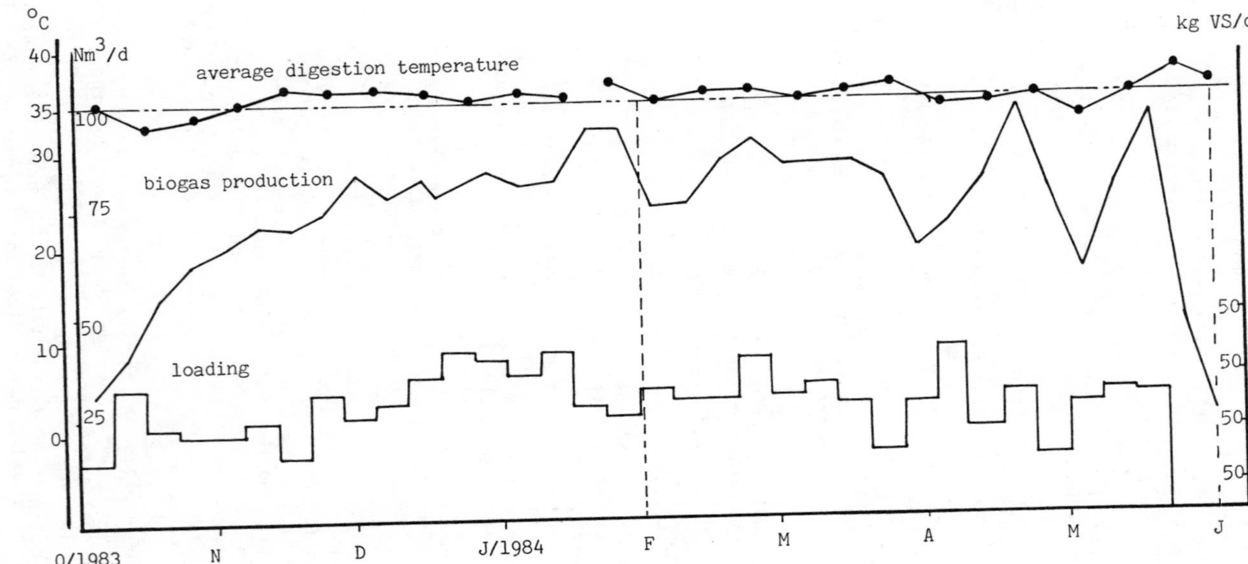

Figure 4 Diagram of the average digestion temperature (°C), the gross biogas production (Nm^3/d) and the loading (kg VS/d) from 1st of October 1983, to the last of June 1984.

production. Experience shows that a process temperature lower than 35°C determines a much lower specific biogas yield.

The feedstock manure has an average total solids (TS) content of 131.3 g/l (107.0 - 143.4 g/l), and an average total volatile solids (VS) content of 107.0 g/l(85.6 - 116/1 g/l) corresponding to about 80% of the TS. Figure 5 shows the decrease in concentration of TS, VS and VA related to HRT. The graph shows clearly the plug-flow behavior of the plant and points out that hydraulic retention times below 20 days are able to exploit more than 90% of the feedstock biogas potential.

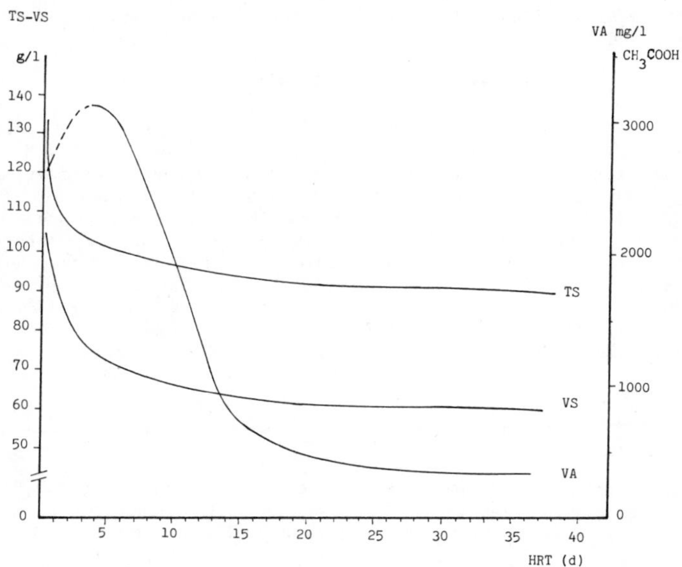

Figure 5 Trend of the concentration of total solids (TS), total volatile solids (VS), and volatile acids (VA) in relation to hydraulic retention time (HRT).

The average gas production rate over the whole trial period, from October 1983 to June 1984, was 0.365 Nm^3/kg VS added. Over the last 115 days, after the stabilization of high gas production performance, the average rate was 0.41 Nm^3/kg VS added. The average total volatile solids destruction efficiency was about 40%, with loading rates ranging from 3 to 4.5 kg VS/m^3. The specific gas production rate reached a peak of 1.5 vol/vol of digester calculated on weekly basis; pH values range from 7.4 to 7.6 with no significant difference along the length of the reactor. Therefore, even high VA concentration, like in the first part of the reactor, is not able to change the pH because of the very high alkalinity, ranging from 20,000 and 16,000 mg/l $CaCO_3$.

Total nitrogen in the feedstock ranges from 3 to 4 g/l with 30% N-NH_4^+. In the effluent there is no change in NTK concentration but ammonia grows up to 50%. After the opening of the rubber cover, observations done from the bull's eyes were confirmed. No crust or foam were found on the liquid surface, except for a very little amount of floating material in the first part of the reactor (only 50kg after one year). The presence of a thin sediment layer was observed only under the heating pipes. Its thickness decreased in the shallowest points of the pipes, a phenomenon probably due to convective movements. The results obtained through the prolongation of the HRT allowed us to calculate the biodegradable fraction of total volatile solids (VSb), which averages around 43% with this kind of waste.

Assuming an ideal plug-flow behavior, we could apply the following kinetic equation.

$$VSb\ (effluent) = VSb\ (influent)\ e^{-kHRT}.$$

The calculated kinetic constant (k) value at 30d HRT is 0.077; it is very close to the 0.073 value found by Jewell et al. (1978) for the Cornell plug-flow system. But if we calculate the k value at short or very short retention times, we have much higher values (0.4 at 5 d HRT). Therefore this equation does not fit very well to our process. It seems likely that it should be divided into 2 separate stages, the first very rapid and the second slower. These data correspond to the observation made on the production delays corresponding to given loading. The first effect can be noticed after a few days (1-5 d), and the second after a longer period (10-15 d). More analyses and more detailed figures are needed to confirm these observations.

CONCLUSIONS

The first year of operation of the plug-flow plant at the Pellerano farm gave us good results that allowed us to go ahead in optimizing the design, the reliability and the efficiency of the plant.

Capital costs could be reduced again using self-construction instead of buying a turn-key plant. In countries warmer than Northern Italy, costs can be further reduced by decreasing the insulation, the dimension of heat exchanges, and the power of the boiler.

We are now designing a new project -- a daily cattle housing in which a plug-flow plant is already enclosed. This plug-flow design could be suitable for developing countries, for community plants or for modern farms equipped with manure collection systems, for its low capital cost and relatively high gas production rate. In these countries a plug-flow can represent a substantial technological development in comparison to the traditional simplified biogas plants; in spite of the greater technological difficulties in the construction phase, this plant can overcome some of the problems that limited until now the diffusion of biogas technology.

ENEA will publish design and construction manuals with general layouts, advised construction machinery, and materials.

Other two similar plug-flow plants projected by ENEA are now under construction in Italy.

REFERENCES

American Public Health Association. (APHA). 1975. Standard Methods for examination of water and waste water. 14th edition. APHA. New York.

Bonazzi, G., L. Cortellini, S. Piccinini, and A. Tiche. 1984. The Biogas Project in Emilia-Romagna (Italy). Paper presented to the Bio-Energy 84 Conference, June 18-21. Goteborg. Sweden

Chiumenti, R., F. De Poli, P. Gabbi, A. Mazzi, and A. Tiche. 1983. A Data Acquisition System for the Monitoring of Anaerobic Digesters: Design and Application, in Proceedings of the Anaerobic Waste Water Treatment Symposium. Noordwijkerout. 23-25 November. 559-572.

Howard, K., and E. De La Fuente. 1983. The Anaerobic Digester at the Marindale Dairy, Novato, California, Report to the California Energy Commission. Landal Institute, Sausalito. 24 pages.

MULTICRITERIA OPTIMIZATION DESIGN OF METHANE FERMENTATION SYSTEMS

R.Chamy , S.Videla , and E. Navarrete
Chemical Engineering Department, Universidad de la Frontera,
Casilla 54-D Temuco, Chile

ABSTRACT

Existing design approaches of methane fermentation systems employ unicriteria optimization methods, constraining the problem by using a single objective function. Multiple functions of anaerobic digestion represent a complicated and complex problem. The purpose of this paper is to present an optimization approach employing a mathematical model which includes two objective functions, subject to physical contraints. One objective function represents energy balance, and the other corresponds to total investment. Optimal solution is a trade-off curve defined by a non-inferior solution set. The non-inferior trade-off curve presented a non-convex behaviour, so it was characterized by using the k-th objective $\mathcal{E}$-constraint problem. Some preliminary results are presented to indicate the validity of the technique employed.

INTRODUCTION

Many papers have been published over the past five years on design of methane fermentation systems. Though some significant efforts have been devoted to the development of an optimization design procedure (Hashimoto, 1981; Lavagno, 1983; Hill, 1984), there is still a lack of a complete and exhaustive engineering optimization methodology.

An optimum methane fermentation system is determined by plant design and by plant operating characteristics. Fermenter volume, volatile solids (VS), loading concentration, hydraulic retention time (HRT) and operating temperature are the most important variables (Hill, 1984).

Gas production rate, which is affected by such selection of variables, is one of the main objectives of economic evaluation. Furthermore, energy conservation represented by the energy balance, and resource conservation expressed by the size of equipment, are the other two main objectives.

The existing design approaches are only unicriteria optimization methods constraining the problem by using a single objective function. In a recent paper, Hill (1984) used in his study a parameter which was defined as unit energy production cost (UEPC) defined by:

$$UEPC = \frac{\text{Total annual cost value-Value of energy produced}}{\text{Total methane - Methane used internally}}$$

It is clear that UEPC represents a unique objective function which intends to resolve the conflict between resource conservation and energy conservation.

There are, broadly, three classes of conflict between objectives design which correspond to the following multicriteria optimization problems:

a) Maximization of Biogas Production and Minimization of the organic content of the discharge, (Economic and Ecological Conflict).

b) Maximization of Biogas Production and Minimization of Heat losses, which corresponds to the unicriteria decision problem defined by maximization of the energy, (Economic and Energy Conservation Conflict).

c) Maximization of Energy Production and Minimization of Resources (Energy Conservation and Resource Conservation Conflict).

In a recent work, the authors of this paper applied multiobjective analysis to resolve b-class conflict (Navarrete et al., 1984). Other authors have analyzed similar problems (Nishitani, 1983). In these three cases, the optimal solution is a trade-off characterized by a non-inferior solution set (Chankong, 1983). In accordance with some decision criteria of the decision agent a preferred solution can be obtained on this trade-off curve.

This paper is concerned with this new approach to resolve these three classes of design problems by using multiobjective analysis. The main purpose is to present a multicriteria decision problem, which corresponds to the Energy Conservation and Resource Conservation Conflict defined by the biogas plant design.

MATHEMATICAL TECHNIQUE EMPLOYED

Consideration of two criteria gives rise to a two-objective problem, (TOP), given as follows:

$$\min (f_1 (x), F_2 (x))$$
$$\text{Subject to } g_i (x) \quad 0, \; i = 1 \ldots\ldots m$$

Where x represents an N-dimensional vector of decision variables, System constraints are defined by $g_i (x)$, $i = 1 \ldots\ldots m$

To solve a TOP is to find its set of non-inferior or Pareto-Optimal solutions, given by the following definition:

X^x is said to take a non inferior solution of a TOP, if there exists no other feasible x such that $f_j (x) \not\geq F_j (X^x)$, j=1,2 with strict inequality holding for at least one j.

A common strategy to solve a TOP is to characterize non-inferior solution in terms of optimal solutions of some appropriate scalar optimization problem. In order to avoid false characterization due to the presence of "gaps" in the set of non inferior solution, we choose the 'kth-objective constraint problem' to characterize optimal solution (Chankong, 1983).

PROCESS MODELING

The main assumptions on which the process modeling is based are as follows:

1. A continuous stirred tank reactor is selected as a digester unit.
2. A heat exchanger heats the influent of the digester.
3. A heat exchanger consisting of coils of pipe heats the liquid in the digester.
4. The volumetric methane production rate can be quantitatively described by the Chen and Hashimoto kinetic equation.
5. The physical properties are taken to be constant at an average temperature.

A scheme of the biogas plant used as an illustrative model for the design optimization is shown in Figure 1. We select this simplified model of biogas plant in order to show the advantages of multi-objective analysis applied to biogas digester design.

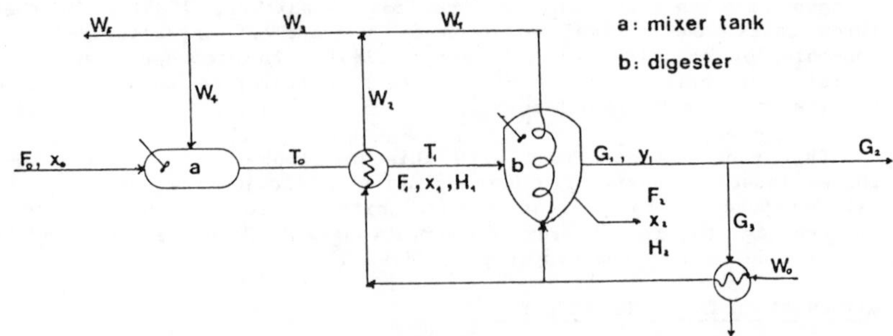

FIGURE N°1

Energy conservation is represented by the energy balance of the system,

$$\text{Exergy} = \text{Energy produced} - \text{Heat losses}$$

Energy produced corresponds to the biogas production rate expressed by the Chen and Hashimoto kinetic equation. Heat losses are the results of two factors: the heat losses of the digester and the energy necessary to heat the feed stream.

Resource conservation is expressed by the total investment cost. This cost can be estimated by adding the investment cost for each of the main equipment components of the biogas plant. We use the well-known function:

$$I = \sum_{j}^{n} C_j \, (S_j)^{q_j}$$

Table 1 shows the cost values, the corresponding exponent q_j, and function size, S_j for the main equipment of the biogas plant.

Table 1 (Chamy, 1983)*

ITEM	C_j	q_j	S_j
Digester	36.2	0.6	$\pi R^2 H$
Mixer tank	21.51	0.6	$\pi R^2 H$
Propeller	29.8	0.54	$\pi R^2 H$
Coil pump	73.1	0.11	F_2
Water pump	29.6	0.11	G_3
Heat exchanger	0.67	0.6	G_3
Coil pipe	0.09	0.6	E

* All data correspond to Chilean situation. The cost data are expressed in thousands of Chilean $ (1982).

With the above assumptions, the multicriteria decision problem can be written as follows:

A TWO OBJECTIVE PROBLEM

Exergy balance

Investment Cost

MAX $G_2 = G_1 - G_3$ MIN I

Where:

$$G_1 = \frac{Y_v \pi R^2 H_1}{\% \, CH_4}$$

$$Y_v = \frac{B_0 X_1}{\theta}\left(1 - \frac{K}{u_m \theta - 1 + K}\right)$$

$$u_M = 0.013T - 0.129$$

$$\theta = \frac{\pi R^2 H_1}{F_1}$$

$$G = \frac{2\pi R H_1 (T-T_b)}{\frac{l_1}{k_1} + \frac{l_2}{k_2} + \frac{b}{h_b}} + \frac{\pi R^2 (T-T_t)}{\frac{l_1}{k_1} + \frac{l_2}{k_2}} + \frac{\pi R^2 (T-T_b)}{\frac{l_2}{k_2} + \frac{l_3}{k_3} + \frac{1}{h_a} + \frac{1}{h_b}} + \frac{2\pi R H_2 (T-T_b)}{\frac{l_1}{k_1} + \frac{l_2}{k_2} + \frac{1}{h_G} + \frac{b}{h_b}}$$

$$+ F_1 p_1 C_p (T-T_0)$$

$$H_1 = 0.9\ H \ ; \quad H_2 = 0.1\ H$$

The two objective problem is subject to:

$$25 < T < 65$$
$$R > 0$$
$$H > 0$$
$$0 < 0 < 60$$
$$0 < X < 100$$

All necessary data are summarized in Table 2.

Table 2 Design Data for a Biogas Plant

l_1	= 0.2 m	T_0	= 15°C
l_2	= 0.1 m	B_0	= 0.25 m^3 CH4 /kg VS
l_3	= 0.0765 m		
k_1	= 0.96 W/m °K	K	= 1.4
k_2	= 0.035 W/m°K	b	= 0
k_3	= 6.5 W/m°K		
T_t	= 15°C		
T_b	= 10°C		
h_b	= 7.9 W/m^2°K		
h_G	= 9.6 W/m^2°K		
C_p	= 4.184 X 10^3 j/kg°K		

RESULTS AND DISCUSSION

The non-inferior solution presented 'gaps' so it was characterized by the following primal scalar program (see Kitagawa, 1982) and the kth-objective ε - constraint problem Chankong (1983),

Min I (X)
Subject to $-G_2 < \varepsilon$ *
V (X) $\in$ S

Where S is the normal set of restrictions and X the decision vector.

A FORTRAN program that implemented the augmented Lagrangian algorithm was used to optimize it for different values of exergy in order to obtain the trade-off curve in the objective speces (Figure 2). This trade-off curve shows us the cost that each decision agent must pay for incrementing energy. For all cases analyzed, the optimal temperature of non-inferior points was 25°C (the minimum of the model). This illustrates that producing more gas by increasing temperature does not pay off for the loss of energy.

Obviously, this conclusion is affected by the structure of the model used. Nevertheless, we think that the preliminary results are

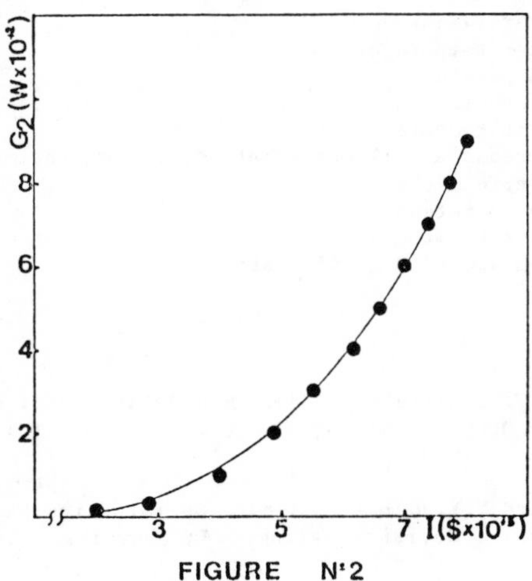

FIGURE N°2

indicative of the validity of the multicriteria optimization method presented in this paper.

Table 3 shows values of the decision vector for some points of the trade-off curve. In all cases, the input was selected as F_0x_0=126 kgVS/day. Optimal solutions gave a constant R/H ratio which was found to be 0.5 for this illustrative example.

Table 3 Decision Vector

Non inferior points		Decision vector				
($x1000) I	(W) G_3	R (m)	H (m)	θ (day)	T(°C)	F_1(m^3/day)
801.8	9000	1.635	3.271	33.1	25.0	0.740
703.7	6000	1.455	2.911	14.4	25.0	1.198
392.8	1000	0.845	1.680	6.1	25.0	0.554

NOMENCLATURE

B_0 = ultimate methane yield
C_p = specific heat
C = investment cost for each equipment
F = volumetric substrate utilization rate
G_1 = energy corresponding to biogas production
G_2 = exergy (net energy)
G_3 = total energy requirements
H = digester height
l_1 = digester thickness
l_2 = insulating thickness

l_3 = top thickness
R = digester radius
T = digester temperature
T_b = air temperature
T_t = land temperature
T_0 = influent temperature
X = biodegradable influent substrate concentration
Y_v = volumetric methane yield
0 = Hydraulic retention time
μ = specific growth rate
μ_m = maximum specific growth rate

REFERENCES

Chamy, R., and G. Schaffeld. 1983. Simulation Model of a Biogas Process. 2nd Interuniversitary Symposium on Energy. Santiago, Chile.

Chankong, V., and Y.Y. Haimes. 1983. Optimization-based methods for multiobjective decision making. An overview. Large Scale Systems 5:1-33.

Hashimoto, A.G., and R.L. Hruska. 1983. Conversion of straw-manure mixtures to methane at mesophilic and thermophilic temperatures. Biotechnology and Bioengineering, 15:185-200.

Hashimoto, A.G., and Y.R. Chen. 1981. Economic Optimization of Anaerobic Fermenter Designs for Beef Production Units. In: Livestock Wastes: A Renewable Resource. ASAE. Pub. No. 2-81.

Hill, D.T. 1984. Economically optimized design of methane fermentation systems for swine production facilities. ASAE 27:525-529.

Kitagawa, H., H. Watanabe, Y. Nishimura, and M. Matsubara. 1982. Some pathological configurations of non inferior set appearing in multicriteria optimization problems of Chemical Engineering. Journal of Optimization Theory and Applications 38:541-563.

Lavagno, E., P. Ravetto, and B. Ruggeri. 1983. An analytical model to study the performance of an anaerobic digester. Agricultural Wastes 5:37-50.

Navarrete, E., S. Videla, and R. Chamy. 1984. Multiobjective Analysis Applied to Biogas Digester Design. 3rd. Mediterranean Congress on Chemical Engineering. Barcelona, Spain.

Nichitani H., and E. Kunugita. 1983. Multiobjective analysis for energy and resource conservation in process systems. Journal of Chemical Engineering of Japan 16:235-240.

A VILLAGE SCALE BIOGAS PILOT PLANT STUDY USING HIGH RATE DIGESTER TECHNOLOGY

S. K. Vyas, Punjab Agricultural University
Ludhiana, India

ABSTRACT

In this paper a study of a village-scale biogas plant for producing more biogas during the winter seasons in the Northern Indian Plains is presented. Generally, during many cold nights the ambient temperature goes to 1°C. The average subsoil temperature along the plant reaches 18°C to 20°C and slurry temperature attains 15°C to 17°C. The technology for heating the slurry, reducing the heat losses from a plant and stirring slurry suited to the rural areas has been termed high-rate digester technology. In this study, heating of slurry from solar heated water and reduction of heat losses by insulating gas holder and composting all around the walls of the digester to a meter depth is reported. The increase in gas production is discussed. Suggestions for various other methods of treating slurry to be studied at the University Center are also made.

INTRODUCTION

Traditionally the cooking fuel in rural India has consisted of cow dung cakes, firewood and crop residues. However, with increasing population and the depletion of forests, firewood is becoming scarce and the most needed organic matter for enriching Indian soils is being burnt in the form of dung cakes for cooking fuel (Patel, 1981). One solution to the imbalance of resource use is to derive the cooking fuel energy from the fresh cattle dung in the form of biogas through anaerobic fermentation. The effluent received from digesters after anaerobic fermentation is rich in nitrogen, micronutrients, and organic matter can be applied to soils to improve their fertility.

According to an estimate (Vashist and Boss, 1981) the fresh cattle dung produced annually is 1000×10^6 tons, out of which, one third is used in dung cakes and two thirds for composting to improve soil fertility. This large amount of fresh dung has the tremendous potential of $3{,}800 \times 10^6$ m^3 of biogas as cooking fuel and 750×10^6 tons of manure. To tap the large potential, a large number of successful village-scale and family size biogas plants have to be installed in the shortest possible time.

HIGH RATE DIGESTER TECHNOLOGY

The success of a village-scale community biogas plant depends on uninterrupted and sufficient biogas supply generated through anaerobic fermentation of fresh cattle dung collected daily by the community. By adopting the existing conventional technology in the construction of a community digester, optimum conditions of temperature and stirring for anaerobic fermentation in the mesophilic range cannot be realized. It has been observed in the Northern Indian plains that the winter temperatures of slurry in conventionally designed plants attains low temperature of 14°C to 18°C resulting in poor production of gas during winters when gas needs are greater. In one technical committee review meeting at Delhi (1983) on the causes of failure of community biogas plants, it was revealed that the poor production gas during winter contributes greatly to the failure of the community biogas program. To increase the biogas production, it is required to maintain slurry temperature in the digester above 30°C along with periodical stirring. The slurry temperature above 30°C could be maintained by minimizing heat losses, both from digester and gasholder, through insulation, and heating the slurry in the digester from the renewable energy sources or biogas produced in the digester, or waste heat energy obtained from engine generating set, or a combination of these sources. The technology required to maximize biogas production is termed high rate digester technology. It is also essential that the high rate digester technology being developed to improve the biogas production should be inexpensive and easy to maintain and manage with the skills available in rural areas of a developing country such as India.

Digester

The digester was designed so that the retention period of two tons of slurry fed daily worked out to be the 50 days required for conventional design in winter. The retention period could be reduced by increasing inputs of fresh dung slurry. The difference in levels of inlet and outlet was kept as 0.8 m to ensure free flow of slurry through the digester. The detailed dimensions of plant are shown in Figure 1. For circulating hot water through the plant to heat the slurry, pipe lengths 55 m each were installed on either side of the dividing wall of the plant. The digester walls had to be taken 1.8 m above ground due to the difficulty of digging in coarse sand layer. There is one enclosure well to the digester wall in which pipe outlets are provided for taking slurry sample from different depths and points along the diameter. There are two additional enclosures for installation of two stirrers. A concentric ring of masonry at a distance of one meter to a height of one meter above ground was erected for composting.

Insulation of Digester and Gasholder

The exposed surface of well wall was covered with alternate layers of dung and straw 15 cm thick, each at 70% moisture level provided in annular space for composting. This, however, did not cover the enclosures for stirrers and sampling. The details of insulation of gasholder are shown in Figure 2.

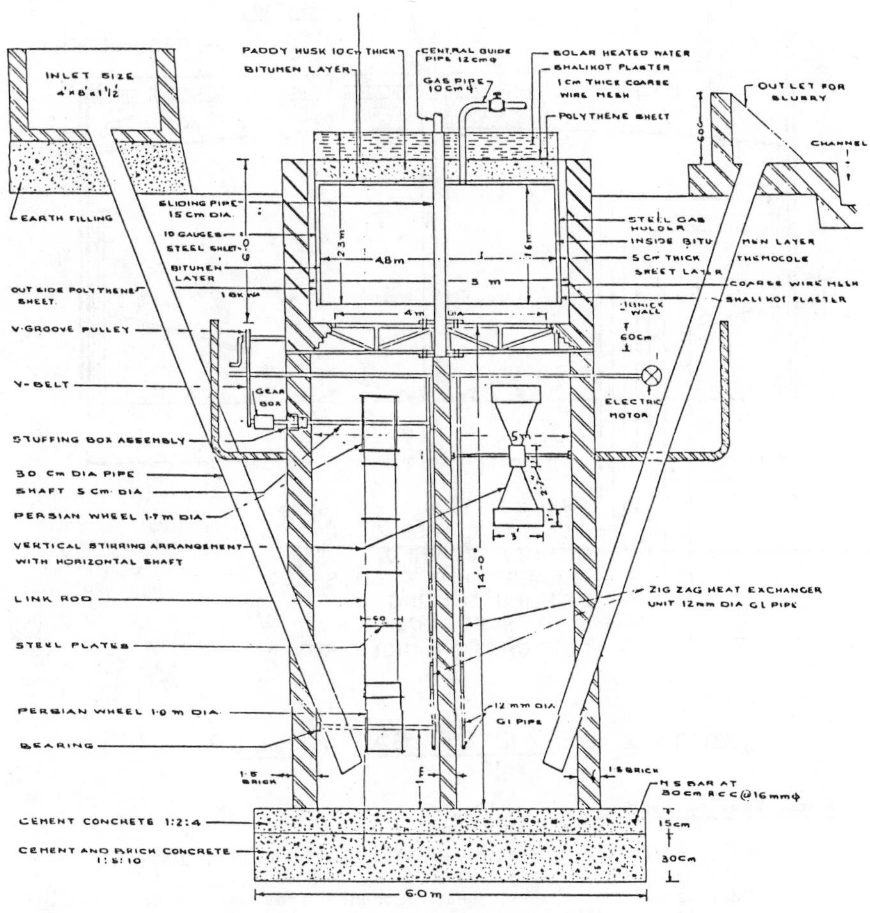

FIGURE 1 COMMUNITY BIO GAS PLANT AT P. A. U.

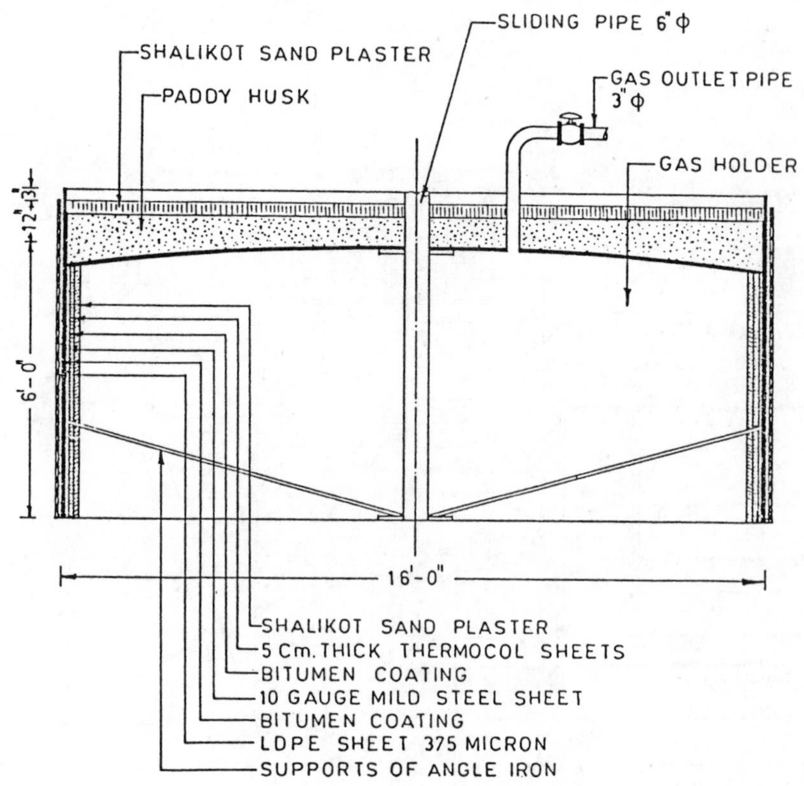

FIGURE 2 DETAIL OF GAS HOLDER

Slurry Heating Systems

In 1983, the following systems of heating slurry were proposed:

1) Heating slurry through heat exchange from hot water obtained from 500 l solar water heater.

2) Waste heat from engine generating set of 25 kW capacity.

However, only the first system could be installed during the period under report.

OPERATION OF PLANT AND PRESENTATION OF DATA

The digester of the biogas plant was charged in November 1982. Regular daily feeding of the plant with 500 kg of fresh cattle dung was

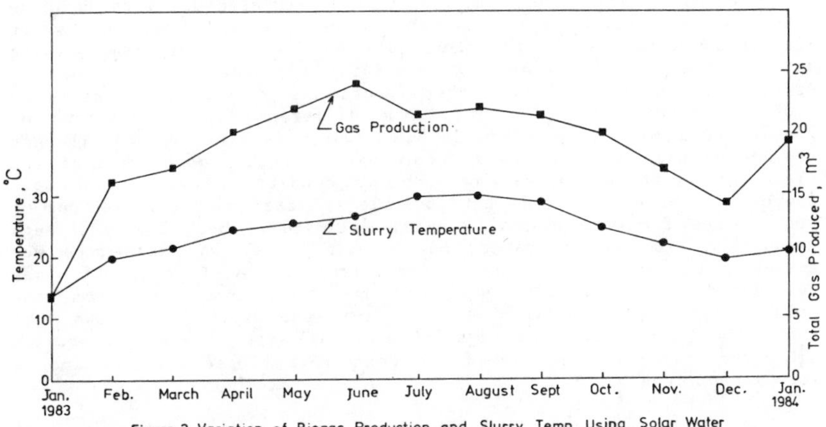

Figure 3 Variation of Biogas Production and Slurry Temp. Using Solar Water Heating over the Period from Jan.1983 to Jan. 1984.

started in January 1983. The slurry temperature was lowest at 14°C in January 1983 and then showed a continuous rise to 30°C in July 1983. Biogas production showed a rise from 7.1 m^3 in January 1983 to 24.2 m^3 in June 1983, but a slight decrease later. Both gas production and temperature declined continuously from 24.2 m^3 to 14.3 m^3, and from 30°C to 19°C respectively. It should be noted that heating of slurry with solar heated water through the heat exchange process was started in the second week of December 1983, when slurry temperature reached 17°C. In the last week of December, daily feed to the digester of fresh dung was increased from 500 kg to 1,000 kg. Both the production of biogas and the temperature of the slurry have shown an increase in January 1984.

DISCUSSIONS OF RESULTS

Because of continued fall in ambient temperatures during the winter of 1982-83 and no installation of any device for heating slurry, the average temperature was 14°C during January 1983. Subsequently the biogas production increased due to improvement in slurry temperature with the increasing ambient temperatures from February 1983 and onwards till July 1983. The biogas production during corresponding period showed increase until June but a slight decline during July, August, and September 1983. This is attributed to a lower feeding of solid contents to the digester. As the feed to the digester was measured in terms of baskets, the cattle dung at the higher moisture contents from July to September 1985 resulted in lower inputs of solids to the digester. The average moisture contents of dung increased from 85% to 90% from June to September 1985.

The solar-heated hot water circulated through the installed pipelines in the digester lost 30°C temperature through heat exchange in transit. A total amount of 600 l was circulated daily through the plant on all sunny days. It should be realized that hot water circulation could arrest the continuous decline of slurry temperatures as it happened in December 1983, without circulation of hot water. Besides, circulation of hot water recorded a rise in temperature by 4.0°C in about six weeks time. The difference in temperatures of January 1983 and January 1984 is 5.6°C which is significant. The gas production has also increased accordingly. As the boundaries of the plant are exposed to varying ambient conditions and varied soil temperature profiles, it is not possible to make exact calculations of heat losses from the plant. A rough estimate of heat losses through conduction from the digester and the gas holder was made on lower ambient temperature of 1°C and subsoil temperature of 18°C, which was calculated to be 110×10^3 kcal per day. The input heat from the solar water heater was only 18×10^3 kcal per day. This shows that heat input was low. The cost of a solar water heater to provide 110×10^3 kcal may even exceed the cost of a biogas digester and as such is not feasible.

The alternative method of heating digester slurry with waste heat obtained from 25 kW engine generating set will be carried out this year. The purchase of the engine generating set is in process. Besides this heat energy from agrowaste glass-house effect, etc. will also be tried at this center to maximize the biogas production by attaining optimum conditions of heating and stirring required for anaerobic fermentation.

CONCLUSIONS

1) Heating of digester slurry during winter with solar heated water increased gas production from 0.014 to 0.028 m^3 per kilogram of fresh cattle dung fed to the plant.

2) A suitable technology for design, fabrication and installation of stirrers was developed so that the stirrers could be easily operated and maintained by a rural blacksmith.

3) The cost of solar water heater capable of maintaining 30°C during winter is equal to that of the biogas plant digester. Other alternatives of heating slurry with waste heat from engine generating set, heat energy of a part of biogas produced, heat radiations received from greenhouse effect, and heat from agrowaste need to be tried individually or in combination.

REFERENCES

1. Patel, J.J. 1981. Biogas in India, Proceedings of the National Seminar on Biogas Technology. USG Ludhiana.

2. Vashist V.N. and P.R. Bose. 1981. Energy Alternatives for India Prospects and Problems. Proceedings of the National Seminar on Biogas Technology. USG Ludhiana.

3. Smith P.J., R.L. Fehr, J.A. Miranowski, and E.R. Pidgeon. 1977. Proceedings, Cornell Agricultural Waste Management. Ann Arbor Sciences.

COMPOST-HEATED SMALL SCALE FARM DIGESTER APPROPRIATE FOR KOREAN CONDITIONS

Chong Joon Hong
Biomass Division
KOREA INSTITUTE OF ENERGY & RESOURCES (KIER)
P.O. Box 339, Daejeon, Chungman, Korea

ABSTRACT

Livestock manure is a potentially important biomass energy resource in Korea. Its effective use for biogas production depends, among other things, on the ability to maintain a suitable fermentation temperature during the harsh winter season. This paper describes a viable compost-heated system for producing biogas year-round. The system design, operability, and relevant results are presented and discussed. Economic feasibility is also assessed. Results indicate that the tested system is viable from both the technical and economic points of view.

INTRODUCTION

The amount of animal wastes produced by Korean livestock farms has increased greatly year after year, just as the scale of livestock farms has become larger and yielded greater income. Anaerobic digestion of these wastes is essential for pollution control. It is also an important energy source.

In order to retain a constant fermentation temperature, methods using vinyl house or solar energy collecting sheets or indirect coil heat are employed. The application of these methods poses several problems during the harsh winter season. Furthermore, the optimum temperature of a fermentation tank may not be maintained by means of installing it underground or in vinyl houses. The efficiency of using the collecting sheet of solar energy is too low for the immense expense of plant construction. The indirect coil heat method using the produced gas as a heat source demands almost 70 percent of the total produced methane gas with low economical feasibility.

Therefore, in order to solve these problems, a compost-heating fermentation system was designed to produce methane gas year-round. This system utilizes the compost heat itself, negating the need for the conversion to indirect heat with the consequent expense of plant construction.

SYSTEM DESIGN AND CONSTRUCTION

The digester has a cylindrical shape and consists of a 10-cubic-meter fermentation section and a 3-cubic-meter gas space. It is made up of Fiber Resin Plastics (FRP). The material has such merits as strength, durability, ease of repair and maintenance, and light weight for installation and removal without using tools.

Pig manure is used as the feeding material. In order to increase the temperature inside the digester, the manure from pig sheds mixed with rice straw is piled on to compost around the digester. The structural diagram of the FRP tank is shown in Figure 1 and the system is depicted in Figure 2.

BIOGAS GENERATION FROM PIG MANURE

In this work, the input volatile solids (VS) concentration of pig manure was controlled to maintain a pH 7 and the temperature was held at 35°C. The cultivated seed solution was added and amounted to 10 percent of the total content. It consisted of pig manure anaerobically fermented for 40 days in 10 percent volatile solids at 40°C.

The compost was piled 2 meters wide and high on the digester to maintain a constant temperature inside. A one-millimeter-thick vinyl cover was placed on top of the compost. Also the underground floor was sealed with the cover to prevent airflow.

It took around 30 days from initial stage to the normal activated conditions. The pH value was maintained between 7.0 and 7.2. The compositions of produced gases were 60-63% CH_4, and 37-40% CO_2 respectively.

A thermo sensor was installed at the compost loading and digester. The sensor recorded the temperature change on a Temperature Recorder (TR). From that recorder, a curve was obtained to show the anaerobic fermentation temperature of the compost as well as the proper method of using the compost heat.

In order to monitor the operating conditions, the pH change was measured by the hydrogen ion concentration guage (HM-1K), attached with glass electrodes. A gas chromatograph (GC-9A) was used to analyze the biogas composition, especially the ratio of CH_4 to CO_2 gas. The total amount of gas was measured by a gas flowmeter.

RESULTS AND DISCUSSION

Organic Material Loading Conditions

From the time of initial operation, the biogas composition, the hydrogen ion concentration, and the gas production changes are shown in Figure 3, Figure 4, and Figure 5 respectively.

From Figure 3, it can be seen that the CH_4 content decreased for the first 12 days during the initial organic material loading but it

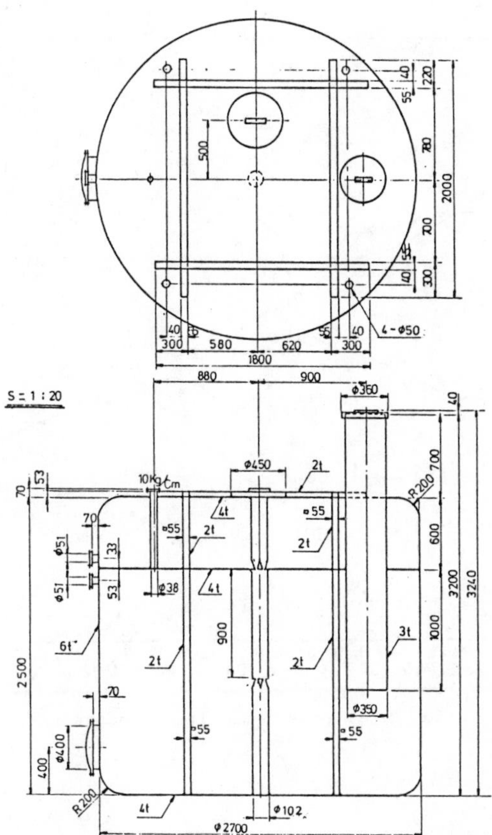

Fig, 1. Methane Fermentation Tank Design.

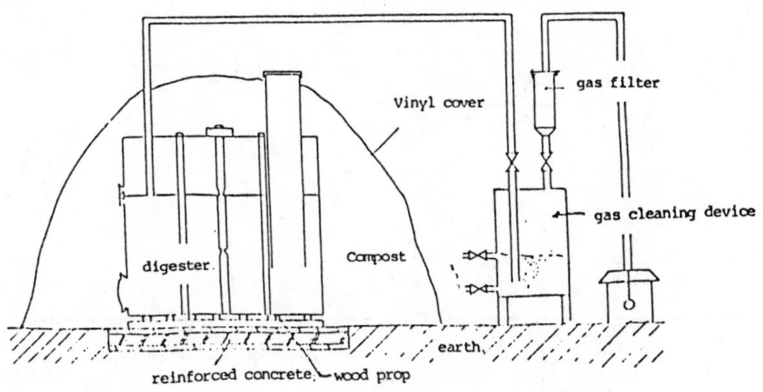

Fig, 2. Methane Gas Installation System.

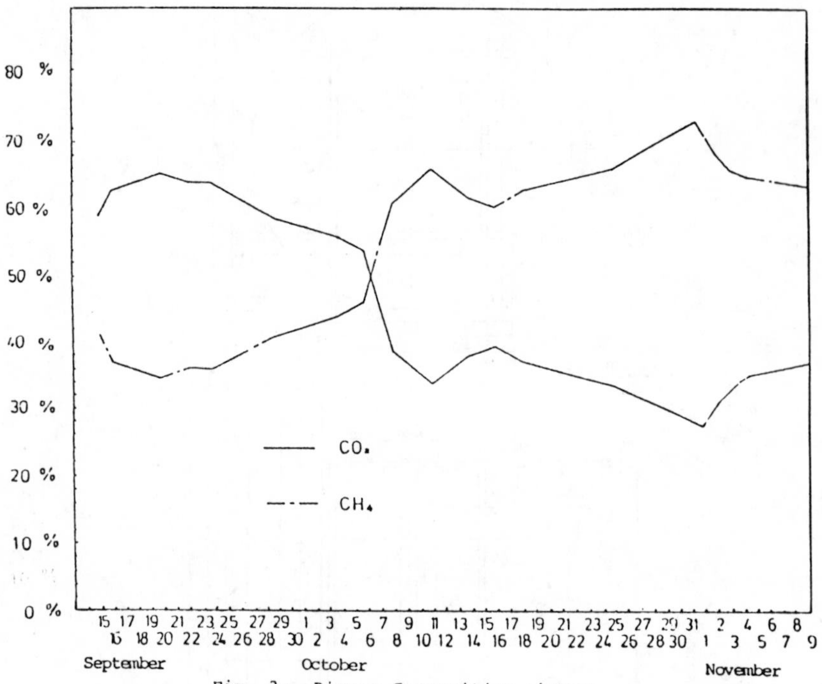

Fig, 3: Biogas Composition change

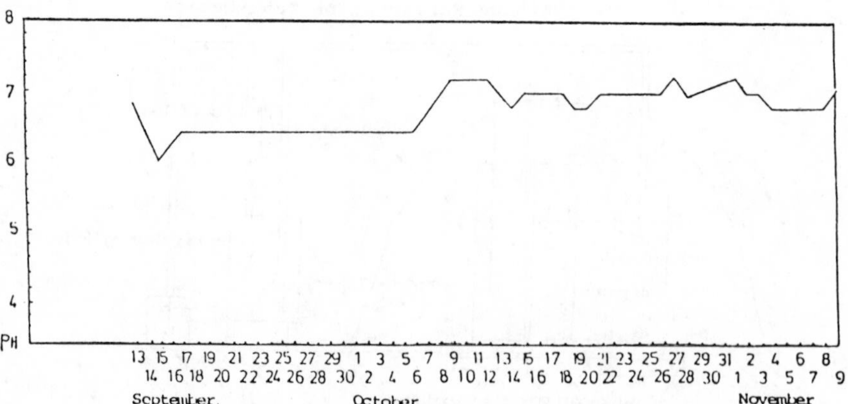

Fig, 4: Hydrogen Ion Concentration Change

recuperated and increased sharply after 30 days. This sort of CH_4 change shows a transition from the acid-fermentation phase through the acid-reduction phase to the alkali fermentation phase. Similarly, in Figure 4, the pH changes from an acid state at the initial stage, maintains a normal state for 30 days, then increases.

In Figure 5, the gas production exhibited an increasing trend toward a stable plateau. On an average, a gas rate of 4 to 4.5 cubic meters per day was obtained at stable conditions.

Maintaining High Temperature By Compost Heat

The compost heating effects by compost loading on the digester are shown in Figure 6. As shown, there is an initial sharp temperature rise for the first 10 days of the aerobic fermentation phase. The next 20 days produce essentially a heat-insulation effect. After 30 days,

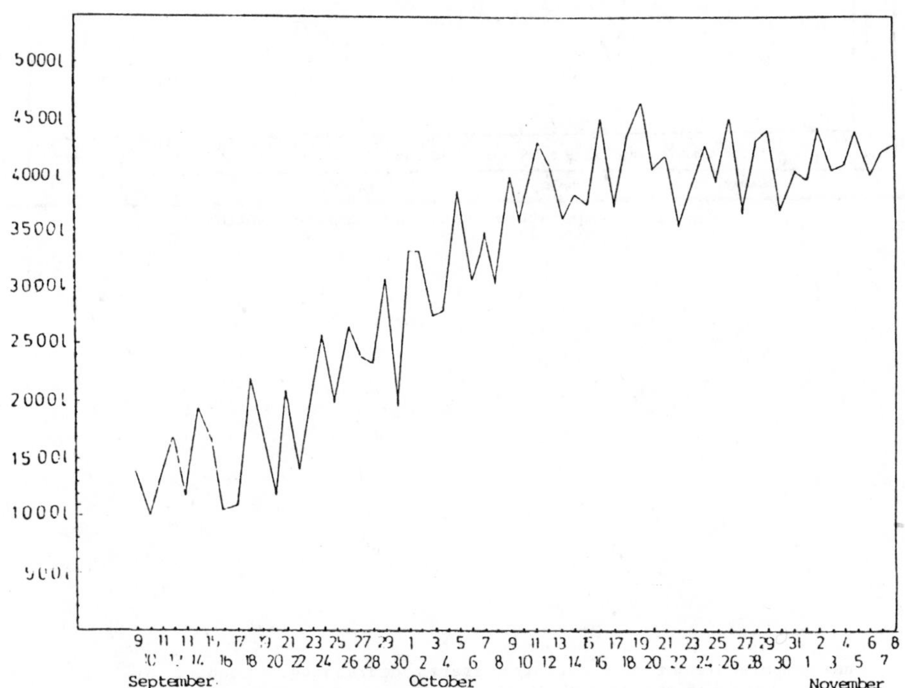

Fig, 5: Biogas Generation Estimate

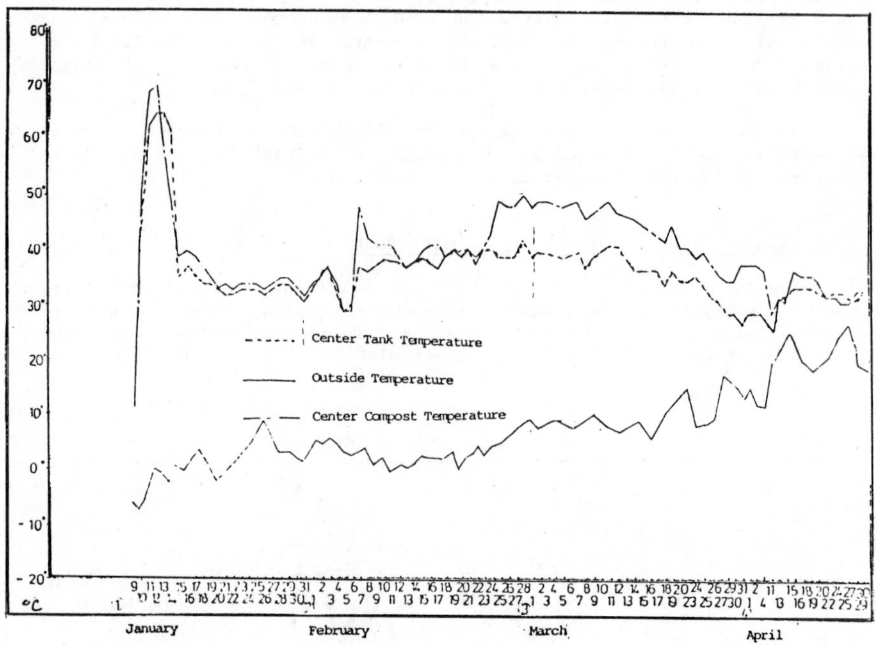

Fig, 6. Temperature change by Compost Loading

the temperature of the digester starts decreasing. Loading of a new compost batch affects a second steep temperature rise.

In order to find a more natural method, no further compost loading was applied and resort was made to solar heating through the plastic cover, with the result that the compost temperature decreased while the open air temperature increased as shown in the graph. Ultimately, however, the optimum fermentation temperature of 35°C could be fully retained, utilizing the compost heating solar-energy combination system (as noted in the last part of the graph).

Heat Loss Comparison

As can be seen in Figure 7, heat losses in the compost-heating system have three components Q_1, Q_2, and Q_3. System study and analysis indicated that these component heat losses are according to the order: $Q > Q_3 > Q_2$. Total heat losses can be estimated from the following calculations:

$$Q_1 = U_1 A_1 (T_2 - T_3); \; U_1 = \frac{1}{\frac{1}{h_t} + \frac{1}{h_o}}$$

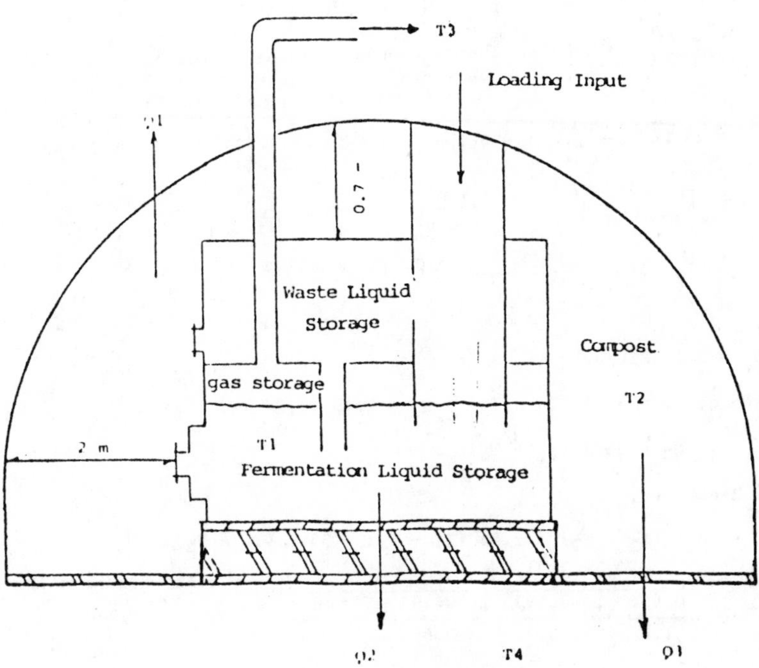

Fig, 7. Places for Heat Loss Measure

$Q_2 = U_2 A_2 (T_1 - T_4);$ $\qquad U_2 = \dfrac{1}{\dfrac{x_1}{k_1} + \dfrac{x_2}{k_2} + \dfrac{x_3}{k_3} + \dfrac{x_4}{k_4}}$

$Q_3 = U_3 A_3 (T_2 - T_4); \; U_3 = h_t$

A_1 = The surface area of the hemisphere

A_2 = The surface area of the digester floor

A_3 = The area of the compost floor

$h_t = 0.35 \times 0.6 + 0.043 \times 0.4 = 0.1$ Btu/hr x ft^2 x °F
(Compost of 60% humidity)

$h_o = 2.1$ Btu/hr x ft^2 x °F

$x_1 = 0.656$ ft $\qquad x_3 = 0.197$ ft

$x_2 = 0.197$ ft $\qquad x_4 = 1$ ft

$x_1, x_2, x_3, x_4,$: The digester floor, cement mortar, nitre stone, pebble thickness

$k_1, k_2, k_3, k_4,$: Thermal conductivities of the materials

$A_1 = 2x\,\pi \left(\frac{300}{30.4}\right)^2 = 611.6 \text{ ft}^2; \; A_2 = \left(\frac{130}{30.4}\right)^2 = 18.3\text{ft}^2$

$A_3 = \pi \left(\frac{300}{30.4}\right)^2 - \pi \left(\frac{130}{30.4}\right)^2 = 351.7 \text{ ft}^2$

$U_1 = \dfrac{1}{\dfrac{1}{0.1} + \dfrac{1}{2.1}} = 0.095$ Btu/hr x ft^2 x °F

$U_3 = 0.1$ BTU/hr x ft^2 x °F

Therefore, $Q_1 = 58.1 (T_2 - T_3)$, $Q_2 = 2.43 (T_1 - T_4)$

$Q_3 = 35.2 (T_2 - T_4)$

Total Heat Loss,

$Q = 58.1 (T_2 - T_3) + 2.54 (T_1 - T_4) + 35.2 (T_2 - T_4)$

$U_2 = \dfrac{1}{\dfrac{0.656}{1.0} + \dfrac{0.197}{0.75} + \dfrac{0.197}{0.05} + \dfrac{1}{0.43}} = 0.139$ Btu/hr x ft^2 x °F

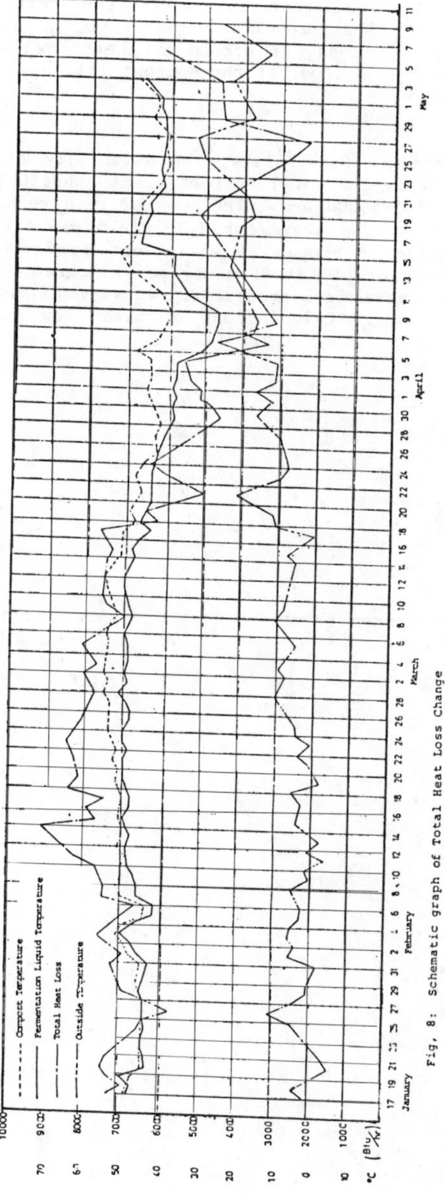

Fig, 8: Schematic graph of Total Heat Loss Change

According to the results of comparative analysis in Figure 8, the Q_1 value was highly dependent on the outside temperature, since it was almost exposed to the outside. In order to reduce the heat loss Q_1, it is desirable that economical heat insulation material be used to cover the compost. To reduce Q_3 heat loss, an economical heat insulation sheet should similarly be put on the ground floor of the compost loading.

Economic Feasibility

In studying the economic feasibility, the durability of the FRP digester was estimated as 30 years. The total cost of plant construction with a volume of 10 cubic meters was 2,195,000 won. The cost of repairs and maintenance was estimated at 125,000 won. The average biogas yield was 4.5 cubic meters per day. The calorific value of biogas is equivalent to 5,500 kilocalories/m^3. Applying the retail price of LPG at 92 cents per kilogram, the rate of return is around 24% on the whole investment. A recovery of input costs can be achieved in around 4 years. Calculations of cash flow and discounting are given in Tables 1 and 2.

Table 1 - Estimated Cash Flows

Unit: KOREAN WON

(Rate: 800won/US$1)

Year No.	Cost	Benefit	Cash Flow
0	-2,195,000		-2,195,000
1	-	557,073	557,073
2	-	"	"
3	-	"	"
4	- 125,000	"	432,073
5	-	"	557,073
6	-	"	"
7	- 125,000	"	432,073
8	-	"	557,073
9	-	"	"
10	- 125,000	"	432,073
11	-	"	557,073
12	-	"	"
13	- 125,000	"	432,073
14	-	"	557,073
15	-	"	"
16	- 125,000	"	432,073
17	-	"	557,073
18	-	"	"
19	- 125,000	"	432,073
20	-	"	557,073
21	-	"	432,073
22	- 125,000	"	557,073
23	-	"	"
24	-	"	"
25	- 125,000	"	432,073
26	-	"	557,073
27	-	"	"
28	- 125,000	"	432,073
29	-	"	557,073
30	-	"	"

Table 2 - Discounted Cash Flows

Year No.	Cash Flow	Discount rate 22%	25%
0	-2,195,000	-2,195,000	-2,195,000
1	557,073	456,658	445,658
2	"	375,276	356,526
3	"	306,783	285,221
4	432,073	195,037	176,977
5	557,073	206,116	182,541
6	"	168,948	146,032
7	432,073	107,408	90,612
8	557,073	113,509	93,461
9	"	93,040	74,769
10	432,073	59,150	46,392
11	557,073	62,510	47,852
12	"	51,238	38,282
13	432,073	32,574	23,753
14	557,073	32,421	24,500
15	"	26,575	19,600
16	432,073	17,939	12,161
17	557,073	18,598	12,544
18	"	15,539	10,035
19	432,073	9,879	6,227
20	557,073	10,440	6,422
21	"	8,557	5,138
22	432,073	5,440	3,188
23	557,073	5,749	3,288
24	"	4,712	2,630
25	432,073	2,996	1,632
26	557,073	2,595	1,683
27	"	2,127	1,346
28	432,073	1,650	835
29	557,073	1,171	862
30	"	960	689
Total	-	199,914	- 74,142

STRUCTURAL BEHAVIOUR AND STRESS CONDITIONS OF FIXED DOME TYPE OF BIOGAS UNITS

M. Amaratunga, Department of Civil Engineering,
University of Peradeniya, Sri Lanka

ABSTRACT

Brickwork and small precast concrete blocks, which can be easily handled manually, seem to provide the most suitable type of construction material for biogas digesters with fixed dome gasholders. Even large domes can be constructed using these small blocks; the use of formwork, which is often expensive, can be eliminated by adopting special techniques of construction. The ability of ordinary masons to quickly master these techniques makes this type of biogas unit very useful for popularization in rural areas.

Many problems have arisen both during construction and during use of fixed dome biogas units, due primarily to a lack of understanding of the engineering principles involved. There is a possibility that the brickwork, or blockwork, which is expected to be self-supporting, could collapse at some stage of the construction. The internal gas pressure could cause tensile cracks to develop in the dome which would result in the leakage of gas, unless suitable precautions are taken in the choice of dimensions and the external loads to ensure that adequate initial compressive stresses are induced.

This paper explains the technique of biogas dome construction using bricks or small blocks without resorting to the use of formwork. The stress conditions generated in spherical domes at various stages of construction and use are discussed and the principles to be followed for the successful construction of fixed dome type of biogas units are outlined.

INTRODUCTION

Bricks, or small precast concrete blocks which can be easily handled by an individual, offer the most convenient and perhaps the most economical means of construction of biogas units in rural areas. Considerations of the economy of the use of materials and labor and the convenience of construction result in the choice of a circular cylindrical shape for the fermentation chamber and a spherical dome shape for the cover of the gas storage section.

Failures due to deficiencies in design and construction are, however, frequently encountered and these tend to hinder efforts made to popularise the use of biogas technology. An understanding of the structural behaviour of the unit as well as its individual components will help to avoid much failures and help in the maintenance of these units. The essential structural components of a biogas unit are identified in this paper and the manner in which these behave under the various loads imposed on them during construction and subsequent use is explained and the precautions to be observed to minimize possibilities of failure are discussed.

The essential components of a fixed dome type of biogas unit are shown in Figure 1. The more significant structural components are:

i. the base of the fermentation tank;
ii. the wall of the fermentation tank;
iii. the dome of the gasholder.

For construction in brickwork, or plain concrete blockwork, structural shapes and construction techniques which do not give rise to excessive tensile stresses need to be used. These are described in the subsequent sections.

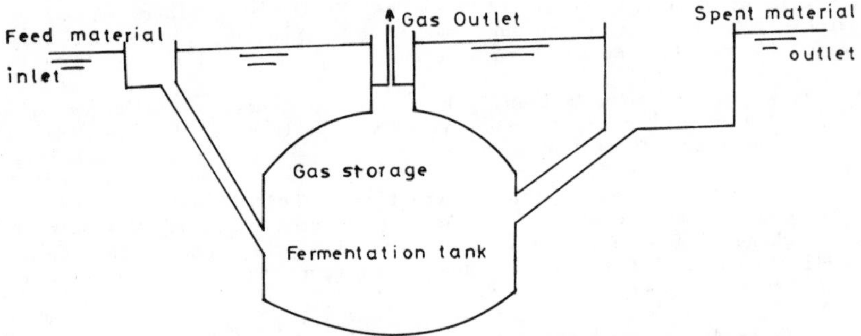

Figure 1 Cross-sectional view of typical fixed-dome biogas unit

Base of Fermentation Tank. An inverted shallow spherical shell provides a structurally desirable shape for the base of the fermentation tank. A flat circular plate is sometimes used for ease of excavation and construction.

Failure of this component shows up in the form of loss of water from the slurry in the fermentation chamber and this will affect the performance of the unit. Such failure can be due to excessive permeability or to cracking caused by shrinkage or differential settlement of the base. Sometimes roots of trees can cause forces on the base which might result in cracking. In areas which might become water logged, there is a possibility of a flat base failing due to flexural cracks caused by the upthrust of water.

The use of an adequate thickness of brickwork, usually the thickness of two bricks, together with suitable internal plastering and rendering prevents failure due to excessive permeability. Shrinkage strains are not of much significance when the tank diameter is less than about 4 m; if a larger unit is envisaged, reinforced concrete construction would be more appropriate but the use of several smaller brickwork units could offer better reliability than the single large unit.

Differential settlement too is not a serious problem in most locations. The weight of excavated material will generally be greater than the weight of the components of the unit and therefore the imposed pressure on the soil will be smaller than the pressure under normal conditions. It would, however, be desirable to avoid construction on soil of non-uniform nature or on freshly filled unconsolidated soil. In such situations, the weaker soil may be removed and replaced by a rubble-sand mix. The earth must be well compacted before the base is built upon it, and the bricks should be in good contact with the soil. The use of a thin layer of lean concrete screed, although this adds to the cost, is good constructional practice.

In sites subjected to waterlogging, the use of an inverted spherical shell shape for the base is more suitable in view of the compressive nature of the stresses caused by the net upthrust forces. The connection between the base and wall is a potentially critical region due to the possibility of the development of tensile stresses caused by localized bending effects, but the precautions mentioned in this section and the following one would ensure that these effects are not significant in small units. In larger units, the magnitudes of the bending stresses can be reduced by increasing the thickness around the interconnections.

Wall of Fermentation Tank. The circular cylindrical shape for the wall of the tank lends itself to easy construction. The wall will be subjected to external forces due to lateral earth pressure and to hydrostatic loading under submerged conditions. These will induce compressive hoop stresses in the wall, which can be easily resisted by brickwork. The slurry inside the tank will exert internal pressure which will tend to cause tensile hoop stresses in the wall (See Figure 2).

The external pressure will be greater than the internal pressure and therefore the net effect will be compression in the wall, provided

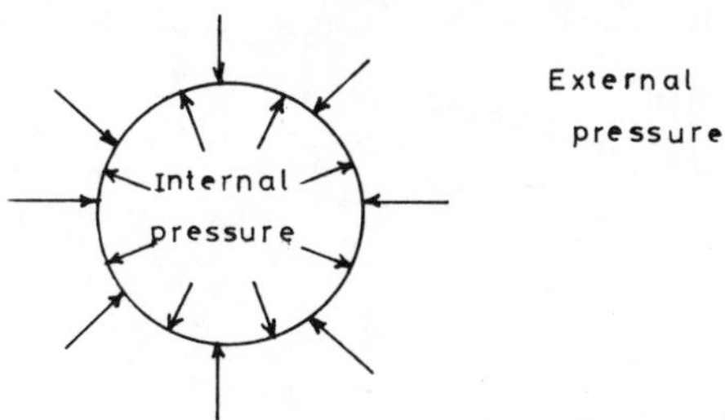

Figure 2 Cross-section of tank wall showing loading conditions

that the full external pressure acts. In order to ensure this, backfilling of earth against the external face of the wall should be carried out with every 150 mm to 200 mm rise in the height of the wall as the construction progresses commencing at the interconnection of the base and the wall. This backfill should be well compacted at a suitable moisture content so that the tank wall will be subjected to the external lateral earth pressure throughout.

If any likelihood of differential settlements has been eliminated, as described in the previous section, the tank wall will not cause problems of a serious nature. Wall permeability can be eliminated by effective internal plastering. Small leaks, should they exist, tend to get sealed after some months of usage, but the rich fertilizer could attract roots of nearby trees which might cause eventual localized damage to the tank. The regions around the connections of the inlet and outlet pipes tend to leak at times, but these can be rectified without much difficulty.

Dome of Gasholder. The spherical dome and its interconnections to the cylindrical wall and the skylight are the most critical regions of the biogas unit. At the same time, these provide the most interesting portions from the point of view of construction and analysis. The basic concept of the construction of the dome without using supporting formwork is described below; the loading and stress conditions which arise at various stages of construction are also discussed.

Construction Technique. The spherical shell dome can be constructed in brickwork without the use of shuttering by adopting simple techniques. The basis of these is to construct the dome as a series of rings, each of which will be stable as soon as it is formed--even before the mortar bonding the bricks together had hardened. These rings can be constructed with the aid of three sticks of wood fitted with protrusions, such as nails, to provide temporary support to a brick during placing (See Figure 3). Once three bricks of a ring have been placed in position supported by three sticks and jointed by mortar, the middle stick can be removed. The second brick, although now unsupported, will continue to be stable due to the wedging action of the other two bricks. The stick so removed can now be used to support the fourth brick of the ring. This procedure can be continued, keeping the first stick in position and alternately moving the other two until the ring is completed, when the whole ring will be self supporting. Variations of this technique are possible. For instance with the aid of an additional stick and another mason, construction of the ring can

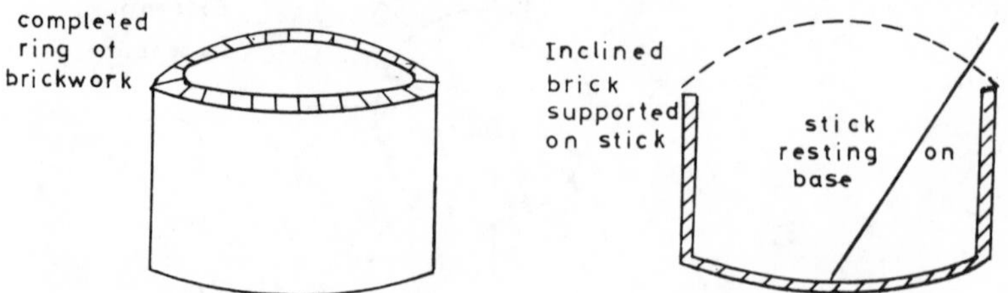

Figure 3 Construction of dome without using formwork

proceed in both directions. Once the dome and its accessories, such as the skylight and cover lid, are complete, plastering will be necessary in order to avoid gas leaks. Several thin coats of plaster, internally and externally, are required.

There is a possibility of the dome collapsing during construction if there is some impact force which acts before the mortar hardens. Failure of the dome to perform its expected function of storing gas can be due to a number of reasons. If the plasterwork has not been done satisfactorily gas could permeate out. The use of candlewax rubbed on the portion coming in contact with gas is a valuable remedy. Cracking of the dome, particularly due to localized bending effects around sections of interconnections occurs sometimes. A common cause of cracking of the dome in other regions is due to tensile stresses developed due to excessive internal pressure. This is due to inadequate depth of soil overburden which is an essential load expected in the design. These are further discussed in the analytical section below.

Analytical Considerations. The dome is an axi-symmetric shell, subjected to axi-symmetric systems of loading. During construction, the shell loading may be approximated to that shown in Figure 4 (a) with a self-weight load uniformly distributed along the surface and

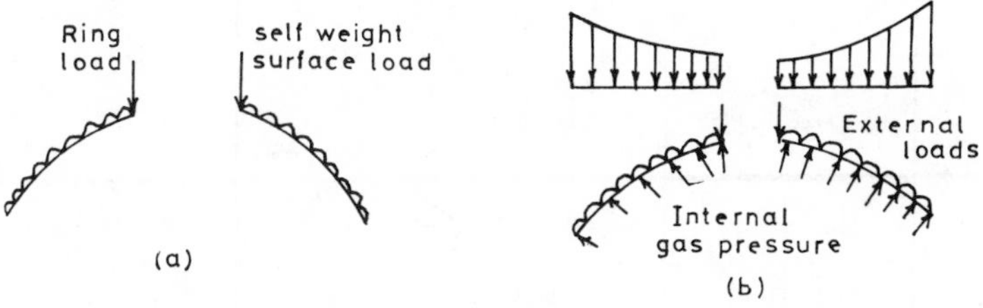

Figure 4 Loading conditions on gas-holder dome

skylight ring load from the freshly constructed brickwork rings. During use, the load systems may be represented by those indicated in Figure 4(b) comprising the self-weight surface load, the earth load, the ring load and the internal gas pressure.

Structural failure can occur due to these load systems. A somewhat common cause of failure is due to the lack of adequate external loading due to earth as shown in Figure 4 (b) when the gas pressure builds up a few days after initial charging of the fermentation chamber.

In order to obtain an idea of the nature of stresses induced in the spherical dome, as well as the base, some standard cases of loading are indicated in Figure 5. Membrane behavior may be taken as a satisfactory approximation for this type of structure. Hoop and meridian stress components are indicated, and the limiting semi-vertical angle of the dome which makes both stress components compressive, i.e. the position of the ring of rupture, is also indicated for the two appropriate cases.

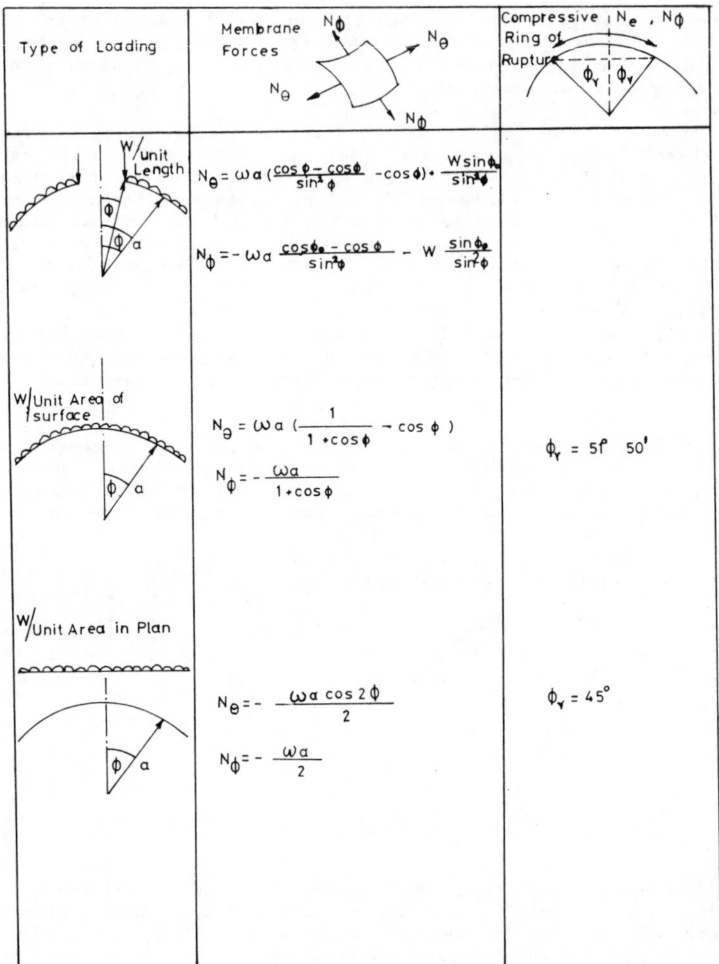

Figure 5 Membrane Stress States in Spherical Domes

If a semi-vertical angle less than 45° is selected for the spherical dome, the membrane stresses developed will be entirely compressive under earth loads and self weight loads, the combination of which may be taken as equivalent to a uniformly distributed load in plan. Such compressive stresses can be carried by brickwork and if the magnitude of the stresses induced by these external loads are greater than those generated by internal gas pressure, there will be no cracks developed in the gasholder. This will help to prevent gas leaks.

Very good compaction of earth around the dome-wall connection is necessary, and in larger units, the provision of a reinforced concrete ring beam to carry the horizontal thrust would be desirable.

Structural Testing of Biogas Unit. Once the unit has been completed, it would be desirable to test it for structural serviceability before charging the fermentation chamber, because it would be simpler to take remedial action at this stage in the event of shortcomings. This can be done by filling the fermentation chamber with water up to a suitable height and pressuring the air in the gas storage section up to about 1.5 m head of water. If no water or gas leaks occur in 24 hours, the structure has no defects. If water or air leaks occur, their positions can be isolated by having different levels of water before pressurizing and remedial action are taken.

CONCLUDING REMARKS

Fixed dome type of biogas digesters made of brickwork offer a meaningful and economical solution to the propagation of biogas technology. Most failures are due to:

i. inadequate compaction of earth below the base, causing cracking due to differential settlements;
ii. inadequate compaction of backfill outside the fermentation chamber wall causing tensile cracking of wall due to internal fluid pressure;
iii. the use of semi-vertical angles greater than 45° for spherical gasholder dome;
iv. the provision of too small a depth of soil over the gasholder dome;
v. unsatisfactory plastering and connection of pipes, giving rise to water and gas leaks.

The behavior of the components described here will help to understand the likely reason for any structural failure and enable precautionary or remedial measures to be taken.

FERROCEMENT GASHOLDER FOR TWO 60 M^3 DIGESTERS

G. Frettlôh

Regional Biogas Extension Program CDB-GATE

ABSTRACT

Ferrocement has many advantages to commend its use for the construction of the digester gasholder. Corrosion resistance, the high gas pressure realized through the weight of the ferrocement gasholder, and the relative low cost were the foremost reasons leading to the construction of two $20m^3$ gasholders in Barbados. This paper summarizes procedures for their construction and recommends their extended use, as they offer a very good chance for cost reduction.

INTRODUCTION

Ferrocement is a highly versatile construction material consisting of cement mortar matrix reinforced with layers of welded mesh and chicken wire mesh. These can be moulded into any desired shape and can be precast in mass scale or cast at the site itself. The ready availability of semiskilled and unskilled labor in developing countries can be used with advantage in ferrocement construction. The major applications are: water-storage, tanks, reservoirs, grain silos, pipes, shipping vessels, etc. It has the distinct advantage of being mouldable and of one-piece construction. Other major advantages are its low cost, nonflammability, high corrosion resistance, and, ease of repair. Considering the above advantages, ferrocement has been adopted for the construction of gasholders for biogas plants.

What led us to the decision to choose ferrocement for the construction of two $20m^3$ gasholders in Barbados?

The most obvious reason is the location of the dairy farm; only 500 meters away from the Atlantic shore, a constant wind blows upcountry and fills the air with salt. Any steel, even with good protection, that is exposed to the atmosphere will corrode in no time.

Apart from the high corrosion resistance, the weight of the floating gasholders offered another advantage for the project--the high gas-pressure (150-200 mm water). The distance from the biogas plant to the gas consumer is about 100 meters. Instead of a 1 1/2" pipe, a 1" pipe was installed. Due to the higher pressure, the conversion of the already existing equipment run by LPG (gas stove, water heater, and generator) was much easier.

Finally, it was the lower cost of ferrocement construction, compared to other materials, such as steel or fiber glass, that led to this decision. The success of implementing biogas technology depends in nearly all cases on the cost.

ROCEDURES FOR CONSTRUCTION OF A 20m^3 GASHOLDER

(1) Construction of the Mould

The cylindrical part of the mould consists of twelve equal segments made of plywood -- 20 mm thick for the rim and 4 mm thick for the surface. Each ring has one key segment with 45 ^{o} ends. This makes the dismantling of the mould much easier and also enables it to be used again. A cross at the top, and another at the bottom, keep the center pipe of 45 diameter in position. The dome of the mould is composed of wooden beams covered with plywood and the final shape was given with plaster. Then, the whole mould was painted with mould oil after the plaster had completely dryed.

2) Reinforcement

The square welded mesh (75 mm x 75 mm) was laid around the cylindrical part of the mould and the joints were welded together. The upper end of the mesh was bent according to the shape of the mould. In this way, the edge of the mould was covered. The concentric steel dome has a distance of 120 mm between one ring and the next. On top of this layer, other rods with the same diameter were fixed radially from the center towards the perimeter. On this "welded mesh basket," chicken wire mesh (12.5mm hole) was wrapped around and fastened with binding wire.

3) Plastering

The plastering was done in two shots and vibrated in with an orbital-action electric sander. The whole outside plastering needed four hours work (four masons and one laborer). Two days later the mould was removed so that the masons could proceed with the inside plastering.

4) Gas-Tightness

To eliminate gas leakages, the inside was painted twice with "Bitumen Emulsion." Before the second coat of bitumen was dry, aluminium foil was stuck onto it with an overlapping of approximately 50mm. To protect the foil, another coat of bitumen was painted on.

5) Inner-Steel Structure

There are two reasons for the inner-steel structure: first, to give a stirring effect inside the digester when the gasholder is moving up and down or when it is being rotated; second, to prevent cracking when the gasholder is being raised or lowered by a crane.

CONCLUSIONS

On completing the construction of the biogas plant, we found that ferrocement was not only the solution for that particular plant but that it could be used successfully for biodigester construction in developing countries because of its lower costs. Nearly all over the world, access can be found to the necessary materials -- cement, sand, wiremesh, and steel-rods. All that is needed is the basic knowledge of construction and skills. Apart from materials, the decisive factor is the price. Compared to steel or fiberglass gasholders, ferrocement gasholders are much less expensive. Thus, they offer a very good chance for cost reduction in the overall cost of the biogas plant.

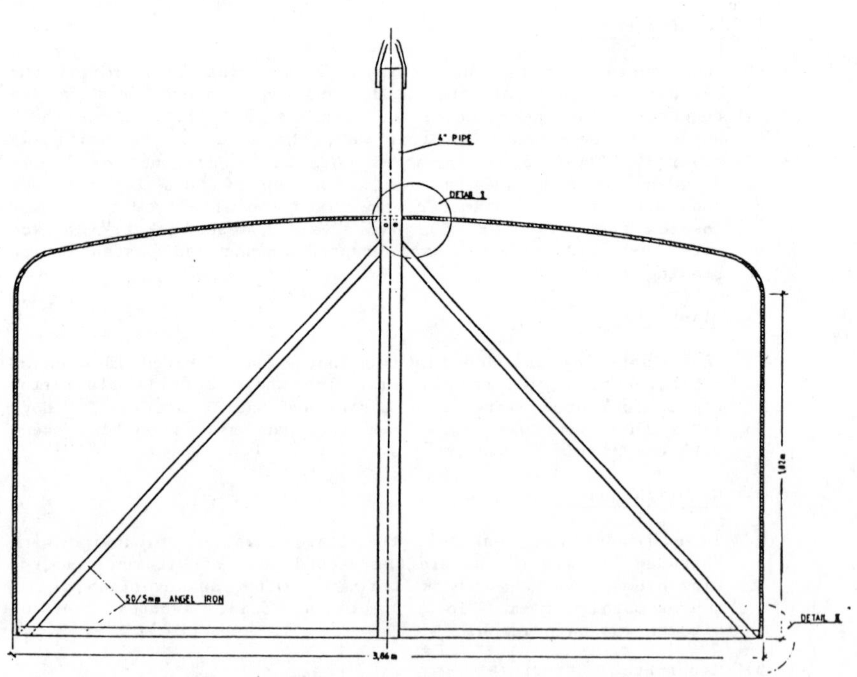

20 m³ ferrocement gasholder

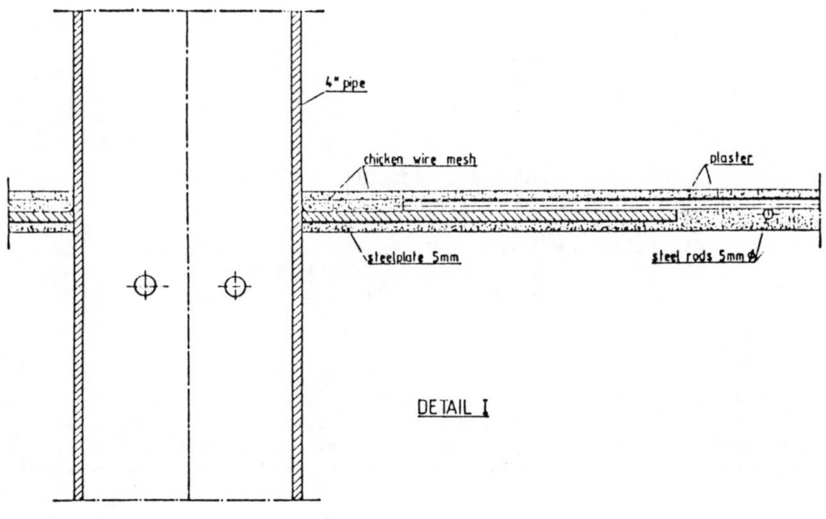

DETAIL I

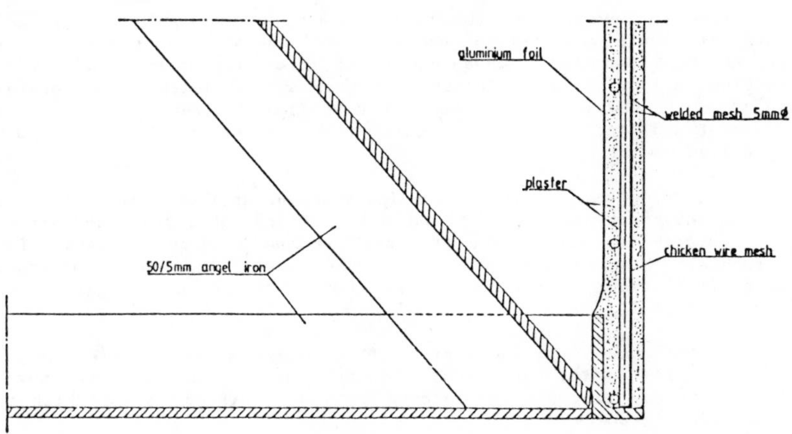

DETAIL II

SIMPLIFIED ANAEROBIC DIGESTERS FOR ANIMAL WASTE

P. Balsari, P. Bonfanti, E. Bozza, and F. Sangiorgi
Istituto di Ingegneria Agraria
Via Celoria, 2 - 20133 Milano - Italy

ABSTRACT

This paper describes four relatively simple anaerobic digestion systems suitable for farm-scale uses. Tests on a small electric generator set fueled by biogas are also outlined. Finally, a preliminary economic evaluation of the four tested systems is presented.

FOREWORD

Anaerobic digestion plants have not attained the expected diffusion into Italy's animal husbandry because of poor reliability, the need for constant supervision and specialized labor, and the high cost ($400-500 per m^3 digester) and consequent long payback period of 15-20 years.

Since the purpose of the biogas plant is to supply energy to a farm, it should be sized to meet the farm's demand. In certain cases, it may be economically advantageous to feed user points such as the milking parlour which, although they have a peak demand, are constant in time. It should also be kept in mind that Italian agriculture has low heat and electricity requirements (60% of farms have less than 3 kW installed power).

To tackle these problems, the Institute of Agricultural Engineering of the University of Milan, financed by CNR and ENEA, has been carrying out research to define simplified and low cost anaerobic plants. Four farm-scale plants (one plug-flow, two batch, and one "continuous expansion") and a system for the recovery of slurry and biogas storage were installed.

At the same time, a series of experiments was carried out in collaboration with the ACME Motori Company on a low power, biogas-fueled, electric generator set to be operated jointly with the anaerobic digesters.

BATCH DIGESTER PLANT

The plant was built around an existing manure pit of some 30 m^3 capacity (Figure 1).

The anaerobic digester was produced by covering the pit with a reinforced plastic sheet, sealed by fastening the sheet to the bottom of a 35 cm deep ditch.

The gas leaves the digester through a central opening, flows through a plastic hose, and is stored in a pillow-type gas holder (capacity 30 m^3 approx.) made entirely of reinforced plastic sheet.

The digester was fed with manure from the permanent litter of the heifer barn. Straw consumption is some 4 kg per head per day; the litter is changed monthly.

Results

The chemical analysis of the substrate fed to the digester during the various test cycles showed a total solids content (TS) in the range from 16 to 24%, and the (VS/TS) ratio to average 0.79 (VS = volatile solids).

Fermentation caused VS reduction of 19 to 30%.

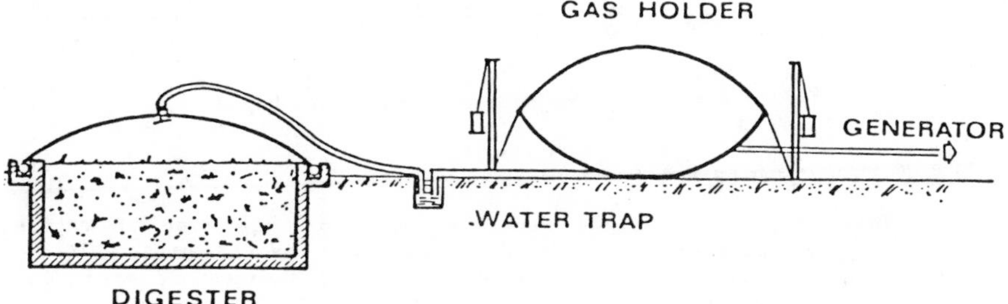

Figure 1 - Schematic View of the Batch Digester.

The temperature readings during the test cycles were found to closely correlate to ground temperatures and ranged from 6°C in winter to 25°C in summer.

Retention time depended on the seasonal average temperature. In the test period, it went from a minimum of 40 days in summer to a maximum of 120 days in winter.

Usually, the first gas formed some 48 hours after sealing the digester and contained more than 70 percent CO_2; methane content increased later on and levelled off at 52 to 58 percent.

The digester's overall production during the test cycles ranged from 50 to 700 Nm^3/month for a daily rate of 0.05 and 0.65 Nm^3 per m^3 digester respectively (Figure 2). Higher rates were achieved in the summer months, with high ground and ambient temperatures.

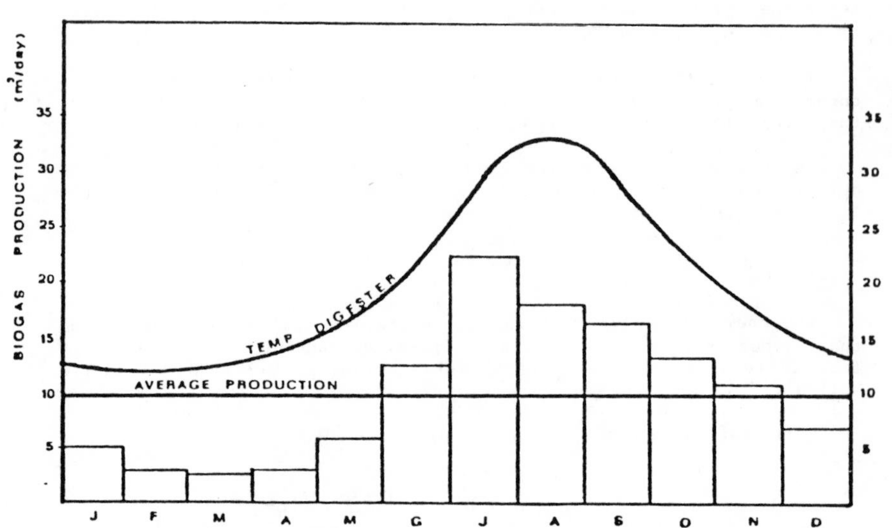

Figure 2 - Biogas Production of the Above Digester.

PLUG FLOW DIGESTER PLANT

This plant was designed for the anaerobic digestion of cattle slurry from paddocks or from slotted floor covered pits. It is composed of a collecting pit, a grinding pump, a digester, a pit for effluent storage and a gas holder (Figure 3).

The concrete and rubberized fabric digester has 12 m^3 effective operating capacity and is able to treat the excreta of 16 to 20 adult head.

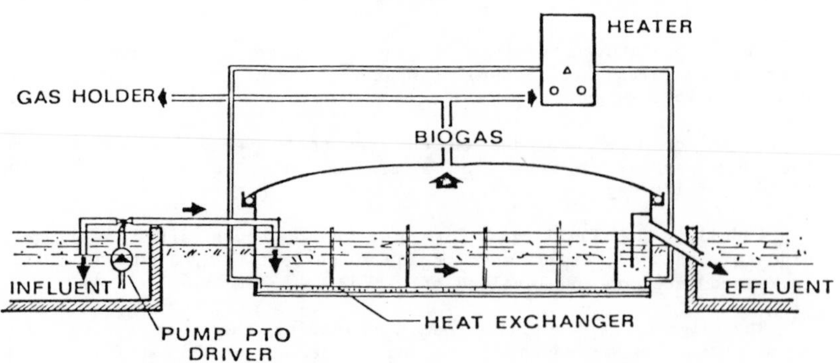

Figure 3 - Schematic View of the Plug Flow Digester.

The heater is a 15 m^2 rubber exchanger, as used in solar collectors, connected to a boiler fired with biogas from the plant.

Biogas is stored in a 30 m^3 container built of plastic sheet,

Results

The plant was run at 30°C temperature, with 20 days hydraulic retention time, and a daily load of 0.6 m^3.

After an initial period of some 15 days, the daily biogas production rate was 1.1 Nm^3 per m^3 digester and methane content levelled off at 50 to 53 percent (Figure 4).

The chemical analysis of the slurry during this first test cycle showed total solids to be 9 to 12 percent.

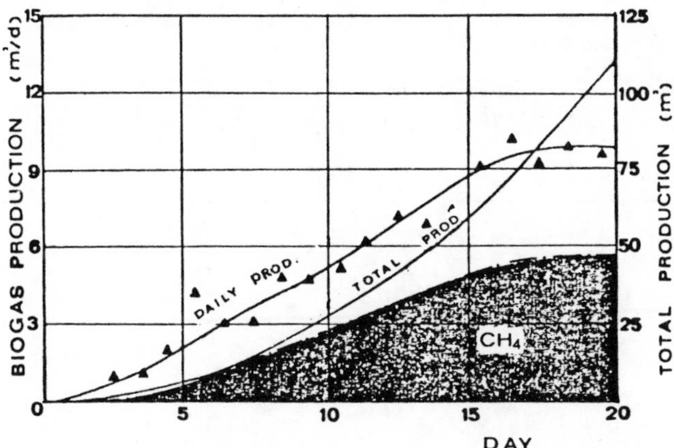

Figure 4 -Biogas Production of the Plug Flow Digester.

In the successive stages, significant non-uniformities of the fermentation process occurred. Specifically, it was found difficult to reduce stratification because of variations in the slurry's solid content (from 4.6 to 11.2 percent) and consequent uneven specific load. This poor homogeneity can be ascribed to the stabling system which is affected by weather and fodder variations from season to season.

A mechanical stirrer was installed in the digester to do away with mass stratification; however, controlling the excreta's total solids to even out the specific load was difficult. For this reason, the plant did not run smoothly; specific production rates ranged from 0.65 to 0.40 m^3 biogas per m^3 digester per day, and methane content from 50 to 58 percent.

COVERED LAGOON BIOGAS SYSTEM

During the first stage, a 10 m^2 floating collector made of reinforced Hypalon rubber was installed over a 3000 m^2 pit downstream of the anaerobic digestion plant on the Manerbio farm. This was done to evaluate the areas of the lagoon having the highest biogas production and to determine the effects of temperature and slurry level on the production.

The slurry coming from a one-stage digester, of 750 m^3 volume, enters the pit at a rate amounting to 100 m^3 per day at a temperature of 30°C.

The slurry has a TS content of 0.5 percent and a VS/TS ratio of 65 percent. During the test period (January 10 to September 10, 1984), the slurry retention time in the pit averaged between 22 and 37 days (Table 1).

Table 1 Average Seasonal Values of Daily Biogas Production in a Swine Slurry Storage Lagoon (m^3 biogas/m^2 covered surface)

Season	Slurry Level (m)	RT (d)	Temperature bottom (°C)	surface (°C)	ambient (°C)	Biogas yield (m^3/m^2)
Winter (19/1-21/2)	1.5	37.5	8.5	7	5	0.15
Spring (16/4-28/4)	1	25	20	20	17	0.21
Summer (11/7-3/9)	0.9	22.5	22	25	25	0.23

Results

On the basis of the preliminary tests carried out, the following observations were made:

- the temperatures inside the pit ranged between 6 and 22°C depending on the season;
- the gas output had no correlation with the slurry level in the pit and ranged from 0.1 to 0.3 Nm^3 per m^2 of covered surface;
- the highest output was obtained near the slurry entry (Figure 5); about 30 percent of the pit surface was involved in the anaerobic digestion process;
- the CH_4 percentage averages from 40 to 60 percent.

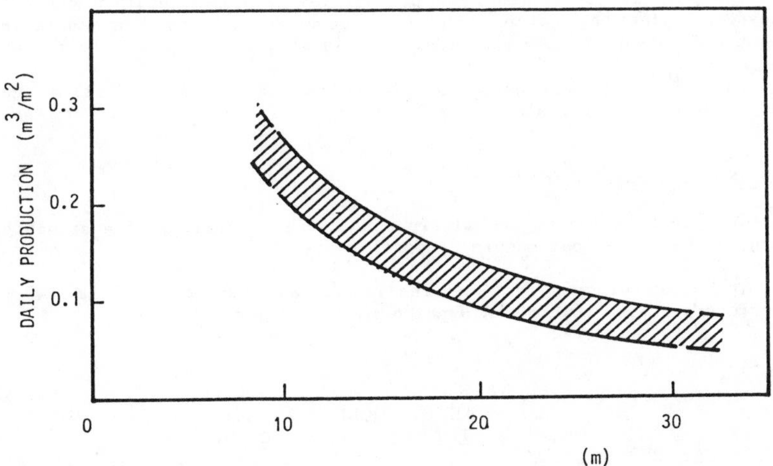

Figure 5 - Daily Specific Biogas Production in Relation to the Distance from the Slurry Inlet in the Covered Lagoon System.

Following these results, a 400 m^2 cover was made with the same material (Figure 6).

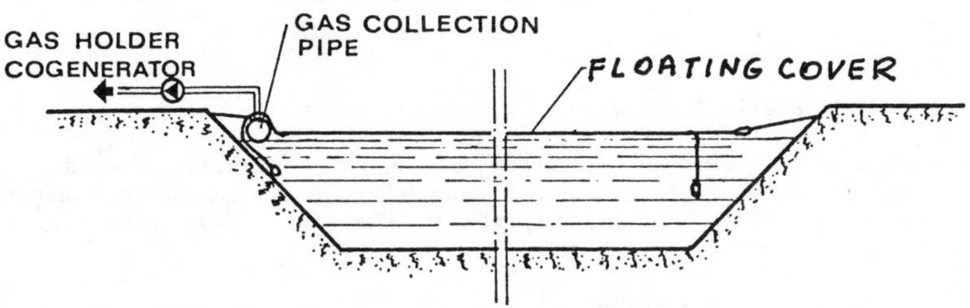

Figure 6 - Schematic View of the Covered Lagoon Biogas System

The lagoon cover is made of 1.2 mm fabric reinforced black Hypalon. A polystyrene section (150 mm x 300 mm), enclosed in 0.7 mm Hypalon, is placed underneath to allow the cover to float over the liquid. The gas flows into a 160 mm bore perforated PVC pipe laid along the wall under the cover.

Gas is removed from under the cover by a piston pump rated 10 m^3/h, driven by a 1 kW electric motor.

The pump is controlled by a pressure switch to maintain a constant pressure under the cover. The gas output is fed to a cogeneration plant and converted into electricity and heat.

In the first two weeks the plants had an output of 0.25 m^3 gas per m^2 of covered surface.

CONTINUOUS EXPANSION DIGESTER

The plant is fed by the slurry produced in a dairy-cattle stanchion barn with a mechanical manure disposal system.

The digester was made by insulating and waterproofing a 140 m^3 concrete pit with reinforced Hypalon rubber (Figure 7).

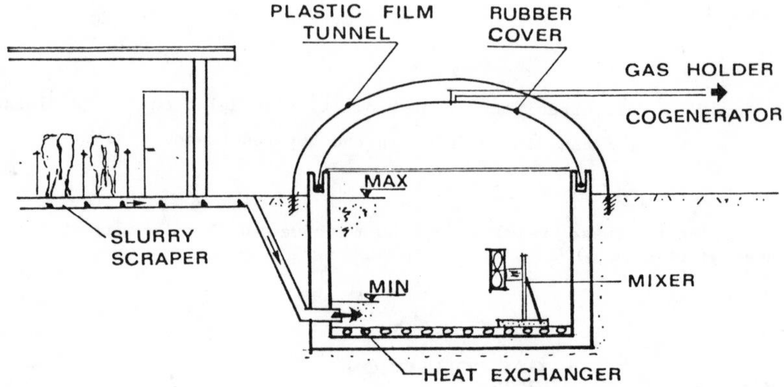

Figure 7 - Schematic View of the Continously Expanded Digester

The heating system consists of a heat exchanger made of polyethylene pipes laid on the pit bottom, which maintains a constant temperature of 30°C. A continuously running stirrer is installed at pit bottom.

The cover of this digester, as that of the batch plant, is a plastic film similar to that used to reduce heat losses from protected crops.

The plant, which is currently being started up, will receive twice daily the slurry produced by 30 lactating cattle (about 1.2 m^3), and the effluent will be unloaded after a retention time of 90-120 days, depending on the field's fertilizer requirements.

The slurry has an average 11.5% TS content and 9.4% VS content. Till now no data on production is available.

TESTS ON A SMALL ELECTRIC GENERATOR SET FUELLED BY BIOGAS

To ensure that the electric generation system being developed is of interest to a wide range of users, it was decided to experiment in the 3 to 5kW range.

A first test cycle was run with a generator set composed of a 3.5 kVA three-phase generator and a single cylinder Otto engine of 327 cm^3 capacity and 5.5 kW rated output.

The engine modifications involved the carburetor, replacing the standard head in order to increase the compression ratio, and a new control panel.

Results

During the tests, the engine was run for 300 h on biogas, and a series of tests under variable electric load and various types of fuel with different CH_4 content were carried out, as well as a series of bench tests on the engine only.

The tests were run on a farm in order to highlight the aspects of the use of an electric generating set under actual conditions.

This called for monitoring performance relating to the following:

- fueling with gas whose pressure and CH_4 content vary with time;
- fuel consumption;
- actual electric conversion efficiency;
- engine performance with time.

Therefore, a test code similar to OCDE's directive for small engine testing was adopted.

The engine was initially fueled with gas at a pressure of 0.002-0.03 bar (according to how full the gas holder was). At minimum pressure values, this was insufficient to ensure proper engine operation.

To overcome this difficulty, a low rated (40 W) induction compressor capable of delivering biogas at constant of 0.04 bar pressure (approximately) is needed. The generator set ran correctly when the CH_4 content ranged from 50 to 60%. Higher methane precentages posed no problems but are hardly attainable with the plant and substrate used for this experiment.

On the other hand, when the methane content drops to 45-50%, combustion is difficult and the engine runs poorly; below 45% the engine cannot be operated at all.

Specific consumption with gas fueling (see Table 2 for gas characteristics) and under 0.75 to 2.5 kW electric load was 2.8 to 1.1 Nm^3kWh (Figure 8), for an overall efficiency of 7-17.5%. The low yield is to be ascribed to poor carburation and to the electric generator's efficiency (70%). During the tests no engine trouble or failure was recorded. Servicing was limited to an oil change every 50 hours as specified by the manufacturer.

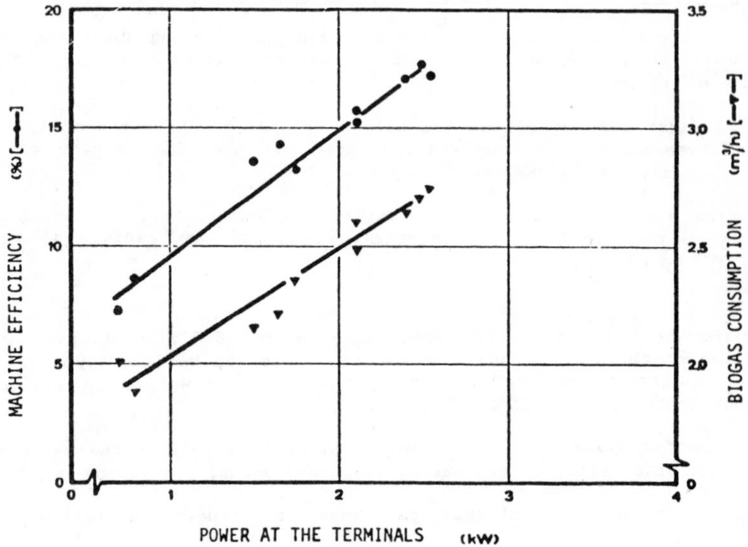

Figure 8 - Efficiency and Consumption of the Biogas Fuelled Generating Set

AN ECONOMIC EVALUATION OF THE PLANTS

Plant cost was evaluated by not accounting for any existing facilities but including retrofitting costs. The corresponding capital outlay is US$30 per m^3 (or approximately US$80 per head) for a batch digester; for a plug-flow digester, the outlay is US$250 per m^3 (or US$150 per head).

Based on daily gas production rates, if the value of the biogas output is calculated from its heat value and current fuel oil price in Italy, the payback period is 2 years for a batch plant and 6.5 years for a plug-flow plant. If no adequate facilities exist on the farm, their added cost extends paypack periods to some 5 years for a batch plant and 10 years for a plug flow plant.

The covered lagoon biogas system has a comprehensive cost (pit excluded) of about $15 per m^2. Referring to the pilot plant biogas production, a payback period of not more than 2 years is expected.

The continuously expanding digester costs are about $80 per m^3; considering the expected production, the payback period could be approximately 4 years.

Table 2 - Characteristics of the Biogas Used for the Tests

Component	Quantity
Hydrogen	0.01% by gas volume
Oxygen	0.53% by gas volume
Nitrogen	4.86% by gas volume
Methane	56.80% by gas volume
Carbon dioxide	37.80% by gas volume
Gross heat value at 0°C	6.28 kWh/Nm^3
Gross heat value at 15°C	5.95 kWh/Nm^3
Net heat value at 0°C	5.66 kWh/Nm^3
Net heat value at 15°C	5.36 kWh/Nm^3
Density at 0°C	1.2232 kg/Nm^3
Density at 15°C	1.1595 kg/Nm^3

It should be pointed out that these evaluations assume that the biogas output is fully utilized. This condition is hardly met on a livestock farm, especially in the summer months when biogas output is high and energy demand is low.

CONCLUSIONS

The results of the experiments indicated the feasibility of simplified, low-priced anaerobic digesters.

It should be emphasized, however, that a plant's cost efficiency does not depend only on its low cost, but chiefly on the utilization of the energy output which is likely to be higher for small or very small plants.

The data gathered for the batch plant indicate that it should be suited for the small to medium farms of south-central Italy, where, because of higher ambient and ground temperatures and low energy demand, the capital payback period can be less than 2 years.

The digester-generator combination appears especially interesting for small and medium farms and to meet the constant electric load of larger farms. A point worth mentioning is that the generating set should be rated from two to three times the farm's installed power to cope with startup surges.

The results of the plug-flow plant tests again prove that anaerobic digestion is a process depending on a number of parameters which could hardly be optimized on a livestock farm without drastically changing the farm organization. Thus, plug-flow plants are more suitable for livestock farms where animal feeding is uniform.

The covering pit system seems to be particularly interesting as it does not release odors and it produces a biogas output which allows a payback period of 2 years.

REFERENCES

Sangiorgi, F., P. Balsari, and P. Bonfanti. 1981. Low Cost Biogas Installations for Italian Agriculture. Poster Session, 2nd International Symposium on Anaerobic Digestion. Travemunde.

Balsari, P., P. Bonfanti, and F. Sangiorgi. 1981. Utilizzazione della digestione anaerobica come fonte energetica nelle aziende agricole: primi risultati ottenuti da impianti sperimentali. Regione Lombardia, 2° rapporto informativo, 39-63.

Balsari, P. and G. Riva. 1983. Primi risultati di prove su un piccolo generatore di corrente alimentato con gas biologico. Conferenza Internazionale Energia e Agricoltura 2:100/1-100/22. Milano.

Balsari, P. and P. Bonfanti. 1983. Impianti semplificati per la digestione anaerobica: possibilit di inserimento nelle aziende agricole e risultati di due anni di prove. Conferenza Internationale Energia e Agricoltura 2:102/1-102/19. Milano.

Balsari P., P. Bonfanti, F. Sangiorgi, and G. Riva. 1984. Results of Test Carried Out with Simplified Anaerobic Digester Coupled to a Small Generator Set. 10th International Congress on Agricultural Engineering 3:164-171. Budapest.

COLD CONDITION BIOGAS

S. B. Pradhan
Soil Science & Agriculture Chemistry Division
Department of Agriculture
HMG, Nepal

ABSTRACT

Biogas plants insulated with straw and plastic covering increased the gas production throughout the year. Insulated plants with feeding materials at 35°C and at ambient temperature produced 28 and 40 percent more gas respectively as compared to plants without insulation. Between two insulated plants, treatment with warm feeding material showed positive effect only during the cold period.

INTRODUCTION

Nepal is a land-locked, mountainous country situated between China and India. It has an area of about 141,000 sq. km and the population is approximately 16 million. Around 17 percent of the total area comprises flat lowlands, with the rest being hills and mountains, including the highest mountain in the world. The climate varies from tropical to alpine within the distance of 175 km. Ninety-three percent of the population is engaged in agriculture and 66 percent of gross domestic product is based on it.

The country has not yet found any proven deposits of fossil fuel, such as gas, oil and coal. Fuels for transport and industry must be imported. Firewood is the chief source of energy for the domestic sector, although kerosene and gas are also used, to a certain extent for cooking and heating purposes in urban areas.

Rapid population growth has forced the people to bring more land, including the forests, under cultivation. The increasing need for firewood is leading to massive deforestation. The prices of fossil fuels and firewood are increasing as supplies become scarce and expensive. People have started to use cow dung, formerly applied in the fields, as a cooking fuel, thus depriving the soil of necessary nutrients.

Although Nepal has high potential for hydropower, only a small fraction of it has yet been tapped. There are several reasons for this underexploitation of the hydropower, one of them being the high capital investment.

To supplement the domestic fuel demand, a biogas program was introduced in 1975. This could reduce the forest cutting and cow dung burning. More compost and humus will thus be available to the field, which will help to improve soil fertility. In 1978, Gobar (cow dung) Gas Tatha Krishi Yantra Vikas (P) Ltd. was established to work toward the promotion of biogas.

There are now more than 1,500 biogas plants in the country and the number is increasing every year. However, the biogas users have not been fully satisfied due to reduced gas production in winter, during which the demand is at peak. To meet the winter requirement, the farmer has to install the larger plants but the amount of feeding material available to them limits plant size.

Many sophisticated techniques for raising the slurry temperature have been reported by different countries. South Korea, for example has developed the technique of blowing hot air through a pipe called "bubble gun" into the slurry inside the digester. Similarly, Gobar Gas Tatha Krishi Yontra Bikas Company uses the method of circulating hot water coming out of the engine cooling system through a pipe dipping in the slurry. These techniques used sophisticated equipment, that require heavy initial investment, compared to the cost of less sophisticated plants installed by the farmers. In addition, the spare parts for sophisticated equipment are not readily available in the villages where most of the family size plants are situated, and the maintenance is a handicap.

Therefore, it has become very important to work out simple as well as cheap technology for increasing gas production during the cold conditions, so that there will be less problems in day-to-day operation, and more chances for widespread adoption of the technology.

METHODOLOGY

Three identical dome type biogas plants, each of 4 m^3 capacity digester were constructed at Khumaltar, Kathmandu (Altitude 1320 m). One of the plants was left to operate at ambient temperature (T_1), whereas the other two were covered with paddy straw and white plastic sheet. One of two insulated plants was being charged with feeding material at ambient temperature (T_2), whereas the warm feeding material at 35°C was used in another (T_3).

Each of these three biogas plants was charged with equal amounts of cow dung and water mixed in a 1:1 ratio. To enhance the gas production, fresh dung was mixed with the digested slurry during the initial period. This mixture started producing gas after 15 days. Daily gas production from three plants was measured with gas meters. Maximum and minimum temperatures of the slurry were recorded at about 0.8 m depth in the outlet. Similarly, maxima and minima of air and the temperature of the feed, just before charging, were also recorded. In case of insulated gas plant (T_2), a dial thermometer having a long stem was being used to record the daily fluctuation of slurry temperature.

RESULTS AND DISCUSSION

In the beginning, gas production was not stabilized. Therefore, in the following discussion the results for the initial three months (February-April, 1982) are not considered.

The highest monthly average gas production was observed during the month of August from insulated plants (T_2 and T_3), whereas in the case of the plant operating at ambient temperature, it was observed during July (Table 1). Effect of treatments, as shown by increase in gas production over T_1, was maximum during October and November for T_2 and T_3 respectively (T_2 - T_1 and T_3 - T_1 in Table 1). Gas production was minimum during February for all the treatments, whereas the increase in gas due to treatments was observed lowest during the month of June.

Highest reading of maximum and minimum temperature of the air was recorded during the months of June and July, respectively; whereas lowest reading for both was recorded during January (Table 2). Similarly, highest reading of maximum and minimum temperature of slurry for the treatments T_2 and T_3 was recorded during the months of July and August respectively. In the case of T_2 only minimum temperature was recorded with dial thermometer, and the highest reading was observed during September.

The lowest reading of maxima and minima temperatures of air, slurry, and feed was recorded during January.

The maximum temperature of the slurry seems to be effected by air temperature after one month, whereas the minimum was during the same period (Table 2). Gas production seems to be in accordance with the slurry temperature. Increase of gas due to treatment was maximum as the temperature of the air started decreasing rapidly. During that period, slurry temperature of the insulated plants decreased slowly.

In terms of percentage, increase in gas production due to both treatments was highest during November (Table 1). Overall increase in gas production due to treatments T_2 and T_3 for the entire period, was found to be 32% and 25% respectively. Insulation of T_2 and T_3 was removed during a warm period in 1984 (March-June). Considering the increase in gas production, during the insulated period, the effect of treatments T_2 and T_3 was found to be 40% and 28% respectively. Effect of T_3 over T_2 was found to be positive only during the cold period.

Acknowledgement

The author is grateful to FAO, particularly Organic Recycling Project RAS/75/004 for providing financial support for this work and to Mr. B. K. Thapa, Joint Secretary, Ministry of Agriculture, for his guidance during preparation of this paper. Last, but not least, the cooperation of Mr. B. N. Regmi, Assistant Soil Scientist, and Mr. N. K. Shrestha, Junior Technician of Soil Science and Agriculture, Chemistry Division, are acknowledged.

Table 1 Monthly Average Gas Production and Effect of Treatments

	Gas Production, m^3			Increase in Gas Production			
Months	T_1*	T_2*	T_3*	T_2-T_1	%	T_3-T_1	%
February 1983	0.105	0.221	0.221	0.116	110	0.116	110
March	0.312	0.464	0.501	0.152	49	0.189	61
April	0.335	0.550	0.513	0.215	64	0.178	53
June	0.521	0.658	0.551	0.137	26	0.030	6
July	0.595	0.794	0.628	0.199	33	0.033	6
August	0.586	0.807	0.681	0.221	38	0.095	16
September	0.338	0.773	0.682	0.235	44	0.144	27
October	0.440	0.679	0.553	0.239	54	0.113	26
November	0.337	0.525	0.496	0.188	56	0.159	47
December	0.340	0.450	0.466	0.110	32	0.126	37
January 1984	0.320	0.350	0.393	0.030	9	0.073	23
February	0.293	0.323	0.355	0.030	10	0.062	21
March	0.372	0.379	0.410	0.007	2	0.038	10
April	0.424	0.468	0.515	0.044	10	0.091	21
May	0.460	0.476	0.496	0.016	3	0.036	8
June	0.497	0.520	0.521	0.023	5	0.024	5

*Treatments

T_1 uninsulated

T2 insulated, ambient temperature feeding

T3 insulated, warm feeding (35°C)

Table 2 Monthly Average Temperature, °C

			T_1			T_2		T_3	
	Air		Feed	Slurry		Feed	Slurry	Slurry	
Months	Max.	Min.		Max.	Min.		Min.	Max.	Min.
February 1983	19.99	14.87	15.9	-0-	12.43	-0-	12.4	-0-	13.78
March	27.50	13.3	18.4	18.9	15.3	-0-	15.28	-0-	17.6
April	30.09	15.86	20.62	22.0	16.66	20.3	16.4	-0-	18.4
June	36.27	24.09	27.23	25.23	22.82	27.1	22.6	25.8	23.3
July	35.50	26.0	27.5	25.77	23.6	27.9	23.6	24.6	
August	34.7	25.8	27.3	25.5	23.9	28.0	23.9	26.1	24.7
September	33.3	25.0	27.1	25.0	23.7	27.6	24.2	25.1	24.0
October	30.6	19.0	24.6	23.2	21.8	24.6	21.8	23.5	21.9
November	25.7	12.2	19.7	19.7	16.2	20.6	17.6	20.5	18.5
December	20.8	6.9	14.7	16.6	14.6	15.8	13.8	17.2	15.4
January 84	19.5	5.7	13.4	14.6	12.4	13.9	10.7	16.1	14.3
February	22.8	8.3	16.1	15.2	13.6	16.2	11.9	16.6	14.7
March	26.7	14.0	21.1	17.2	15.6	21.6	16.5	21.3	17.4
April	32.4	16.8	22.3	19.6	17.6	22.8	19.4	23.6	20.3
May	33.1	21.8	24.2	21.7	19.9	24.4	21.5	23.2	21.1
June	31.9	25.1	25.9	24.2	22.4	26.7	23.8	24.4	23.0

A MODIFIED ANAEROBIC FERMENTER FOR EGYPTIAN AGRICULTURAL RESIDUES

A. Zakaria Ebrahim
Department of Chemical Engineering, Faculty of Engineering and Technology, Minia University, Minia, Egypt

ABSTRACT

Anaerobic digestion of local agricultural residues was investigated in a laboratory fermenter. The efficiency of the mixing devices (modified paddle mixers) was assessed. The three-phase fluidized bed techniques was also investigated.

It was found that the suggested modifications in design and technique had decreased the detention time by about 20% and the incubation period by about 15%, while an increase in the rate of evolution of gaseous products, by a factor ranging from 0.23 to 0.27, was detected. The fluidized bed technique was also found to be quite advantageous.

INTRODUCTION

Agitation of the fermenter contents is often recommended to ensure intimate contact between the microorganisms and their food and to increase the rate of decomposition by releasing small trapped gas bubbles from the microbial cell matrix. It also helps to break up scum. Nevertheless, the Public Works Research Institute of Japan found that continuous mixing produced only 5% more biogas than once a day agitation.

Mixing methods may be affected by: daily feeding of the digester; effective design and manipulation of the inlet and outlet arrangements to ensure mixing; creating a flushing action of the slurry through a flush nozzle; creating a mixing action by flushing the slurry so that the flow is tangential to the digester contents; installing wooden conical beams that cut the surface when moving up and down[1]; gas recirculation[2]; and finally, installing mixing devices that can operate manually or mechanically.

This work aims at evaluating the efficiency of the mixing devices installed in a laboratory scale fermenter, and also at choosing a design which is more adequate and whose results may be compared with those obtained by other methods of mixing.

On the other hand, in the techiques of fermentation processes as energy generating systems, three-phase fluidized particle operations had been rarely used as a mean for increasing fermentation efficiency. Thus, another aim of this work was to carry out trials to test the merits of fluidized bed techique in the fermenter operation and design.

APPARATUS

Choice of a Laboratory Fermenter

A paddle wheel horizontal fermenter was chosen, the features of which are shown in Figures 1, 2, 3[3]. This type of fermenter is suitable for the production of methane gas from agricultural, animal, and human wastes. Also, it is easy to design, manipulate, and operate.

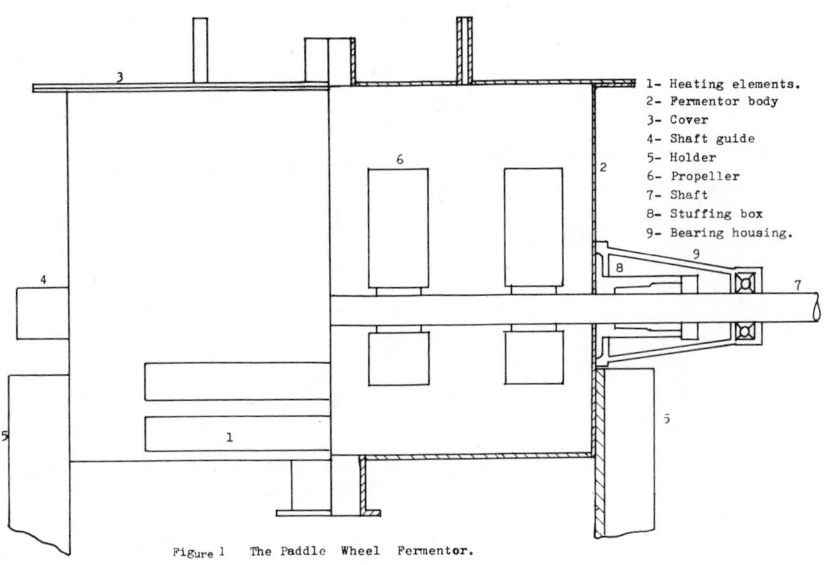

Figure 1 The Paddle Wheel Fermentor.

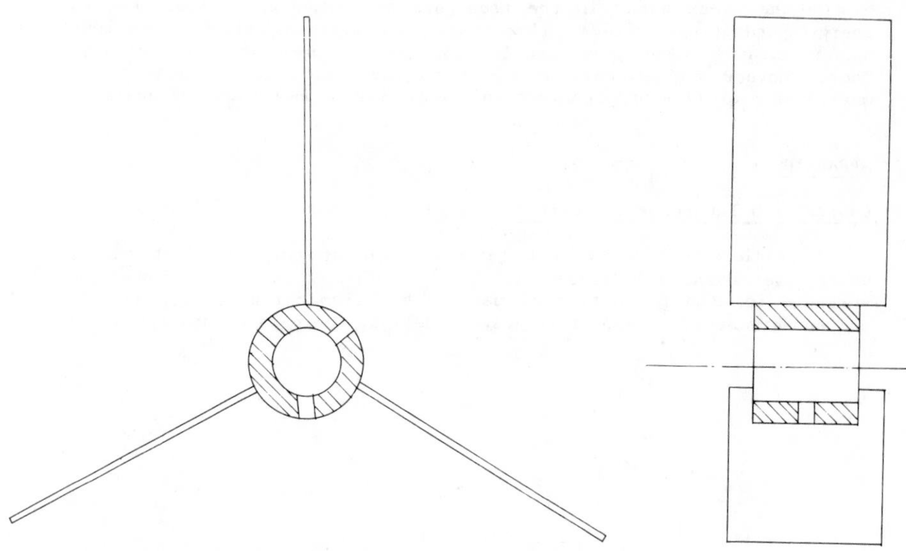

Figure 2 Solid Plade Impeller

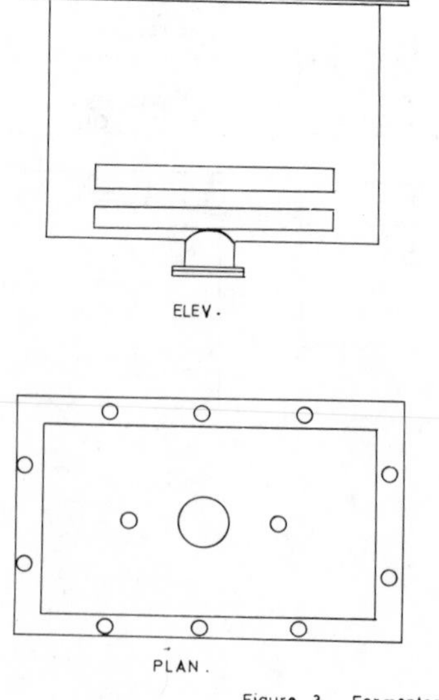

Figure 3 Fermentor (General Views)

On a laboratory scale, a fermenter capacity of about 10 l is sufficient to give information about its performance, and the results obtained would be valuable for scaling-up purposes. Length to diameter ratio is chosen to be about 2. Stainless steel is the material of construction of all parts in contact with the substrate. To ensure better gas production efficiency, the charge used was kept as shallow as possible. Temperature, concentration, density, viscosity and other physical constants are easily controlled and measured in this chosen fermenter.

The original design data are as follows:

- Fermenter body (stainless steel): 350 mm length, 200 mm bottom diameter and 250 mm height.
- Four impellers of the blade type 1mm thickness, 40 mm width and 80 mm length each. Each impeller has three arms connected to the impeller hollow hub by riveting.
- Power supply is 0.125 hp motor.
- Revolution reduction system consisting of:
    a. a gearbox (worm type) reduction ratio 1:40.
    b. two pulleys (D_1 125 mm, D_2 425 mm) connected together by a V-belt to reduce the rotation of the fermenter shaft to about 10 r.p.m.

Within the scope of this work, impellers have been the main concern. Changes have been carried out trying not to affect any of the other design features of the fermenter, thus allowing comparisons to be made between the original impeller's design and the proposed ones.

The Proposed Impeller Design

The factors and considerations that determine the most suitable type of impeller for a given duty are often dependent on those influencing agitation characteristics [4,5]. In the applications of agitation the primary effects are concerned with one or more of the following physical processes: mass transfer at an interface, heat transfer, and the dispersion of solids, liquids or gases. On the other hand, the degree and rate of mixing, as well as its efficiency, are dependent to a large extent on characteristics pertaining to rotating impeller, such as its shape, speed, dimensions and its position in the vessel. The physical properties of the material to be agitated, besides the shape and dimensions of the containing vessel, are all of non-ignorable influence on the rate and degree of mixing in a biochemical reactor. Internals immersed in the reactor, such as fittings and connections, have also an effect.

Taking the above statements into account, it is not so simple to select an impeller design specified for a certain reactor job, especially when its contents are to be sensitive to sudden changes in temperature, concentration, and acidity or alkalinity (pH value). Thus, the choice of an impeller which operates smoothly without sudden impacts on the charge is of importance. Figure 4 represents a proposed modified design for the paddle impeller discussed. This is expected to be more consistent with biochemical processes. It is analogous in shape to the original design but differs in replacing the solid blades by a frame containing free rotating (swinging) sub-blades for which the

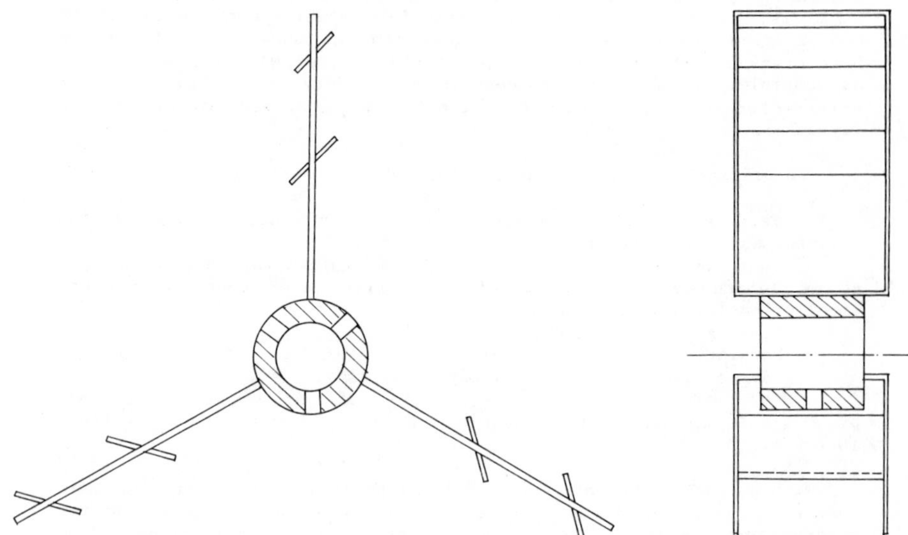

Figure 4 Modified Impeller (Swinging Type)

angle of inclination with respect to the plane of the frame can be changed and fixed in place from 0° to 90°, i.e. from being in plane with the frame itself to the extent of being perpendicular to it.

This design for the impeller had proved to be flexible, giving different modes of mixing, depending on the type of the charge, by merely changing the angle of inclination of the sub-blades as well as by changing the number of frames used.

Three-Phase Fluidized Bed

Another proposal is the use of three-phase fluidization for a double acting purpose as a means of complete mixing and also as a means of gaining the merits of fluidized beds to increase the fermenter efficiency.

A perforated tube, lined by a circular screen in its length, (to minimize the escape of the solids contained in the substrate to the inside of the tube) is inserted in the bottom of the fermenter. This is externally connected to gas mains. These may be branches from the main gas exit from the fermenter as a recirculated part at a slow flow rate. Beside the mixing effect of these circulating gases, the solid contents in the fermenter becomes highly fluidized in a medium containing gases, liquids and solids. The contact between the microorganism and the solids becomes highly efficient as will be seen later.

EXPERIMENTAL TECHNIQUE

To start a given experiment, the fermenter is cleaned and all connections and accessories are prepared. The chosen type and number of blades for mixing are then connected. The constituents for a run ready for use in the fermenter are next fed carefully and well homogenised. The fermenter is finally closed; its temperature is regulated and controlled.

The gas measuring system is connected to the fermenter outlet. This consists of a measuring vessel full of a displacement salt solution (sodium chloride/citric acid) as shown in Figure 5, by which the volume of solution collected is a measure of the volume of gas evolved from the fermenter outlet in a certain time.

Samples of the gas evolved are periodically injected in a IGC-112 gas chromotograph to identify the constitution of gases and to verify qualitatively the presence of methane and carbon dioxide gases. At the end of an experiment, the fermenter is opened and emptied from its residual contents. Samples of these residuals are taken for analysis to evaluate for use as a fertilizer[6].

Experiments were carried out using agricultural residues (such as rice straw and tree leaves), waste paper (treated), and water weeds. In this research, stress is made on the results obtained from the water weeds, especially water hyacinth whose high moisture content is an asset for a fermentation operation and thus requires no dewatering. This is economically advantageous [7]. Also water hyacinth provide the nutrients required for the methane-producing bacteria, such as nitrogen, potassium and phosphorus, in quantities and proportions adequate for good growth of the gas-producing bacteria.

RESULTS AND DISCUSSIONS

Effect of Using the 3-Phase Fluidisation Technique

In the course of this research, nitrogen from compressed cylinders was used at a low flow rate as the fluidizing gas. Figure 6 shows the results obtained. It is obvious that the use of fluidized bed technique had improved the efficiency of the fermenter. The retention time had decreased by about 25%, the total volume of gases generated had increased, while the incubation period had decreased.

These improvements are due to the great advantages of the fluidized bed techniques, which is characterized by an excellent heat and mass transfer, as well as good mixing. This improves the contact between the microorganisms and the solids, and increases its activity in action.

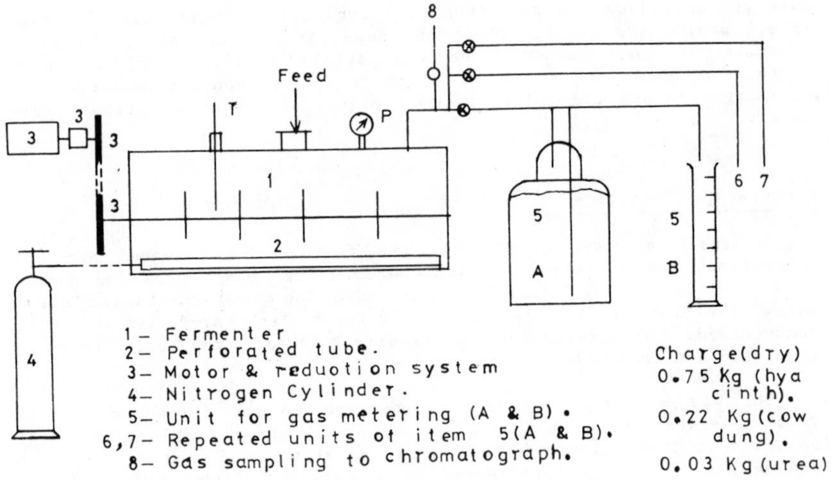

Figure 5 Metering System and Connections.

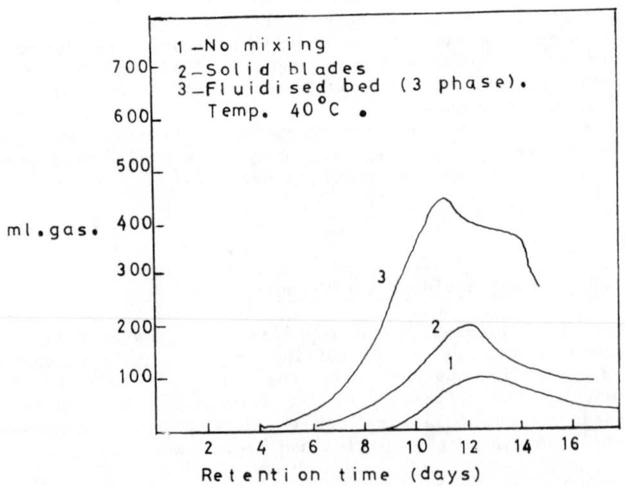

Figure 6 Effect of the Fluidised Bed Technique.

The decreased incubation periods, on the other hand, are due in most part to the flushing action of nitrogen gas which expells air from the fermenter, and thus the anaerobic fermentation begins earlier.

Effect of the Modified Paddle Mixer

Figure 7 presents the results obtained. The swinging blade mixer gives higher total volume of gases for a 45° inclination to the plane of the frame, while this volume is lower for 30° and 60° inclinations. However, these last values are higher than those of the solid blade paddle itself.

This effect may be in greater degree attributed to the efficient mixing characteristics of the swinging blade paddle than to the solid blade, especially at an inclination of 45°. The effect of mixing depends to a large extent on the angle proximity of the mixer blade with the surface of the substrate; smoother angles give better results. In this latter case microorganisms get in better and efficient contact with the solid particles[5]. It is also essential to note that smoother contacts prevent local over-heating and permit temperature homogeneity in the substrate, thus allowing efficient bacterial action.

Effect of Type and Duration of Mixing

Figure 8 represents a comparative study between two different types of mixing. Curve 1 represents the results where the mixing is carried out by the swinging blade paddle mixer once a day, while curves 2, 3 and 4 are the values for twice, three-time and four-time-a-day mixing, each for a period of 15 minutes (at about 10 r.p.m.).

It is clear that there is an obvious advantage for repeating the mixing operation three times a day, while the fourth one has no effect. Such enhancement may be attributed to the good contact which happens between the micro-organism and the renewable solid surfaces as a result of repeated mixing.

Effect of Temperature

To study the effect of temperature, some experiments were carried out using water hyacinth. The results are shown in Figure 9.

As expected, increasing temperature increases the productivity of the fermenter. Yet, the interesting observation is that the productivity of the fermenter at lower temperature, using the modified swinging blade mixer, is higher than that observed when using the ordinary solid blade mixer at higher temperatures. This adds to the available data on the efficiency of the swinging blade mixer in the fermenter. Figure 10 represents this comparative observation.

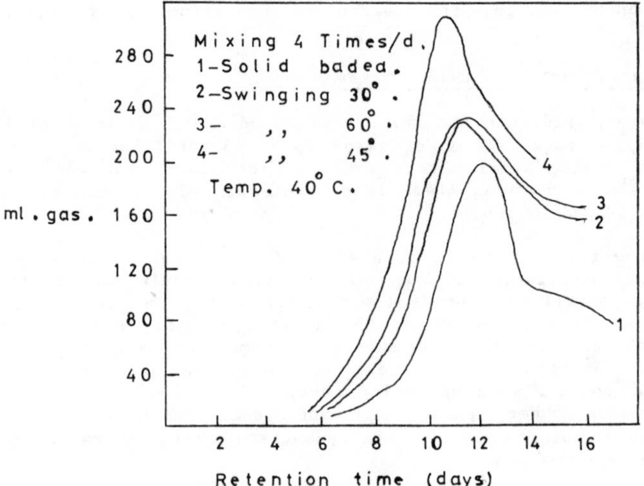

Figure 7 Effect of the Swinging Paddle Mixer

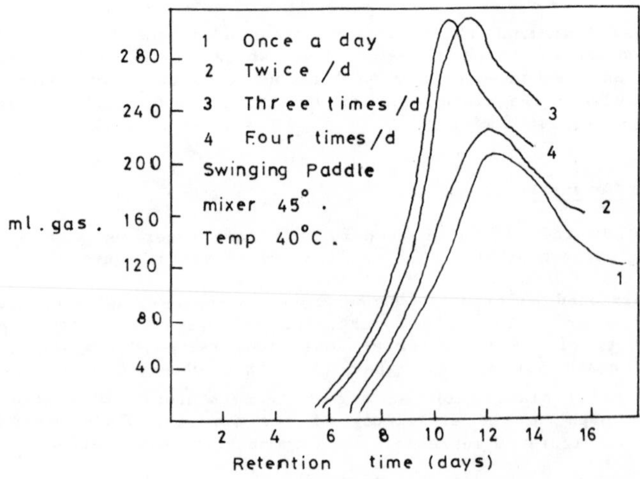

Figure 8 Effect of Repeating mixing Operations

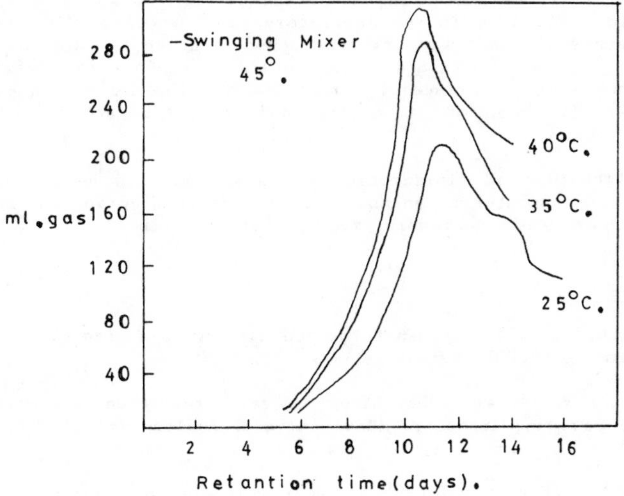

Figure 9 Effect of Temperature.

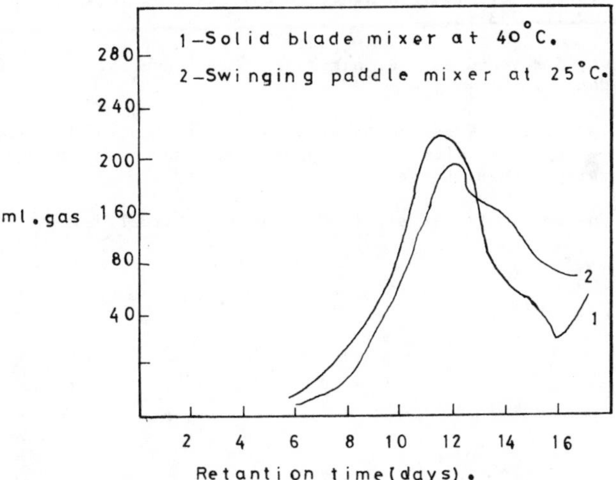

Figure 10 Double Effect of Both Temperature & Type of Impeller.

CONCLUSIONS AND RECOMMENDATIONS

In spite of the increasing advantage of using a wheel paddle mixer with solid blade arms in the laboratory scale wheel paddle fermenter, it has been shown in this work that a modification in this solid blade to a swinging form contained in an outside framebody is advantageous. The retention time decreased by about 20% and the incubation period by about 15%. The rate of gas evolution increased by about 25% (23% to 27%).

The technique of fluidization was also found to be advantageous. Flushing the fermenter by an inert gas before running an experiment had proved successful in decreasing the incubation period.

REFERENCES

1. Joppich, W. 1957. German farms too use fuel gas plants. Indian Farming. 6(11), 35-40.

2. Patha, B.N., et al. 1956. Effect of gas circulation in a pilot scale cow dung digester. Environmental Health (India) VII, 208-212 (1956).

3. Sewilam, F.F., et al. 1981/82. Design of a Biogas Fermenter. Faculty of Engineering and Technology. Minia University. Minia, Egypt.

4. Uhl, V.W., et al. 1967. Mixing, Theory and Practice. Academic Press. V.11. 293-96.

5. Coulson, J.M., and J.F. Richardson. 1968. Chemical Engineering. V.2, Chap. 18.

6. Ebrahim, M.Z.A., Research in progress, Department of Chemical Engineering. Faculty of Engineering and Technology, Minia University, Minia, Egypt.

7. N.P. Cheremisinoff, et al. 1980. Biomass. Marcel Dekker. P.26.

MODULAR BIOGAS PLANT PARTICULARLY ADAPTED TO TROPICAL RURAL AREAS: A SHORTER COMMUNICATION

D. Compagnion, D. Rolot, H.P. Naveau and E.J. Nyns
Unit of Bioengineering, University of Louvain,
1/9, Place Croix du Sud, B-1348 Louvain la Neuve, Belgium

and

V. Baratakanwa, D. Nditabiriye, J. Ndayishimiye and P. Niyimbona
Study Centre of Burundi on Alternative Energies (C.E.B.E.A.),
BP 745, Bujumbura, Burundi

A large experience on biomethanation in rural areas has been acquired from two classical methane digesters, the Chinese type on the one hand and the Indian type on the other hand. Whereas the operation of these digesters is rather simple, their building is more complex and in many cases their performances (daily rates of biogas production and biogas yields from starting material) are low (Chen Ru-Chen, 1983).

A biogas plant, adapted to rural tropical areas, is being developed as part of a biomethanation development project in Burundi (Africa). It combines reinforced concrete to insure a liquid-proof bottom part of the methane digester, with flexible polymeric material to insure lightness of the gas of the upper part of the methane digester and of the separate biogas holder.

The shape of the methane digester is a flat parallelopiped which insures a large surface area to minimize the hazards of scum formation. Mixing occurs manually through a simple mechanical device near the exit end of the methane digester. Operation of the methane digester is semi-continuous. Loading occurs through an inlet on one of the narrow sides of the methane digester, the outlet being located at the oppositve narrow side. The bottom of the methane digester is slightly inclined from the exit end towards the inlet end to promote the accumulation of active biomass in the methane digester.

To maintain the temperature of 30°C in the digestion mixed liquor, the methane digester is only partly buried in the ground and as much as possible exposed to the ambient warm atmosphere. The color of the polymeric material which constituted the upper part of the methane digester is black. Wherever necessary, a complementary greenhouse effect is obtained through transparent panels organized around the methane digester.

Several specimens of this biogas plant are presently in operation in Burundi. The feed for biomethanation consists of different kinds of animal manure and of slaughterhouse residues.

This work is subsidized through the Belgian Administration for Development Cooperation (ACCD) in a project fully identified elsewhere (Compagnion et al., 1983).

REFERENCES

Chen Ru-Chen. 1983. Up-to-date of anaerobic digestion technology in China. Proc. IIId Int. Symp. Anaerobic Digestion, Boston.

Compagnion, D., D. Rolot, H.P. Naveau, E.J. Nyns, V. Baratakanwa, D. Nditabiriye and P. Niyimbona. 1983. Development, popularisation et integration de la biomethanisation au Burundi. Tropicultura.

GAS-USE DEVICES

GAS-USE DEVICES

OVERVIEW

Most of the papers presented at the conference discussed family-sized biogas plants producing gas primarily or exclusively for domestic cooking. Several papers described the substantial benefits derived from the use of biogas as a cooking fuel. They noted that biogas provides a quicker, easier, cleaner, and healthier fuel than the fuels it usually replaces. By reducing the time needed for fuel collection and preparation, the biogas system frees the housewife for more productive activities. The use of biogas reduces eye disease by eliminating smoke from the burning of biomass fuels. If dung cakes have been burned, the use of biogas eliminates the odor, disease, and insect problems associated with the preparation of the dung cakes. If biogas replaces wood fuel, it helps to limit deforestation. It can also save straw and crop stalks for use as fodder and silage.

Several papers described the use of gas from larger biogas plants to provide heat energy for larger installations and institutions. Alaa El-Din, et al. in their paper, "Biogas Production form Kitchen Refuses of Army Camps...," described the use of biogas in an army camp in Egypt, while Edelmann's paper, "Technical and Social Problems for the Propagation of Biogas in Rwanda," reviewed a system providing cooking gas for a school in Rwanda. Obias's paper, "A Biogas System for Developing Countries," reported on the use of biogas as process heat for a meat packing plant. Tilche, et al. in their paper, "An Improved Plug Flow Design for the Anaerobic Digestion of Dairy Cattle Waste," outlined a plant producing heat energy for a large dairy farm in Italy. Presentations from Brazil, Philippines, Thailand, Sudan, and Costa Rica referred to the present or planned use of biogas from stillage to provide process heat for distilleries.

Eight papers covered the use of biogas as a fuel for internal combustion engines. Two of these are presented in this section of the proceedings. An overview of the production of "Mechanical and Electrical Power from Biogas in Developing Countries" is provided in the following paper by D. Mahin. The subsequent paper by K.V. Mitzlaff describes experience in Tanzania with the "Performance of a Small Diesel Engine Operating in a Dual Fuel Mode with Biogas." In addition, six papers in other sections of these proceedings include descriptions of the use of biogas in engines. These include the papers of Hamad et al., "Design Parameters Affecting Success and Failure of Biogas Systems," Fathy et al., "An Integrated Energy System," Ghosh, "Novel Processes for High Efficiency Biodigestion of Paticualar Feeds," Bellamy et al., "Preliminary Assessment of Farm-Scale Digester in Relation to Possibilities of Operating an Integrated Biogas System," Henning, "Biogas Plant in the Ivory Coast," and Khatibu, "Biogas Technology for Water Pumping in Botswana."

MECHANICAL AND ELECTRICAL POWER FROM BIOGAS IN DEVELOPING COUNTRIES

Dean B. Mahin
Editor, Bioenergy Systems Reports
P.O. Box 591
Front Royal, VA 22630 USA

ABSTRACT

Experience with the production of mechanical and electrical power from biogas has been developing rapidly in recent years in North America, Europe, and China, and there is some experience with use of biogas in engines and generating sets in about a dozen developing countries. The current situation, limitations and solutions to the use of biogas-fueled engines are outlined in this paper which is based on a report prepared for the U.S. Agency for International Development's Bioenergy Project on mechanical and electrical power from biogas in developing countries.

INTRODUCTION

There are two basic types of systems using biogas in engines. In one type the biogas-fueled engine provides mechanical power directly to some kind of machinery using a shaft connection, gear drive, or belt drive. Almost all of these biogas-fueled engines are in developing countries. More than 800 small engines fueled with biogas are providing direct mechanical power on farms in China; most of these are diesel engines producing less than 10 horsepower.

In China and elsewhere the most common use of shaftpower from biogas-fueled engines is for pumping water--for crops, for livestock, and for people. In Botswana a system with a 75 cubic meter Indian-type digester is used with an 8 horsepower pumpset at a borehold; the system delivers about 4,300 liters per hour from a depth of 60 meters and provides water for 2,500 animals and more than 40 families. In the unique Maya Farms complex in the Philippines, four large biogas-fueled pumpsets deliver 580 gallons of water per minute (131,000 liters per hour) to a meat-packing plant.

Biogas-fueled engines are also used in China and some other countries to operate stationary machinery for rice hulling, grain threshing, chaff cutting, flour milling, and feed grinding. At Maya Farms four-cylinder and eight-cylinder spark ignition engines fueled with biogas operate corn grinders and feed mixing equipment which produces up to 100 metric tons of animal feed every day.

Around the world, however, the majority of the biogas-fueled engines drive electric generators. Much of this electricity is ultimately used in electric motors to provide mechanical power. Although there is some loss of efficiency involved in this indirect production of mechanical power, a biogas-fueled generating set provides much greater flexibility in the use of the available energy to meet the highest priority energy needs of each day and season.

Precise figures are not available, but there may be 100 biogas-fueled electric generating sets on farms in North America and perhaps about the same number on farms in Europe. In the U.S. many of the systems provide the electrical power needed on large farms, and the surplus power is sold to the grid. In 1982 there were 1,100 biogas-fueled generating sets in China; most of these units produced from 5 to 7 kW, but some larger Chinese farms had 12 kW systems. Two biogas-fueled generating sets provide lighting for about 40 households in a village in Sri Lanka. Seven biogas-powered generating sets provide all of the electric power needed by the 50,000 hog farm and meat packing plant at Maya Farms. Eight other large hog farms in the Philippines and at least two large hog farms in Singapore are meeting most of their electric power needs with biogas. A Belgian-built biogas system provided up to 28 kW hr per day at a rural institution for children in Rwanda between 1981 and 1983. A German-built system in the Ivory Coast provides up to 280 kW hr per day for a large slaughterhouse.

ENGINES MODIFICATIONS FOR BIOGAS USE

Several misconceptions may be limiting interest in developing countries in the use of biogas in engines. One is that extensive engine modifications are required. In fact, the one essential modification of a spark-ignition or diesel engine is the addition of a simple mixer to add biogas to the engine's air intake stream. A stoichiometric mixture of air and biogas consists of six to seven volumes of air for each volume of biogas, so it is only necessary to add about 15% biogas to the intake air. Several types of air/gas mixers have been used; they are easily installed between the air filter and the air intake pipe. A spark-ignition engine operates very nicely on only this mixture of air and biogas. In a diesel engine up to 90% of the liquid fuel can be displaced with biogas, but some liquid fuel is necessary as a pilot fuel to ignite the biogas. Changes in the timing and/or compression ratio have been made in some engines, but there is no consensus as to the need for such changes. Most such modifications have been made in developed countries in the large spark-ignition engines which are normally used with biogas in those countries, such changes have rarely been made in the small diesel engines which are most often used in biogas/engine systems in developing countries. In China, the Sichuan Institute of Agricultural Machinery concluded that changes in the injection timing or in the compression ratio of single-cylinder diesel engines operating with biogas would have a negative net effect on the performance of these engines.

PERFORMANCE OF BIOGAS-FUELED ENGINES

The technical literature contains numerous references to substantial reductions in power output in engines fueled with biogas; these also apply primarily to large spark-ignition engines, rather than to small diesels. A stoichiometric mixture of air and average biogas contains about 85% of the energy in a stoichiometric mixture of air and gasoline and about 90% of the energy in a stoichiometric mixture of air and diesel fuel. In a dual-fueled diesel engine the total fuel energy and power output is influenced by the extent of the diesel fuel displacement. Most reports have indicated that the power output of dual-fueled diesel engines equaled or even exceeded that with diesel fuel only. However, a power drop of at least 15% can be expected in a spark-ignition engine.

There have been serious corrosion problems in some of the biogas-fueled engines in developed countries due to the hydrogen sulphide--H_2S--and other sulfur compounds in the gas. The gas is usually scrubbed by passing it through a chamber of iron oxide, but these scrubbers have not eliminated all of the corrosion problems. In most developing countries, however, both the level of H_2S in the biogas and the extent of sulfur corrosion problems have been rather limited. I have discovered several theories which may explain this situation.

One is that in many areas in developed countries sulfur concentrations in the soil--and thus in the manure--are higher due to crop dusting with sulfur for alkalinity reduction or for insect control and/or to the sulfur deposited on the soil as a result of the burning of high-sulfur fossil fuels in power plants and factories. Another theory is that the higher protein level in the livestock feed in most developed countries leads to higher concentrations of organic sulfur compounds in the manure.

Developing countries seem to have lower sulfur levels in the soil and in the manure. Also, developing countries are in the broad tropical zone which has laterite soils containing a high percentage of iron oxide. At most plants some soil is inevitably scraped up with the manure; the iron oxide in the laterite soil may be reacting chemically with the H_2S in the digester just as in the iron oxide scrubbers used in developed countries. In addition, most of the biogas/engine systems in developing countries use diesel engines. Sulfur corrosion problems are more limited in diesel engines, due apparently to the built-in resistance to the sulfur in diesel oil and to the continuous lubrication of internal parts by the remaining percentage of diesel oil needed for dual-fuel operation.

MAIN FACTORS LIMITING USE OF BIOGAS-FUELED ENGINES AND PROSPECTIVE SOLUTIONS

We now come to the real problems. Most of these are related to the production of sufficient gas for the operation of an engine.

The most important problem is the design of a digester which is productive enough to meet the estimated gas requirement of the proposed biogas/engine system. A typical biogas-fueled engine uses about 0.45 m^3 of gas per horsepower per hour. If you are going to run a 10 horsepower engine for six hours a day, you will probably need about 27 m^3 of gas per day, which is several times the daily output of most family-sized plants. Due to the small size and low gas production rates of the commonly used types of family digesters, none of them is very well suited for use in a biogas/engine system. As a result of severe gas leakage problems and low productivity, the typical Chinese-type displacement digester is apparently being deemphasized even in China. Larger versions of the typical Chinese plants and interconnected sets of several smaller plants provide gas for over 2,000 engines and generating sets in China, but this type of digester does not seem to have been used with an engine in any other country. A few larger versions of the typical Indian plant with floating steel gasholders have been used with engines in India, Botswana, Sri Lanka, and perhaps elsewhere. However, due to the high cost of the steel gasholders, these plants are very expensive and also present construction and maintenance problems. A few plastic bag digesters have provided gas for engines in Taiwan and elsewhere, but the commercial availability of these bag digesters is rather uncertain at present. There is very little experience with heating any of these digesters, and the gas production from all three of these types of unheated digesters is very low.

Insulated above ground tanks of steel or concrete are widely used as heated digesters in Europe, but the high cost of these units will probably limit their use in developing countries. Fortunately, two other types of digesters are more promising for biogas/engine systems in developing countries. One is the underground reinforced concrete digester with a solid concrete roof and a separate steel or plastic gasholder. This type of digester is used at Maya Farms and at a number of other large hog farms in the Philippines. The other promising type of biogas plant has a below-surface digester pit with an open top which is covered by an inflatable flexible gasholder. The pit can be a rectangular concrete tank resembling a swimming pool, a V-shaped concrete trough, or just an earthen pit, with or without a lining of plastic or synthetic rubber. The German-built plant in the Ivory Coast uses a 400 m^3 unlined pit dug in porous laterite soil. The organic matter in the slurry nearly sealed the pit after a few days; thereafter the loss of liquid from seepage has been only equivalent to about 5% of the daily slurry loading. The flexible cover is attached to stays inside a water-filled concrete trough which surrounds the pit. A unique modification of this type of system is being used at a large pig farm in California. Wastes are flushed to a large anerobic lagoon; a synthetic rubber cover, supported by floats and polypropene ropes, covers a section of the lagoon and captures biogas for use in a 75 kW generating set.

Another set of real problems also greatly limits the use of biogas in engines in developing countries at present. I refer to the almost total use of unheated digesters in these countries, to the low gas production per unit of volume from these unheated digesters, and to the high capital costs of the large digesters which are needed to provide sufficient gas for engines at these low volumetric gas production rates.

Methane-forming bacteria are very sensitive to temperature. The optimum temperature for the mesophilic group of methane-forming bacteria is about 35°C. However, the unheated and predominately underground digesters in developing countries are operating at temperatures which are near the temperatures of the earth and are thus 10°C to 20°C lower than the optimum temperature for anaerobic digestion. The most serious temperature problems are in areas with rather cool winters, which unfortunately include some of the best livestock-raising country and the best sites for larger biogas plants. Earth temperatures between 12°C and 18°C have been recorded in winter in Central China, northern India, southern Africa, and here in Egypt. In latitudes near 25°C--North or South--gas production in unheated digesters in winter is usually only one-third to one-half the output in summer. However, even in summer the temperatures in these unheated digesters are well below the optimum. The highest summer gas rates I have been able to find are around 0.3 m^3 per m^3 of digester per day for Chinese-type digesters and around 0.5 for Indian-type digesters. Both the plastic bag digesters and the below-surface digesters with inflatable flexible gasholders absorb some solar heat through the plastic or synthetic rubber, and somewhat higher gas rates have been recorded for these plants. Gas production rates of more than 1.0 m^3 per m^3 of digester per day are common in heated digesters in the United States and Europe, and some of the daily rates go up to nearly two cubic meters of gas for each cubic meter of digester. A very interesting and unique heated 50 m^3 digester has been built by El-Halwagi and his National Research Centre (NRC) colleagues at a poultry farm at Shubra Kass; it is producing between 1.0 m^3 and 1.2 m^3 of gas per m^3 of digester per day.

If the available manure supply is quite large, the gas production rates per volume of digester can be increased by increasing the daily loading rate. For example, at the German-built plant in the Ivory Coast the volumetric gas rate increased from 0.26 when the 400 m^3 digester was fed with manure from 150 cows to about 0.60 when it was fed the wastes of 300 cows. However, in most situations in developing countries it is necessary to obtain the maximum gas from a limited quantity of manure. One way to do this is to provide a longer retention time by increasing the volume of the digester; this has usually been a viable option for family-sized biogas plants producing cooking gas, although some plants have had larger volumes and longer retention times than necessary. However, larger digesters are more expensive digesters, and very large digesters with low gas output rates do not seem an economically viable option for many biogas/engine systems. In China, where the average gas rate has been only 0.15 m^3 per m^3 per day, 180 cubic meters of digester capacity are needed to produce the gas used by a 10 hp engine operating six hours a day. In the borehole system in Botswana the Indian-type digester produces an average of about 0.30 m^3/m^3/day; 75 m^3 digester volume is needed in order to operate a 8 horsepower pumpset for at least six hours a day. These very large digesters involve a very substantial investment which is not economically feasible in very many situations in developing countries.

The other way to increase digester output is by digester heating. Many European biogas plants and some units in the United States burn some of the biogas in a boiler or water heater and circulate the hot water through a heat exchanger in the digester. Even though this use of biogas reduces net biogas production, the net output is still much higher than that of unheated digesters. However, if the biogas is used in an engine or generator set, the greatly preferable means of digester heating is with waste engine heat. In a biogas-fueled generator set only about one quarter of the energy in the biogas can be converted to electricity. In a water-cooled unit about half of the waste heat is dissipated through the radiator and the other half goes out the exhaust pipe. In such units the easiest way to heat a digester is to circulate the engine's cooling water through a coil in the digester before it passes through the engine's radiator. At the NRC demonstration site at Shubra Kass village, a simple heat exchanger is used to extract heat from the exhaust stream of a small air-cooled generator set and uses it for digester heating; similar heat exchangers around the engine's exhaust pipe have been used in the U.S., Europe, Nepal, and elsewhere.

Digester heating has been widely perceived in developing countries as too complex or expensive. This perception is undoubtedly valid for most family-sized plants, but I believe that digester heating could greatly increase the technical and economic feasibility of many potential biogas/engine systems. The additional capital costs of digester heating with waste engine heat are modest and certainly much less than the substantial savings in digester costs which could be obtained by using smaller but heated digesters. There is a clear need for additional experiments in developing countries with the heating of moderate-sized digesters using waste engine heat.

CONCLUSION

In summary, biogas systems of only moderate complexity can provide an important source of mechanical and electrical power in many rural situations in developing countries. At present the wider use of such systems seems to be inhibited primarily by the very limited international sharing of experience with such systems, by the lack of consensus as to the most effective designs for digesters and other components for biogas power systems, and by the low productivity of the most widely used unheated digesters. All of these problems could be resolved through more effective international communication and cooperation.

PERFORMANCE OF A SMALL DIESEL ENGINE OPERATING IN A DUAL FUEL MODE WITH BIOGAS

Klaus von Mitzlaff, and Moses H. Mkumbwa
Faculty of Engineering
University of Dar es Salaam, Tanzania

ABSTRACT

This paper investigates the performance of a small diesel engine of 8 kW which was modified to operate in a dual fuel mode with biogas. The performance was studied under different operating conditions and was compared with operation using diesel fuel only.

The power output was found to be slightly higher in dual fuel mode than for diesel fuel operation. The efficiency is inferior especially in the lower power ranges but the savings of diesel fuel range between 60 percent at higher engine speeds and up to 90 percent at lower engine speeds.

INTRODUCTION

Small stationary diesel engines are widely used in developing countries for various purposes such as water pumping, grain milling, and electricity generation. Some of these countries face problems to supply enough diesel fuel especially to the more remote areas. On the other hand these countries often have big resources in biomass from animal and plant materials.

Tanzania has ample supply of biomass and other renewable energies[1] and at the same time lacks sufficient availability of liquid fuels in the "up country" areas.

In the light of these constraints, fuels like producer gas or biogas can provide a share in helping to ease the prevailing fuel problem. Research on the use of various gases in internal combustion engines is nothing new. However, some of the reports dealing with the use of biogas in diesel engines provide only few data on the actual performance of the engine in all possible operation conditions,[2-4] others go further but either include more sophisticated modification of the diesel engine or discuss engines specifically designed for the use of biogas or natural gas.[5, 6]

OBJECTIVES OF THE RESEARCH

The objective of this research is to operate and test a single cylinder diesel engine running in a dual fuel mode using diesel fuel and biogas. The only modification in this case is a mixing chamber connected to the intake manifold of the engine. The mixing chamber is to be of a simple design in order to facilitate its manufacture even locally. All other elements of the engine remain unchanged so that the engine can at any time be used on lower biogas/diesel ratios or on diesel fuel alone if necessary without reversing the modifications or causing any adverse effect on its original performance.

The engine shall be tested in all possible modes of operation so that conclusions can be drawn on that range of operation in which the engine can run satisfactorily and with an optimum in efficiency and diesel fuel savings when using a mixture of diesel fuel and biogas.

THE TEST UNIT

Figure 1 presents a schematic view of the test unit. The biogas was generated in a plant of about 60 m^3 capacity with an average daily biogas production of about 30 m^3/d. The gas was led via a conventional watertrap to a gas volume flow meter (1) where the time used to pass a volume of 0.1 m^3 of biogas was measured using a stop watch. The flow meter was also connected to a U-tube manometer (2) to establish the actual biogas pressure. From the gas flow meter the biogas entered the mixing chamber (3) of the engine through a hand operated gate valve (4).

The engine used (5) was of a type commonly used in Tanzania, i.e., a single cylinder diesel engine of about 1,000 cm^3 displacement volume. The model used here was a Hatz E 108 U air cooled engine with a displacement of 1,007 cm^3 and a rated power of 10 kW at 2,000 rpm. The engine was connected to a hydraulic dynamometer (6) (Type Froude DPX) for the control of its shaft speed and the measurement of its torque. The shaft speed was measured using a tachometer (7).

Diesel fuel consumption was measured using a measuring cylinder (8) of a fixed volume (50 cm^3) and a stop watch to measure the time necessary for the consumption of this volume.

Exhaust gas temperatures were measured with an electronic thermometer (9) at the outlet of the silencer (10) to obtain at least qualitative data on the change of exhaust gas temperatures. A constant flow CO gas analyzer (11) was used to determine the volume percentage of carbon monoxide in the exhaust gas to draw some conclusions in the combustion. (Type: Bosch ETT).

FUELS USED

The diesel fuel used is the one available in Tanzania with a mean density of 0.85 kg/l and a mean lower heating value of 42,700 kJ/kg.

The biogas used was produced from cow dung and was found to contain 38 percent CO_2 and 62 percent methane (CH_4) leaving aside other

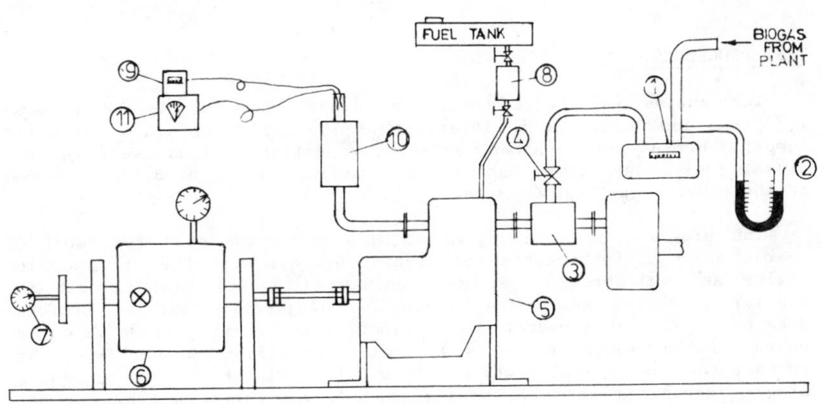

Figure 1 The Test Stand

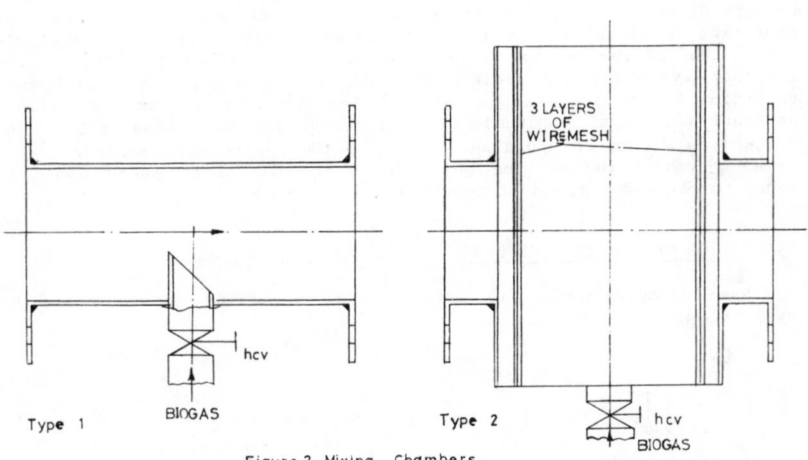

Figure 2 Mixing Chambers

traces of gases. The tests were run at Arusha Tanzania at an altitude of about 1,350 m above sea level where the average ambient atmospheric pressure was at 850 mbar and the average ambient temperature was 25°C with only little variations during the test period. The above data result in a lower heating value of 17,224 kJ/m^3 of the biogas where the methane was considered to be the only component contributing to the heating value.[7]

TEST PROCEDURE

The engine was first tested on diesel fuel at speeds between 1,300 rpm and 2,500 rpm in intervals of 200 rpm to obtain a basis for comparison. At each constant speed, five different loads were set and diesel fuel consumption, exhaust temperature, and CO percent by volume in the exhaust gas were measured.

The program was then repeated twice using the dual fuel mode of operation, i.e. introduction of biogas together with the air into the engine and injection of a small amount of diesel fuel through the regular injection system to ignite the biogas/air mixture. (Biogas does not ignite at temperatures obtained by compression alone in diesel engines.) One series of tests was run using mixing chamber Type 1 and another test using mixing chamber Type 2 (see Figure 2). The amount of biogas introduced into the mixing chamber was controlled by the hand operated valve. The point of maximal ('optimal') biogas consumption was determined by observing the smoothness of running and the CO percentage in the exhaust gas where values below 0.2 percent were considered to be still acceptable.

For assessment of the function of the governor some tests were conducted whereby a point of operation was set while the engine was operating on diesel fuel only. The biogas valve was then opened slowly until the engine began to show signs of unsteady ignition. The governor was obviously capable of reducing the amount of diesel fuel according to the increase in biogas. The engine speed remained almost constant with a slight tendency to increase (maximum 3 percent). The value lies within the tolerance given by the engine manufacturer. With a minor correction of the governor lever, the shaft speed could be reset to the exact speed if required.

EVALUATION OF THE TEST RESULTS

The following relations were used to determine the data for the evaluation:

a) Power

$$P = \frac{M \quad n \, \pi}{30{,}000} \qquad \text{kW} \qquad (1)$$

b) Diesel Fuel flow rate

$$V_D = \frac{50 \text{ ccm}}{\text{time (s)}} \quad 3.6 \qquad \text{l/h} \qquad (2)$$

c) Biogas flow rate

$$V_{BG} = \frac{0.1\ m^3}{time\ (s)} \quad 3{,}600 \qquad m^3/h \qquad (3)$$

d) Energy flow provided by biogas

$$Q_{BG} = V_{BG}\ H_{u\ BG} \qquad kW \qquad (4)$$

e) Energy flow provided by diesel fuel

$$Q_D = V_D\ \rho_D\ H_{uD} \qquad kW \qquad (5)$$

f) Brake thermal efficiency

$$M_e = \frac{P}{Q_D + Q_{BG}} \qquad (6)$$

g) Specific fuel consumption diesel fuel

$$sfc_D = \frac{V_D}{P} \qquad 1/kWh \qquad (7)$$

h) Specific fuel consumption biogas

$$sfc_{BG} = \frac{V_{BG}}{P} \qquad m^3/kWh \qquad (8)$$

i) Percentage of diesel fuel saved

$$S = 1 - \frac{sfc_D\ \text{dual fuel mode}}{sfc_D\ \text{diesel fuel only}} \qquad (9)$$

Out of a total of about 180 tests, 110 were used for evaluation.

DISCUSSION OF TEST RESULTS

Power Output

Figure 3 shows that the maximum power for the dual fuel operation is slightly higher than for diesel fuel operation. This can be explained by the fact that diesel engines usually operate with an excess air ratio of about 1.5, i.e., there is enough air for additional biogas to be combusted and thus increase the power output. However, this is limited especially at higher speeds as the time for complete combustion of the biogas in the cylinder is not sufficient due to the slow burning velocity of methane.

Exhaust Gas Temperature and Combustion

The exhaust temperature rises significantly in dual fuel operation (see Figure 4).* The lower gradient of the curve for dual fuel

*Sundaram[3] reports lower exhaust temperatures for dual fuel mode than for diesel fuel.

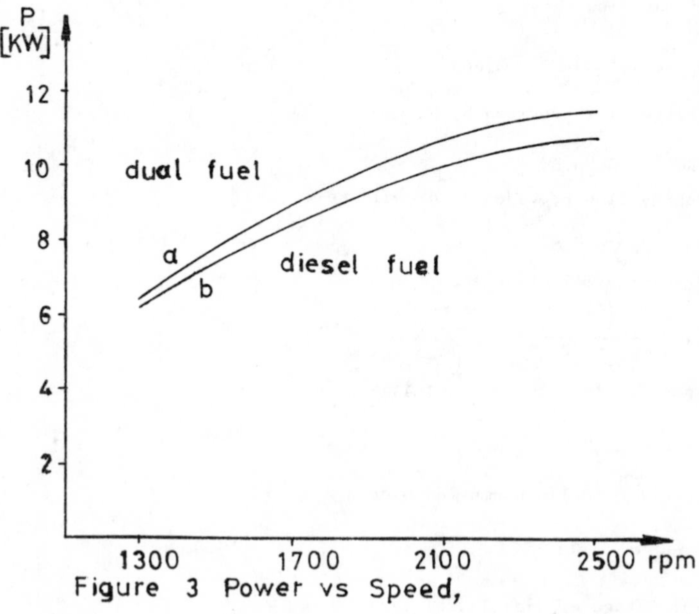

Figure 3 Power vs Speed,

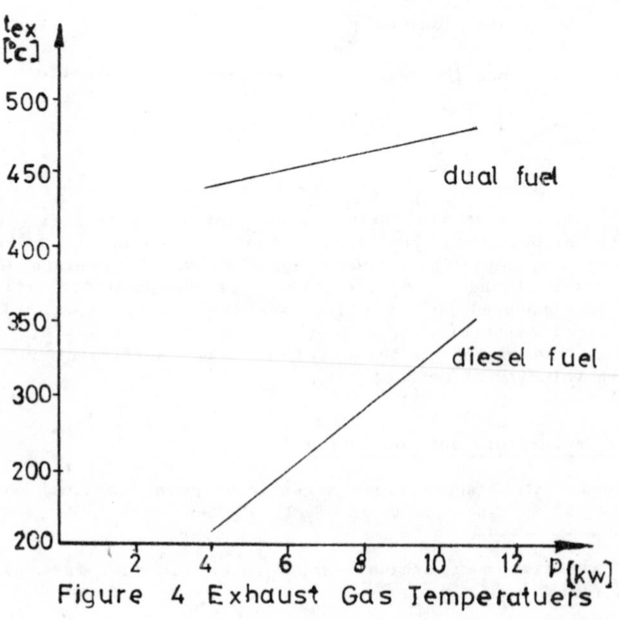

Figure 4 Exhaust Gas Temperatuers

operation results from the fact that the exhaust temperature was measured in the outlet of the silencer which was acting as an exhaust gas cooler. The temperature drop between exhaust valve and point of measurement increases with increasing exhaust gas temperatures. At higher engine speeds, temperatures at the outlet valve can be expected to be about 150°C higher than in diesel fuel operation at the same conditions. This may have an adverse effect on the lifespan of the exhaust valve.

The higher exhaust gas temperatures relate to the lower efficiency in dual fuel operation and indicate that the gas is still burning while the exhaust valve opens. The burning velocity of methane is known to be comparatively slow especially at higher pressures.[7]

Another effect of the reduced combustion speed in dual fuel operation is that the running of the engine appears much smoother than in diesel fuel operation. The typical knocking of these engines disappears immediately when biogas is introduced into the mixing chamber. The pressure increase in the cylinder seems to be smoother than for diesel fuel operation and the peak pressure appears well after TDC. Ortiz-Canavate[5] recommends advancement of injection point by about 4°C for better efficiency. The smoother combustion can be expected to have a positive effect on the engine's lifespan.

Specific Fuel Consumption and Fuel Savings

Figure 5 to 10 show the specific fuel consumption (sfc) of diesel fuel and biogas for both regimes, the brake thermal efficiency, and the amount of diesel fuel saved in dual fuel operation for three selected speeds. It is obvious that in medium shaft speeds (1300-1900) conditions seem to be optimal because:

- -- time for complete combustion is sufficient;
- -- conditions for even ignition are favorable;
- -- the excess air ratio is high enough to avoid necessary air for combustion to be displaced by biogas in the engine intake;
- -- exhaust gas temperatures are still tolerable (about 500°C).

At lower speeds and lower loads diesel can only be replaced by biogas to a smaller extent as a minimum of diesel fuel is required for adequate ignition. Oversupply of biogas caused unstable running.

At higher speeds the slow combustion of biogas contributes to the drop in efficiency. In dual fuel operation the specific diesel fuel consumption is higher at higher engine speeds and lower at lower speeds. This is explained by the fact that the biogas flow displaces the air in the engine intake to an extent that the excess air ratio decreases to values below 1.0. This was also indicated by the higher CO values in the exhaust gas. However, it cannot be excluded that at speeds above 2,000 rpm the biogas inlet pipe to the mixing chamber was slightly undersized (3/4") as the control valve remains fully open at higher loads.

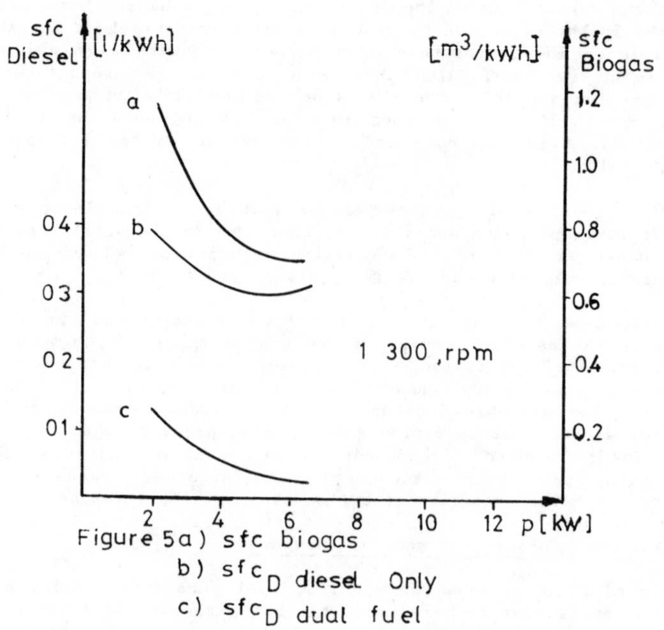

Figure 5 a) sfc biogas
b) sfc_D diesel Only
c) sfc_D dual fuel

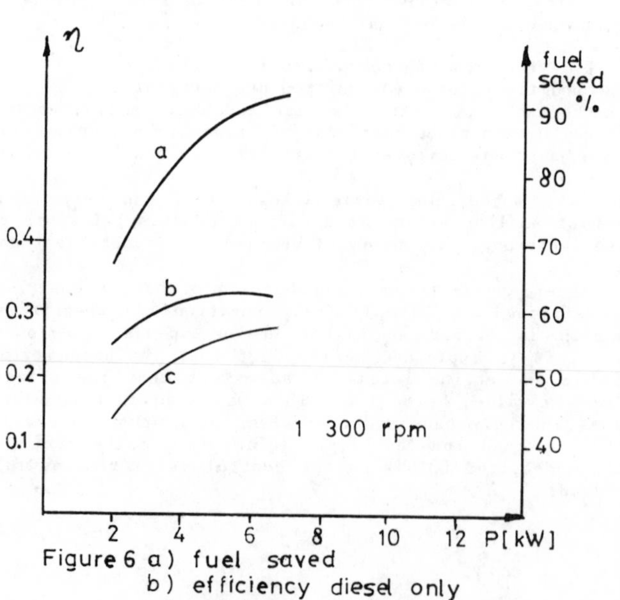

Figure 6 a) fuel saved
b) efficiency diesel only
c) efficiency dual fuel

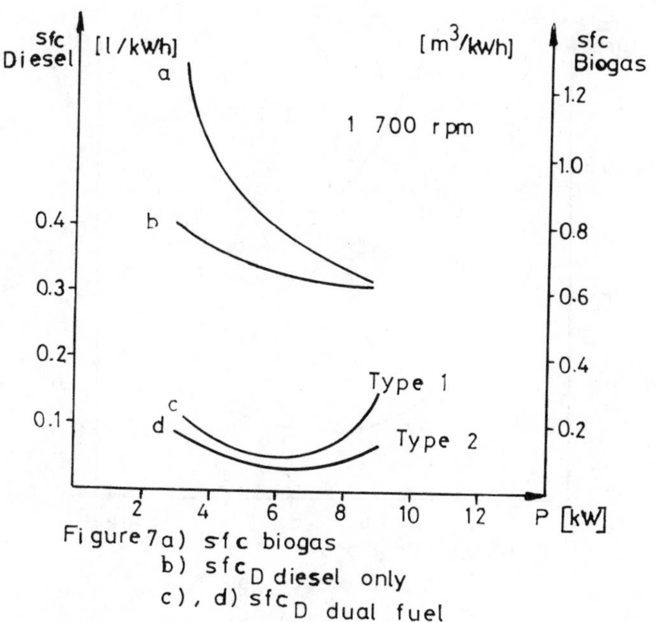

Figure 7 a) sfc biogas
b) sfc_D diesel only
c), d) sfc_D dual fuel

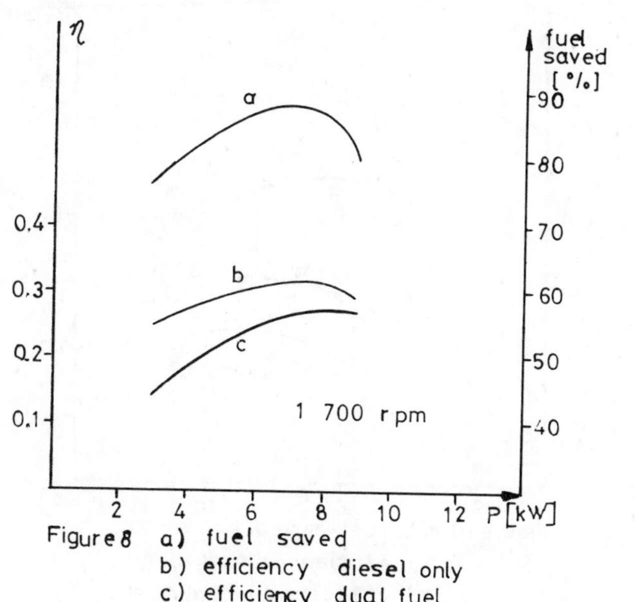

Figure 8 a) fuel saved
b) efficiency diesel only
c) efficiency dual fuel

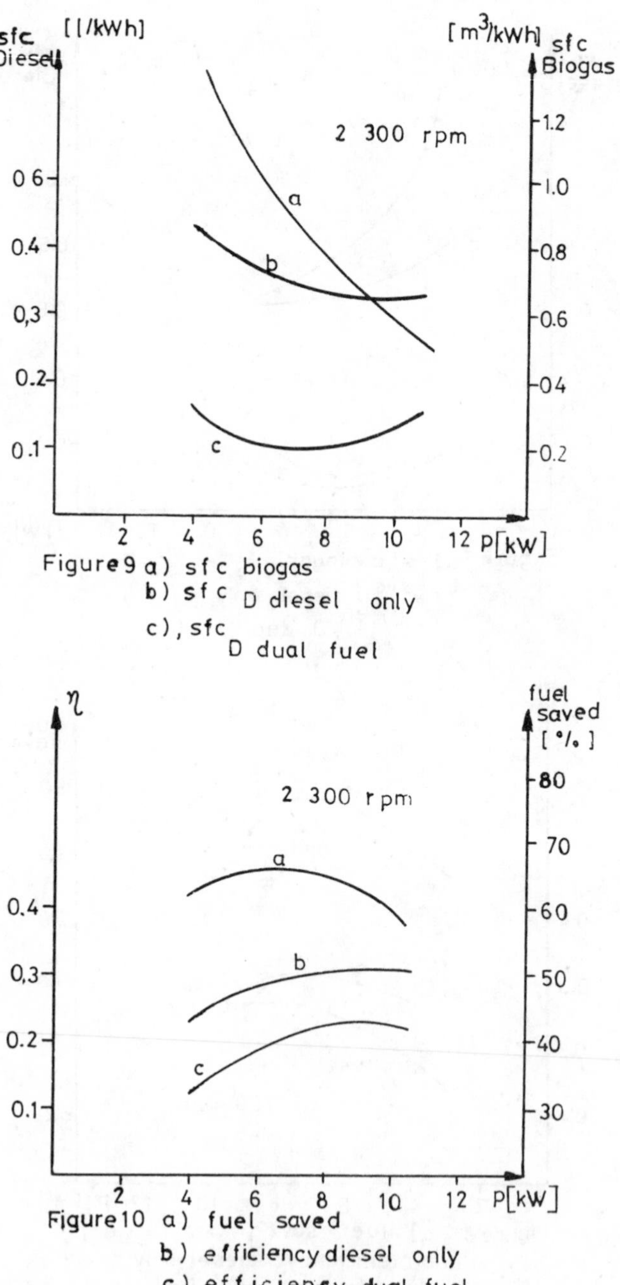

Figure 9 a) sfc biogas
b) sfc_D diesel only
c), sfc_D dual fuel

Figure 10 a) fuel saved
b) efficiency diesel only
c) efficiency dual fuel

Efficiency

The brake thermal efficiency in dual fuel operation was observed to be lower than in diesel fuel operation (Figures 6, 8, and 10). This does not only result from the incomplete combustion during the combustion stroke of the engine but also from the fact that about 40 percent of the biogas is CO_2 which does not contribute to the combustion but consumes energy while being processed through the engine.

However, the lower efficiency should not be overvalued as long as considerable diesel fuel savings can be achieved.

Comparison of Mixing Chamber Types

Figure 2 presents schematic drawings of the mixing chamber Types 1 and 2. As can be seen from Figure 7, Type 2 obviously provides a larger intake of biogas into the engine. This is due to the larger diameter of the chamber. The layers of wire mesh and the larger chamber volume facilitate producing a more homogeneous mixture of biogas with air.

However, in low speed operation the difference in performance of the two mixer types allows Type 1 to be a possible and easy to manufacture alternative.

In Figure 7 the specific fuel consumption of diesel fuel in dual fuel operation is plotted for both mixing chamber types. The curves in all other graphs relate to results obtained from tests with mixing chamber 2.

CONCLUSIONS AND RECOMMENDATIONS

- Small diesel engines can easily be modified to operate on a mixture of biogas and diesel fuel. The modifications do not require special expertise and can be undertaken locally.

- Modifications of the governor of the engine do not seem to be required as long as abrupt changes in biogas flow can be avoided.

- The performance of the modified engine is very satisfactory at shaft speeds below 2,000 rpm. Diesel fuel savings of 70 percent to 85 percent can be achieved.

- Overloading and running at maximum shaft speeds should be avoided in dual fuel operation as this may cause adverse effects on the engine's lifespan.

REFERENCES

1. Lwakabamba, S. and S. Kitutu. 1983. The Development of Biomass/ Biogas Technology as an Alternative Source of Energy in Tanzania. ANSTI-Seminar paper. Dakar, Senegal.

2. Mathur, H. B. 1981. Utilization of Biogas and Alcohol Fuels in Diesel Engines. Research and Industry, Volume 26.

3. Sunderam, P. S. N. 1981. Alternate Fuels for Engines. Research and Industry, Volume 26.

4. Sichuan Institute of Agricultural Machinery. 1979. Application of Biogas on Farm Combustion Engine. 1979. Biogas Technology and Utilization. Chengdu Seminar. China.

5. Ortiz-Canavate, J. et al. 1981. Diesel engine modification to operate on biogas. Transactions of the ASAE.

6. Motorenwerke, Mannheim. 1981. The Diesel Gas Engine (in German). Technical report. Mannheim.

7. Technical Data on Fuel. 1977. Edited by J. W. Rose and J. R. Cooper. The British National Committee, World Energy Conference. London.

BIOCONVERSION AND EXPERIENCE IN USING ALTERNATIVE RAW MATERIALS

MICROBIOLOGY AND BIOCHEMISTRY OF METHANOGENESIS, A SELECTED REVIEW

Robert A. Mah, Professor Environmental Microbiology
Division of Environmental and Occupational Health Sciences
School of Public Health, UCLA
Los Angeles, CA 90024

ABSTRACT

This selected review of the methanogenic fermentation encompasses a brief examination of the difficulties encountered by early microbiologists as background for discussing current concepts of microbiological interactions between methanogenic and non-methanogenic organisms. Some of the basic findings uncovered by modern workers help to explain culturing difficulties as well as problems associated with the types of presumed methanogenic substrates chosen. The importance of "interspecies H_2 transfer" as it relates to the mixed fermentation typical of anaerobic digesters is reviewed. The final section is devoted to a review of the unique cofactors and enzymes characteristic of the methanogenic bacteria.

EARLY HISTORY

Natural gas is an important energy source in many parts of the world. It is comprised of about 95% CH_4 and probably originated from decaying vegetation and other organic matter which was fermented to CH_4 over the span of geological time. Over 4000 years ago in Southwest China, methane was piped up from underground deposits using long bamboo poles and was combusted for heating brine during the process of salt production. This usage of methane as a fuel was obviously preceded by an earlier unrecorded observation of the existence and occurrence of a combustible gas in nature. In the western world, the modern discovery of methane formation came from reports in Italy of its association with decaying vegetation during the early 1800s. Its biological origin was also established during this period.

Early Isolation Attempts

The first real attempt to isolate methane-producing bacteria came from the studies of Omeliansky (1904), who thought that cellulose-decomposing bacteria were methanogenic. However, when he pasteurized a subculture of a methane-producing cellulose enrichment, it only produced hydrogen. Omeliansky found he could maintain one set of enrichments which produced only CH_4 from cellulose and the other (following pasteurization) only H_2. He concluded that he had succeeded in separating two kinds of cellulolytic bacteria: one which produced H_2 and one which produced CH_4. We know today that a cellulolytic methane-producing organism in all likelihood does not exist. Thus, it is not surprising that, in retrospect, he had no further success in his isolation attempts and did not obtain pure cultures. Maze (1903) was only slightly more successful in his culture attempts on methanogenic bacteria but made some interesting microscopic observations that leave no doubt regarding the methanogen he was studying. He observed a large microorganism which he termed a "pseudosarcina" and successfully subcultured an enrichment of this organism into medium containing acetate or butyrate as the major methanogenic substrate but was not able to isolate pure cultures of the pseudosarcina. A few years later, Soehngen showed that H_2 and CO_2 supported a methanogenic fermentation and even-numbered fatty acids up to caproic acid also served as substrates for a methanogenic enrichment. Because Soehngen (1906) was not successful in isolating any methanogenic bacteria, he concluded that these organisms could not be obtained axenically, perhaps because of their "slow growth rates or peculiarities of the organisms." In fact, we now know that methanogens do grow slower than most bacteria and that the biochemical and chemical properties of this group mark them as distinctly separate from all other prokaryotic and eukaryotic cells. (Balch et al., 1979)

The Basis for Early Isolation Problems

There were two physiological properties of the methanogenic organisms which undoubtedly made it so difficult in the early culture attempts:

(1) the anaerobic nature of the methanogens, and
(2) the limited range of methanogenic substrates which can actually be used directly by these bacteria.

Although Soehngen recognized the first property by incubating his cultures in an oxygen-free dessicator evacuated with H_2, he could not successfully culture any colonies of these organisms, probably because of inadequately low oxidation-reduction conditions which his anaerobic system failed to satisfy. This technical problem was only solved much later by using stricter anaerobic techniques (Schnellen, 1947; Stadtman and Barker, 1951) and eventually by application of the Hungate roll-tube method (Smith and Hungate, 1958; Hungate, 1984) and now also by use of the anaerobic glove box. The nature of the second property, the range of methanogenic substrates, proved to be much more complex, and its study is intimately linked with the history and study of the microbial ecology of the methane fermentation.

Early Concepts of the Methanogenic System

Before 1947, when Schnellen first reported the isolation of pure cultures of methanogens, all studies on the methanogenic fermentation were accomplished with mixed or enrichment cultures in which certain physical and chemical selective pressures were imposed to favor growth of certain methanogens but did not exclude growth of other, especially heterotrophic, organisms. Early reports of cellulolytic methanogens were unconfirmed and were replaced with the hypothesis that methane was formed from the fermentation products of cellulose produced by nonmethanogenic cellulolytic and associated bacteria. This hypothesis set the stage for later workers, especially engineers interested in anaerobic digesters, to adopt the "two-phase or two-stage" hypothesis which has been a focus for many engineering studies of methane fermentation. The concept has been restated many times during the course of study of methanogenesis.

Methanogenic Substrates

Several fermentation products were tested as direct methanogenic substrates including, formate, acetate, and H_2-CO_2 in addition to ethanol, propionate, and butyrate. Other potential fermentation products, including longer chain volatile fatty acids and alcohols were also reported as methanogenic substrates, but the above products are the ones of main interest because of their historical and/or actual importance in the methane fermentation. We know now that only a very limited number of compounds serve as methanogenic substrates. They are: H_2/CO_2, HCOOH, CH_3OH, CH_3COOH, and the methylamines (mono-, di-, and trimethylamine). CO may be used to slight extent by some isolates, but it is not an important quantitative source of CH_4.

EARLY CULTURES

The idea that low molecular weight fermentation products could serve as methanogenic substrates formed the basis for many attempts to isolate specific organisms from enrichments containing specific substrates. One of the first attempts was reported in 1916 by Omeliansky who published photographs of an ethanol-oxidizing methanogen. Barker later (1936) inoculated similar enrichments containing ethanol and obtained a mixed culture containing similarly shaped organisms; he named the methanogen present in this mixture *Methanobacterium omelianskii*, in honor of Omeliansky. In 1940, Barker reported its isolation. From this point in the history of methanogenesis until 1967, the metabolism and biochemistry of *M. omelianskii* (Barker, 1956) was studied more than any other methanogen since it was easy to grow and thought to be an axenic culture. This culture, which is still available since that time, eventually served as the focus for developing our present understanding of the ecology and interrelationships between methanogenic and nonmethanogenic bacteria in the methane fermentation.

The culturing of *M. omelianskii* was, in retrospect, one of the milestones in the history of methanogenesis because the choice of ethanol as a methanogenic substrate indelibly sealed the outcome of the culture process and mandated the development of a co-culture: *M. omelianskii*, as isolated by Barker in 1940 and still available to this day, was in actuality a monoxenic (two-membered) co-culture and not an axenic culture. This finding was reported in 1967 by Bryant et al. who demonstrated, in a series of elegant experiments, that the culture was composed of a chemoheterotrophic non-methanogen, the so-called S organism, and a methanogen, Methanobacterium bryantii, and each could be grown axenically.

Methanobacillus omelianskii, A Co-culture

Barker described M. omelianskii as a strictly anaerobic bacterium which obtains energy for growth by oxidizing ethanol to acetate and reducing CO_2 to CH_4 by the following reaction:

$$2\ CH_3CH_2OH + CO \text{ ---- } CH_3COOH + CH_4 \quad (1)$$

The electrons generated by the oxidation of ethanol are used to reduce CO_2 to CH_4.

The organism also catalyzes the reduction of CO_2 by the oxidation of H_2 according to the following equation:

$$4H_2 + CO_2 \text{ ---- } CH_4 + 2H_2O \quad (2)$$

In reality, the methanogenic partner (later named Methanobacterium bryantii) used H_2 and CO_2 (but not ethanol) to produce methane (equation 2) while the other or "S" organism oxidized ethanol and formed H_2 (equation 1) but was inhibited by exposure to 0.5 atm H_2. In fact, addition of H_2 completely inhibited the "S" organism and ethanol could not be metabolized. When Methanobacterium ruminantium, a methanogen which uses only formate or H_2, is combined with the "S" organism, good growth and a lot of methane are produced just as expected (and obtained) with the combination of M. bryantii and "S" organism.

These studies by Bryant et al. (1967) established the physiological interdependence of the two organisms when ethanol served as the energy source. The oxidation of ethanol to acetate and H_2 was dependent upon the removal of H_2 by reduction of CO_2 to form CH_4 by the methanogen. The "S" organism was apparently sensitive to its own H_2 end product and furthermore could not oxidize ethanol in the presence of H_2; in turn, the methanogenic partner was dependent on the formed H_2 as an energy source since it could not use ethanol as a substrate. Hence, the two organisms established a syntrophic mutualistically beneficial relationship in order to utilize ethanol. Neither isolate was capable of growing on ethanol by itself. This type of physiological interaction is characteristic of the methane fermentation and helps to explain some of the problems of earlier investigators in their attempts to isolate methanogens from methanogenic conversion of substrates other than H_2-CO_2, formate, methanol, acetate and mono-, di-, and trimethylamine. This work, and the work of Wolin (1976) set the stage and direction for work on methanogen-nonmethanogen interaction studies that has led to the current understanding of the methane fermentation.

METHANOGENIC PARTNERSHIPS

Development of Concept of Interspecies H_2 Transfer

The hypothesis that a physiological partnership existed between a non-methanogenic chemoheterotroph and a methanogen was first predicted from the rumen microbial ecology studies of R.E. Hungate. Hungate found that axenic cultures of carbohydrate-using rumen chemoheterotrophs often produced reduced end products not found in the natural mixed rumen system. In other words, mixed volatile fatty acids were the chief end products found in the rumen, and they consisted mainly of formate, acetate, butyrate, and propionate; yet, in axenic cultures, lactate, succinate, and even ethanol were common end products. Furthermore, CH_4 and CO_2 were the chief gases formed in the rumen; in axenic cultures, only H_2 and CO_2 were formed by the fermentative chemoheterotrophs. Some years before, Smith and Hungate (1958) isolated the first methanogenic rumen bacterium, Methanobacterium ruminantium, which oxidized H_2 and reduced CO_2 to form CH_4 and demonstrated that it was the most numerous rumen methanogen. Since H_2 is not a major rumen gaseous fermentation product, it seemed reasonable that H_2 may be present at low concentrations because of its rapid turnover through utilization by H_2-oxidizing methanogens. Because of these observations, Hungate postulated that the electrons formed from the oxidation of fermentation substrates were diverted from the formation of fermentation end products to the reduction of CO_2 to CH_4 in the rumen. This would explain why axenic cultures of chemoheterotrophs from the rumen formed reduced end products not found in the mixed rumen itself. Since interspecies electron disposal by methanogenic bacteria was not available in the axenic culture, formation of end products by reduction of fermentation intermediates served as the final mechanism for electron disposal and resulted in formation of lactate, succinate, ethanol, etc. On the other hand, in the mixed rumen system (or any other natural anaerobic mixed culture system), the direction of fermentation-generated electrons is channelled towards proton reduction; consequently, H_2 formation becomes the main electron sink. Of course, this proton reduction reaction to form H_2 may occur only if the appropriate hydrogenase enzymes were present and only if the H_2 is removed by methanogenic bacteria. The key biochemical reaction involves NAD (or ferredoxin) according to the following stoichiometry:

$$NADH + H^+ \quad ----- \quad NAD^- + H_2 \quad (3)$$

delta G = +18.0 kJ
at °1 atmosphere H_2

(NAD = nicotinamide adenine dinucleotide)

It is apparent that at 1 atmosphere of H_2, the free energy change (delta G) does not make this a thermodynamically feasible reaction. It becomes thermodynamically feasible only if H_2 were reduced to vanishingly low concentrations by removing it from the system. A discussion of this relationship between the partial pressure of H_2 and delta G was presented by Wolin (1976), who calculated the delta G for H_2 at several partial pressures of H_2 less than one atmosphere. The results showed that the free energy becomes more and more favorable at lower and lower partial pressures of H_2.

Interspecies H_2-transfer, Experimental Evidence:

In mixed culture systems (e.g., the rumen or the digester), the task of H_2 removal is accomplished by methanogenic or other

H_2-using non-methanogenic bacteria. This phenomenon, termed "interspecies transfer of H_2" by Iannotti et al. (1973), was first experimentally demonstrated in a co-culture of two chemoheterotrophs fermenting glucose. Only organism 1 could metabolize glucose; when growing alone, it produced acetate, ethanol, H_2 and CO_2. In co-culture with organism 2 and in the presence of added fumarate, which organism 1 cannot metabolize, fumarate served as the terminal electron acceptor and succinate was formed. Under these co-culture conditions, the end products of organism 1 shifted completely to acetate and CO_2. Since organism 2 obtained energy by H_2 oxidation and fumarate reduction to form succinate, the acetate and CO_2 products of organism 1 must have resulted from an interspecies transfer of H_2, by-passing reduction of acetaldehyde to form ethanol in favor of fumarate reduction by organism 2. The net consequence of these reactions may lead to a higher cell yield, metabolism of a greater quantity of substrate by the co-culture, formation of a greater quantity of oxidized end product, namely acetate, and greater net energy production.

The Methanogenic Interaction

In natural anaerobic mixed culture systems, H_2-oxidizing methanogens are essentially completely dependent on non-methanogenic fermentative bacteria for their source of energy (i.e., H_2). However, the methanogens also play a very important role in the fermentative process in mixed culture systems by maintaining the H_2 concentration at very low partial pressures. In fact, in any anaerobic fermentation where CH_4 is actively formed, the partial pressure of H_2 is low. Under these conditions, the fermentative bacteria are able to degrade the starting substrates completely to CO_2, H_2O, and acetate.

Effect of Methanogens on the Fermentation Products

During the fermentation of organic substrates by the chemoheterotrophs alone, the products are determined by the final organic electron acceptor compounds that serve as the electron sink for that culture system. Thus, many types of organic end products including formate, acetate, butyrate, propionate, ethanol, acetone, butanol, and propanol may be formed. H_2 and CO_2 may also be common end products of these reactions. However, when H_2-oxidizing methanogens are present, proton reduction to form H_2 becomes the major electron sink for the chemoheterotrophs and acetate is the major organic end product. In fact, reduced end products may be formed only at low concentrations or not at all. If they are produced, these reduced end products may, in turn, be converted by methanogenic interspecies H_2 transfer to acetate by a second more specialized group of chemoheterotrophs. This further oxidation of the reduced end products results in the overall conversion of the starting substrates (regardless of whether they were carbohydrates, fats, proteins, or other organic compounds) primarily to acetate, CH_4 and CO_2.

The Importance of Acetate

What happens to the acetate is of primary importance in the

man-made anaerobic digester system and in most natural aquatic ecosystems where acetate is the chief precursor of methane via a unique acetate-splitting (aceticlastic) reaction in which the methyl group of acetate is converted directly to methane and the carboxyl group to CO_2. Most of the methane in anaerobic digesters comes directly from the splitting of acetate into CH_4 and CO_2. Early labeling studies showed that the methyl group of acetate is converted to CH_4 and the carboxyl group to CO_2. The chief aceticlastic methanogens responsible for this conversion in continuously stirred tank reactors (CSTRs) belong in the genus, Methanosarcina.

Interspecies H_2-transfer, the Concept

It is now clear that the utlization of H_2 by methanogenic bacteria may exert a direct effect on the metabolism of chemoheterotrophic bacteria in the biodigester. In the biodigester, the nonmethanogenic chemoheterotrophs involved all share the characteristic of H_2 formation by proton reduction. They may produce H_2 as their sole means of electron disposal (e.g., some homoacetogenic chemoheterotrophs) or, more commonly, as a supplement to the formation of reduced organic end products. The term, "interspecies H_2-transfer", has been used to describe the coupled oxidation-reduction reactions (mediated by H_2) between two or more interacting anaerobic bacteria during fermentation of one initial substrate. The concept may apply to any two partners: one partner must be capable of proton reduction during fermentation of the starting substrate; the other must be capable of H_2 oxidation. This latter H_2-oxidizing partner may be a methanogen or a nonmethanogen. Thus, interspecies transfer of H_2 provides mechanisms (Mah et al., 1977) by which:

1. Relatively unfermentable substrates, usually low molecular weight such as propionate and butyrate may be used as carbon and energy sources under anaerobic conditions.

2. Some carbohydrate-fermenting bacteria can carry out more complete oxidations of their substrates.

3. Methanogenic bacteria may obtain H_2 and CO_2 from substrates such as carbohydrates, proteins, or fatty acids, which they are otherwise not able to use. (Any fermentative reactions leading to formation of H_2 and CO_2 from complex substrates may also satisfy this general requirement of methanogens for energy substrates).

The Fermentation of Low Molecular Weight Compounds

The anaerobic metabolism of low molecular weight compounds other than the methanogenic substrates, formate, methanol, acetate, and mono-, di-, and trimethylamine is mediated by the joint efforts of proton-reducing, acetogenic chemoheterotrophs and H_2-oxidizing methanogens. Ethanol, butyrate, propionate, and even benzoate have all been shown to undergo an interspecies H_2 transfer type of fermentation. The following section will illustrate how these types of partnerships operate and why the anaerobic metabolism of these and other common non-methanogenic fermentation products cannot be carried out by any one single organism.

Ethanol, Other Alchohols, and Fatty Acids

In addition to the "S" organism isolated from the co-culture called *Methanobacillus omelianskii* (previously described), there are other anaerobic bacteria which may also oxidize ethanol, provided the reaction can be made thermodynamically favorable by linking proton formation to proton oxidation via methane formation or sulfate redcution. In the experiments of Bryant et al. (1977), two sulfate reducers were each tested for their ability to metabolize ethanol in partnership with *Methanobacterium bryantii*. Either of the sulfate reducers metabolized ethanol when grown alone in the presence of added sulfate as a terminal electron acceptor. If, however, no sulfate were added, no growth occurred unless the methanogenic partner were inoculated in co-culture with the sulfate reducer. When a limiting amount of sulfate was added to an ethanol-using co-culture, the methane yield was reduced by an amount approximately equal to the reducing equivalents used for reduction of the added sulfate. If sulfate were added in excess to the co-culture, no methane was produced. From these experiments, it was clear that sulfate was the preferred electron acceptor over CO_2 reduction by interspecies H_2 transfer. However, if sulfate were completely used up, oxidation of ethanol could only continue via the methanogenic partnership. Many other chemoheterotrophs in addition to the sulfate reducers may also engage in this type of mutualistic association as long as they make the appropriate enzymes. For example, *Thermoanaerobium brockii* may oxidize ethanol in partnership with *Methanobacterium thermoautotrophicum*.

The Generalized Pattern of Metabolic Interaction

Higher alcohols such as *n*-propanol, *n*-butanol, isobutanol and *n*-pentanol were also used by partnerships similar to the ethanol-oxidizing co-cultures. In fact, methane formation from historically important substrates once thought to be directly methanogenic, including all alcohols higher than methanol and all volatile acids higher than acetate, conform to the general picture of interspecies H_2 transfer. The general metabolic pattern of the nonmethanogens growing alone vs. the nonmethanogen-methanogen co-cultures involves a shift in end product composition from a more reduced to a more oxidized form (Mah, 1982) because of electron flow towards H_2 formation via proton reduction. This increased electron flow toward H_2 formation by the nonmethanogens is maintained by the oxidation (and hence removal) of the H_2 by methanogenic bacteria to form CH_4 by CO_2 reduction. Thus, not only more methane but also greater quanitities of acetate, instead of formation of end products more reduced than acetate, would result. In addition, low molecular weight alcohols and acids which methanogens cannot metabolize are dissimilated by combinations of chemoheterotrophs and H_2-oxidizing methanogens. In several cases, the ubiquitous sulfate reducers growing axenically are able to metabolize the substrates, provided sulfate is present in the culture medium. Without sulfate, a H_2-oxidizing methanogenic partner must be co-cultured with the sulfate-reducer. Other organisms (such as *Selenomonas ruminantium* or S organism), which are not sulfate reducers, may also metabolize some of these low molecular weight substrates. However, these organisms usually exhibit only limited growth when cultured by themselves; they may only metabolize the substrate significantly when co-cultured with a methanogen.

Propionic and Butyric Acids

The metabolism of propionic and butyric acids is of primary importance in methanogenesis of biogas systems because they are the two main compounds through which carbon flows when acetate, the main precursor of methane, is produced. Smith and his collaborators (1973, 1980) examined the role of propionic and butyric acids in enrichment cultures and domestic sludge and demonstrated their importance by radioactive turnover studies of each acid. They showed that the fermentation of both propionate and butyrate involved the formation and utilization of H_2 and that significant quantities of H_2 were evolved during metabolism of these acids. When either the propionate or butyrate fermenting system was sparged with H_2, methanogenesis from these substrates was completely inhibited. This indicated that the oxidation of propionate or butyrate was probably dependent on H_2 removal, as might be expected during interspecies H_2 transfer.

The Importance of Hydrogen

Hydrogen formed from the oxidation of propionate or butyrate comes from electrons generated at a much higher potential than ethanol or lactate. Thus, if these reactions are to be thermodynamically feasible, the partial pressure of H_2 must be maintained at even lower levels for propionate (10^{-5} atm) and butyrate (10^{-4} atm) than for ethanol (10^{-3} atm) at substrate concentrations of 10^{-4}.

Evidence for Propionate- or Butyrate-oxidizing Co-cultures

The actual existence of propionate and butyrate-oxidizing, acetogenic, obligate proton-reducing bacteria was verified by growth of the following co-cultures that have been reported by Boone and Bryant (1980). The co-cultures consisted of two different pairs of organisms capable of metabolizing either propionate or butyrate. Isolation in axenic culture of either chemoheterotroph was not possible because of the absolute and inescapable requirement for interspecies H_2 transfer. These studies were pioneered in the laboratory of Marvin P. Bryant at the University of Illinois; they stemmed from the earlier findings by Bryant and his co-workers on *Methanobacillus omelianskii* and mark the stage for future studies of the microbial ecology of the methane fermentation.

The Butyrate Oxidizer

The first co-culture of a strictly obligate proton-reducing organism was the buytrate-oxidizing organism reported by McInerney et al. (1979). Although it was not possible to isolate the butyrate oxidizer in pure culture, the organism was named *Syntrophomonas wolfei*. Its co-culture confirmed the hypothesis (first stated in the *Methanobacilus omelianskii* paper by Bryant et al., 1967) that proton-reducing, acetogenic bacteria were responsible for the oxidation of traditional methanogenic fatty acids greater than two carbons in length.

Studies on the butyrate-oxidizing couplet by McInerney et al. (1979, 1980, 1981) have led to the co-culture and description of four strains of S. wolfei from two different parts of the world: from Goettingen, F.R.G., and from Urbana, U.S.A. Apparently, S. wolfei is ubiquitously distributed. It was isolated from butyrate enrichments inititated from several primary habitats including anaerobic digester sludges, creek sediment, and bovine rumen fluid; it has also been observed in butyrate enrichments in several laboratories from California to Florida.

The Propionate Oxidizer

The sulfate-reducing bacteria are among the most commonly occurring anaerobes, even if sulfate is absent from the habitat but especially abundant if sulfate is present. They have also been implicated in the oxidation of propionate and may account for part of the priopionate oxidation in anaerobic systems, provided that sulfate is present. The types of sulfate-reducing species involved appear to differ from the usual vibrio morphology and belong to new genera such as Desulfobulbus propionicus, only recently described by Widdel. This, or a similar organism, was present in low numbers (100-1000 nl inoculum) in the roll-tube dilution experiments of Boone and Bryant (1980).

Extended incubation of roll-tube dilutions for 6 weeks led (Boone and Bryant, 1980) to the development of large (>1 mm), isolated, dark-centered colonies in dilutions receiving only 10 and 1 ml of inoculum. These dark-centered colonies were present only when a lawn of Desulfovibrio sp. was introduced to allow interspecies H_2-transfer to occur between the propionate-oxidizing proton reducer and the H_2-oxidizing sulfate reducer lawn. In the absence of propionate, no dark-centered colonies formed. These colonies yielded an organism which they named Syntrophobacter wolinii. The inability to co-culture S. wolinii and M. hungatei without Desulfivbrio has still not been resolved. The generation time of the dixenic co-culture of S. wolinii, methanogen, and sulfate reducer was about half as fast (161 +/- 18 h) as the monoxenic sulfate-containing culture of S. wolinii and Desulfovibrio (87 +/- 7 h). Although the selection of Desulfovibrio as a syntrophic partner was based on the thermodynamic advantage of sulfate over CO_2 as a terminal electron acceptor, it appeared that some other, as yet unexplained, factor(s) may be involved in co-culturing the propionate oxidizer at the monoxenic level with a methanogenic partner. In spite of these difficulties, it was clear from these experiments that propionate-oxidizing acetogenic organisms, other than sulfate reducers, do exist and may be linked to methanogenesis in nature. The relationship is, in fact, too intimate to yield to current culture attempts at the axenic level.

BIOCHEMISTRY OF THE METHANOGENS

Metabolism

The oxidation of H_2 and reduction of CO_2 to CH_4, as well as the aceticlastic dissimilation of acetate to CH_4 and CO_2, involves a biochemistry that is unique to the methanogens. Interest has focused on this reaction sequence since Barker's 1956 hypothesis regarding an unknown methyl carrier involved in CH_4 formation. However, progress was necessarily slow because of inadequate knowledge about the biology and growth conditions of the methanogenic bacteria themselves. Recent advances have developed much more rapidly in the last 10 years because of the isolation and characterization of many new methanogens and the development of better anaerobic and other modern biochemical methods for studying these systems.

Coenzyme M Taylor and Wolfe (1974) identified a new co-enzyme, Coenzym M (CoM) (2-mercaptoethanesulfonic acid; $SH\text{-}CH_2CH_2SO_3^-$), involved in the methyl transfer reaction in methanogenic bacteria. This cofactor is required (McBride and Wolfe 1971) in the transfer of methyl compounds in methanogens. It is the smallest of all known coenzymes and has an exceptionally high sulfur content and acidity. CoM is required by methyl-CoM reductase, an enzyme present in all methanogens and responsible for the terminal steps in the reduction of CO_2 to CH_4. CoM was not detected in any other procaryotic or eucaryotic cells. In contrast, it is found in high concentrations in all methanogens available in pure culture. An average intracellular concentration of 0.2 to 2 mM CoM was reported. CoM apparently does not play a general role in other methyl transfer reactions; its singular importance is in the terminal methyl transfer to form methane by methanogenic bacteria.

The methyl reductase of *M. thermoautotrophicum* (R.P. Gunsalus) required a H_2 atmosphere, ATP, and Mg^{2+}, as well as three additional components fractioned from the cell-free extract. Component *A* has hydrogenase activity and is a large protein; Component *B* is a heat-stable oxygen-sensitive co-factor with a molecular weight of about 1000; Component *C* is a heat-sensitive oxygen-stable protein. A number of compounds with terminal sequences similar to that of CoM were tested for their ability to serve as methyl donors for the methylreductase system but none was active. However, when an ethyl moiety (CH_3CH_2-S-CoM) was substituted for the methyl group] (CH_3-S-CoM) of CoM, ethane was formed at about 20% of the rate at which methane was formed from methyl-S-CoM. The propyl analog was inactive.

RPG Effect The stimulation of CO_2 reduction to methane by CH_3-S-CoM is known as the RPG effect. Gunsalus and Wolfe observed that addition of CH_3-S-CoM to cell extracts of *M. thermoautotrophicum* incubated under a H_2/CO_2 atmosphere stimulated the rate of methanogenesis 30X! Neither HS-CoM nor $(S\text{-}CoM)_2$ replaced CH_3-S-CoM. The requirements for the RPG effect are the same as those of the methylreductase reaction: ATP, Mg^{2+}, H_2, CO_2, and CH_3-S-CoM. This reaction demonstrated for the first time that the methylreductase reaction is coupled to the activation and reduction of CO_2: under a H_2 atmospnere only, CH_4 was formed in stoichiometric proportion to the addition of CH_3-S-CoM; however addition of CO_2 greatly stimulated the formation of CH_4. The effect could be re-initiated again and again in the presence of CO_2 by the addition of CH_3-S-CoM. No additional ATP was ever needed. Thus, evidence strongly favors CoM as the C_1 carrier originally proposed by Barker for the transfer of the methyl-group to methane.

Coenzyme F_{420} and F_{430}. F_{420}, another unique cofactor found exclusively in methanogenic bacteria, is a flavin mononucleotide analog containing a glutamyl-glutamic acid side chain attached to a phosphodiester linkage; it is a 5-deazaisoalloxazine derivative. Tzeng et al. (1975) showed that F_{420} participates as an electron carrier in the nicotinamide adenine dinucleotide phosphate-linked hydrogenase and formate dehydrogenase systems of methanogens. F_{420} is reduced by coenzyme A-dependent pyruvate and alpha-ketoglutarate dehydrogenases in extracts of Methanobacterium thermoautotrophicum, a reaction normally mediated by ferredoxin in most bacteria. Finally, F_{420} also acts as a direct electron donor of the methylcoenzyme M reductase system, the final step of methanogenesis. In addition to F_{420}, chromophoric factors F_{342}, F_{350}, F_{340} and F_{430} were also described. F_{430} contains substantial amounts of nickel and lacks other metals commonly associated with molecules of biological origin (Whitman and Wolfe, 1980). F_{430} appears to be a tetraphyrrol with nickel as an essential central component of the tetrapyrrol structure; iron, cobalt, and molybdenum are not involved. It is the first nickel-containing, biological compound of low molecular weight to be reported. F_{430} may be the prosthetic group of methyl CoM reductase. Cytochromes, menaquinones, or ubiquinones have not been detected in methanogenic bacteria so far.

F_{342}, F_{350}, and F_{340}, the Methylpterins

Methanogens contain a novel class of pterins, the 7-methylpterins, whose structure and possible function have only just been worked out. The simplest representative is 7-methylpterin itself. This compound was previously described as F_{342}, F_{350} and presumably as F_{340}. Its occurrence in methanogenic bacteria is the basis for the practice of using epifluorescence microscopy to identify methanogens presumptively. The structure of 7-methylpterin was established by comparison of native purified 7-methylpterin obtained from Methanobacterium autotrophicum with a synthetic compound. Because of its novel occurrence in methanogens, this compound is now called methanopterin.

The methanopterins were first reported by Keltjens and Vogels in 1981. Initially 7-methylpterin was identified as a pterin on the basis of the UV-visible light absorption spectrum and other spectrographic (NMR) and biochemical oxidation methods. It has since been confirmed by much more elaborate methods. During short-term labeling experiments with cells of M. thermoautotrophicum, $^{14}CO_2$ was incorporated into a compound, YFC (yellow fluorescing compound), later identified as a reduced derivative of methanopterin. Cell suspensions of M. barkeri incubated in the absence of hydrogen, incorporated $^{14}CH_3OH$ into a similar compound. YFC was identified as a tetrahydromethanopterin (THMP) derivative, 5,10-methenyl-5,6,7,8-tetrahydromethanopterin. Two additional methanopterins have been desribed by Vogelset al. (1982).

Methanopterin stimulates methanogenesis in cell-free extracts of methanogens. The identification of methenyl-THMP as a one-carbon carrier derived from methanopterin and the similarity of the structures of methanopterin and folic acid suggest the participation of a folate-like biochemistry for methanopterin during methanogenesis. The evidence suggests that methenyl-THMP is converted to 5,10-methylene-THMP and 5-methyl-THMP by the successive participation of methylene-THMP dehyrogenase and methylene-THMP reductase.

CO_2 Reduction to Methane. In addition to NAD(P) and FAD, other coenzymes are involved in the electron transfer reactions of methanogens. Coenzyme F_{420} is present in all methanogens, and a cytochrome of the b-type is present in methylotrophic methanogens. Coenzyme F_{420} and the cytochrome are different in physical-chemical properties from related coenzymes in eubacteria and eucaryotes. However, these electron carriers are not uniquely confined to methanogens; F_{420} is found (but with a different function) in Streptomyces griseus and in Anacystis nidulans.

Acetate as Methanogenic Substrate

Conversion of acetate to methane was an early focus of radioactive C^{14} studies by Buswell and coworkers, who demonstrated that the methyl-group of acetate was converted directly to methane and the carboxyl-group to carbon dioxide. This reaction appeared contrary to the ideas of unity in biochemical pathways for formation of methane by oxidation of H_2 at the expense of CO_2 reduction. It seemed logical that acetate ought first to be oxidized to H_2 and CO_2 with the subsequent formation of CH_4 conforming to the CO_2, reduction model. Subsequent experiments by Barker and his group, however, confirmed the earlier isotopic labeling studies of Buswell. The evidence for this unusual aceticlastic reaction was documented by using CD_3COOH and showing its intact conversion to CD_3H. Nonetheless, since enrichment cultures were used in these studies, final confirmation of this reaction awaited a repetition of the C^{14} labeling studies on axenic cultures of Methanosarcina barkeri. This was accomplished by the growth of an acetate-adapted strain of Methanosarcina strain 227 in medium completely devoid of added H_2 (Mah et al., 1978).

Intermediary Metabolism

The pathway of CO_2 fixation into cell carbon in methanogens is still not established. Certain key oxidoreductases of the TCA cycle, except for isocitrate dehydrogenase, have been demonstrated in Methanobacterium thermoautotrophicum. However, certain enzymes present in one methanogen were deficient in others. Thus, either additional enzymes remain to be detected in each organism or novel pathways of CO_2 fixation may be operative. Acetate is assimilated into cell carbon by many methanogens, providing up to 60% of total cell carbon in cells grown in the presence of acetate. Its assimilation into alanine, aspartate, and glutamate in M. thermoautotrophicum was examined by Fuchs et al. Pyruvate (and hence alanine) may be synthesized from 1 C_2 compound and 1 CO_2; oxaloacetate (aspartate) from 1 C_2 compound and 2 CO_2; and alpha-ketoglutarate (glutamate) from 1 C_2 and 3 CO_2 via oxaloacetate, malate, fumarate, and succinate. Acetate assimilation into cell carbon in Methanosarcina barkeri, however, differed: synthesis of alpha-ketoglutarate occurred via oxaloacetate, citrate, and isocitrate from 2 C_2 compounds and 1 CO_2. These differences may reflect basic differences due to the methylotrophic pathways present in M. barkeri.

Autotrophic CO_2 assimilation by M. thermoautotrophicum proceeds via a unique pathway in which acetyl CoA, formed by an as yet unknown mechanism (instead of 3-phosphoglycerate as in the Calvin Cycle), is the key intermediate. Acetyl CoA is not regenerated within the assimilatory sequence as is true for the Calvin Cycle. The hypothesis is based on enzymes shown to be present in M. thermoautotrophicum. The key enzymes of the Reductive Pentosephosphate Cycle and the Reductive

Tricarboxylic Acid Cycle for CO_2 assimilation, as well as the Serine Pathway and Hexosephosphate Pathway for formaldehyde fixation could not be detected in cell-free extracts of methanogens.

REFERENCES

Balch, W.E., G.E. Fox, L.J. Magrum, C.R. Woese, and R.S. Wolfe. 1979. Methanogens: reefaluation of a unique biological group. Bacteriol. Revs. 43:260-296.

Barker, H.A. 1936. Studies upon the methane-producing bacteria. Arch. Mikrobiol. 7:420-438.

Barker, H.A. 1940. Studies upon the methane fermentation. IV. The isolation and culture of *Methanobacterium omelianskii*. Ant. van Leeuwenhoek. 6:201-220.

Barker, H.A. 1956. Biological Formation of Methane. *In* Bacterial Fermentations. pp. 1-27. Wiley. New York.

Boone, D.R. and M.P. Bryant. 1980. Propionate-degrading bacterium, *Syntrophobacter wolinii* sp. nov. gen. nov., methanogenic eco-systems. *appl. Environ. Microbiol.* 40:626-632.

Bryant, M.P., E.A. Wolin, M.J. Wolin, and R.S. Wolfe. 1967. *Methanobacillus omelianskii*, a symbiotic association of two species *of bacteria. Arch. Microbiol.* 59:20-31.

Bryant, M.P. L.L. Campbell, C.A. Reddy, and M.R. Crabill. 1977. Growth of *Desulfovibrio* on lactate or ethanol media low in sulfate in association with H_2-utilizing methanogenic bacteria. Appl. Environ. Microbiol. 33: 1162-1169.

Hungate, R.E. 1984. Development of Ideas on the Nature and Agents of Biomethanogenesis. First Symp. on Biotechnol. Adv. in Processing Municipal Wastes for Fuels and Chemicals.

Iannotti, E.L., D. Kafkewitz, M.J. Wolin, and M.P. Bryant. 1973. Glucose fermentation products of *Ruminococus albus* grown in continuous culture with *Vibrio succiongenes: changes* caused by interspecies transfer of H_2. *J. Bacteriol.* 114:1231-1240.

Keltjens, J.T. and G.D. Vogels. 1981. Novel Coenzymes of Methanogens. *In*: H. Dalton, ed. Microbial Growth on C-1 Compounds, pp. 152-158. Heyden. London.

McBride, B.C. and R.S. Wolfe. 1971. A new coenzyme of methyl transfer, coenzyme M. Biochemistry, Wash. 10, 2317-2324.

McInerney, M.J. and M.P. Bryant. 1980. Syntrophic Associations of H_2-Utilizing Methanogenic Bacteria and H_2-Producing Alocohol and Fatty Acid-Degrading Bacteria in Anaerobic Degradation of Organic Matter. In Anaertobes and Anaerobic Infections (ed. G. Gottschalk). pp. 117-126. Gustav Fischer Verlag. Stuttgart. New York.

McInerney, M.J., M.P. Bryant, and N. Pfennig. 1979. Anaerobic Bacterium that degrades fatty acids in syntrophic association with methanogens. Arch. Microbiol. 122:129-135.

McInerney, M.J., M.P. Bryant, R.B. Hespell, and J.W. Costerton. 1981. *Syntrophomonas wolfei* gen. nov. sp. nov., and anaerobic, *syntrophic, fatty acid*-oxidizing bacterium. Appl. environ. Microbiol. 41, 1029-1039.

Mah, R.A. 1982. Methanogenesis and Methanogenic Partnerships. Phil. Trans. R. Soc. London. B 297. pp. 599-616.

Mah, R.A., D.M. Ward, L.Baresi, and T.C. Glass. 1977. Biogenesis of methane. Ann. Rev. Microbiol. 31:309-341.

Mah, R.A., M.R. Smith, and L. Baresi. 1978. Studies on an acetate-fermenting strain of *Methanosarcina*. Appl. Environ. Microbiol. 16:358-361.

Maze, M. 1903. Chimie biologique. Sur la fermentation formenique et le ferment qui la produit. C.R. hebd. Seanc. Acad. Sci. 137:887-889. Paris.

Omeliansky, V.L. 1904. Centralbl. F. Bakteriol. Abt. II, Orig. 11, 369.

Omeliansky, V.L. 1916. Fermentation methanique de l'alcohol ethylique. Annals Inst. Pasteur. 30:56-60. Paris.

Schnellen, Ch. G.T.P. 1947. *Underzoekingen over de methaanigisting*. Dissertation, *Techniscke Hoogeschool, Delft*.

Smith, P.H., and R.E. Hungate. 1958. Isolation and characterization of *Methanobacterium ruminantium* n. sp. J. Bact. 75: 713-718.

Smith, P.H. and P.J. Shuba. 1973. Terminal anaerobic dissimilation of organic molecules. *In* Proc. Bioconversion Energy Research Conf. pp. 8-14.

Smith, P.H. 1980. Studies of Methanogenic Bacteria in Sludge. EPA-600/2-80-093.

Soehngen, N.L. 1906. *Het onstann en verdivijnen van watersstof an methaan onder den invloed van het organische leven.* Dissertation, *Technische Hoogeschool, Delft*.

Stadtman, T.C., and H.A. Barker. 1951a. Studies on the methane fermentation. VIII. Tracer experiments on fatty acid oxidation by methane bacteria. J. Bact. 61:67-80.

Stadtman, T.C. and H.A. Barker. 1951b. Studies on the methane fermentation. X. A new formate-decomposing bacterium, Methanococcus vannielii. J. Bact. 62:269-280.

Taylor, C.D.and R.S. Wolfe. 1974. J. Biol. Chem. 249, 4879 and 4886.

Tzeng, S.F., R.S. Wolfe, and M.P. Bryant. 1975. Factor 420-dependent Pyridine nucleotide-linked hydrogenase systems of *Methanobacterium ruminantium*. J. Bacteriol. 121:184-191.

Vogels, G.D. and J.T. Keltjens, T.J. Hutten, and C. Van der Drift. 1982. Coenzymes of methanogenic bacteria. Zbl. Bakt. Hyg., I. abt Orig. C 3: 258-264.

Whitman, W.B. and R.S. Wolfe. 1980. Presence of Nickel in Factor F_{430} from Methanobacterium bryantii. Biochem. Biophys. Res. Commun. 92:1196 - 1201.

Wolin, M.J. 1976. Interactions between H_2-producing and methane-producing species. In Symposium on Microbial Production and Utilization of gases. pp. 141-150. Göttingen.

METHANE PRODUCTION FROM FARM WASTES

Sambhunath Ghosh
Institute of Gas Technology
Chicago, Illinois

ABSTRACT

The energy production potential from farm wastes could be significant. Because of their high moisture contents, several types of farm wastes are ideal feeds for anaerobic digestion. Apart from its ability to produce methane energy, this process provides a usable residue and affords a number of public health and environmental benefits. Anaerobic digestion is clearly preferable to thermal processes in cases of small farms and when a heterogeneous mixture of several wastes has to be processed for the dual purpose of waste stabilization and energy production.

There was an upsurge of interest in applying anaerobic digestion for energy production from farm wastes in Germany, France, India, and a few other countries after World War II. However, the digestion process proved to be uneconomical because of falling fuel prices. There was a great renewal of interest in anaerobic digestion in the 1970's after the oil embargo. This time a large number of countries became involved in farm waste digestion R&D to improve the energy production aspect of this process. Examination of the published data seems to indicate that, except in the Peoples Republic of China, digestion of farm wastes may not be economically attractive for small installations when energy production is the only goal. Systems which were complicated and included several unit operations were uneconomical even on a large scale. The economic feasibility of farm digestion could be improved significantly by simple, high-SRT (solids retention time) digesters which can process concentrated farm wastes to provide high methane yields; simple cost-effective feed pretreatment techniques could be developed to this end. There are a number of barriers to widespread application of anaerobic digestion to farm wastes; these must be removed. Last, it is important that ways be found to obtain economic credits for many of the intangible benefits that accrue from application of anaerobic digestion to farm wastes, which could otherwise be a major source of pollution.

INCENTIVES FOR INSTALLING FARM DIGESTERS

There is considerable potential for energy production from farm wastes by anaerobic digestion. For example, it has been estimated that farm wastes can provide 10 to 20 percent of the agricultural energy needs in West Germany. Similarly, methane gas from farm digesters could satisfy 60 percent of the cooking-energy demands of the rural population in Tanzania. In the United States, conversion of all animal manures could produce energy equivalent to about 4 percent of the country's crude oil consumption.

Farm digesters produce a high methane-content gas which, without cleanup, can be used for water or space heating, lighting, steam production, or to meet other thermal energy demands. It is also technically feasible to utilize methane gas as automobile or tractor fuel. Digester gases can be used to drive internal combustion (I.C.) engines to produce mechanical power. Electrical power can be produced by coupling the I.C. engine with a generator. Alternatively, the methane gas can be reformed and used as feed for a fuel cell.

Energy production is only one beneficial aspect of the farm digesters. Production of a useful residue could be equally important. The digested residue is a sanitary fertilizer; it can be used as a soil conditioner, as animal bedding, as fish or worm feeds, as animal feed or for re-feeding.

A number of intangible benefits are derived from application of anaerobic digestion to process farm wastes. These include: odor reduction; improved sanitation; reduction of enteric pathogens, viruses, and parasitic eggs; reduction of fly and mosquito breeding; reduction of water, land, and air pollution; and improved public health. Installation of farm digesters may be justified on the basis of intangible benefits alone, although these are commonly ignored in evaluating economic feasibility.

HISTORY OF APPLICATION OF FARM DIGESTERS

Post World War II Developments

There was a flurry of activity in developing various farm digester designs and in installing farm digesters during and after World War II because of fuel and fertilizer shortages. Several digester system designs evolved in Germany, France, and India. The most popular German designs included the Darmstadt system (heated and stirred covered pit plug-flow digester, $15 m^3$ in size, in which fermentation proceeded for weeks to exhibit gas production rates [GPR] between 0.3 and 0.5 vol/culture vol-day), the fully mechanized Schmidt-Egersgluss (also known as the Bihugas process, consisting of heated [30° to 35°C] above-ground concrete silo digester operated at a hydraulic residence time [HRT] of 18 to 20 days to exhibit a GPR of 0.6 to 0.9 vol/vol-day and methane yield of 0.16 m^3/kg VS added), and the Harnisch system (continuous-flow horizontal cylindrical tank, heated [30°C], housed within a greenhouse type of structure, and equipped with a wind-powered agitator). The German digesters proved to be uneconomical with falling fuel prices and 80 percent of the biogas plants were converted to "biodung" storage plants by 1956.

As if not to be outdone by the Germans, the French built thousands of biogas plants after World War II. Perhaps the most popular design was the Ducellier-Isman system, which consisted of a battery of gas-mixed batch digesters with gas holders. The waste to be digested was first aerated to raise the temperature, at which point it was flooded with liquid manure and water to produce the digester feed. The digester was insulated with manure and straw. It exhibited a GPR of about 0.4 vol/vol-day. The system had a 10-year payback period. The other French systems included the unmixed, heated (30° to 35°C in summer and 25°C in winter) double-wall (with cork, mineral wool, or glass wool between the walls) Salubra digesters, the Betur tank, the Baudot-Hardoll and OFTA system, and the Samogaz system.

Several Indian digesters were in operation in the 1940's and 1950's. These included the Indian Agricultural Research Institute (IARI) pilot plant in New Delhi, the 1800-2300 kg manure/day plant in Walchanduagar Industries Estate, the oil-drum digesters in Haringhatta, West Bengal, and others. These digesters were the forerunners of the present-day Gobar-gas digesters.

Post-1970 Developments

There was a renewal of interest in farm digester construction in the 1970's after the oil embargo. Overly optimistic estimates of fuel production were made, and farm digesters became popular in about 40 developed and developing countries. This was the age of back-yard digesters and pseudo-experts advocating the construction of various types of "simple" systems.

A number of digester designs evolved during the 1970's in India, China, the United States, and many other countries. The Gobar gas digesters, with a floating-cover gas holder, have received the widest publicity. Several other models, including the KVIC (Khadi Village Industries Corporation), the fixed-dome Janata, the RKM (Rama Krishna Mission), and the IARI digestion systems are now available. The GPR of these digesters is about 0.25 vol/vol-day. The feasibility of family digesters in India is in question. In a department of Science and Technology study, the following were concluded:

Availability of manure and water could be crucial to the success of a plant.

The economic feasibility of a plant is dependent not only on the use of the gas, but credits received for use of the digested residue are quite important.

Institutional and community plants are easier to manage and are more feasible than family plants.

In still another study by the Center for Science for Villages Study, it was concluded that the fixed-dome Janata digester design was more efficient and maintanance-free than the floating-dome design.

A significant development in farm digester design is the evolution of the water-pressure fixed-dome Chinese digesters which are available in 6-, 8-, 10-, 12-, 50-, and 100-m^3 sizes. There is a considerable body of literature on the design, operation, performance, and application of the water-pressure digesters. These digesters are simple in design and operation, and are now applied outside of China in Nepal, Africa, and South America. These digesters have no moving parts and are inexpensive and economical. Disadvantages of these systems include low loading rates, long HRT's, low gas production rates, and gas loss by leaks under pressure. Construction of the digesters requires skilled labor.

American Farm Digesters

A large number of farm digesters were installed in the United States after the oil embargo of the 1970's. Notable among these are the following plants:

Brawley & Imperial Valley, California
Monroe, Washington (200 dairy cows)
Bartow, Florida (10,000 cattle)
Bedford, Virginia (100 dairy)
Green Bay, Wisconsin
Ludington, Michigan
University of Illinois, Urbana, Illinois (3,000 pigs)
Gettysburg, Pennsylvania (1200 dairy cows)
Oahu, Hawaii (50 hogs)
Stevensville, Texas (2700 hens)
Henniker, New Hampshire (100 dairy cows)
Middlebury, Vermont (350 dairy cows)
Rice Lake, Wisconsin (100 livestock)
Guymon, Oklahoma

As exemplified in Figures 1-3, the American farm digestion system designs are complicated and capital-intensive. Except for a few systems, like the one in Oahu, Hawaii, most American systems are uneconomical or at best, are break-even facilities (Table 1).

TECHNICAL PROBLEMS

A wide variety of materials including green crops, waste vegetables, weeds, husks, grasses, straw, stalks, bagasse, manures, trees, sawdust, aquatic plants, domestic wastes, etc. are available for charging the farm digesters. Some of these materials have low biodegradabities, and exhibit low methane yields (Table 2) even when operated at long HRT's (frequently over 20 days) and low organic loading rates of 1 to 2 kg VS/m^3- day. Methane production rates under these conditions are also low (less than 0.5 vol/vol-day). These operating conditions and performances could lead to unfavorable economics.

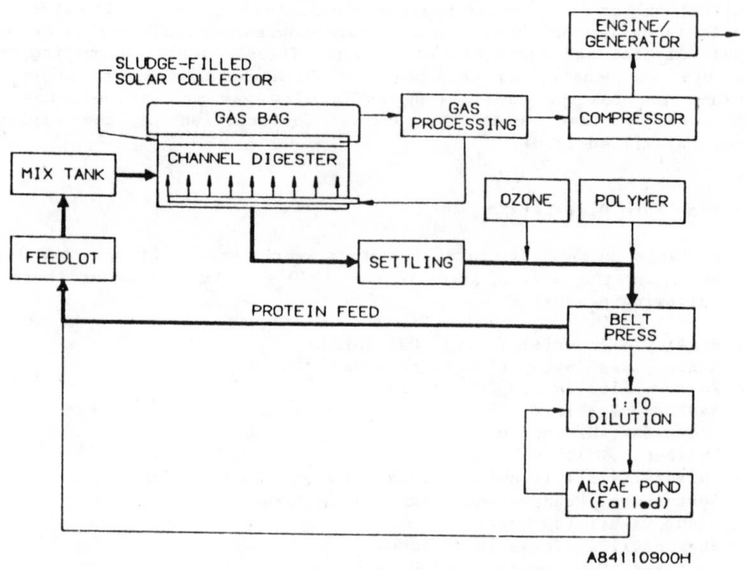

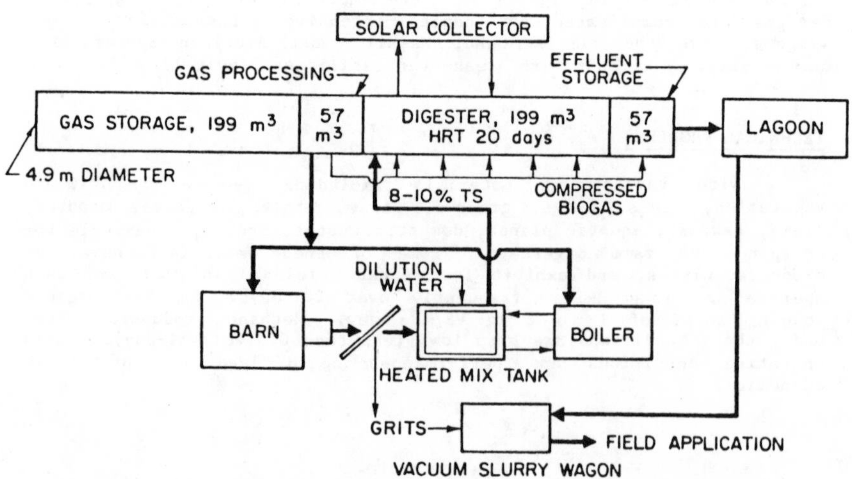

Figure 2 University of Illinois Swine Waste Digestion System

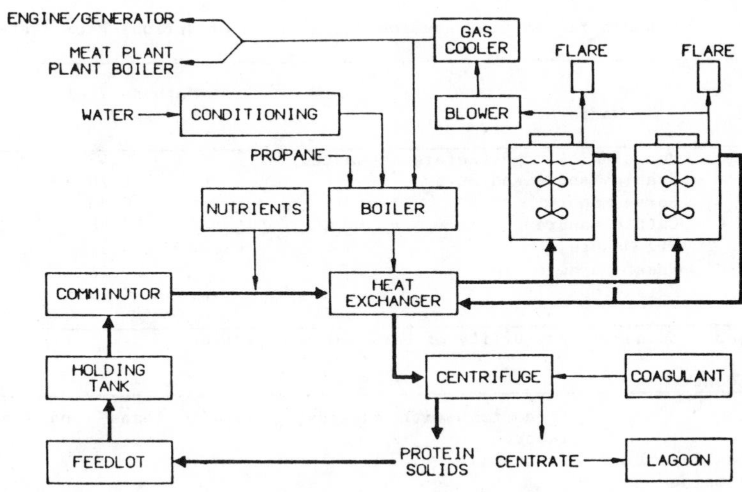

Figure 3 Hamilton Standard Thermophilic Digestion System

Table 1 Case Histories of Digestion Plants on American Farms

Location	Plant Features	Capital Cost in $1,000s	Payback
Oahu, Hawaii (50-hog farm)	20 ft x 6 ft rubber-bag plug-flow insultated digester with condensate and iron sponge gas traps (3 v CH_4/v-day). Biogas for heating, water pump, lighting	3.9	2 years
Stevensville, Texas	7000-gallon horizontal gas-mixed heated digester (0.6 v CH_4/v-day)	97	Uneconomical
Henniker, New Hampshire	40 ft x 12 ft x 8 ft plug-flow digester, heated by engine cooling water. Biogas for electric power. Fertilizer	45	Uneconomical
Middlebury, (350 head dairy)	Same as Henniker	45	Break-even facility
Rice Lake, Wisconsin (100 head livestock)	Plug-flow digester with generator	116	Uneconomical

Table 2 Methane Yields from Anaerobic Digestion of Various Farm Wastes

Waste	Methane Yield, m^3/kg VS
Municipal Sludge (Reference Waste)	0.47
Stable Manure and Straw	0.26
Horse Manure	0.33
Cattle Manure	0.26
Pig Manure	0.34
Wheat Straw	0.26
Maize Tops	0.43

Table 3 Economic Feasibility of Farm Waste Digestion

Country	Status
India	Economical with subsidy, favorable loans, and free labor.
Peoples Republic of China	Economical
Finland	Economical only for large farms (100 animals)
Switzerland	Uneconomical
West Germany	Economical only when oil price is higher than 0.75 DM/l.
Netherlands	Central plants are marginally feasible for farms with more than 100 cows or 1500 pigs with tax credits and favorable financing.
New Zealand	Commercially available 20,000 gallon systems are feasible.
United States	Poultry waste digestion is economical with more than 30,000 birds. Cow manure digestion is economical with more than 100 animals. Cornell Study projects a 2 year payback for 500 cow plug-flow digesters; 6 year payback for a 25 cow system. U.S. Environmental Protection Agency study specifies a 250 cow minimum size for an economical plant. Hamilton Standard specifies 8,000 cattle as a minimum size with a 5 year payback.

ECONOMIC FEASIBILITY OF FARM WASTE DIGESTION

The economic feasibility of farm digesters is dependent on a multitude of factors, which include price of the methane, credit for use of the residue, the nature of financing, labor cost, capital cost, plant size, available subsidies, digester design, and rate of return or payback period acceptable to tne plant owner. Thus, whereas a plant may be economical in one location of a country because of favorable economic factors operative at that location, it may be economically unattractive in another location of that country or in another country. Table 3 provides an idea of the status of economic feasibility of farm digesters in several countries. In general, economic feasibility is enhanced as the size of the plant increases. One factor that may significantly improve economic feasibility of farm digesters is the salability of the digester residue as fertilizer, animal bedding, or protein supplement.

BARRIERS TO APPLICATION OF ANAEROBIC DIGESTION TO FARM WASTES

There are a number of reasons why anaerobic digestion of farm waste is not economically attractive. As noted in Table 4, there are a number of barriers. These could be overcome by developing suitable material handling systems, cheaper storage vessels, higher-efficiency conversion equipment, and more cost-effective dewatering systems. Alternate uses of the biogas and digested residue should be developed.

TECHNICAL APPROACHES TO SYSTEM IMPROVEMENT

A number of approaches can be envisioned to improve the efficiency of farm waste digestion. Unconventional fermentation modes and digestion reactor designs could be utilized to conduct digestion of high-solids or undiluted feeds to enhance net energy production and to reduce digester size. Work in India, Ireland, and the United States indicates that two-phase digestion with plug-flow, complete-mix, and upflow (packed or unpacked bed) bioreactors provides high methane yields and production rates with small bioreactor capacities.

Research at the Institute of Gas Technology (IGT) and the Indian Institute of Technology (IIT) in New Delhi, has shown that a blend of various wastes provides higher methane yields than one type of waste.

Considerable work has been done by the U.S. Department of Agriculture, the Israeli Kibbutz Industries, and Hamilton Standard, Inc. to indicate that thermophilic digestion could provide a higher net energy production than mesophilic digestion. However, thermophilic digestion may not be feasible for small digestion installations.

RESEARCH NEEDS

The following recommendations may be considered to improve farm digester performance and economics:

- Develop simple feeding and effluent withdrawal systems with a minimum of moving parts.
- Use concentrated feeds.
- Develop simple high-SRT bioreactors and apply novel fermentation modes to process concentrated feeds without digester upsets.
- Develop optimum feeding schedules to obtain gas production rates matching energy demands.
- Develop cost-effective and simple effluent dewatering processes.
- Develop simple and inexpensive feed pretreatment techniques.
- Develop alternative uses for the biogas.
- Develop cost-effective storage vessels for the effluent and the biogas.
- Develop a system of cash credits for digester owners to pay for the intangible benefits that accrue from waste stabilization by anaerobic digestion.

Table 4 Barriers to Widespread Application of Farm Digesters

Barrier	Probable Solution
Materials handling problems feeding, effluent withdrawal	Need efficient feed system design
Inert plug digesters	Develop effective separation techniques
Net biogas production does not match variation of farm energy use	Need low-cost waste-storage fermentation system, and gas storage
Biogas-to-electric power conversion is inefficient, uneconomical, maintenance-intensive, marginally attractive	Need more efficient and low cost gas-to-power system
Use of biogas as truck or tractor fuel not feasible	Develop centralized system so that such uses are feasible
Other uses of biogas not developed	Investigate methanol or ammonia production or other uses
Effluent processing and disposal are problematic	Develop low-cost dewatering processes. Develop alternate uses of residue for aquaculture and algal growth. Need governmental subsidy/tax subsidy as credits for intangible benefits.
Process needs maintenance and expert service, quality control	Develop low-maintenance fool-proof system designs. Organize service groups.
Economy of scale unfavorable for family digesters	Organize community plants.
Gas production low at high loading rates	Need improved high-efficiency high-solids digestion systems. Need cost-effective chemical pretreatment techniques.
Low social and cultural acceptance	Reduce failures, increase reliability, provide assurance of service and safety.

OPTIMIZATION OF BIOCONVERSION OF SOLID AND LIQUID RESIDUES

M.J.T. Carrondo, I. Coutinho, J. Lampreia, J. LeGall, A.R. Lino, I. Moura, J.J.G. Moura, M.A.M. Reis, M. Teixeira and A.V. Xavier

Centro de Quimica Estrutural INIC, Universidade Nova de Lisboa, FCT, 2825 Monte da Caparica, Portugal

and

Dept. of Biochemistry, University of Georgia, Athens, Georgia 30602, USA

ABSTRACT

This paper presents strategies to scale up, start up, and operate a bioconversion process based on microbiological, biochemical, and engineering studies in an interdisciplinary program recently funded by the Program in Science and Technology Cooperation, United States Agency for International Development.

Two main objectives in this project are:

i) Optimization of the biogasification process of solid and liquid residues (agroindustrial, crop, and forestry) by using phased two-stage anaerobic fixed film reactors.

Two presumably different strategies will be adopted for the acid phase reactor:

- Operation of the acid phase reactor to yield maximum production of volatile fatty acids for chemicals production;
- Operation of the acid phase reactor to yield the best mixture of substrates for the methanogenic reactor.

In the case of molasses distillation slops, it is further expected that the acidogenic reactor will eliminate most of the sulphide produced before it can become "toxic" to the methanogens.

In the case of the methanogenic reactor, emphasis will be placed on achieving maximum methane content of the gas (e.g., by seeding and fixing or adding immobilized hydrogen-producing bacteria) to reduce or eliminate the gas purification stage.

ii) The microbiological biochemical effort will entail screening and characterization of electron transfer proteins and enzymes relevant for the metabolic pathways involved (e.g., hydrogen and methane production). Also, quantification of cofactors relevant to the operation of the digesters will be examined.

INTRODUCTION

Although anaerobic digestion presents many advantages for residue biogasification, the "black box" approach used until recently gave rise to many operational difficulties, namely during the start up and occasionally when complete failure occurred.

Recent basic knowledge of the kinetics, microbiology, and biochemistry, the use of fixed film reactors and phase-separated operations should allow an increased expectation in overcoming large scale operational difficulties, with a higher stability of biogas production as well as elimination of later gas purification stages.

The screening and characterization of the electron transport proteins, as well as the enzymes involved in the metabolic production of methane and hydrogen, and the quantification of cofactors, could conceivably allow for the development of novel start up and operational techniques.

This paper is a brief presentation of the background knowledge which allowed our group to develop a multidisciplinary research strategy, recently funded by USAID.

TECHNOLOGICAL ASPECTS

In traditional digestion systems, one single reactor, generally with suspended growth by stirring, carries out the whole anaerobic treatment. Sometimes, e.g., in domestic wastewater treatment, a secondary digester exists, whose role is mainly that of a solid-liquid separator.

More recently, three process layouts or their combinations have been considered (Figure 1).

1. Parallel operation - after an initial soluble-suspended phase separation, both lines of treatment carry out the acid and methanogenic steps;

2. Series (stages) operation - two or more anaerobic reactors exist in series, each one with acid and methanogenic steps;

3. Phased operation - acid and methanogenic phases are separated in different reactors.

An important factor in deciding which lay-out to choose can be described by the solubility index[1] of the effluent stream. Residues with average solubility indexes (0.2-0.8) might benefit from the parallel operation, separating the soluble from the suspended fraction, whereas low (0-0.2, almost entirely suspended matter) or high (0.8-1,

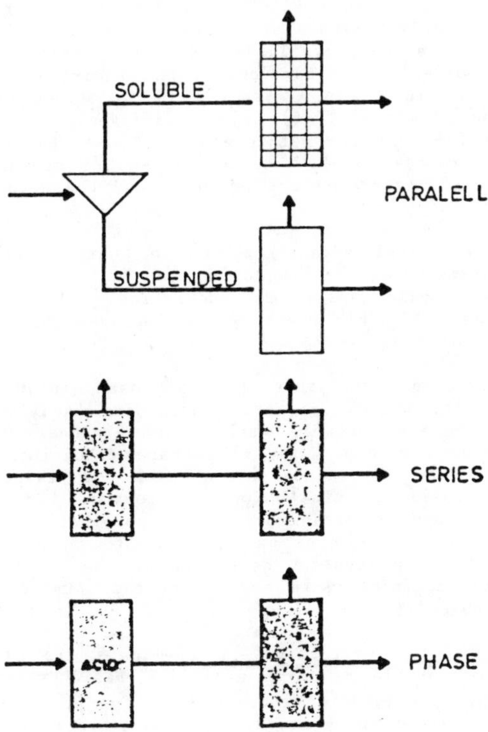

Figure 1 Anaerobic Processes

almost entirely soluble matter) solubility indexes should be processed in only one process line, with either one reactor or series or phased operation.

Advantages of Phased Separation

Phased separation for anaerobic digestion has both theoretical and practical advantages.

From the kinetic viewpoint, since the growth rate of the acidogenic microorganisms is much higher than the methanogens, different wash out velocities will take place and more economical dimensioning of the reactors is possible. Nutritionally, the methanogens can only operate if metabolites from the acidogenic microorganisms are already available. Physiologically, their behavior is also rather diverse. In particular, the importance of controlling pH within 6.6-7.4 in a single reactor is in effect an optimization between the low pH optima (5-6) of the acidogenic organisms[2] and the high pH optima (7-8) of the methanogens[1]. Those reasons substantiate the conceptual proposal of Pohland and Ghosh[3] of operating the first reactor until acids (acetate) are produced and the second reactor for methane production.

Further theoretical advantages of two phase operation are: i) larger toxic resistance, if the toxins can be confined in the first reactor [sulphate, pulp, and paper xenobiotics [1,4]]; ii) the lower pH available in the acidogenic reactor also improves the hydrolysis step which precedes acidogenesis[5,6].

The engineering advantages of two-phase separation include: optimization of the overall process, increasing stability with easier process monitoring and control, and ease of disposal of fast growing acid forming sludge without loss of methane producing bacteria[3,7,8]. Operationally, phase separation is normally maintained by kinetic control (short residence time) chemical control (low pH values or addition of methanogenic inhibitors) or a combination of both[5,7-10]. Optimization of such a process may concentrate on acid formation or hydrolysis in the acid reactor, depending on the substrate or on the slow-growing methanogenic system and the contact time of the effluents from the first phase[5,7].

Economically, the extraction of chemicals from the acidogenic reactor may be feasible if more expensive residues are used which require chemical or enzymatic pretreatment. The products which can be obtained range from organic acids, to their salts or esters[9,11-13]. These processes will be strongly dependent upon the separation operations needed, e.g., liquid-liquid extraction, adsorption-esterification, or membranes[9,12,14].

Fixed Film and Suspended Growth Reactors

Conceptually one might consider five types of fixed film reactors and three types of suspended growth reactors [of which some might become fixed film if the floc shows a tendency to granulate throughout its operation[1,15]].

The types of anaerobic fixed film reactors are (Figure 2):

I - Fixed bed - uses various filling materials and can be operated up- or down-flow[15]; usually operated without recirculation as a large plug flow system even though gas bubbling mixes it to some extent; recirculation might be needed if there is a need to control biofilm thickness, toxicity or if the pH at the inlet is too low;

II - Expanded bed - introduced by Jewell and Switzenbaum[16] these utilize slightly larger solid supports than fluidized beds, expanded by the upflow rate obtained with

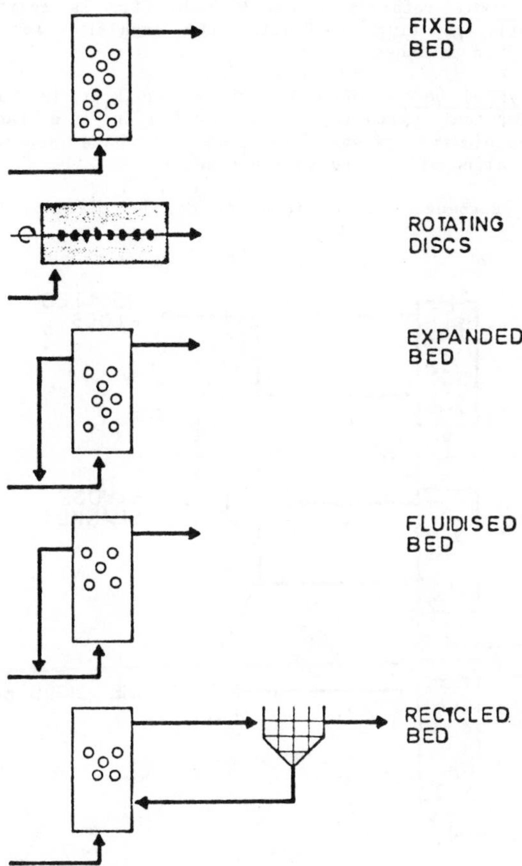

Figure 2 Fixed Film Anaerobic Reactors

recirculation. The particles keep their position within the bed and the thickness of biofilm is controlled by physical contact.

III - Fluidized bed [17] - operated at higher upflow rates with faster recirculation rates; the particles are kept within a reasonably small "parking" space. Biofilm thickness is controlled by the bed regeneration strategy and size and density of the inert materials in relation to the upflow rate.

IV - Anaerobic rotating discs [18] - biofilm is formed on discs slowly rotating in liquid; the angular velocity controls biofilm thickness.

V - Recycled bed - inert materials are kept in suspension by mechanical agitation or gas bubbling[19], a large part of the biomass exists as flocs; a phase separator allows recycling of the bed to the reactor.

Anaerobic suspended growth reactors can be classified as follows (Figure 3):

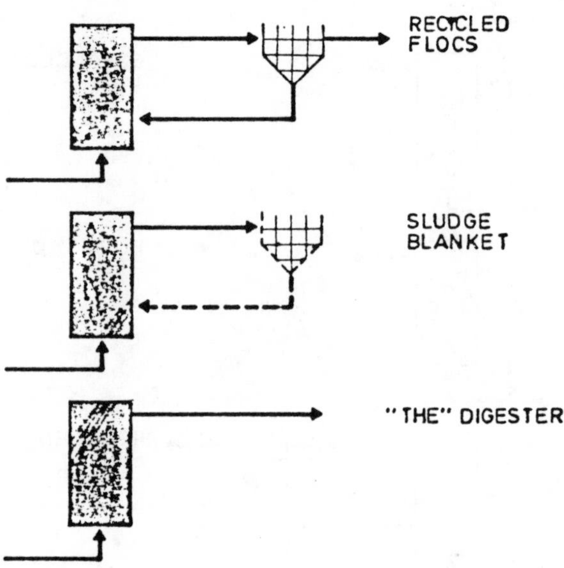

Figure 3 Suspended Growth Anaerobic Reactors

A - Contact or recycled flocs[20] - conceptually similar to the recycled bed mentioned above, without explicit addition of inert solids apart from those existing in the residue or due to floc "granulation".

B - Fluidized flocs or sludge blanket[21] - conceptually similar to the contact reactor except upflow and gas bubbling are responsible for fluidization; a solid profile exists along the reactor.

C - "The Digestor" - the classical reactor used for wastewater sludges high in suspended solids, operating under similar hydraulic and solids retention time thus requiring long hydraulic retention times and consequently large reactor volumes.

When compared with suspended growth reactors, the fixed film types are less susceptible to washout, can be operated at much shorter hydraulic detention time, are more likely to cope with shock loadings and sustain active microbial cultures even after long periods of starvation; given their long mean cell residence time, they have higher conversion efficiencies and reduced nutrient requirements[1,15]. Furthermore, their plug flow operation (unless high recirculation rates are used) produces some phase separation, perhaps increasing the stability and possibly also increasing the methane production rate and final methane concentration in the gas [1,22].

Choice of Process and Reactor Type

Where no phase separation is envisaged, the solubility index criteria allows us to decide for or against parallel operation[1]. Fixed film reactors should be considered the optimum choice for effluents with a high solubility index; the choice between fixed film and other systems will depend on size, energy efficiency required, and acceptable level of complexity.

If phase separation is chosen, (as in our project) the different limiting steps and engineering behavior of the reactors determine that[1-11, 15-25]:

a. For soluble, easily fermentable substrates from the agro-industries, the acid stage can probably be operated at very high loading rates as hydrolysis should not be rate- limiting; much of the engineering optimization strategies can thus focus on the quality of intermediary metabolites, either as methanogenic substrates or as chemicals. Thus, for the acidogenic phase, either expanded or fluidized beds seem appropriate, given their high loading rate capabilities and smaller requirement of sludge settling properties; we propose to utilize an expanded bed reactor as it is easier to operate, uses less energy and lower effluent recycle rates, thus being closer to plug flow than the fluidized bed.

b. For particulate feeds, such as those arising from crop or forestry residues less information is available. It looks as if long solids retention times, as well as high solids concentrations, and as low pH as reactionally feasible will be helpful for enzymatic action and further increase hydrolysis. Almost certainly much of the engineering strategies will focus on increasing the rate of hydrolysis for which culture seeding and breeding or cell immobilization of interesting bacteria might have to be contemplated. The most probably correct choice of reactor lies with the recycled bed type of the upflow solids blanket to which the former ultimately reduces, given the internal settling stage and the fact that part of the solids will operate as inert-floc support. We intend to use the sludge blanket reactor as it reduces the amount of equipment required and does not add a settling stage which also has to be kept anaerobic. The acid production rate and concentration distribution will, to a certain extent, have to yield to the first strategical priority of guaranteeing extended hydrolysis.

c. The methanogenic reactor will have similar constraints for both the soluble and particulate original feeds high biological solids retention times, average hydraulic retention times (larger than is usual for the acidogenic reactor for soluble substrates), feed almost completely made up of solubilized substrate. The most appropriate reactor type for the job is generally recognized to be the upflow fixed bed or anaerobic filter, although the expanded bed might still be a good choice, but will probably yield lower methane concentrations in the gas; we therefore selected the former.

MICROORGANISMS

At the moment, we are characterizing the following microorganisms: Desulfovibrio gigas (NCIB 9932); Desulfovibrio vulgaris (Hildenborough) - this last one being a very interesting bacterium, because its hydrogenase is extremely active (specific activity of about 3000 micro liter H_2 evolved per minute per miligram protein, as compared to about 400 micro liter H_2/min mg in D. gigas); Desulfovibrio desulfuricans (Berre Eau), a sulfate reducing bacterium which is able to fix N_2; and Methanosarcina barkerii (DSM 800 and 804), the most versatile mesophylic methanogenic bacterium, as it can grow in methane acetate, CO_2+H_2 and methylamine.

BIOCHEMICAL STUDY OF THE PROCESS

Screening of the Electron Transfer Proteins and Enzymes

Purification Processes The purification processes used will depend on the organism and the cell quantity. In some cases, depending on the enzyme purified, the purification scheme may be altered. For instance, in the purification process of the D. gigas hydrogenase, an extremely active fraction can be obtained if the periplasmic space fraction (corresponding to the cell washing) is previously treated, thus making unnecessary some of the cromatographic steps, with an overall decrease of purification time.

All the purification steps are carried out at 4°C , in cooled chambers and columns, and the cromatographic elution processes are done with appropriate buffers of controlled molarity and pH.

The cells are either washed or broken in a French-Gaulin pressure cell, to obtain the periplasmic fraction or the crude extract respectively. In the last case, DNase is added to the medium to decrease its viscosity. The fractionating of the cellular extract is achieved through a series of cromatographic steps (in batch or column), using mainly sephadex, ultragel, DEAE-sephadex, alumina, silica gel, hydroxylapatik, DEAE, and CMC as cromatographic materials.

Bacterial Control of the Digestor through Cofactor Analysis

Previous studies of enzyme and cofactor contents present in some bacterial species lead to the following conclusions:

- The methanogenic bacteria have unique cofactors, which can be used as their activity indicators (e.g. F 420, F 430 (containing nikel) and corrinoid B-12).

- One of the largest components of the enzymatic apparatus of sulfate-reducing organisms is the dissimilatory sulfite reductase, which is used as a taxonomic label [26] of the bacteria _D. gigas_, _D. Salexigens_ and _D. Vulgaris_ (desulfoviridin), _D. desulfuricans_ (Norway 4) and D. 9974(desulforubidin), and _D. desulfotomaculun_ (P582). The chemical treatment of these enzymes with acetone/HCl produces an extract containing sirohydrochlorine (a demetalized siroheme), which presents a characteristic fluorescence.

  The analysis of these cofactors may be relevant in the control and identification of the bacterial populations present in the system, when other experimental parameters (such as the pH and the [$SO_4^=$] present in the medium) are changed.

ACKNOWLEDGEMENTS

The authors acknowledge the INIC, JNICT, and Calouste Gulbenkian Foundation, Portugal, and the Agency for International Development, United States, for financial assistance.

REFERENCES

1 HENZE, M. and P. HARREMOES. June 1982. Anaerobic treatment of wastewater in fixed film reactors. Literature review in Proc. IAWPR Seminar, pp. 1-90. Copenhagen.

2 ZOETEMEYER, R.J. et al., pH influence on acidogenic dissimilation of glucose in an anaerobic digestor. Water Res. 16:303-311.

3 POHLAND, F.G. and S. GHOSH. 1971. Anaerobic stabilization of organic wastes: two phase concept. Env. Letters. 1:255-266.

4 CARRONDO, M.J.T., et al. 1983. Anaerobic filter treatment for molasses fermentation wastewater. Water Sci. Technol. 15:117-126.

5 GHOSH, S., J.R. CONRAD, and D.L. KLASS. 1975. Anaerobic acidogenesis of wastewater sludge. J. Water Pollut. Control Fed. 47:30-45.

6 EASTMAN, J.A. and J.F. FERGUSON. 1981. Solubilization of particulate organic carbon during the acid phase of anaerobic digestion. J. Water Pollut. Control Fed. 53:352-366.

7 VERSTRAETE, W., L. De BAERE and A. ROZZI. 1981. Phase separation in anaerobic digestion: motives and methods. Trib. Cebedeau, No. 453-454, 367-375.

8 COHEN, A., et al. 1979. Anaerobic digestion of glucose with separated acid production and methane formation, Water Res. 13:571-580.

9 LEVY, P.F., J.E. SANDERSON, and D.L. WISE. 1981. Development of a process for production of liquid fuels from biomass. Biotechnol. Bioeng. Symp. 11:239-248.

10 MASSEY, M.C., and F.G. POHLAND. "Phase separation of anaerobic stabilization by kinetic control. J. Water Pollut. Control. Fed. 9(m978): 2004-2222.

11 GHOSH, S. 1981. Kinetics of acid-phase fermentation anaerobic digestion. Biotechnol. Bioeng. Symp. 11:301-313.

12 DATA, R. 1981. Production of organic acid esters from biomass-novel processes and concepts. Biotechnol. Bioeng. Symp. 11:521-532.

13 CLAUSEN, E.C., R.B. SHAH, G. NAJAFFOUR, and J.L. GADDY. 1982. Production of organic acids from biomass by acid hydrolysis and fermentation. Biotechnol. Bioeng. Symp. 12:238-248.

14 OMSTEAD, D.R. et al. 1980. Membrane controlled digestion: anaerobic production of methane and organic acids. Biotechnol. Bioeng. Symp. 10:247-258.

15 YOUNG, J.C., and P.L. McCARTY. 1969. The anaerobic filter for water treatment. J. Water Pollut. Control. Fed. 41: R 160 - R 173.

16 SWITZENBAUM, M.S. and W.J. JEWELL. 1980. Anaerobic attached film expanded bed reactor treatment. J. Water Pollut. Control Fed. 52: 1953-1965.

17 SHIEH, W.K. 1980. Suggested kinetics model for fluidized biofilm reactor. Biotechnol. Bioeng. 22:667-676.

18 TAIT, S.J. and A.A. FRIEDMAN. 1980. Anaerobic rotating biological contactor for carbonaceous wastewaters. J. Water Pollut. Control Fed. 52:2257-2269.

19 MARTENSSON, L. and B. FROSTELL. June 1982. Anaerobic wastewater treatment in a carrier assisted sludge bed reactor. In: Proc. IAWPR Seminar on Anaerobic Treatment, pp. 179-192. Copenhagen.

20 SCHROEPFER, C.F. and N.R. ZIEMKE. 1959. Development of the anaerobic contact process. Sew. Ind. Waste 31:164-190.

21 LETTINGA, G. et al. 1980. Use of the upflow sludge flanked reactor concept for biological wastewater treatment, especially for anaerobic treatment. Biotechnol. Bioeng. 22:699-734.

22 CHYNOWETH, D.P. 1981. Microbial conversion of biomass to methane. 8th Ann. Energy Conf. and Exposition, p.25. Washington, March 1981.

23 DATTA, R. 1981. Acidogenic fermentation of corn stover. Biotechnol. Bioeng. 23:61-77.

24 JEWELL, W.J. et al. 1978. Anaerobic fermentation of agricultural residues: potential improvement and implementation. Final U. S. Department of Energy Report. EY-76-5-02-2981-7.

25 JERGER, D.E., J.R. CONRAD, K.F. FANNIN, and D.P. CHYNOWETH. 1982. Biogasification of woody biomass. In: Energy from Biomass and Wastes VI, p.27. Florida.

26 LeGALL, J. and J.R. POSTGATE. 1973. The sulfate reducing bacteria. In: Advances in Microbial Physiology (A.H. Rose and D.W. Tempest (eds.), pp. 81-133. Academic Press. London.

STUDY OF THE OPTIMUM OPERATING CONDITIONS FOR THE ANAEROBIC DIGESTION OF SOME ORGANIC WASTES

S.I. EL-Shawarby, N.A. Shabaan, F. EL-Gohary* and M.M. EL-Halwagi
Pilot Plant Laboratory, National Research Centre, Egypt

ABSTRACT

Bench-scale studies using batch and semi-continuous fermenters have demonstrated the feasibility of producing reasonably high methane rates from three different organic farm wastes.

The effect of initial solids concentration, retention time, temperature, and mode of operation on the percentage reduction of the total volatile solids production rates have been studied. Cattle manure, poultry-droppings, and poultry-droppings/maize stalks mixtures have been used as the organic substrates.

It seems that an initial total solids concentration of eight percent, a hydraulic retention time of five days, and a temperature of about 40°C are the optimum operating conditions for digesters using cattle manure within the levels of variables studied in this work.

On the other hand, the optimum conditions, within the ranges studied in this investigation, for poultry-dropping digestions appear to be an initial total solids concentration of about seven percent and a hydraulic retention time of 10 days at 50°C.

For poultry-droppings/maize-stalks mixtures, the TS figure was kept constant and the C/N ratio varied experimentally. The optimum C/N ratio considered in these experiments lies between 20 and 30.

INTRODUCTION

The general characteristics and the controlling factors of anaerobic digestion have been extensively studied and well documented by many investigators[1-4]. However, the rate of fermentation for a specific set of conditions, including those relevant to the quality and type of local substrates, should be determined experimentally under controlled conditions[5]. Thus, the main objective of this work is to assess the performance of both batch and semi-continuous systems using

* Water Pollution Control Laboratory, NRC

the various operating conditions and some organic substrates that are likely to be utilized in rural Egyptian settings. Such studies constitute an indispensable stage in the process of designing and optimizating full-scale digesters.

DESIGN OF THE EXPERIMENT

A schematic of the experimental setup is shown in Figure 1. Five-liter aspirator bottles were used as fermenters. They were operated in both the batch and semi-continuous modes. Mixing was performed by manual shaking twice a day. The biogas produced was measured daily by collection and analyzed for methane content by gas chromatography[6]. Each type of feed stock was stored in a refrigerator (4°C) to minimize degradation in the laboratory

A total of thirty four experimental runs were conducted (excluding replicates). In addition to the two different modes of operation, three substrates (cattle manure, poultry droppings, and poultry droppings/maize stalks mixtures) in different concentrations and over a temperature range varying from mid to high levels were covered. Temperature was elevated gradually and slowly, about 3-5°C/5 days, starting from room temperature to the desired level. The effect of each variable on both the total volatile solids (TVS) reduction and gas production was assessed. TVS, carbon (0.47TVS), total solids (TS) and total nitrogen were determined using standard methods[7].

RESULTS AND DISCUSSION

1. Effect of the initial total solids (TS) concentration on:

A. TVS reduction

Results indicate that for batch experiments, no significant reduction in TVS took place by increasing the initial TS of cattle manure from about 6 to 10 percent. The values varied between 30.57 to 29.1 percent. However, increasing the TS to 14 percent led to a significantly smaller reduction of TVS to 24.38 percent. For the semi-continuous fermentations at a retention time (θ) of 30 days, the TVS-reduction assumed values varied between 28.18 and 31.61 percent. These results are similar to those reported by Hayes et al[8], Singh et al.[9], Stuckey[10], Bryant et al.[11] and Jewell[2,12].

For poultry droppings the results indicate that, the reduction in TVS decreased gradually from 68.55 to 66.41 percent as the initial TS increased from about five to 14 percent. These results agree well with those of Gramms et al.[13] and Hill[14].

The reduction in the percentage removal of TVS may be attributed to overloading. It should be mentioned that the percentage reduction in TVS was much higher for poultry droppings than for cattle manure.

For poultry-droppings/maize-stalk mixtures, the C/N ratio would be preferable to the TS concentration as a basis of comparison, since all experiments had the eight percent TS as an initial concentration chosen. Results show that, as the C/N ratio increased from about 12 to

19, the percentage reduction in TVS slightly increased from 56.77 to 63.02. Further increase in the C/N ratio up to about 43 was accompanied by a decrease in TVS-reduction to 45.36 percent. Hashimoto[15] and Hills[16] reported very similar trends.

B. Biogas and methane

The results show that the average rates of biogas obtained over the studied range of TS varied from 0.15 to 0.23 l/l/day with methane percentage of about 60 percent for cattle manure in batch systems. Although there is a slight tendency for the biogas to level out between six and 10 percent TS[18], the 10 percent TS appears to be the optimum where methane production rate reached 0.136 l/l/day. When the TS increased to 14 percent, the methane production rate decreased to 0.088 l/l/day. This could be attributed to organic overloading[17]. Methane rates obtained in this work are slightly lower than some reported values[2,13] probably due to the lower biodegradable fraction of the manure used[18]. On the other hand, the results show that the best initial TS concentration for semi-continuous systems was eight percent where the maximum methane rates of 0.67 l/l/day was attained for 20 days retention time. These rates agree with those reported by Singh et al.[9] and are comparable to those of Jewell[2]. The results also show that the average volumes of biogas ranged between 0.24 to 0.56 l/l/day for poultry-droppings with methane content ranging between 25 and 64 percent in batch systems (18). A slight increase in methane rate was noted as the TS increased from about five to seven percent. Further increase in TS was accompanied by a sudden decrease of biogas as well as methane rates. Ammonia inhibition probably accounts for this decrease[18].

While Hassan et al.[19] concluded also that the optimum solids concentration lies between 7 and 7.5 percent, other investigators[17] reported similar biogas and methane rates. On the other hand, the semi-continuous fermenters exhibited much higher biogas rates of 0.8 to 2 l/l/day and methane rates of 0.28 to 0.96 l/l/day. Hill[14] published similar order of magnitude.

In the semi-continuous fermentations of the mixtures, the results show that the biogas rate increased from 1.65 to 1.92 l/l/day as the C/N ratio increased from about 12 to 19 as shown in Figure 2. Increasing C/N ratio to about 43 caused the rate to decrease. Methane rates showed similar trend. This may be attributed to ammonia inhibition[20] at lower C/N ratio and too high percentages of lignin at higher C/N ratios. These results are reinforced by other investigators findings that the optimum C/N ratio lies between 20 and 30.

2. Effect of hydraulic retention time (O) on:

A. TVS reduction

The results show that for the initial TS concentration of eight percent the decrease in the retention time from 30 to five days caused a decrease in the reduction of TVS of cattle manure from 31.61 to 11.31 percent. Similar trend and order of magnitude have also been reported by Converse[21], Jewell et al.[2], Gramms et al.[13] and Singh et al.[9].

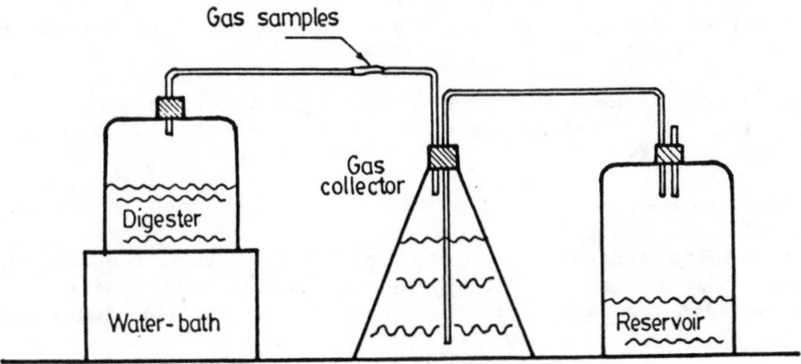

Figure 1 Schematic Diagram of Digester.

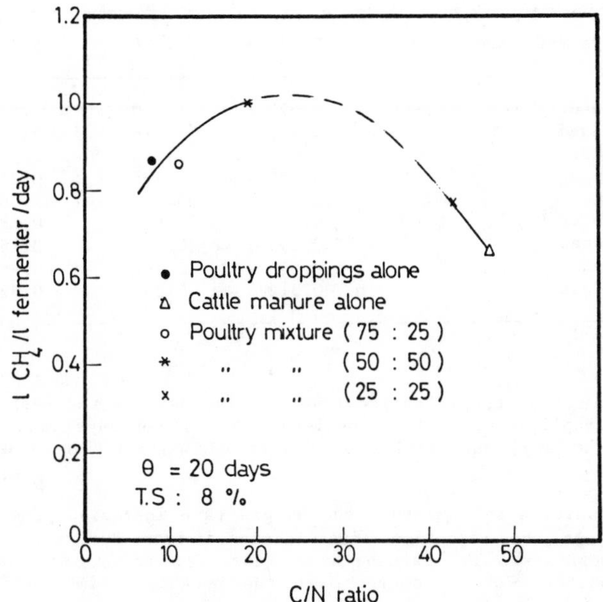

Figure 2 Variation of Methane Production Rate with C/N Ratio for S.C. Anaerobic Fermentation of Different Wastes at 35 °C.

For poultry-droppings the results show that as θ decreases from 30 to five days, the reduction in TVS decreased from 59.26 percent to 24.97 percent. The order of magnitude and the trend of this reduction agree well with Huang et al.[22] and others[13,23].

Generally, it seems that the hydraulic retention time has a greater effect than the loading rate on the lowering of TVS-reduction percentages.

B. Biogas and methane

The results indicate as shown in Table 1, that the biogas rate increases from 0.575 to 2 l/l/day as the retention time decreases from 30 to five days for cattle manure. The methane rate also increases.

Table 1 Variation of Biogas and Methane Produced with Hydraulic Retention Time for Cattle Manure 8 percent TS (Anaerobic Fermentation at 35°C).

θ (Days)	30	20	10	5
Loading Rate gm VS/l/Day	2.010	2.880	5.660	11.31
l gas/l Fermenter/day	0.575	1.250	1.875	2.000
l gas/gm VS added	0.286	0.376	0.331	0.177
l gas/gm VS dest.	0.905	1.498	1.798	1.563
l gas/gm TS	0.230	0.275	0.253	0.134
% CH_4	68.00	53.56	47.00	57.00
L CH_4/l Fermenter/Day	0.391	0.670	0.881	1.140
l CH_4/gm VS added	0.194	0.201	0.156	0.101
l CH_4/gm VS dest.	0.615	0.802	0.845	0.890
l CH_4/gm COD dest.	0.312	0.331	0.329	0.259

It seems that the retention time of five days is the best, but for practical applications, it is believed that 10-15 days would allow a margin of safety, especially for wastes of higher fibrous or lignin content.

The results also show that the biogas rate increased from 1.125 to 2 l/l/day as the retention time decreased from 30 to 10 days for poultry-droppings. The methane rate decreased from 58 to 48 percent simultaneously. Further decrease in the retention time to five days caused a drastic decrease in both biogas and methane rates. The increase in biogas and methane rates as retention time decreases could be attributed to the increase in the loading rate; the following decrease in the rates is believed to be due to ammonia inhibition[18]. Hobson[17] reported comparable biogas rates.

3. Effect of temperature on:

A. TVS reduction

The data indicate that as the temperature gradually increased from 16 to 60°C,the reduction in TVS increased from 15.2 to 29.1 percent for cattle manure. The same trend was noticed in case of poultry-droppings, for which the TVS-reduction ranged from 29.44 to 62 percent. This trend is in agreement with the findings of Golueke[24] for sewage sludge, Chen[25] for cattle manure, and Nelson et al.[26] for some agricultural wastes.

B. Biogas and methane

Table 2 shows that as the temperature increased from 16°C to 60°C the biogas increased from 0.625 to 1.42 l/l/day for cattle manure wastes. However, above 40°C the methane content decreases significantly. Converse et al.[21], and Hobson et al.[17] found similar effects.

Table 2 Effect of Temperature on Biogas and Methane in S.C.* Digestions

Temp °C	Type of Substrate	Cattle Manure		Poultry-Droppings		
	1 Biogas/l/Day	lCH_4/l/D	%CH_4	1 Biogas/l/Day	lCH_4/l/Day	%CH_4
16	0.625	0.336	53.70	0.55	0.341	62.00
28	0.900	0.427	47.56	0.80	0.496	62.00
35	1.250	0.670	53.56	1.50	0.875	58.30
40	1.250	0.733	58.80	1.64	0.818	53.50
45	1.350	0.631	46.75	1.53	0.948	62.00
50	1.400	0.621	44.32	1.80	0.985	54.72
60	1.420.	0.576	40.59	2.05	0.910	44.38

* S.C. = Semicontinuous.

Experiments with poultry-droppings showed the same trend, and the biogas rate ranged between 0.55 to 2.05 l/l/day. Regarding methane rates, it seems that 45 and 50°C were the optimum temperatures for poultry-droppings fermentations. Although, Hassan et al.[19] concluded that 35°C was the best for biogas yield, Havang et al.[27] claimed that thermophilic digestion (55°C-60°C) was better for poultry-droppings.

4. Effect of mode of operation on:

A. TVS reduction

The previous results indicate that the percentage reductions in TVS for batch systems are generally higher than the corresponding values for the semi-continuous ones. This could be partially due to the fact that the actual retention time in the latter is much less than the calculated time[28] and also due to the fact that in the batch systems the micro-organisms have the opportunity to grow, acclimatize and efficiently utilize the substrate.

B. Biogas and methane

For cattle manure and poultry droppings, the results show that the biogas production rates as l/l/day were higher in all cases of semi-continuous fermentations (0.5 to 2) than the batch ones. It is also noted that the methane content and consequently the methane production rate is higher for the former case. Boshoff[5] and Hill[14] also showed that continuous fermentation is the most efficient method for gas production using cattle manure and poultry droppings respectively.

Generally, one can attribute this higher efficiency in the semi-continuous systems to the dilution effect of any toxic materials and to the higher capability of the micro-organisms to acclimatize themselves.

REFERENCES

1. Lapp, H.M., D.D. Schulte, A.B. Sparling, and l.C. Buchaman. 1975. Can. Agr. Eng., 17,97.

2. Jewell, W.J., H.R. Capener, S. Dell'Orto, K.J. Fanfoni, T.D. Hayes, A.P. Leuschner, T.L. Miller, D.F. Sherman, P.J. Van Soest, M.J. Wolin, and S.J. Wjcik. 1978. U.S. Dept. of Energy Rep. HCP/T2981-07 Nat. Tech. Inf. Serv., Springfield, Virginia.

3. Lawerence, A.W. and P.L. McCarty. 1969. J. Wat. Poll. cont. Fed., 14,RI.

4. Pfeffer, J.T. 1974. Biotech. and Bioeng., 16:771.

5. Boshoff, W.H. 1965. Tropical Science, 3,155.

6. Association of Official Analytical Chemists. 1975. Official methods of Analysis, 12th ed., AOAC, Washington, D.C.

7. American Public Health Association. 1975. Standard Methods for the Examination of Water and Waste-Water, 14th ed., APHA, Inc., New York.

8. Hayes, T.O., W.J. Jewell, S. Dell'Orto, K.J. Fanfoni, A.P. Leuschner, and D.R. Sheridan. 1979. Anaerobic Digestion, Applied Science Publishers, London, U.K.

9. Singh, R., M.K. Jain, and P. Tauro. 1982. Agr. Wastes, 4,267.

10. Stuckey, D.C. 1983. A state of the Art Review, IRWD, Switzerland.

11. Bryant, M.P., V.H. Varel, R.A. Frobish, and H.R. Iseacson. 1979. Seminar on Microbial Energy Conversion Gotlingen, Germany.

12. Jewell, W.J., H.R. Davis, W.W. Gunkel, D.J. Lathwell, J.H. Williams, Jr., T.R. McCarty, G.R. Morris, D.R. Price, and D.W. Williams. 1976. U.S. ERDA Rep. No. TID 27164, Cornell Univ.

13. Gramms, L.C., L.B. Polkowski, and S.A. Witzel. 1971. Trans. Amer. Soc. Agr. Engrs., 14,71.

14. Hill, D.J. 1982. Poultry Science, 61, 677.

15. Hashimoto, A.G. 1982. SERI/PR - 624.

16. Hill, D.J. 1979. Amer. Soc. Engr., St. Joseph.

17. Hobsen, P.N., S. Bousfield, R. Summers, and P.J. Mills. 1980. Proceedings 1st Int. Symp., Applied Science Publishers, London.

18. Shabaan, N.A. Study of the Kinetics of Anaerobic Fermentation of Some Organic Wastes. 1984. A Ph.D. Thesis, Faculty of Eng., Cairo University.

19. Hassan, A.E., H.M. Hassan and N. Smith. 1975. Energy, Agriculture and Waste Management. An Arbour Science publisher. Michigan, U.S.A.

20. McCarty. P.L. 1964. Public Works, 95.

21. Converse, J.C., J.G. Zeikus, R.E. Graves, and G.W. Ivans. 1977. Trans. Amer. Soc. Agr. Eng., Paper No. 77-0411.

22. Huang, J.J.H., J.C.H. Shih, and S.C. Steinsberger. Proceedings of the 2nd. International Seminar. Oxford, New York.

23. Converse, J.C., G.W. Evans, C.R. Verhoeven, W. Gibbon, and M. Gibbon. 1980. Proceedings of the 4th Int. Symp. Livestock Wastes, Amarillo, Texas.

24. Goluke, G.G.. 1958. Sewage Ind. Wastes, 30, 1225.

25. Chen, Y.R., V.H. Varel and A.G. Hashimoto. 1979. Proceedings of the 2nd. Symp. on Biotechnology in Energy Production and Conservation, Gatlinburg, Tennessee U.S.A.

26. Nelson, G.H. 1939. J. Agr. Res. Washington, D.C., 58, 4.

27. Havang, J.J.H., and J.C.H. Shih. 1981. Biotech. and Bioeng., 23, 2307.

28. Hamad, M.A., A.M. Abdel Dayem and M.M. El-Halwagi. 1983. Energy Agricl, 1,235.

NOVEL PROCESSES FOR HIGH-EFFICIENCY BIODIGESTION OF PARTICULATE FEEDS

Sambhunath Ghosh
Institute of Gas Technology
Chicago, Illinois

ABSTRACT

The development, application and advantages of three advanced anaerobic digestion processes utilizing phase-separated fermentation and novel upflow digesters are described. Unlike single-stage conventional digestion, these processes were suitable for the digestion of solid and concentrated semi-solid feeds which are by far the largest biomass and waste resources available for simultaneous stabilization and energy production. The three novel digestion processes included continuous stirred tank reactor (CSTR) two-phase digestion, upflow two-phase digestion, and leach-bed two-phase digestion. All novel processes were significantly superior to conventional single-stage digestion in terms of methane yield and production rate, solids conversion efficiency, and net energy production. The upflow and the leach-bed two-phase digestion processes effected virtually complete conversion of the biodegradable fraction of the feed; the leach-bed process could accomplish this with a nutritionally deficient highly cellulosic refuse-derived substrate.

INTRODUCTION

Organic feeds that are suitable for methane fermentation by biodigestion could be "liquid," semi-solid or "solid" in nature. From the viewpoint of biodigestion process design, "liquid" feeds could be defined as substrates in which most of the biodigestible organic carbon remains in a soluble form and the volatile suspended solids (VSS) content of the feed is below about 0.5 wt percent. The important liquid feeds are sewage and certain industrial wastes; however, these wastes constitute a small fraction of the overall renewable carbon resource available for biodigestion and at best can only have a small impact on biogas production potential on a national scale. Conversely, feeds that are available in large quantities to individual farmers and communities for biogas production for public benefit are semi-solid or "solid" in nature. "Solid" or "dry" feeds are organic substrates that usually have a solids concentration of more than 15 wt percent and do not exhibit physical properties peculiar to slurries. Semi-solid substrates are slurries of particulate organic matter and usually have solids concentrations of one to 15 wt percent. Conventional "standard-rate" and "high-rate" digestion processes that have been practiced for more than 100 years are suitable for semi-solid feeds having solids concentrations between two and five wt percent. These single-stage conventional anaerobic digestion processes cannot be applied to solid feeds; with semi-solid feeds, substantial reduction in process efficiency occurs at feed solids concentration above five wt percent and further, as the feed flow rate is increased. Consequently, novel process configurations need to be developed to effect high-efficiency rapid-rate anaerobic digestion of solid and concentrated semi-solid feeds.

The above considerations are relevant because the physical nature of the feed, and feed solids consistency are of crucial importance in the selection, development, and application of appropriate fermentation modes and reactor designs. As an example, hydrolysis, which could be the controlling biochemical reaction in the digestion of particulate and solid feeds, is of little importance in the case of liquid substrates. Also, liquid feeds do not present many of the difficult engineering problems that are peculiar to semi-solid and solid feeds. Therefore, as expected, major advances were made during the last decade in developing novel rapid-rate processes for the stabilization and biogasification of sewage and liquid industrial wastes to methane. Examples of such systems include the Anamet process,[1] the Biothane process,[2] the Anflow process,[3] the Celrobic process,[4] the Bicardi process,[5] the baffle-flow digester,[6] the fluidized-bed system,[7] the Anthane-Anodek process,[8] and others. Most of these systems utilize single-stage mixed acidogenic-methanogenic fermentation, and are effective for highly biodegradable liquid substrates of moderate strengths (COD concentrations up to 15,000 to 20,000 mg/l); higher-strength wastes need to be diluted to prepare digester feeds. High-strength liquid wastes can be processed without dilution by the Anthane-Anodek system which utilizes two-stage two-phase digestion providing for separate liquefaction-acidification (LA) and acetogenic-methanogenic (AM) fermentations. Comparable advances have not been made in developing novel and advanced digestion systems to process concentrated organic slurries or solid feeds at lower hydraulic residence times (HRT) and higher loading rates to afford higher methane yields (and stabilization efficiencies) and production rates than those of the conventional digestion processes.

Limitations of Conventional Anaerobic Digestion

By far the major biomass resources available for small- to medium- to large-scale production of biogas are residential and commercial solid wastes, agricultural residues, manures, sludges, and various herbaceous plants (e.g., water hyacinth, algae, grasses, etc.) grown to clean up polluted waters. For conventional digestion, it is a common practice to dilute these high-solids-content feeds to produce a thinner organic slurry suitable for balanced LA-AM fermentation in a single complete-mix digester. This traditional approach is not attractive because it not only requires external water, larger digesters, and higher-capacity feeding and effluent-withdrawal systems, but it also increases the energy requirement for pumping and heating the diluted slurry. In addition, with a diluted feed, a larger volume of digester effluent has to be processed and disposed of.

Conventional stirred-tank digesters are operated at high hydraulic residence times (HRT) and low volatile solids (VS) loading rates (LR) to obtain acceptable methane yields and gasification efficiencies, and to ensure the production of low volatile solids (VS) loading rates (LR) to obtain acceptable methane yields and gasification efficiencies, and to ensure the production of low volatile acids (VA)-content stabilized effluents. Typical operating and performance characteristics of single-stage conventional digesters are exemplified by the IGT data presented in Table 1.

Table 1 shows that satisfactory digester performance, as indicated by good methane yields and low effluent (or residual) volatile acid concentration, were obtained at HRT's above 12 days and at loading rates below 2 kg VS/m^3-day. At the high HRT's and low loading rates, the rates of acid production balanced those of acid conversion to methane with the result that the digester volatile acids concentration was about 100 mg/l, a concentration that is not inhibitory to the methanogenic or the acidogenic bacteria. Such was not the case when more concentrated feeds (e.g., 6.1, 6.6, and 12.6 wt percent), high loading rates, and/or shorter HRT's (e.g., 2.9, 4.3, 5.0, and 6.9 days) were used, as would be evident from the data in Table 2. Under these operating conditions, the rate of volatile acids production exceeded the rate of acids turnover to methane with the result that volatile acids accumulated to high concentration levels that were inhibitory to the LA and AM reactions. Feed VS conversion efficiencies were very low, and "sour" digestion conditions and process failure were experienced. Clearly, unconventional approaches are needed to stabilize and gasify concentrated semi-solid and solid feeds at higher efficiencies than those indicated in Table 1. An engineering approach that has been successful in this regard with bench-scale units and merits further development through pilot-scale operation involves 1) the application of an advanced mode of digestion, two-phase fermentation, and 2) the utilization of a novel, but simple, biochemical reactor design. The remainder of the paper is devoted to the discussion of novel digestion process development for application to concentrated semi-solid and solid organic feeds.

NOVEL PROCESS CONCEPTS

Various approaches can be envisaged to enhance the efficiency, kinetics, and stability of the anaerobic digestion process. Chemical or enzymatic hydrolysis of the particulate feed and utilization of genetically improved microorganisms, for example, could improve the process, but these methods do not seem to be cost-effective or practical at this time. Engineering approaches including application of advanced operating or fermentation modes and utilization of novel reactors are feasible and merit development for digestion process improvement in the near term. Thus, a two-part approach, as outlined below, is needed:

- First, the kinetically dissimilar reaction steps of the overall digestion steps must be optimized in isolated environments or reactor stages because this is not achieved in single-stage mixed-phase digesters as illustrated by the data in Table 2.
- Second, novel digestion reactor designs that provide high substrate and microbial solids retention times (SRT's), must be developed and applied.

Table 1 Performance of a Mesophilic (35°C) Single-Stage Complete-Mix Conventional Digester Operated with a Dilute Feed at High HRT's

Operating Conditions			Digester Performance				
Feed Solids Conc., wt %	HRT days	LR,* kg VS/m^3-day	Methane Content mol%	Methane Yield, SCM/kg VS added **	MPR,+ std vol/vol-day	Effluent Volatile Acids, mg/l as acetic	Feed VS Converted to Gas, %
4.7	19	1.28	73	0.30	0.39	100	41
4.7	18	1.44	70	0.31	0.26	80	46
4.7	17	1.44	68	0.26	0.37	130	39
4.7	13	1.93	70	0.29	0.55	100	42

Table 2 Failure of a Mesophilic (35°C) Single-Stage Complete-Mix Conventional Digester Operated with Concentrated Feeds at Low HRT's

Operating Conditions			Digester Performance				
Feed Solids Conc., wt %	HRT days	LR,* kg VS/m^3-day	Methane Content mol%	Methane Yield, SCM/kg VS added **	MPR,+ std vol/vol-day	Effluent Volatile Acids, mg/l as acetic	Feed VS Converted to Gas, %
12.6	6.9	10.09	66.7	0.21	1.90	2500	33
3.8	5.0	5.45	57.7	0.08	0.41	1370	15
6.1	4.3	9.13	51.1	0.05	0.25	3220	11
6.6	2.9	12.83	64.0	0.06	0.76	3820	10

*LR is loading rate.

**SCM is cubic meters of methane gas at a selected standard temperature of 15.56°C (60° F) and a standard pressure of 762 mm of Hg.

+MPR is methane production rate.

Phase Separation

For the anaerobic digestion process, which is mediated by at least three symbiotic groups of microorganisms, the simplest approach to optimization is to utilize two independently controlled digesters to maximize substrate conversion (and product formation) per unit time per unit digester volume. In this two-stage, two-phase approach, hydrolysis and acidogenesis are dominant in the first-stage digester and aceticlastic methanogenesis is the predominant reaction in the second-stage digester (Figure 1). Methane fermentation by carbon dioxide reduction, which is mediated by hydrogenotrophic methanogens and is faster than the aceticlastic reaction, occurs in both stages. However, the slower-rate aceticlastic reaction, which is recognized to be the primary source of methane in anaerobic digestion, is not dominant in the lower-HRT acid-phase digester. Acetogenesis, which is the process of oxidation of higher fatty acids to acetate, is supposedly a slow reaction because of the unfavorable free energy of reaction, and could be an unimportant conversion step in the first-stage digester. For highly biodegradable liquid substrates which could be rapidly converted by fermentative pathways to acids and molecular hydrogen, acetogenesis may not occur at all in the first-stage digester and is thus completely shifted to the second stage. On the other hand, some acetogenesis may occur in the first stage digester when it is charged with particulate substrates. Acetogenic conversion could also occur in the first-stage digester when higher HRT's or SRT's are used to promote hydrolysis at a higher efficiency.

High-SRT Digesters

The efficiency of substrate conversion, is dependent on the microbial and substrate solids residence times (SRT_m and SRT_s) as indicated by the following equation:[9]

$$\eta = 1 - \frac{K\,(\theta/SRT_m + k'\theta)(\theta/SRT_s)}{S_i\,(\theta\mu - \theta/SRT_m + k'\theta)}$$

where -

K = saturation constant
0 = HRT
k' = microbial death and decay rate constant
S_i = biodegradable feed substrate concentration
μ = maximum specific growth rate constant

The conversion efficiency increases as the SRT's are increased. To achieve a high SRT in a continuously stirred-tank reactor (CSTR) digester, it is necessary to first degas the digester effluent and then to settle the solids in a heated anaerobic settler so that a concentrated stream of microbial mass and substrate solids can be recycled to the CSTR digester. The microbial and substrate solids residence times are dependent on the ability of the anaerobic settler to concentrate these solids, and on the rate of recycling of the concentrated sludge to the CSTR. A system which can accomplish this is

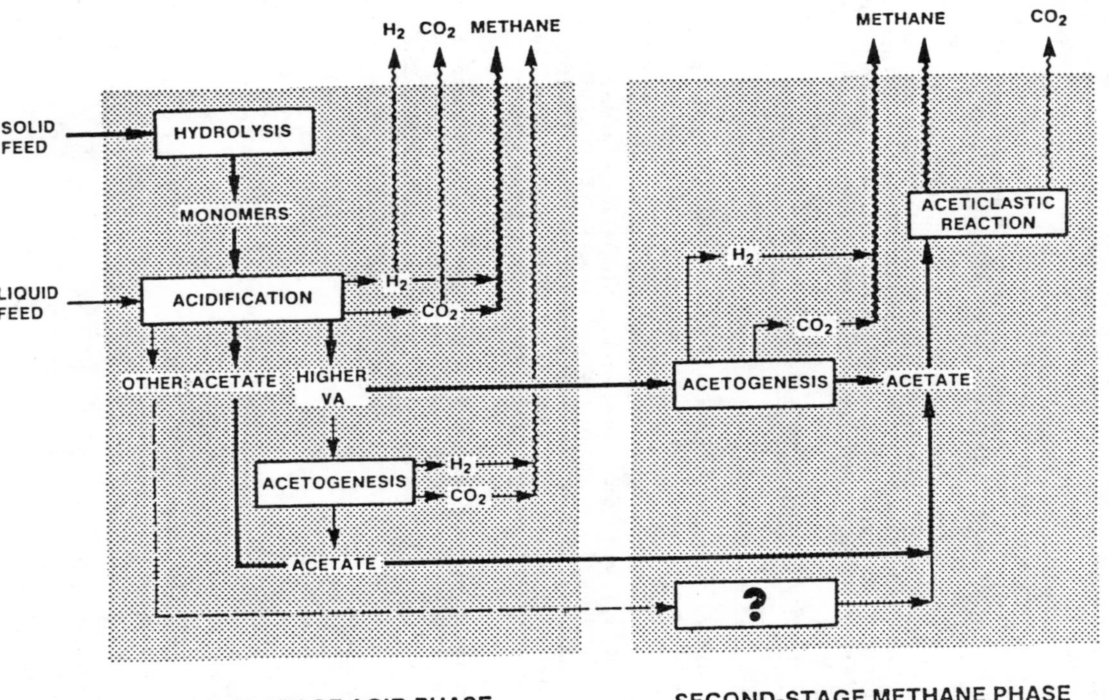

Figure 1   TWO-PHASE ANAEROBIC DIGESTION PROCESS CONCEPT

the anaerobic contact process (also known as the anaerobic activated sludge process); this system is complicated, is capital and energy intensive, and has not worked satisfactorily because of degassing and settling problems with concentrated digested slurries. A simpler and more direct approach to achieving a high SRT is to utilize the digester itself as a clarification tank to selectively retain the solids within the digester. The upflow digester designs discussed in this paper accomplished these objectives as reflected by their superior performance relative to CSTR digesters. We believe that these unconventional digester designs, which are relatively less expensive to build and operate but are superior in performance, would be more suitable for small - as well as large-scale applications.

Three novel digestion processes which are suitable for semi-solid and solid substrates -- these feeds, it should be noted, constitute by far the largest biomass or waste resources -- and which utilize phase-separated fermentation and apply conventional and unconventional reactors, are described in the remainder of this paper.

TWO-PHASE DIGESTION OF SEMI-SOLID FEEDS

Studies with CSTR Digesters

These studies were conducted with a seven wt percent mixed primary and activated Chicago sludge which had a theoretical methane yield of 0.53 SCM/kg VS added based on an elemental analysis of 41.7 wt percent carbon, 6.3 wt percent hydrogen, 18.3 wt percent oxygen, 5.1 wt percent nitrogen, and 1.5 wt percent phosphorus. Single-stage high-rate and two-stage two-phase digestion runs were conducted at 35°C with CSTR digesters (Figures 2 and 3) to demonstrate the benefits of phase-separated fermentation. Results of these studies presented in Table 3 show that the methane yield, the methane production rate, and

Table 3 Conventional and Two-Phase Mesophilic Digestion of Wastewater Sludge

		Two-Stage Two-Phase		
	Single-Stage Conventional	Acid Phase	Methane Phase	Overall System
Operating Conditions				
Feed Consistency, wt % TS	7.0	7.0	-	7.0
HRT, days	6.9	2.0	5.2	7.2
Loading Rate, kg VS/m^3-day	7.7	25.0	9.6	7.1
Performance				
Methane Yield, SCM/kg VS added	0.22	0.09	0.22	0.31
Methane Production Rate, SCM/m^3-day	1.69	2.37	2.06	2.17
Methane Content, mol %	69.5	57.5	67.9	64.5
VS-to-Gas Conversion Efficiency, %	33	18.5	34.3	52.8
Effluent VA, mg/l as acetic	230	1630	110	110
Observed Methane Yield as a % of Theoretical	41	17	41	58

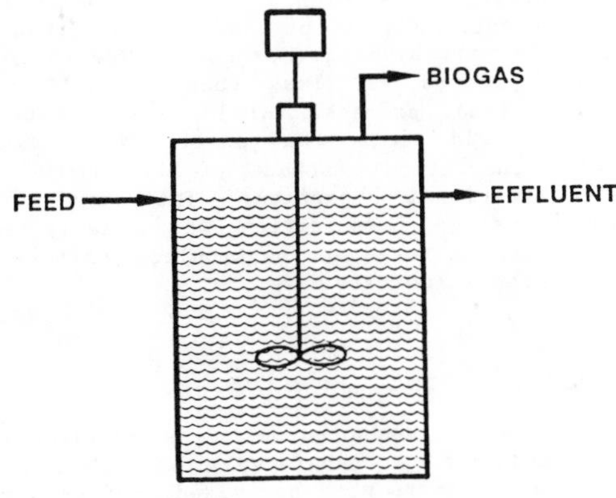

Figure 2 SINGLE-STAGE CONVENTIONAL ANAEROBIC DIGESTION

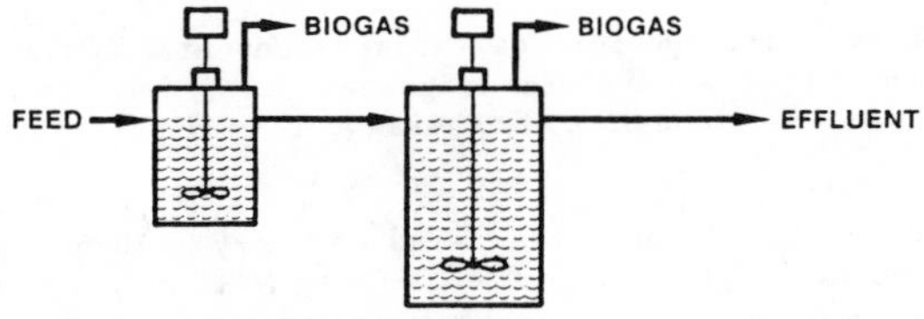

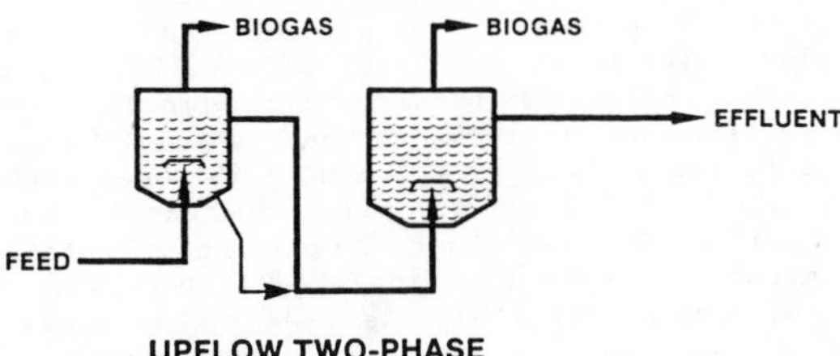

Figure 3 TWO-PHASE ANAEROBIC DIGESTION CONFIGURATIONS

the VS-to-gas conversion efficiency of the two-phase process were 41 percent, 40 percent, and 60 percent higher than those of the single-stage conventional high-rate process. The effluent residual VA of the two-phase process was less than one-half of that of the single-stage conventional process. Also, it should be noted that the observed methane yield from the single-stage CSTR process was 41 percent of the theoretical methane yield compared with 58 percent for two-phase conventional digestion. The above data showed that phase-separated two-stage fermentation was clearly superior to the single-stage process in terms of particulate solids conversion, and methane yield and production rate.

Studies With Upflow Digesters

These studies were conducted with mesophilic (35°C) upflow digesters utilized for separated acid- and methane-phase fermentations. The digesters were not mixed, and both the acid and the methane digesters were similar in design (Figure 3). The particulate feed, concentrated primary sewage sludge, was introduced into the digester at a point above the digester bottom, and digested sludge was withdrawn via a simple overflow at the culture surface. A deflector was placed above the sludge inlet to promote the creation of a bottom zone of denser microbial sludge. The aqueous fraction of the incoming feed and soluble products of microbial metabolism moved upward at velocities higher than those of the particulates so that the SRT could be significantly higher than tne HRT. With upflow digesters of this type, the intensity of gas evolution was highest at the bottom of the digester, and the evolved biogas mixed the digester contents as it moved toward the head space.

Concentrated 12-20-wt percent solids-content bottom sludge was withdrawn from the acid-phase digester and fed to the methane digester. There was no need to withdraw bottom sludge from the methane digester.

The studies were conducted with primary sludge which had a theoretical methane yield of 0.62 SCM/kg VS reacted based on an elemental analysis of 38.4 wt percent carbon, 5.9 wt percent hydrogen, 2.7 wt percent nitrogen, 16.7 wt percent oxygen, and 2.9 wt percent phosphorus. Results of single-stage conventional digestion studies conducted with this sludge and as reported in Table 4, indicated that methane yields of 50 percent , 47 percent, and 34 percent of the theoretical value could be obtained at HRT's of 18, 13, and 6.9 days. By comparison, the upflow two-phase system afforded a methane yield that was 77 percent of the theoretical value (Table 4). Also, comparison of a bioassay yield of 0.5 SCM/kg VS added with the observed methane yield of 0.48 SCM/kg VS added indicated that the upflow two-phase process stabilized about 96 percent of the biodegradable organics. Examination of the data in Table 4 shows that the two-phase process with the novel upflow digesters exhibited methane yields and production rates that were substantially higher than those of the single-stage conventional process.

Table 4 Conventional and Upflow Two-Phase Mesophilic (35°C) Digestion of Wastewater Sludge

Operating Conditions	Single-Stage CSTR			Acid Phase	Methane Phase	Overall System
Feed Consistency, wt % TS	4.7	4.7	12.6	5.8	--	5.8
HRT, days	18	13	6.9	1.3	4.6	5.9
Loading Rate, kg VS/m^3-day	1.4	1.9	10.1	28.9	7.8	6.2
Performance						
Methane Yield, SCM/kg VS added	0.31	0.29	0.21	0.06	0.42	0.48
Methane Production Rate, SCM/m^3-day	0.26	0.55	1.90	1.77	3.27	2.96
Methane Content, mol %	70	70	67	59	70	68
VS-to-Gas Conversion Efficiency, %	46	42	33	12	64	76
Effluent VA, mg/L as acetic	80	100	2500	2830	120	120
Observed Methane Yield, % of Theoretical	50	47	34	10	68	78

Dominant Reactions in the First- and Second-Stage Digesters

During steady-state operation of the upflow two-phase system with a feed flow rate of about 5.3 L/day, about 152 g/day of VSS (volatile suspended solids) was fed to the acid digester; about 59 g/day and 22 g/day of VSS emanated from this digester with the underflow and overflow streams indicating a liquefaction efficiency of 46 percent for the particulate organic matter in the feed. No VSS reduction was observed in the methane digester, indicating little or no solids liquefaction in this digester.

Considering particulate COD (COD of the organic solid particles) inputs and outputs, about 205 g/day of particulate COD (pCOD) was charged to the acid-phase digester with 33 g/day and 60 g/day of pCOD's emanating with the overflow and underflow streams, respectively, indicating a 55 percent reduction of particulate organic matter in this digester. Since methane production from the acid-phase digester could account for only about 1.5 g/day of particulate COD, the observed pCOD reduction was due to liquefaction of the feed organic solids. A pCOD reduction efficiency of 10 percent was computed for the methane digester. This indicated that little liquefaction of particulate organics occurred in this digester. The above observations on VSS and pCOD reductions occurred in this digester. The above observations on VSS and pCOD reductions in the acid- and methane-phase digesters strongly indicated that liquefaction of particulate organic matter was the major reaction in the acid-phase digester. Little solids liquefaction occurred in the methane digester. The data indicated that about 50 weight percent of the particulate organics could be liquefied in an upflow acid-phase digester. This type of acid digester performance has not been reported for particulate feeds.

Measurement of the volatile acids profiles within the culture showed that concentrations of C_3 and higher acids were about constant at all culture depths of the upflow acid-phase digester. Concentrations of acetic, propionic, isobutyric, n-butyric, isovaleric, and n-valeric acids were about 650, 2250, 130, 140, 270, and 80 mg/l, respectively, indicating turnover of the C_2, C_4, and C_5 acids, but accumulation of the C_3 acid; this suggests that little conversion of propionate to acetate occurred in the upflow acid-phase digester. By comparison, acetate and propionate were the only acids present in the upflow methane-phase digester; these acids were present in low concentrations ($<$80 mg/l) and their concentrations decreased with decrease in culture depth, suggesting rapid conversion of the higher VA's to acetate and then to methane within the bottom zone of the methane digester. In particular, propionate conversion was much more prevalent in the methane digester than in the acid digester.

In summary, liquefaction and acidogenesis were prevalent in the first-stage digester. In addition, acetogenic conversion of C_4 and higher acids probably also occurred. By comparison, acetogenic conversion of the C_3 acid and aceticlastic conversion of acetate were predominant in the second-stage digester. Methanogenic conversion of carbon dioxide to methane occurred in both digesters.

Advantages of the Two-Phase Fermentation Mode and the Upflow Digester

Comparison of performances of the single-stage CSTR and the two-phase CSTR digestion processes (see Table 3) suggested that the methane yield (and VS reduction) and methane production rate of the latter process were 40 percent and 30 percent higher, respectively, than those of the former. Thus, the phase-separated fermentation mode enhanced both the efficiency and the kinetics of anaerobic digestion. Similarly, from a comparison of the performances of upflow two-phase digestion and conventional single-stage CSTR digestion under HRT and feed solids concentrations as above, it was projected that the methane yield (and VS reduction) and methane production rate of the former process were 130 percent and 80 percent higher, respectively, than those of the latter. Thus, additional improvements in two-phase process performance relative to that of conventional single-stage CSTR digestion could be achieved by utilizing novel upflow digesters in lieu of CSTR's. Overall, this meant that the methane yield and production rate of the upflow two-phase process were 60 percent and 40 percent higher, respectively, than those of the CSTR two-phase process.

Energetic and Economic Advantages of Two-Phase Digestion

There are three major benefits of the two-phase process. These would be apparent from the information presented in Table 5 for hypothetical conventional and two-phase digestion plants receiving a daily sludge load of 100 metric tons (dry TS basis). The VS content of the sludge was assumed to be 65 wt percent of TS for this analysis. The benefits of an upflow two-phase process relative to single-stage CSTR conventional digestion are compared in Table 5 in terms of gross and net methane productions, digester volume requirement, and VS reduction efficiency. Based on laboratory experience, it was assumed that a conventional process can be satisfactorily operated at a 12-day HRT with a four-wt percent TS-content feed to afford a methane yield of

0.22 SCM/kg VS added. By comparison, the upflow two-phase process was assumed to be operated at a 6-day system HRT with a 7-wt % TS-content feed to afford a methane yield of 0.48 SCM/kg VS added. Mesophilic (35°C) operation was assumed with the feed termperature at 20°C. All digesters were assumed to have a gas space equal to 25% of the culture volume.

Table 5 Benefits of Mesophilic (35°C) Upflow Two-Phase Digestion of 100 Metric Tons/Day (Dry TS Basis) of Particulate Solids

	Conventional Digestion	Two-Phase Digestion
Gross Methane Production, 10^6 kcal/day	129	281
Operating Energy, 10^6kcal/day		
Heating	47	26
Mixing	10	0
Total	57	26
Net Methane Energy, 10^6 kcal/day	72	255
Net Energy/Gross Methane Energy, %	56	91
Digester Volume, 1000 m^3	37.5	10.7
VS Reduction, %	34	78
Residue, metric tons/day	78	49

Table 5 shows that gross and net energy productions from the two-phase process could be 118 percent and 254 percent higher than those of the single-stage CSTR process. Also, the two-phase process digester volume is less than 30 percent of that of the single-stage process, and it produces 37 percent less residue to be processed and disposed of; these benefits considered together translate to substantial savings in plant capital and operating costs.

TWO-PHASE DIGESTION OF SOLID FEEDS

The CSTR or the upflow two-phase digestion systems described above are suitable for concentrated semi-solid feeds, but these processes cannot be applied to gasify and stabilize solid feeds unless they are diluted to form slurries. This approach, although commonly employed, is not attractive for reasons mentioned before. A special process configuration -- leach-bed two-phase digestion -- is more suitable for "dry" or high-solids-content feeds, is simpler than slurry-phase digestion, and is conducted without dilution of the feed, without mixing, and even under ambient conditions. The leach-bed digestion process relies on the basic principles of two-phase anaerobic digestion which has been under development since 1970.[10-13]

Leach-bed solid-phase anaerobic fermentation is particularly attractive for such low-moisture organic feeds as municipal and industrial solid wastes, sludge cakes, manure, agricultural and forestry residues, farm wastes, and other similar organic biomass and wastes. The feasibility of single-stage "solid-culture" or *Koji* fermentation was demonstrated in Japan with *T. viride* and other organisms to saccharify newsprint and other similar dry substrates.[14] More recently, a similar concept termed "dry-fermentation" has been utilized in the United States, West Germany,

France, Hungary, and Guatemala to convert animal and crop wastes to methane.[15] LeRoux et al. in the United Kingdom digested municipal solid waste (MSW) and sludge blends in a similar fashion in a "fixed-bed" digestion vessel.[16] The disadvantages of these systems were that digestion proceeded in an uncontrolled manner and at very low rates, as in a municipal landfill, and it required a long fermentation period to achieve a modest methane yield. The disadvantages of dry fermentation may be attributed to inadequate moisture in the digesting solid bed, mass transport limitations within the organic bed, and the onset of unbalanced acidogenic and methanogenic fermentations occurring with highly concentrated substrates. The leach-bed two-phase digestion process overcomes these difficulties by inducing rapid bioleaching of the solid feed by application of an acidogenic culture, and promoting continued and accelerated liquefaction and acidification of the bed by recirculation of the reactivated culture (Figure 4). The leach-bed approach employs active control of all phases of the overall digestion process. Liquefaction products from the acidosenic leach-bed are moved to an acid-recovery process or are diverted to a separate methane-phase digester for gasification of the volatile acids with recycling of the methane-phase effluent to the leach-bed to conserve the nutrients indigenous to the solid substrate, and thus to eliminate or reduce the need for external nutrient addition. The leach-bed digestion process is superior to the traditional slurry-culture digestion process because of the following reasons:

- Ability to handle "dry" or high-solids-containing feeds;
- A minimum feed processing (e.g., shredding, grinding and separation) and feed pretreatment (e.g., chemical or enzymatic) are necessary;
- Feed slurrification is not necessary;
- Intensive mechanical mixing is not required;
- Addition of external nutrients is eliminated or minimized;
- The process can be applied for in-situ bioconversion of waste deposits (e.g., landfills) and the ultimate disposal of the final residues;
- Ability to conduct fermentation in simple containment vessels; and
- Reduced fermenter volume and energy requirements compared with the conventional dilute-slurry fermentation processes.

The leach-bed two-phase digestion study was conducted by utilizing a 126-l (44.5-cm diameter x 63.5-cm high) acidogenic digester which contained 12 kg (dry) of shredded separated MSW containing 7.6 kg of VS and having an initial moisture content of 43 percent. The cellulosic feed had a C/N ratio of 34:1 and a C/P ratio of 280:1, indicating nutrient deficiency. The leach bed was connected in series with a 12.5-l (9.5-l culture volume) anaerobic filter (methane digester). Both digesters had a temperature of 35°C, and the system was operated in the batch mode.

Leaching of the bed was started by applying an acidogenic culture and recirculating the percolating leachate around the bed at a rate of about 90 l/day. Acidogenic fermentation characterized by volatile acids production in increasing concentrations (Figure 5) prevailed under these conditions. After about one month of operation, the total VA concentration accumulated to a level of about 18,000 mg/l as acetic, (4480 mg/l acetic, 1370 mg/l propionic, 240 mg/l isobutyric, 3210 mg/l

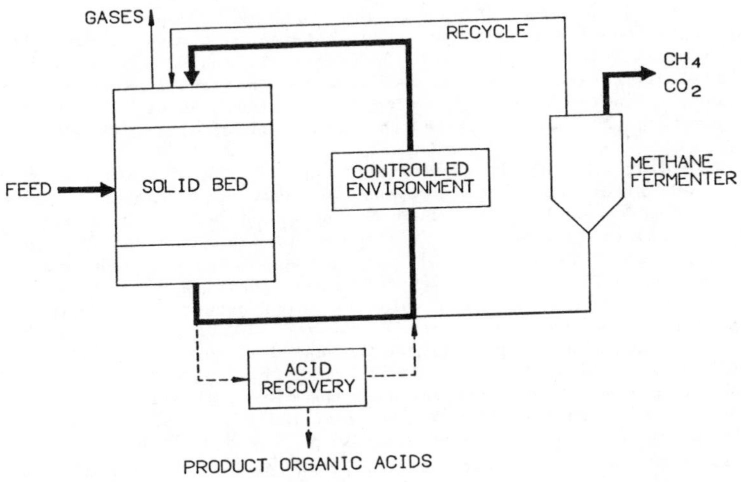

Figure 4. LEACH-BED TWO-PHASE ANAEROBIC DIGESTION

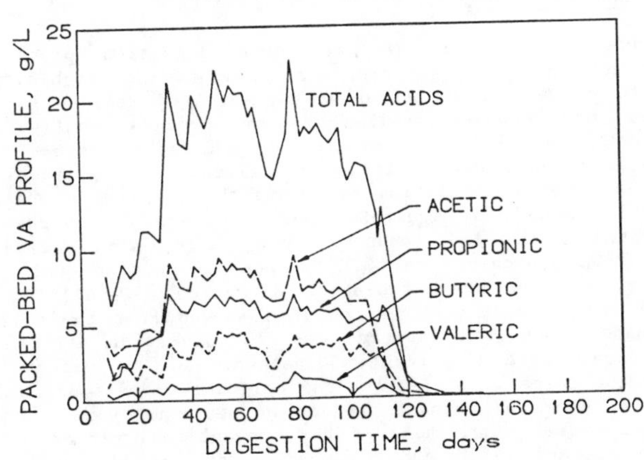

Figure 5 VOLATILE ACIDS PRODUCTION IN MESOPHILIC (35°C) LEACH-BED ACID DIGESTER

butyric, 1530 mg/l valeric, and 400 mg/l caproic) and further accumulation of acids did not occur after this time indicating that the prevailing acids and low pH (5.3) conditions inhibited the liquefaction and acidification processes. As expected, no methanogenic activity was evidenced in the leach bed during this period. About 21 percent of the volatile solids were converted to volatile acids during this time.

Feeding of the leach-bed effluent to the methane digester was started after about two months of leach-bed operation with the methane-phase effluent recycled to the leach bed to inoculate the bed and to reintroduce the feed-indigenous nutrients for utilization by the acidogens. The anaerobic filter methane digester was operated at decreasing HRT's of 19 days, 9.5 days, 4.1 days, and 38 hours. A rapid decrease in the accumulated volatile acids occurred once this mode of operation was instituted, and residual volatile acids in the bed leachate and the filter effluent decreased to zero after about 70 days of operation (Figure 4). Gasification of the bed solids and the accumulated volatile acids followed the classical sigmoidal pattern and 78 percent of the starting VS was gasified after 157 days of two-phase operation affording a methane yield of 0.31 SCM/kg VS added. Since this feed had an ultimate bioassay methane yield of 0.30 SCM/kg VS added in the presence of added nitrogen, phosphorus and other nutrients, the observed methane yield from the leach-bed two-phase system represented complete conversion of the biodegradable organics occurring without addition of supplementary nutrients. Overall, 77 percent of the produced methane evolved from the bed and 23 percent from the anaerobic filter. Methane content of the leach bed gas increased from zero to 60 mol percent, while the methane content of the anaerobic filter gas reached 80 mol percent shortly after its operation was started and thereafter it declined gradually to 56 mol percent at the end of the run.

SUMMARY AND CONCLUSIONS

Conventional single-stage anaerobic digestion processes, which commonly utilize CSTR digesters, are unsuitable for high-efficiency and rapid-rate conversion of solid or concentrated semi-solid substrates. A number of unconventional digestion process configurations are available for stabilization of liquid industrial wastes, but most of these processes are unsuitable for stabilization of solid or semi-solid feeds. Three novel digestion processes that utilize a phase-separated fermentation mode and higher-SRT novel digesters converted solid and concentrated semi-solid feeds at very high efficiencies at short HRT's. These processes are CSTR two-phase digestion, upflow two-phase digestion, and leach-bed two-phase digestion. Parallel single-stage CSTR and two-phase CSTR digestion runs showed that the methane yield, the methane production rate,and the VS conversion efficiency of the latter process were 41 percent, 40 percent, and 60 percent higher than those of the former. Similarly, the upflow two-phase digestion process afforded methane and methane production rate that were 130 percent and 80 percent higher than those of the conventional process. This meant that the methane yield and production rate of the upflow two-phase process were 60 percent and 40 percent higher than those of the CSTR two-phase process. Thus, it was demonstrated that considerable gains in process efficiency and kinetics could be achieved by phase-separated digestion, and further by application of high-SRT upflow digesters.

About 95 percent of the biodegradable VS was gasified by upflow two-phase anaerobic digestion.

Liquefaction (hydrolysis) and volatile acids formation were the predominant biochemical reactions in the first-phase digester; in addition, acetogenic conversion of C_4 and higher VA's also occurred. By comparison, acetogenic conversion of the C_3 volatile acid and aceticlastic conversion of acetate were the major reactions in the second-stage digester.

Net energy production from the two-phase digestion process was projected to be 250 percent higher than that of single-stage conventional digestion. The two-phase process digester volume could be less than 30 percent of that of the single-stage digestion process. In addition, VS reduction of the upflow two-phase process was more than double that of conventional digestion. These benefits of two-phase digestion translate to substantial cost and energetic advantages.

The leach-bed two-phase digestion process is applicable to dry solid feeds, and it can be conducted in simple feed containment vessels without slurrification of the dry feed, without mixing, and without the addition of external nutrients in the case of nutrient-deficient feeds. The leach-bed digester produced volatile acids which accumulated to concentrations up to 20,000 mg/l amounting to a VA yield of 20 percent of the feed VS. Gssification of the bed solids and the accumulated VA could be effected by feeding the leach-bed products to a methane-phase anaerobic filter and recycling the methane-phase effluent to the bed. A methane yield of 0.31 SCM/kg VS added was observed during batch operation of the leach bed indicating complete stabilization of the biodegradable VS. The methane-phase digester could be operated at low HRT's down to 38 hours. Further research is necessary to demonstrate continuously fed leach-bed operation.

ACKNOWLEDGMENTS

The author would like to thank the sponsors of the research discussed in this paper. They are: the Institute of Gas Technology; Executive Offices of Energy Resources, Commonwealth of Massachusetts; Boston Gas Company; Brooklyn Union Gas Company; Bay State Gas Company; Valley Resources, Inc.; Essex County Gas Company; Cogenics Energy Systems; U.S. Environmental Protection Agency. The author also wishes to thank Michael P. Henry and Ash Sajjad for their efforts in carrying out the experimental work.

REFERENCES

1. Frostell, B. 1981. Anamet Anaerobic-Aerobic Treatment of Concentrated Waste Waters. Proc. 36th Ann. Ind. Waste Conf. Purdue University, W. Lafayette, Indiana. May 12-14, 1981.

2. Sax, R.I. 1983. High-Rate Anaerobic Pre-Treatment of Wastewater on the Industrial Scale. Proc. Ind. Waste Symp., 56th Ann. Conf. Water Pollut. Control Fed. Atlanta, Georgia. October 2-6, 1983, WPCF, Washington, D.C.

3. Genung, R.K., et al. 1978. Pilot Plant Demonstration of an Anaerobic Fixed-Film Bioreactor for Wastewater Treatment. Proc. Biotechnol. & Bioeng. Symp. No. 8. Gatlinburg, Tennessee, 329-344. Interscience, New York.

4. Personal Communication. Fischer, W.C. Celanese Chemical Company, Inc. Dallas, Texas.

5. Shea, T.C. et al. July 1984. Rum Distillery Slops Treatment of Anaerobic Contact Process. EPA Report No. 660/2-74-074. Suptd. of Documents, U.S. Govt. Printing Office. Washington, D.C.

6. Bachmann, A., V.L. Beard, and P.L. McCarty. 1982. Comparison of Fixed-Film Reactors With a Modified Sludge Blanket Reactor. Proc. First International Conf. on Fixed-Film Biological Processes. Vol. 1. 1192, Kings Island, Ohio, April 20-23, 1982.

7. Hickey, R.F. and R.W. Owens. 1981. Methane Generation from High-Strength Industrial Wastes With the Anaerobic Biological Fluidized Bed. Symp. Proc. Biotechnology in Energy Production and Conservation. Gatlinburg, Tennessee, May 12-15, 1981.

8. Ghosh, S. et al. 1982. Methane Production From Industrial Wastes by Two-Phase Anaerobic Digestion. Symp. Proc. Energy From Biomass and Wastes VI. Institute of Gas Technology, Lake Buena Vista, Florida, January 25-29, 1982.

9. Ghosh, S. 1981. Kinetic Basis for Optimization of Methane Production by Anaerobic Digestion. Proc. International Gas Research Conf. Gas Research Institute, Los Angeles, California, September 28-October 1, 1981.

10. Ghosh, S. and F.G. Pohland. 1971. Developments in Anaerobic Treatment Processes. In Biological Waste Treatment, Ed., R.P. Canale, Interscience Publishers. New York.

11. Ghosh, S. and F.G. Pohland. April 1974. Kinetics of Substrate Assimilation and Product Formation in Anaerobic Digestion. J. Water Pollut. Control Fed. 46, 4, 748-759.

12. Ghosh, S. et al. 1975. Anaerobic Acidogenesis of Sewage Sludge. J. Water Pollut. Control Fed. 47, 1, 30-45.

13. Ghosh, S. and D.L. Klass. 1977. Two-Phase Anaerobic Digestion. U.S. Patent No. 4,022,665.

14. Takagi, M. et al. February 1977. Symp. Proc. Bioconversion of Cellulosic Substances into Energy, Chemicals, and Microbial Protein. Ed. T.K. Ghose, Indian Inst. of Technol. New Delhi, India.

15. Anon. Chem. Wk., 50, December 15, 1982.

16. LeRoux, N.W. et al. 1979 Conservation and Recycling 3, 165.

BIOGAS FROM ORGANIC WASTE DILUTED WITH SEAWATER

H. Gamal-El-Din

Fayum Faculty of Agriculture
Cairo University, Fayum, Egypt

ABSTRACT

Laboratory glass digesters were used to study the possibility of using seawater to dilute organic wastes for biogas production. The pulverized solid excrement of camel was diluted with distilled water (as a control treatment), synthetic seawater (total soluble salts 40490 ppm), or diluted seawater (20245 ppm) to give 10% total solids and one liter volume. The experiment was continued for 60 days. The highest biogas cumulative volume (20 L) was obtained from the diluted seawater treatment, while the lowest volume (9.5 L) was that obtained from the seawater treatment. However, the seawater treatment showed an increase in the biogas production rate at the end of the experimental period. It seems from the results that seawater can be used as a diluent in the anaerobic digestion of organic wastes if diluted below the inhibitory level, or through the slow acclimation of the micro-organisms involved in biogas production to the seawater conditions.

INTRODUCTION

Recently, there has been an increasing interest in using anaerobic fermentation to produce biogas as an alternative energy source. To utilize animal manures and other low moisture content organic wastes as feedstocks for biogas production, a slurry containing an appropriate solids concentration (5-10%) should be normally prepared. This is usually done by mixing the waste with water. Depending on waste type, considerable amounts of water are needed for slurry preparation.

In some regions, the fresh water sources are either scanty or costly, while plenty of inexpensive and relatively large amounts of seawater or water of similar composition are available. In such a case, the question arises whether seawater can replace fresh water for diluting organic wastes used in the production of biogas.

The aim of the present work has been therefore to answer the above-mentioned question using solid excrement of camels diluted with synthetic seawater for feeding the biogas digester.

MATERIALS AND METHODS

The Organic Waste

Air dried solid excreta of camel were collected, pulverized, and analyzed for: total solids (TS), volatile solids (VS), Kjeldahl nitrogen, and organic carbon (Table 1).

Table 1 Analysis of the Solid Excrement(Camels)

Total Solids (TS)	85.5 %
Volatile total solids	85.2 % TS
Kjeldahl Nitrogen	1.2 % TS
Organic Carbon	48.1 % TS

The Synthetic Seawater

The composition of the syntnetic seawater (Burkholder, 1963) is (in gm): NaCl, 2.348; Na_2SO_4, 0.392; $NaHCO_3$, 0.019; KCl, 0.066; K Br, 0.0096; H_3BO_3, 0.0026; Mg $Cl_2.6H_2O$, 1.061; $SrCl_2.6H_2O$, 0.004; $CaCl_2.2H_2O$, 0.1469 and distilled water to 100 ml.

The Inoculum

The inoculum was prepared from effluent obtained from an actively operating laboratory digester fed with cattle manure. The effluent was passed through layers of cheesecloth to remove undergraded materials.

The Digestion Apparatus

A laboratory digestion apparatus was constructed as shown in Figure 1. It is composed of a one liter brown bottle (A), a gas measuring cylinder (B) equipped with a leveling bulb (C), a gas sampling port (D), and a side arm (E), through which liquid could be withdrawn or added.

Experimental Procedure

The required quantities from the pulverized solid excrement of camel (to give a solids concentration of 10%), were mixed with 100 ml of inoculum and sufficient amounts of distilled water, diluted seawater (total soluble salts:20245 ppm), or seawater (40490 ppm) to give one liter total volume. The distilled water digesters served as a control treatment. A set of distilled water digesters not receiving inoculum was included to study the effect of inoculum on biogas production. The pH of the digesters content was adjusted to pH 7.5. Each treatment was replicated four times.

The digesters were anaerobically incubated at room temperature (25-35°C) for 60 days. No arrangement was made for mixing the digester contents. It was done manually at the time of gas volume measurement.

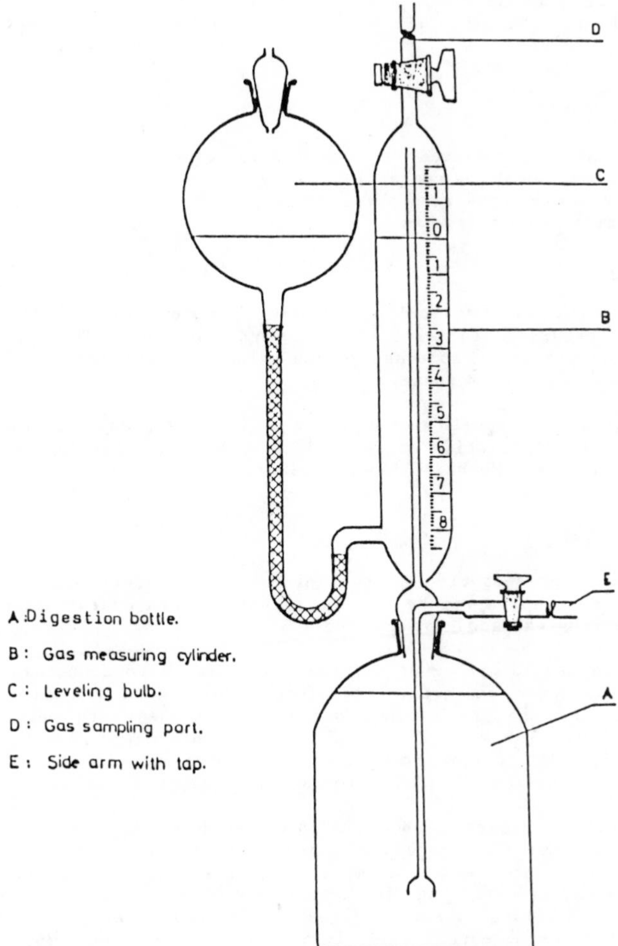

Figure 1 Schematic Diagram of the Digestion Apparatus

The volume of gas produced was measured daily and the gas was assayed for methane percentage every week. At the end of the experimental period, the pH and total volatile acids concentration in the digesters content were determined.

Analytical Procedures

Biogas volume was measured by liquid displacement of a saturated solution of sodium sulphate containing 5% by volume of sulfuric acid. The biogas production rates are expressed as milliliters of gas per day per liter of active digester volume.

Methane Content

Methane content was measured by bringing a volume of biogas into contact with a saturated solution of potassium hydroxide. The loss in volume of biogas was taken as equal to the carbon dioxide content. The balance was assumed to be methane.

Determinations of total and volatile solids, Kjeldahl nitrogen, organic carbon, total volatile acids, and pH were made according to the procedures in "Standard Methods" (Anon., 1980).

RESULTS AND DISCUSSION

A schematic representation of the laboratory digester used in the present study is shown in Figure 1. The digesters were operated easily and reproducibility was good between replicates.

The air-dried solid excrement of camel is presumed to contain a low number of methanogenic bacteria. Therefore, addition of the inoculum not only shortened the lag period, but also increased both rate and total volume of gas produced (Figures 2 and 3). This may be due to the presence of actively growing methanogenic bacteria and some essential substances carried over from the cattle manure digester.

Data plotted in Figure 2 show that the gas production rates from both diluted seawater and seawater digesters, as compared to the control digesters, were lower during the initial priod of digestion. About 25 days after starting the experiment, the rate of biogas produced from the diluted seawater digesters increased markedly, while that of the seawater digesters slightly increased after about 36 days. However, production rates of both the control and diluted seawater digesters decreased at the end of the digestion period, while the increase in the rate of the seawater digesters continued. This increase in gas production rates is likely due to toleration and acclimation of bacteria involved in biogas production, particularly the methanogenic bacteria, to the high salt concentration. Kugelman and McCarty (1964) indicated that the tolerable limits of Na^{+} in the anaerobic digesters are 3500-5500 mg/L. On the other hand, Patel and Roth (1977) found that different strains of methanogenic bacteria varied in their ability to produce methane and multiply in the presence of different sodium chloride concentrations. *M. thermoauto - trophicum* was inhibited by NaCl in excess of 15.2 mM, while *Methanobacterium* MOH was independent of NaCl up to the maximum concentration tested (263.7 mM). The requirement for NaCl, or susceptibility to it, may be related to the source of the organism. For example, the Na^+ concentration in the rumen ranges between 60 and 120 mM (Caldwell and Hudson, 1974), and therefore rumen bacteria may tolerate or even require high concentrations of NaCl for optimal growth. However, the source of the inoculum used in the present study was a cattle manure digester. In addition, the toxicity of the high NaCl concentration may be antagonized by other salts already present in the seawater. Such phenomenon was observed in mixed cultures of anaerobes (Kugelman and McCarty, 1964).

Regarding the total gas produced during the 60-day period, the results (Figure 3) show that the highest biogas cumulative volume (20L) was produced from the diluted seawater digesters, while the lowest volume (9.5 L) was that produced from the seawater digesters. This result indicates that not only the salts concentration in the diluted seawater (20245 ppm) was not inhibitory to the digestion process, but also it may contain some nutrients required for the methanogenic bacteria. However, it seems from results (Figure 2) that the volume of gas produced from the seawater digesters may be expected to rise with continued operation.

The percentage of methane in the biogas produced is shown in Table 2. It is of interest to note that during the initial phase of gas

Table 2 Percentage of Methane in Biogas Produced from Different Treatments

Period Weeks	Treatment			
	Distilled Water		Diluted Seawater	Seawater
	- inoculum	+ inoculum		
1	42.5	64.4	72.2	69.4
2	51.3	63.6	70.5	69.0
3	67.7	65.2	65.9	70.5
4	66.2	57.4	64.6	69.7
5	70.3	57.3	58.7	65.2
6	51.6	53.8	57.1	55.0
7	49.4	56.0	55.5	51.8
8	49.2	54.4	56.2	51.9
9	50.1	54.8	55.0	51.9

evolution, the percentage of methane in the biogas produced from both diluted seawater and seawater digesters was relatively higher than that in the gas produced from the control digesters. However, in course of time, after about 30 days, the percentage of methane generally decreased and ultimately became constant.

The mean values for total volatile acids concentration and pH in the effluents of different treatments at the end of the 60-day period (Table 3) were within the acceptable levels for anaerobic digestion.

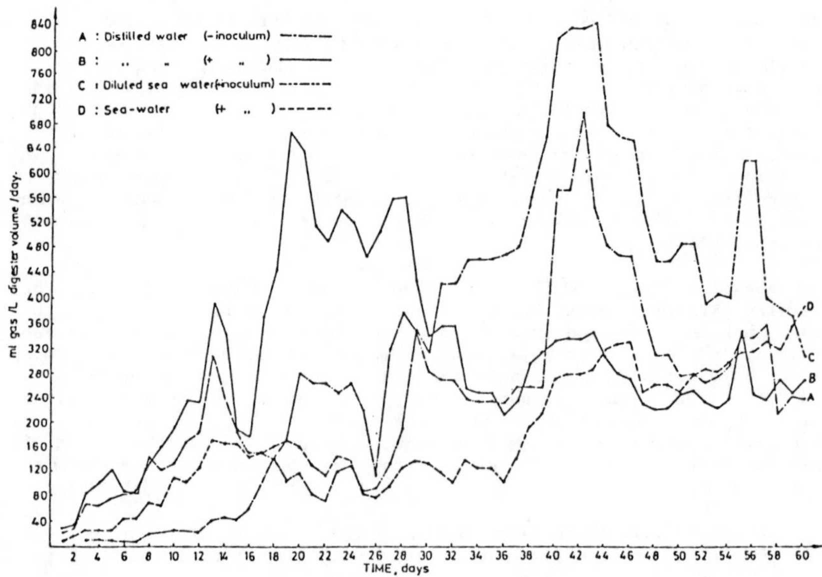

Figure 2 Gas Production Rates from 10% Solid Excreta of Camels

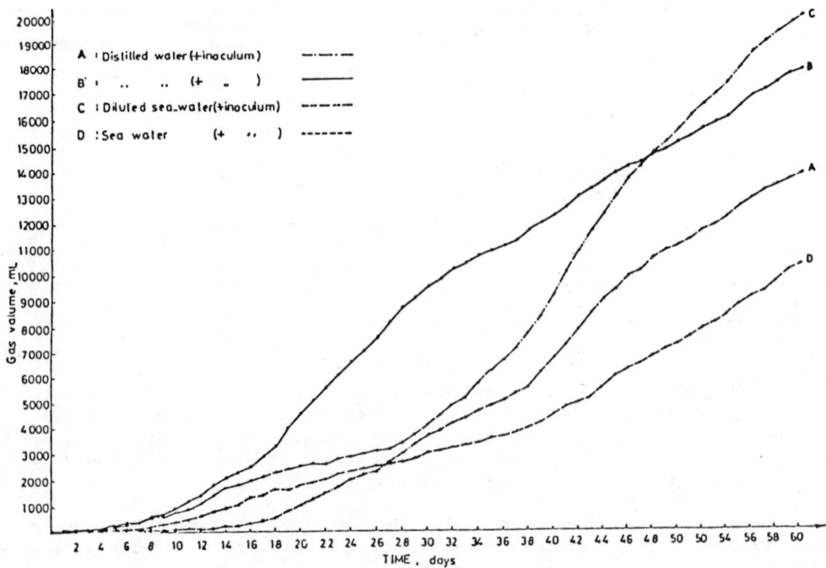

Figure 3 Cumulative Gas Production from 10% Solid Excreta of Camels.

Table 3 pH Values and Total Volatile Acids (TVA) Concentration in the Effluents after 60 Days of Digestion.

Treatment		pH	TVA, mg/1 as acetic acid
Distilled water:	- inoculum	6.9	145
	+ inoculum	7.1	87
Diluted seawater		7.0	146
Seawater		6.9	669

However, the relatively higher volatile acids concentration in the effluent of the seawater digester could be due to the fact that the acid-producers are much less sensitive to stress conditions than methanogenic bacteria.

CONCLUSION

From the results obtained in the present research, the conclusion can be drawn that biogas production from seawater digesters seems quite possible. This can be done by diluting seawater or water of similar composition below the inhibitory salts concentration, or through the slow acclimation of the responsible microorganisms to the saline water conditions.

REFERENCES

Anon. 1980. Standard Methods for the Examination of Water and Wastewater (15th ed.). American Public Health Association. Washington, D.C.

Burkholder, P. 1963. In: Symposium on Marine Microbiology. C.H. Oppenheimer, ed., pp. 133-150. Thomas, Springfield, Illinois.

Coldwell, D.R. and R.F. Hudson. 1974. Sodium, an obligate growth requirement for predominant rumen bacteria. Applied Microbiology. 27: 549-552.

Kugelman, I.J. and P.L. McCarty. 1964. Cation Toxicity and Stimulation in Anaerobic Waste Treatment II. Daily Feed Studies. Proc. 19th Annual Purdue Industrial Waste Conference. Purdue University. Lafayette, Indiana, pp. 667-686.

Patel, G.B. and L.A. Roth. 1977. Effect of sodium chloride on growth and methane production of methanogens. Can. J. Microbiol. 23 : 893-897.

THE ASSESSMENT OF METHANOGENS BY THE QUANTITATIVE DETERMINATION OF COENZYME F_{420} IN ANAEROBIC SYSTEMS

Mohiy Eldin Abdel-Samie, NRC, Cairo, Egypt

ABSTRACT

Coenzyme F_{420} was extracted from ruminant dungs and digested slurries by boiling water. The extract was treated with acetone to get rid of biopolymers. The fluorescent and excitation spectra were identical to the published ones. DEAE Sephadex (A.25) column chromatography was used to isolate the coenzyme. Preliminary polarographic studies gave $E_{1/2}$ more negative than the reported ones of riboflavin, FMN, FAD, NHD, and NADP. The possibility of using polarography for the quantitative determination of F_{420} is discussed.

INTRODUCTION

In order to study the distribution of methanogens within the three zones of the unstirred digesters (i.e., the scum, the active zone, and the sediment), efforts were concentrated on the determination of the coenzyme F_{420}: (a deazaflavin derivative). This coenzyme is a specific indicator of methanogens (Eirich, Vogels, and Wolfe, 1979). This approach has been tried by Delafontaine et al. (1979) and Van Beelen et al. (1983). The latter authors claimed that reversed-phase High-Performance Liquid Chromatograph (HPLC) should be used in order to separate this coenzyme from other substances that might interfere with its fluorometric determination. They expected that the interference might be caused by substances which absorb light at 420 or 470 nm, the excitation and the emission maxima of coenzyme F_{420}, respectively. However they did not show, experimentally, that the interference did exist. In the present work, the excitation as well as the spectra of extracts of fresh cow dung, anaerobically digested cow dung, and its supernatant and fresh green leaves, were recorded and compared with the published spectra of all the expected interfering materials, e.g., riboflavin, FAD, FMN, chlorophyl. . . etc.), as well as coenzyme F_{420} (Udenfreund 1962, 1969; Eirich et al., 1979). In addition, no polarographic studies were done on this coenzyme.

EXPERIMENTAL

Fresh cow dung was collected from cows fed green matter (young maize leaves) and wheat straw roughage. Conventional anaerobic batch digesters (500 ml bottles) were run and the gases were collected over 20 percent saline water.

RESULTS AND DISCUSSION

The excitation and emission spectra of the extracts of fresh cow dung, digested cow dung and its supernatant, and that of green leaves, obtained by method three, are shown in Figures 1, 2, 3, and 4 respectively. The pH values of the starting materials were between 6.5 and 7.0. It will be noticed that the emission spectra contain the intense peak due to light scattering at the exciting wave length and vice versa for the excitation spectra (Pesce et al., 1971).

The recorded spectra in Figure 1, 2, and 3 are in close concordance with those of coenzyme F_{420} (Eirich et al, 1979). The slight shift in the emission maxima towards 460 nm might be due to the solvent effect (50 percent acetone) (Parker, 1959).

By comparing these figures with those published by Udenfreund (1962, 1969) for all the biological materials expected to come from plant material digested by ruminants, the possibility of interference appears to be very low.

Acetone was used, not to extract the coenzyme F_{420}, but to get rid of the dissolved biopolymers which might cause much scattering and/or light absorption. This would affect the total quantum yield of the fluorescence of the coenzyme. It is known that the coenzyme dissolves in 50 percent acetone.

The column chromatographic separaton of F_{420} was easily achieved by using the acetate buffer as an eluent. The fluorescent band was followed by a high pressure mercury lamp (50W) source and an interference filter emitting at 405 nm. The collected band was polarographed and the $E_{1/2}$ was determined and found to be 1.5V against Ag/Agcl/saturated KCl. This value is quite negative with respect to those of other coenzymes (Dryhurst, 1977).

Work is being carried on to establish a polarographic method for selective quantitative determination of F_{420}.

ACKNOWLEDGEMENTS

The author highly appreciates the sincere efforts of Dr. Silvia Kocova, Mrs. Amal M. Nasr, and Miss Sohair H. Ali, in performing the excitation and emission spectra published in the present work.

REFERENCES

van Beelen, P., A.C. Dijkstra, and G.D. Vogels. 1983. Eur. J. App. Microbiol. Biotechnol. 18:67-69.

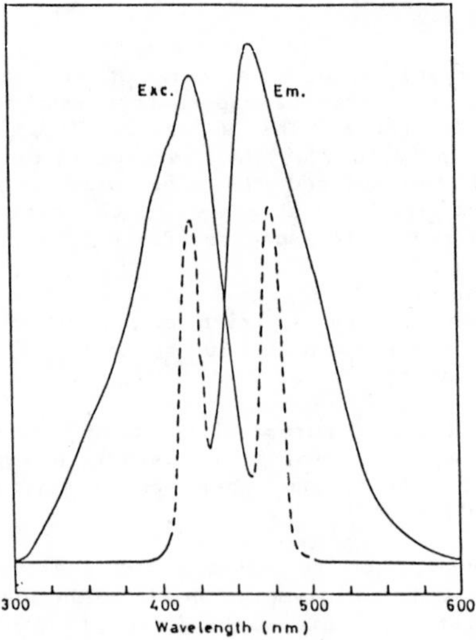

Figure 1 Fresh Dung

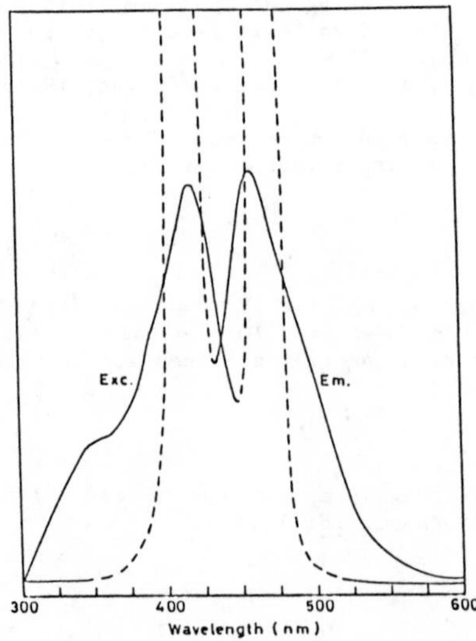

Figure 2 Digested Dung

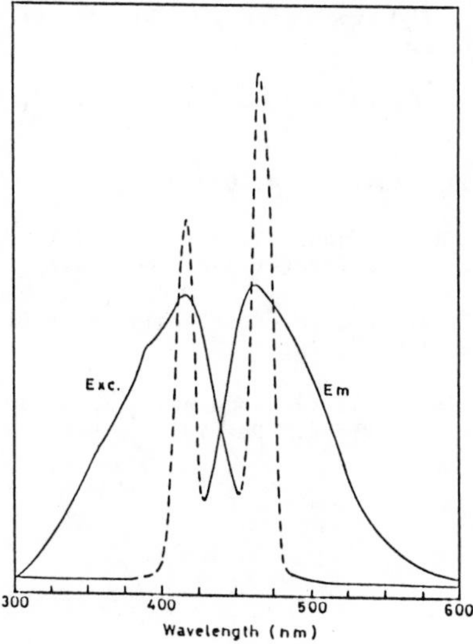

Figure 3 Supernatant of Digested Dung

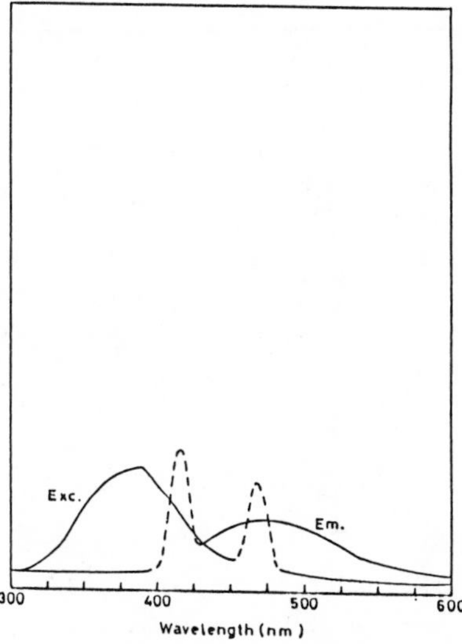

Figure 4 Extract of Green Leaves

Dryhurst, G. 1977. Electrochemistry of Biological Molecules. Academic Press. New York.

Eirich, L.D., G.D. Vogels, and R.S. Wolfe. 1979. J. Bacteriol., 140:20-27.

Parker, C.A. 1959. Analyst 84:1002.

Pesce, A.J., C.G. Rosen, and T.L. Pasby. 1971. Fluorescence Spectroscopy. Marcel Dekker Inc., New York.

Udenfreund, S. 1962. Fluorescence Analysis in Biology and Medicine. Academic Press. New York.

Udenfreund, S. 1969. Fluorescence Assay in Biology and Medicine. Vol. II. Academic Press. New York.

FUEL GAS PRODUCTION FROM ORGANIC WASTES BY LOW CAPITAL COST BATCH DIGESTION

Donald L. Wise, Alfred P. Leuschner,
Ralph L. Wentworth, and Mostafa A. Sharaf
Dynatech R/D Company, Cambridge, Massachusetts

ABSTRACT

The technical background is reviewed on energy recovery from biomass--i.e., all organic wastes, especially municipal solid wastes, but also including agricultural residues and crops grown specifically for energy recovery. This review places special emphasis on biogas production from "controlled landfilling." Some of the problems and techniques of biogas production via this low capital cost batch digestion method are discussed. It is concluded that a very large potential exists for developing countries to utilize biomass as an energy resource when production is carried out using controlled landfilling technology.

BACKGROUND ON "CONTROLLED" LANDFILLING

Controlled landfilling (Augenstein et al., 1976) is the managed, batch anaerobic digestion of municipal solid waste (MSW) to enhance fuel gas production. In controlled landfilling, a suitable inoculum, nutrients, buffer, and moisture are added to MSW, which allows the microbial action of anaerobic fermentation to proceed at a substantially higher rate than in untreated MSW. In this practice, the advantage of high stabilization rates of the organic fraction of MSW, like those obtained in continuous stirred tank reactor (CSTR) systems, following sewage sludge digestion practice, is combined with the advantages of the lower capital and operating costs of MSW landfilling.

Anaerobic digestion refers to the ability of certain classes of microorganisms to grow on a number of different organic substrates in the absence of air, converting them to methane and carbon dioxide. Historically, digestion has been used for the treatment of various relatively dilute liquid wastes or suspensions, such as sewage wastes, packing house wastes, and sulfite liquors. Its function has been primarily to render the wastes inoffensive. Fuel gas generation from municipal wastes has traditionally been of secondary importance, but is now viewed with more interest.

As consequence of widespread application of microbial digestion to the treatment of the usual soluble or suspended wastes over the past several decades, standard digester designs and operating procedures have evolved. When this conventional technology is applied to solid digestible organic substrates, which are of interest for their fuel potential, considerable expense and energy consumption accrue from the necessity to grind and slurry the substrate, stir reactor contents continuously, separate solid and liquid components at the end of the digestion process, and dispose of process effluents, particularly the liquids.

The controlled landfilling concept focuses on circumventing the necessity of continuously stirring the digestion unit, and observing careful, slow, and frequent feed schedules in conjunction with this stirring to prevent what is known as "souring."

In conventional digestion, stirring brings the substrate into intimate contact with the microorganisms. Diffusional rates of mass transfer for these microorganisms may be demonstrated to be adequate to maintain digestion at high rates without mixing, providing the microorganisms, nutrient, and substrates are well distributed throughout the system.

"Souring" occurs when the class of rapidly growing microoganisms which converts substrates to organic acids produces organic acids more rapidly than the organisms which convert these acids to methane can assimilate them. When this occurs, the pH drops, the methane-forming organisms are inhibited, and methane formation is severely impaired or ceases. However, the addition of a buffer, specifically calcium carbonate, is uniquely suited for controlling pH of anaerobic systems, resulting in optimal methane production. The concept requires a system of high solids, as exists in landfills, to form a matrix which supports the calcium carbonate, keeping it dispersed as required. Finally, necessary nutrients (mainly nitrogen and phosphorus) and an inoculum of anaerobic microorganisms, can be added as another waste, sewage.

The main objectives of the controlled landfill concept for enhanced fuel gas production are: a) to increase the amount of energy produced in the digestion process, relative to the amount of energy required to run the process, b) to remove constraints on current digestion technology, enabling lower capital cost waste stabilization and/or fuel gas generation from municipal solid waste, c) to allow digestion of municipal solid waste to occur at much higher solids concentration than allowable or attainable with current technology, d) to permit MSW to be digested with little or no preseparation of nonbiodegradable materials, and e) to enable economic digestion by using longer retention times in a landfill.

Work at Dynatech investigated the microbiological degradation of municipal solid waste to convert a higher fraction to a fuel gas in a shorter time than is required by conventional landfilling practice (less than one year as compared with up to 100 years for conventional landfills). Landfill conditions were simulated in laboratory test cells and the following parameters were studied: moisture, nutrient/inoculum addition, compaction (amount of waste material per volume of landfill), particle size, buffer addition, temperature, and leachate recycle. The results of this study enabled the ranking of the relative significance of each parameter based both on methane produced and on the fraction of waste converted. Large scale controlled landfill projects are now underway based on this work at Dynatech. Also, a U.S. Agency for International Development (A.I.D.) Project involving ten countries has been carried out evaluating controlled landfilling of agricultural residues.

PROCESS DESCRIPTION

It has long been known that methane gas is produced in landfills; however only recently have landfills been viewed as methane producers whose energy is economically harvestable. Subsequently, the art of landfill design has evolved to the point where consideration is being given to designing landfills with energy recovery in mind. Applying landfill gas recovery concepts for energy production from agricultural residues in developing countries holds many significant advantages:

1. Landfill construction is not capital intensive. In the construction of a landfill, the soil is used to contain the residue. This eliminates the need for expensive digesters constructed from steel or concrete. By lining the bottom and top of a landfill with a material such as compacted clay, anaerobic conditions are maintained. In addition, the liner serves as a barrier to liquid migration to the groundwater and migration of methane gas to the atmosphere.
2. Landfills require no residue pretreatment. Conventional digestion systems, whether CSTR, two stage or plug flow digester, all require some form of residue pretreatment. Initially the residue would have to undergo particle size reduction via grinding or chopping. Secondly, the residue would have to be slurried so that it could be pumped into and out of a digester. Landfilling of the residue requires no pretreatment of this form.
3. Landfills are labor intensive, low technology systems. The construction of a landfill requires manpower for placement and covering of residues. Once the residue is in place only minimal attention is required to keep the system operating, and no sophisticated technology is needed to maintain gas production, or recover the gas. In comparison, conventional digestion, or the production of other energy forms from agricultural residues (such as alcohols), requires a higher degree of technical sophistication.
4. No residue disposal is required. With other energy discovery systems the residue after fermentation requires some form of liquid/solid separation and solids disposal. In this approach the residue is already in place in a landfill requiring no further disposal.

Dynatech has pioneered the development of various landfilling techniques designed to improve both the ultimate methane yield from organic material and the rate at which methane is produced from organics. Recently, Dynatech has also explored the feasibility of utilizing a landfill approach for the production of methane from agricultural residues for developing countries. The following review of landfill gas recovery serves as a description of how this technology is applied to municipal solid wastes, as well as agricultural residues.

CONVENTIONAL LANDFILL GAS RECOVERY

The world's first commercial landfill methane recovery facility started in 1971 when the Los Angeles County Sanitation District constructed wells to prevent gas migration to adjacent residential properties at the Palos Verdes landfill in California. Since that time a number of other projects have been initiated. In addition to the

Palos Verdes landfill, the ones in commercial operation include the Mountain View landfill (Mountain View, California), the Azusa-Western landfill (Azusa, California), and the Sheldon-Arleta landfill (Sun Valley, California). Other landfills on line include Ascon, California; City of Industry landfill, California; Operating Industries landfill, California; and Cinnamansion landfill, New Jersey (Yoshioka, 1980).

The first step in recovering methane from landfills is to drill down into the landfill and install gas extraction wells. After drilling, the gas extraction wells are typically lined with a perforated casing and packed with sieved gravel. Collection manifolds connect the wells, bringing the "biogas" to a common site. This method has proved to be generally satisfactory.

Removal of gases from existing landfills has been accomplished by the same principle that is involved in the extraction of ground water. Pumps are manifolded to the wells, creating a pressure gradient within the landfill which draws the gas through the collection systems. It is necessary not to draw on the wells at such a rate as to draw air back down into the landfill. Oxygen is detrimental to the anaerobic process occurring and can severely retard methane production.

After extraction, the gas may be upgraded. A typical system consists of dehydration by compression and cooling, pretreatment for hydrogen sulfide and free water removal in molecular sieve towers, followed by another series of molecular sieve absorption towers to remove carbon dioxide. The resulting methane gas (35 GJ/m^3) is pressurized to 2.5 MPa and injected into a pipeline distribution system. Corrosion problems have been encountered in field installations. However, it is expected that corrosion can be eliminated with the injection of inhibitors into the biogas before processing.

As noted earlier, evaluations have been made of the extraction of methane from existing uncontrolled landfills for the Mountain View Project which is in the range 3.8-14.3 (Blanchet, 1979), the project at Palos Verdes, and by the Los Angeles Bureau of Sanitation (Bowerman, et al., 1976). The low rate of methane production results in high costs for the gas. Typical gas production from existing uncontrolled landfills is in the range 3.8-14.3 m^3/Mg year of deposited wastes, with a methane content above 50%. Costs were based on a 20-year life of a gas recovery system at an existing landfill.

Without enhancement, a landfill undergoes active degradation over a period of many years, the time primarily depending on the moisture content of the fill. Thus, problems associated with a given landfill (e.g., methane migration, ground water contamination) are likely to continue far beyond the termination of refuse placement, and indeed may not appear for some time after fill completion. Post construction maintanence and monitoring of a completed landfill site, to assure that it maintains its integrity and does not become a source of pollution, can therefore be a long-term source of concern. Another disadvantage of slow degradation is the long period of time required for gas extraction and the low annual methane yields per mass of refuse. Significant cost saving could be realized by increasing the rate of degradation within existing landfills. Further, research to date has indicated the higher ultimate gas yields are possible in controlled landfills.

The concept to enhancing methane production in landfills, controlled landfilling as it is sometimes known, has been developed by considering the landfill as a large batch anaerobic digestion system in which optimum conditions for methane production are provided. Urban refuse, which may have been separated, shredded, or baled is combined with nutrients, buffer and innoculum before its deposition into the landfill for the purpose of sustaining high reproductive rates of bacteria during decomposition. The landfill is constructed to allow for a gas recovery system and a moisture control system, and to optimize refuse cell size and geometry. The gas is extracted once the refuse cell is anaerobic and decomposition begins (DeWalle, et al., 1978).

As with the other digestion processes, the composition of the refuse directly affects the rate of methane production and subsequently the methane yield. It is advantageous for the refuse to have a high concentration of biodegradable materials, such as food, garden wastes, and paper. Sewage sludge mixed with the refuse increases the percentages of biodegradable materials in the landfill, and at relatively low concentrations (75-400 mg/l), stimulates gas production. At higher concentrations, however, inhibition may result.

The nutrients, buffer, and inoculum may be provided either by chemicals or sewage sludge, which are mixed or layered with the MSW before its deposition into a landfill, or by recycling leachate through the landfill, or by a combination of these methods.

Since the bacteria which carry out the biodegradation process grow best within a narrow pH within the landfill should be controlled. The optimal pH range of 6.25 to 7.5 has been controlled by the addition of calcium carbonate, a buffering agent, in simulated landfill cells (Augenstein et al., 1976). Based on a bacterial cell formula of $C_5H_7NO_2$, about 12.4% by weight nitrogen is needed for cell growth, while one-fifth of that value of phosphorus is required (Augenstein et al., 1976). The nutrient value of recycled leachate depends on the landfill composition. Evidence suggests that leachate material could provide only a portion of the nutrient requirements; thus addition of sewage or artificial nutrients may also be necessary. However, it has been shown by others that macro nutrient supplementation, i.e., N, P, S, is unnecessary (Le Roux and Wakerly, 1978). The application of sewage sludge and recirculation of leachate in controlled landfills is a potential health hazard for spreading bacteria and viruses, and a potential odor nuisance to nearby residents.

The two most important factors affecting methane production rates in landfills are temperature and moisture content. Methane production is severely limited at temperatures below 15°C, but increases with increasing temperature to an optimal temperature of 30-40°C. This parameter cannot be easily controlled in landfills, but many researchers believe that even in colder climates significant temperatures can be reached due to thermal insulation from surrounding soil and refuse (Rees, 1980). High temperature areas are expected to be non-homogeneous in distribution.

The refuse moisture content should be at least 50% and preferably about 80%, for high methane yields (Augenstein, et al., 1976). Methane production increases exponentially with increases in moisture content in batch digestion (DeWalle, et al., 1978). Moisture content of refuse (normally 25%) may be increased by the addition of water, sewage sludge, industrial wastes, or leachate material before deposition into the landfill. Alternatively, the natural ground water supply could provide water during digestion, but this may contaminate the water as discussed above. Although methods for water addition have been evaluated, large scale implementation of moisture control systems is needed.

Whatever the source of moisture, the landfill must be designed to retain moisture and to obstruct the flow of polluting leachate materials. Impermeable or low-permeable barriers, such as certain types of soils or synthetic liners, can be placed on the bottom and sides of the landfill area; these barriers are commercially available. The landfill should slope towards a point where leachate material can be collected for treatment or collection. If the topography of the landfill site does not provide for natural drainage, recontouring the land, probably at great expense, is necessary.

The enhancement concept appears to be a potentially effective method for producing methane from municipal solid waste. Experimentation using laboratory test cells has provided information on the effects of controlled conditions on methane production, and at present field scale demonstration facilities employing this concept are in operation at the City of Mountain View landfill in California and at the Binghamton, New York landfill.

APPLICATION OF ENHANCEMENT TO AGRICULTURAL RESIDUES

In a study at Cornell University (Jewell, et al., 1981), the principles of microbial innoculation, buffer addition, nutrient addition and moisture requirements were assessed for digestion of wheat straw and corn stover. The effect of these parameters on the anaerobic degradation of these agricultural residues was assessed in reactors similar to a landfill type environment. It was the attempt of this study to show that digestion of agricultural residues could be achieved under minimal moisture conditions (i.e., not having the residues in a slurry) under the proper conditions. The need for adequate microbial inocula and buffer additions to achieve rapid start-up of digestion and efficient conversion of these residues was found to be crucial in this study. More recently, Dynatech carried out a project for A.I.D. dealing with the controlled landfilling of agricultural residues in which the results were similar to those for MSW.

STATUS OF LANDFILLS AS FUEL GAS SOURCES IN THE UNITED STATES

The practice of landfilling to dispose of municipal solid waste in the United States has created many hundreds of accumulations of digestible material. Indeed, like natural, plant-derived wastes, the contents of landfills spontaneously undergo anaerobic digestion. The composition of municipal waste in the U.S. generally is 50% by weight paper and paperboard, an ideal substrate for fermentation. Active exploitation of landfills for fuel gas is now underway. In a few cases the prime incentive for removing gas from landfills is the abatement of nuisance or safety hazard, but for the most part recovery of fuel gas is the prime objective.

The first undertaking of this kind was complete in 1975, at Palos Verdes, California. A census of landfill gas recovery plants completed and in operation in early 1984 shows 29 gas recovery plants in operation. At the same time advanced planning or construction is under way on 53 additional sites. Many more than these are being considered for exploitation, but without firm intentions or contracts in place.

California clearly is the leading seat of this activity, with the New York - New Jersey area being second. The quantity and concentration of waste in these particular densely populated areas accounts for this leadership. The rate of fuel gas recovery from the 29 recovery systems in place is about 4×10^8 cubic meters annually (40 million cubic feet daily). These figures are cited in terms of raw gas containing more or less carbon dioxide and having an energy content of approximately 19 megajoules per cubic meter (500 Btu per cubic foot). The majority of these landfill gas recovery plants practice some kind of dehydration and removal of corrosive constituents, but only 6 of the present plants remove carbon dioxide to produce near-pure methane. Raw gas is entirely acceptable for local combustion, either heating or electricity generation.

In term of pure methane the quantity of gas produced through landfill exploitation, about $200 \times 10^6 m^3$/year, is nearly 7 times the amount of methane estimated to be produced by United States digesters surveyed above, i.e., $30 \times 10^6 m^3$/year. Taken together this amount of methane is approximately 0.04% of the annual natural gas consumption in the United States. Clearly, this contribution to fuel supply is not extensive, but growth is anticipated and nearly every individual production source is a financially significant operation to its managers.

It is notable that landfill gas recovery is sufficiently attractive economically that it has attracted private capital investment. Several firms are engaged actively in the acquisition of rights to landfill gas recovery. A reasonable estimate of the potential for such installations in the U.S. is 200. Given an estimate of 30 ± 10 years as the practical lifetime of gas recovery from a landfill it may be appreciated that this is going to be a vigorous industry. Improvements in the technology of landfill management may be expected to improve the economics and significance. Research and development now under way in the U.S. is aimed at optimizing moisture, nutrients, pH, and leachate circulation in landfills so that more complete gas generation at higher rates is obtained than is the case in untouched landfills. The prospects of "on-line" processing of municipal solid waste in this way for gas recovery are likely to insure many decades of useful gas production.

Academic support of research on anaerobic digestion in the U.S.A. continues strong with greatest emphasis currently in microbiological fundamentals. The over-all level of research and development on anaerobic digestion has diminished drastically in the past five years as financial support from the U.S. Department of Energy has been reduced. The reduction in this support, which has included the construction of large demonstration plants, has made the present level of development activity seem relatively slow. There remain in place a number of Federal and State financial incentives which are available to assist implementation of anaerobic digestion systems. Such incentives, an improved climate for private investment, and growing appreciation of the value of anaerobic treatment of industrial wastes will support steady growth in the utilization of this technology.

REFERENCES

Augenstein, D.C., D.L. Wise, R.L. Wentworth, and C.L. Cooney. 1976. Fuel recovery from controlled landfilling of municipal wastes. Resource Recovery and Conservation, 2:103-117.

Augenstein, D.C., D.L. Wise, R.L. Wentworth, P.M. Gallaher, and D.C. Lipp. 1977. Investigation of Converting the Product of Coal Gasification to Methane by the Action of Microorganisms. Final Report No. FE-2203-17 under Contract No. E. (49-18)-220. Dynatech R/D Company. Cambridge, MA.

Blanchet, M. 1979. Start-up and Operation of the Landfill Gas Treatment Plant at Mountain View. Solid and Hazardous Waste Research Division 5th Annual Research Symp., Municipal Solid Waste: Resource Recovery. Orlando, FL.

Bowerman, F.R., N.K. Rohatji, and K.Y. Chen (eds.). 1976. A Case Study of the Los Angeles Sanitation Districts Palos Verdes Landfill Gas Development Project. U.S. Environmental Project. U.S. Environmental Protection Agency, NERC Contract No. 68-03-2143.

DeWalle, F.B., E.S.K. Chian, and E. Hammerberg. 1978. Gas production from solid waste in landfills. J. Envir. Eng. Div. 104:415-432.

Jewell, W.J., J.A. Chandler, S. Dell-Orto, K.J. Kanfoni, S. Fast, D. Jackson, and K.M. Kabrick. April 1981. Dry fermentation of agricultural residues: SERI Report No. XB-0-9038-H6.

LeRoux, N., and D. Wakerly. 1978. The Microbial Production of CH_4 from the Putresible Fractions of Solid Household Waste. Proc. 1st Recycling World Congress, M.E. Henstock, (ed.). Basel, Switzerland.

Rees, J.F. 1980. Journal Chemical Tech. & Biotechnology, 30:458-465.

BIOGAS PRODUCTION FROM WATER HYACINTH [EICHHORNIA CRASSIPES]: INFLUENCE OF TEMPERATURE

H.B. El Amin, Renewable Energy Research Institute, NCR, and
H. A. Dirar, Faculty of Agriculture, University of Khartoum, Sudan

ABSTRACT

Water hyacinth (Eichhornia crassipes) has been screened on mesophilic anaerobic digestion in a 90-day experiment using a laboratory-scale semi-continuous digester. Fermentation at constant temperatures of 25, 37, and 45°C showed 37°C to be the best incubation temperature. Room temperature varying between 32 and 45°C gave higher yield than 25°C and 45°C. Moreover, fermentation at 45°C was very sensitive to any environmental changes. The pH levels settled around 7.0 without the need for external adjustment.

INTRODUCTION

Sudan is essentially an energy-poor country. Ninety percent of the population uses firewood and charcoal as energy sources. However, the utilization of biomass as such is still irrational. The situation is serious and the fuel to cook food is becoming as scarce as the food itself.

Water hyacinth (Eichhornia crassipes (Mart) Solms) on the White Nile and other aquatic weeds are becoming a problem of crisis dimensions. The hyacinth (covering an area of approximately 11,000 hectares, giving 150 - 300 tons fresh biomass) is found to have little economic value as animal feed, as fertilizer, or as a source for paper pulp[1]. Fortunately, the plant is found to be one of the best substrates for biogas production[2]. Using the hyacinth as a substrate for biogas provides supplementary control measures, fuel, and fertilizer.

Microbial methane formation occurs over a wide temperature range, namely from about 0°C to 97°C[3]. However, with respect to temperature, biogas currently is generated under three types of fermentation: (i) thermophilic with optimum temperature range 47°C - 55°C, (ii) mesophilic with optimum temperature range 35°C, and (iii) natural or ambient temperature fermentation factors. All these types of fermentation have advantages and disadvantages. In any case, digestion at regular temperatures needs more energy input, viz. for stirring, pumping, and heating the substrate inside the digester to maintain the desired temperature. Regular temperature fermentation is practiced in temperate climates, where the majority of the energy input is used for heating the substrate inside the digester[4].

In the Sudan, ambient temperature fermentation is implicated because of the tropical climatic conditions. However, summer, winter and day, night temperature fluctuations could be an important factor for unbalanced fermentation. The most prevailing temperature is almost in the mesophilic range. Thus the optimization of the process is directed mainly to maximum gas production and not for net energy production.

This laboratory study was conducted to investigate the effect of different stable temperature fermentation in the mesophilic range, as well as ambient temperature fermentations to: (i) Point out the relevant biological values and efficiencies of methanogenesis reflected by production of total biogas, methane concentration, pH, and volatile fatty acid fluctuations; (ii) Indicate the least fluctuation temperature range so as to eliminate climatic changes that could occur by means of a suitable fermentation method to avoid the need for artificial heating under all circumstances.

MATERIALS & METHODS

Fermenter

Figure 1 shows the basic design of the 4 liter working volume fermenter, together with the brine displacement system used throughout this study. The layout was adapted from Laura and Idnani[5] and van Velson (6). Full description of this design was given elsewhere[7].

Substrate

Complete (shoot and root) fresh, healthy water hyacinth plants were used. The plants were chopped then minced. The slurry was dispensed into labeled screw-capped plastic bottles. Samples from these bottles were taken for total solids (TS) determination. Then the bottles were frozen.

At any fermentation start, the bottles were thawed and the desired total solids level was calculated according to the TS in any bottle.

Inoculum

All digesters were seeded with digested sludge from a digester previously working on water hyacinth slurry; 10% V/V were used.

Analytical Methods

1. Volumetric gas analyses were performed by means of an Orsat-Kleine apparatus No. 1004. The analyses were conveniently done weekly for methane and carbon dioxide only.
2. The cumulative biogas production (ml/day) was recorded by measuring the brine displaced for every 24 hrs.
3. Volatile fatty acids (VFA), total solids (TS) and volatile solids (VS) were analyzed according to the method of the AOAC (1975)[8].
4. pH was measured with Pye model - 79 pH meter.

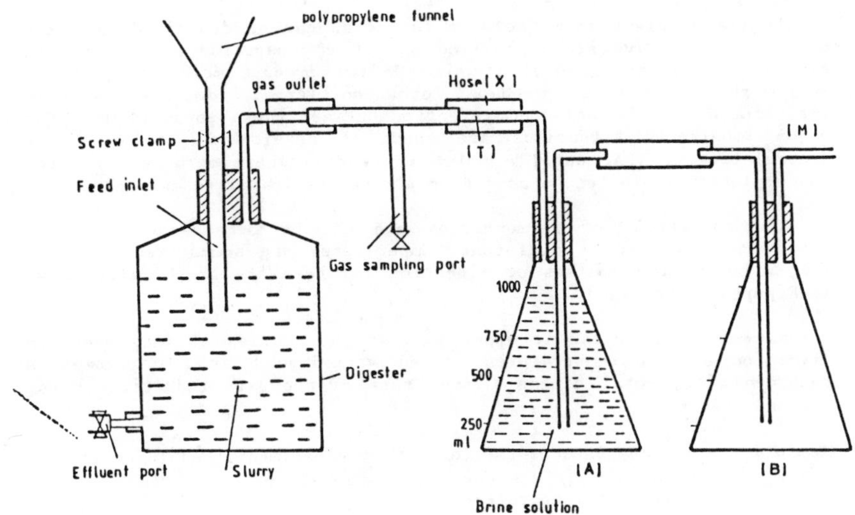

Figure 1 Experimental Setup

Experimental Procedure

Water hyacinth slurry was supplied to the four digesters (designated A_1, A_2, A_3 and A_4) to give a final volume of 4 liters and a total solid concentration of 30 g/l. All digesters were seeded, mixed well, and closed. The digesters were incubated in different water baths maintaining the prescribed temperature (Table 1). Digester A_2 was left at room temperature. The brine displacement system was connected (Figure 1).

All digesters were fed daily. The feeding procedure was as follows. Every day after the determination of the gas production, a prescribed volume (133 ml) (determined by the detention time) of the mixed digester contents was removed; then the slurry was added according to the hydraulic and organic load. Before and after loading, the digesters were mixed for 2 minutes.

RESULTS AND DISCUSSION

In its simplest form, methane formation can be described as three or four consecutive stages carried out by a consortium of interacting synergestic bacteria, most of which do not produce methane as such[9]. Thus, the success of methane formation largely depends on this interaction and the balanced retention between these populations. The major imbalanced situations are known to develop due to a steady accumulation of VFA, until a point is reached where methane formation is completely inhibited, a condition arrived at in silage making[10].

The conditions of fermentation and the performance of the four digesters incubated at different temperatures are summarized in Table 1. Moreover, the courses of digestion for them are graphically shown in Figures 2 through 5.

From these results, it appeared that the courses of the anaerobic digestion at all temperatures tested varied much in all parameters recorded viz., total biogas / day, rate of the gas production, CH_4%, pH, and VFA.

Table 1 Summary Data of the Laboratory Experiments Fed Initially with 30g TS Water Hyacinth Slurry and Conducted at Different Temperatures

Parameter		Digester			
		A_1	A_2	A_3	A_4
Digester working volume (liters)		4	4	4	4
Duration of the experiment (days)		90	90	90	90
Retention time (days)		30	30	30	30
Organic loading rate: kgVS $m^{-3}day^{-1}$		0.779	0.779	0.779	0.779
Incubation temp. (°C)		25	*	37	45
Influent					
TS	g/liter	30.8	30.8	30.8	30.8
VS	g/liter	23.4	23.4	23.4	23.4
Effluent					
TS	g/liter	22.34	17.76	20.1	23.34
VS	g/liter	13.1	7.38	9.4	13.70
Performance					
TS	% destroyed	27.46	42.33	34.93	24.22
VS	% destroyed	44.0	68.40	60.3	41.40
Cumulative gas produced (liters)		67.6	63.6	70.3	58.2
Average CH_4%		59.2	66.7	55.7	56.7

Generally biogas production and methane percent in the gas increase from the outset of the fermentation, and reach a maximum after which they decrease. It was noticed that the acclimatization period, i.e., unbalanced digestion at the start was not detected. This may be due to our previous experimental results which showed that a loading rate of 0.779 kg VS m^{-3} day[11] at 37°C is much better than other loading

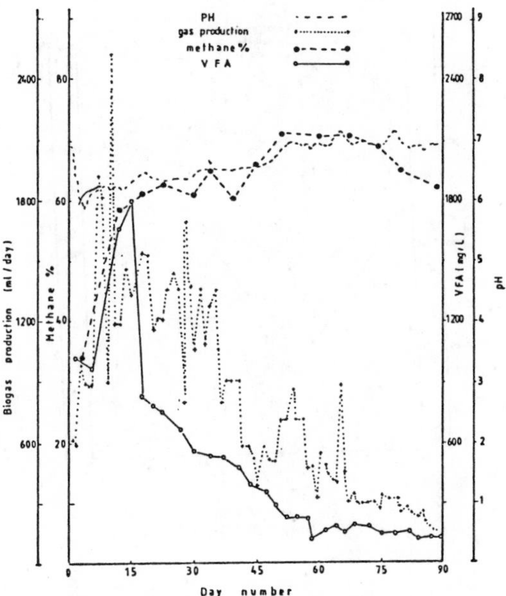

Figure 2 Digestion at 25°C.

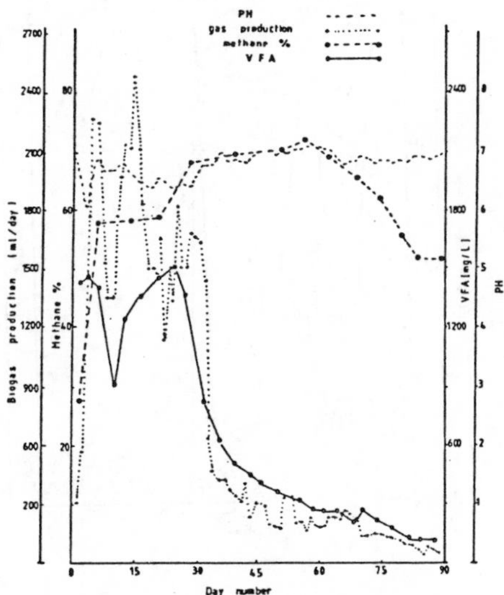

Figure 3 Digestion at Room Temperature.

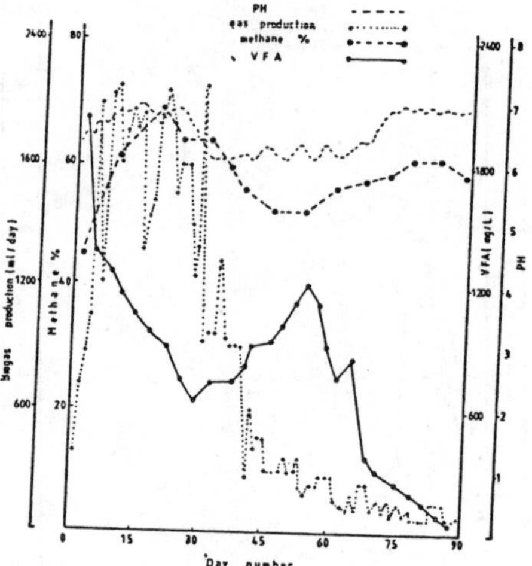

Figure 4 Digestion at 37°C,

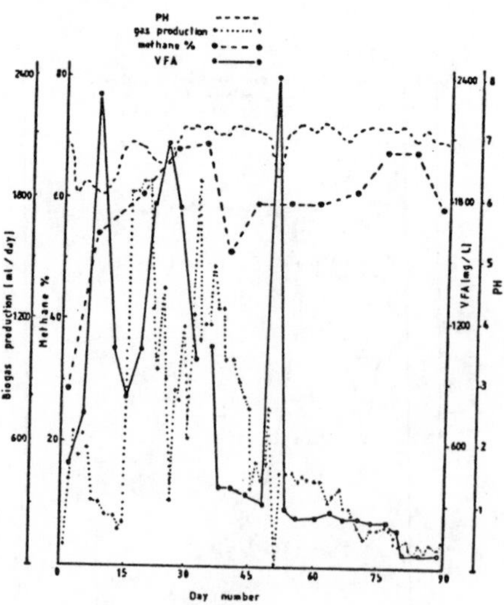

Figure 5 Digestion at 45°C,

rates tested[11]. At the same time, the type of inoculum used is a digested water hyacinth, slurry which is proved to be the best type of inoculum using water hyacinth as a substrate.

Among the constant temperatures tested, namely 25, 37, and 45°C, the performance at 37°C was much better than at the other temperatures. At 37°C cumulative biogas was 70.3 l, while it was 67.6 l for 25°C and 58.2 l for 45°C (Table 1). Generally speaking, while gas production at 37°C followed a steady course, gas production at 45, and to a lesser extent at 25°C, was characterized by strong fluctuation in daily yields. With respect to methane percent in the gas, room temperature was the best, then digestion at 25°C, 45°C, and 37°C, respectively. It is difficult to explain these results because at 37°C maximum cumulative biogas was produced, but the low figure obtained here may be due to erratic sampling procedure.

The idea behind testing a variable room temperature was to test the effect of fluctuations in temperature, which are encountered in practical situations, or digester performance. Results (Table 2) indicate that within the temperature range of 32°C to 42°C (Summer time; inside laboratory with metal roof) the performance of the digester was similar to that of the digester incubated at 37°C. However, the temperature variation did not exceed 7°C, thus it seems possible that stable digestion, high gas production, and methane concentration was due to this narrow range of fluctuation. It was observed[12] that a temperature fluctuation of 5°C can be tolerated for as long as 48 hrs without serious damage to the long-term performance of the system. Furthermore, Zehnder[13] also reported that some strains of methanogenic bacteria show a relatively wide optimum exceeding 5°C .

Table 2 Room Temperature Recorded for Digester A_2.

	June 1983	July 1983	August 1983
Mean maximum temperature °C	42	40	38
Mean minimum temperature °C	35	34	32

Gas production at 25°C seems to be possible since pilot plants in Tawila recorded mesophilic optimum between 25 - 33°C.[14].

Although in this study, temperatures below 25°C were not investigated for reasons beyond our control, it is possible that gas production slowed down during wintertime in the pilot plants, due to drop of temperature below 25°C[14]. Results of van Velson et al.[15] showed that no gas was produced at 13°C. Contrary to these findings, Lettinga[16] found that methane-forming organisms accomodate well to changes in temperture between 10 and 45°C .

Figure 5 demonstrates that digestion at 45°C is quite inferior to the other incubation temperatures. Gas production is characterized by low cumulative gas yields, low methane percentage, and low degree of decomposition of VS. Moreover, the digestion at 45°C was very sensitive to any environmental shock. Electric cuts or laboratory accidents out of our control impede gas production due to high accumulation of VFA (Figure 5).

Fortunately, gas composition seems not to be affected by these off-power days, although by chance, samples were not subjected to analysis. In contrast to these results Hashimoto et al.[17] found that digestion at 45°C gave higher gas yield than at 35°C, but they concluded that methane percentages were not affected by the incubation temperatures in the long run. Results of Wolverton et al.[18] showed that digestion of water hyacinth at 36°C increases not only the rate of biogas production, but also the methane content compared to that at 25°C.

In all temperatures tested we have found that in normal conditions VFA did not exceed 1900 mg/l (as acetic acid). It is commonly stated that a concentration of VFA of 2000 mg/l are toxic[19]. Thus, this level (1900 mg/l) was not toxic to the process of biomethanation, consequently, the pH never fell outside the optimum range (6 to 8).

Since the pH of the fermentation at all temperatures tested was not controlled, and oscillated between pH 6 and 7, the need for artificial adjustment was inconvenient and would not be practical.

At any rate, the results obtained in this study showed that production of biogas in a temperature range between 25°C and 45°C is possible. This range in fact is the prevailing ambient temperature in a tropical country like Sudan. In spite of the fact that winter temperatures in Sudan could slow down gas production, there can be no doubt that a choice of a suitable fermentation method (in our case chosing feed of 30 g/l and long retention time [30 days]) would avoid the need for artificial heat under all circumstances.

REFERENCES

1. Koster, H.J.K., G. Schmalstieg, and W. Siemers. 1982. Densification of water hyacinth - basic data. Fuel 61:791-798.

2. Philipp, O., H.B. El Amin, and V. Leffler. 1978. Some Aspects on the Utilization of Water Hyacinth (*Eichhornia crassipes*) as Supplementary Control Method in the Sudan. Symp. on Crop Post Management. Khartoum.

3. Zehnder, A.J.B., K. Ingvorsen, and T. Marti. 1982. Microbiology of methane bacteria. In: Hughes et al. (Eds.) Anaerobic digestion 1981. Elsevier, Amsterdam.

4. Mills, P.J. 1979. Minimization of energy input requirements of an anaerobic digester. Agric. Wastes 1:57-66.

5. Laura, R.D., and M.A. Idnani. 1971. Increased production of biogas from cow dung by adding other agricultural waste materials. J. Sci. Fd. Agric. 22:164-167.

6. van Velson, A.F.M. 1977. Anaerobic digestion of piggery waste. I. The influence of detention time and manure concentration. Neth. J. Agric. Sci, 25:151-169.

7. El Amin, H.B. 1983. Biogas Production from Water Hyacinth (Eichhornia crassipes). M.Sc. Thesis. University of Khartoum.

8. A.O.A.C. 1975. Official Methods of Analysis (12th ed.) Association of Official Analytical Chemists.

9. Bryant, M.P. 1979. Microbial methane production - theoretical aspects. J. Anim. Sci. 7:437-449.

10. Boshoff, W.H. 1963. Methane gas production by batch and continuous fermentation methods. Tropical Sci. 5:155-165.

11. El Amin, H.B., and H.A. Dirar. 1984. Effect of Loading Rates on Biogas Production from Water Hyacinth (unpublished).

12. Anonymous. 1978. Biogas Production from Animal Manure. Publ. by: Biomass Energy Institute Inc., Cambridge Street Winnipeg, Manitoba.

13. Zehnder, A.J.B. 1978. Ecology of methane formation. P. 349-376 In Mitchell, R. (ed), Water Pollution Microbiology vol. 2. John Willey & Sons. Inc. New York.

14. Philipp, O., W. Koch, and H. Koser. 1983. Utilization and Control of Water Hyacinth in Sudan. Published by the German Agency for Technical Cooperation (GTZ). Eschborn.

15. van Velson, A.F., G. Lettinga, and D. Ottelander. 1979. Anaerobic digestion of piggery waste. 3. Influence of temperature. Neth. J. Agric, Sci. 27:255-267.

16. Lettinga, G. 1979. Direct anaerobic treatment handles waste effectively. Ind. Waste: 25:8.

17. Hashimoto, A.G., V.H. Varel, and Y.R. Chen. 1981. Ultimate methane yields from beef cattle manure: Effect of temperature, ration concentrations, antibiotics, and manure age. Agric. Wastes 3: 241-256.

18. Wolverton, B.C., R.C. McDonald, and J. Gordon. 1975. Biconversion of Water Hyacinth into Methane Gas Part I. NASA. Tech. Memo., (TM X - 72725).

19. McCarty, P.L. 1964. Anaerobic waste treatment fundamentals. III. Toxic materials and their controls. Public Works. 95:91-94.

ANAEROBIC DIGESTION OF ORGANIC FRACTION OF GARBAGE: A PILOT PLANT

F. De Poli(**), F. Cecchi(*), P.G. Traverso(*),
F. Avezzu(*), and P. Cescon(*)

*Universita di Venezia, Dipartimento di Scienze Ambientali,
Calle Larga S. Marta, 2137 - 30123 VENEZIA, Italy

**ENEA (Italian Commission for Nuclear and
Alternative Energy Sources). Dep. FARE, CRE Casaccia,
P.O. Box 2400 - 00100 ROMA, Italy

ABSTRACT

Anaerobic digestion of the organic fraction of municipal solid waste mixed with primary municipal sewage sludge was investigated on pilot scale. Preliminary experimental data are presented. Results indicate that gas yield is close to expectation. Though there are still several operating problems that have to be attended to, the process appears technically feasible.

INTRODUCTION

The overall objective of this paper is to obtain information for large-scale commercial production of methane from the organic fraction of municipal solid wastes (OFMSW). A recent review[1] suggests that the anaerobic digestion of OFMSW is probably to be preferred to other waste treatments. This is still an open question in Italy, however, and some opposing considerations can be found in the literature. We believe that the data from a 3 m^3 working volume pilot digester can be profitably compared with those from full-scale plants. Some preliminary data are shown here.

MATERIALS AND METHODS

The pilot plant is schematically shown in Figure 1. The hand-classified OFMSW is shredded and analyzed for total solids (TS) before being mixed in a homogenizer with the sludge from a municipal wastewater primary settler.

The ratios (OFMSW/Primary Sludge, PS) and TS percentage in the feeding sludge (3 percent) is controlled, and pumped into the storage tank.

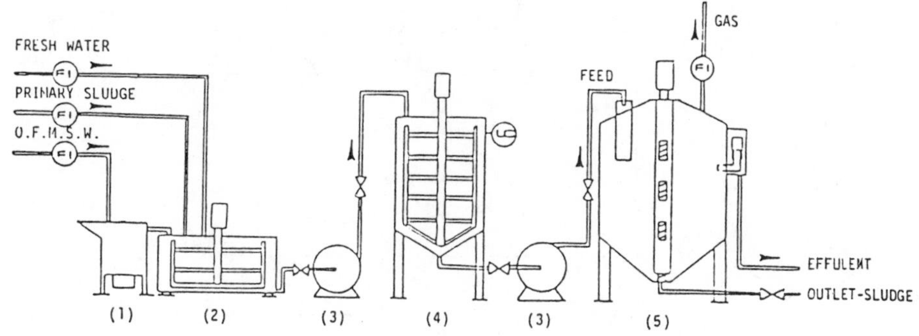

Figure 1 - PILOT PLANT. (1) Shredder; (2) Homogenizer; (3) Pumps;(4) Feed stok;(5) Digester.

The digester is fed discontinuously (three times a day). The TS content in the digester is controlled by discharging the bottom sludge (outlet-sludge).

The liquid level in the reactor is controlled by a hydraulic valve so that the discharge of the effluent is a consequence of the feeding flow. A hydraulic valve in the gas pipe warrants 150-180 mm w.c. pressure of the gas in the digester top. The analytical plan and the parameters which are controlled are in Tables 1 and 2. Furthermore, the MSW, OFMSW, PS, have been characterized and the shredded OFMSW classified.

RESULTS

The seasonal variations of the components of the MSW from a representative quarter in Treviso City are shown in Figure 2; the average values are in Table 3 where they are compared with the related values in the N.W. - Italy[2].

The composition of the OFMSW is shown in Figure 3 from which the time dependence can be deduced too. The results from particle size analysis of the shredded OFMSW are in Figure 4. Two typical periods could be identified: summertime and other seasons. A comparison between the average values from the previous two periods are shown in Figure 5.

The main characteristics of the primary sludge are in Table 4. The particle size classification led to results which are not far from those from OFMSW (other seasons).

The inoculum was drawn off from a full-scale anaerobic digester operating on piggery wastes. Steady-state conditions were achieved in three months by a feed-slurry from the PS. The digester was then operated as shown in Table 5, where the average values and the standard deviations of the operative parameters are shown. The related average values and standard deviations of the analytical data are shown in Table 6.

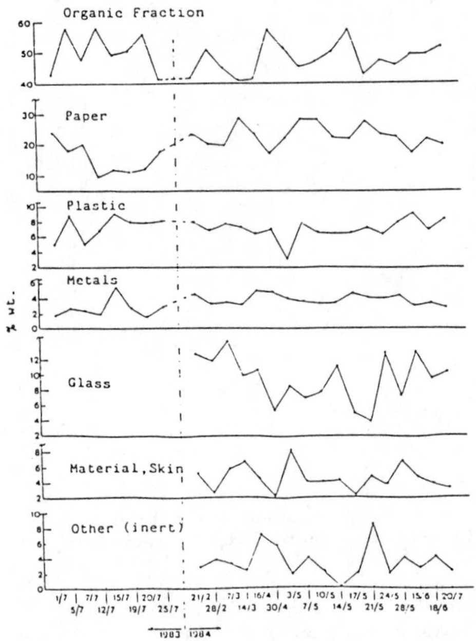

Figure 2 Time Variation of Municipal Solid Waste Components

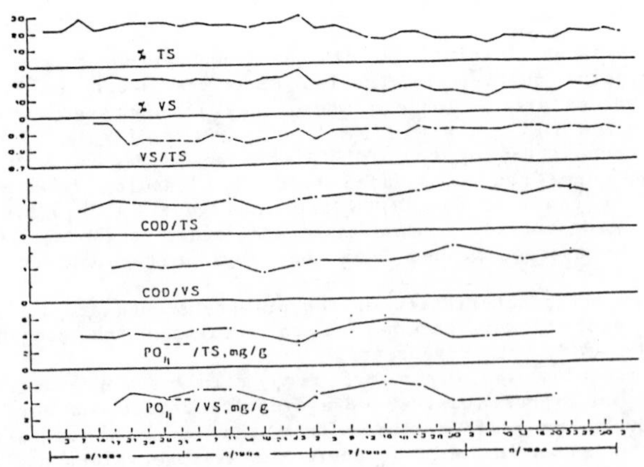

Figure 3 -O.F.M.S.W CARACTERISTICS

CONCLUSIONS

1. The gas yield was close to the expected values.

2. The overflow TS concentration considerably increases as soon as the OFMSW percentage is greater than 35; a persistent scum layer occured, whose depth increases with the OFMSW percentage. We are now solving the problem by modifying the mechanical mixer.

3. The digestion process of proper refuse fraction (16-50 percent) at low organic loading can be carried out without addition of nutrients or of buffers.

4. The gas production was found to be dependent on the refuse loading. This point, however, calls for further experimental results in the absence of the scum layer.

REFERENCES

1. Kispert, R.G. and D.L. Wise. 1981. A review of biconversion systems for energy recovery from municipal solid waste. Part III: Economic evaluation. Resources and Conservation, 6:137-42.

2. Progetto, C.N.R. April 1980. Finalizzato Energetica "Atti del II Seminario Informativo. Utilizzazione Energetica dei Rifiuti Urbani." Padova 21.

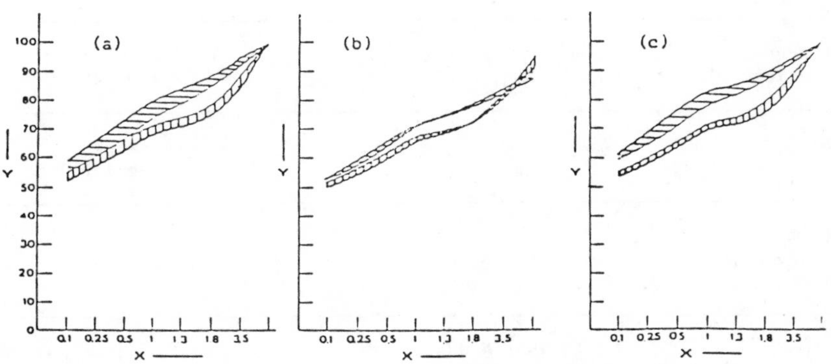

Figure: 4 - Integral description of the results from size classification of the shredded O.F.M.S.W.. X-axis = size (Ø); Y-axis = % wt.of the shredded matter whose size is less than X: a)(TS < X)/(TS); b)(VS < X)/(TS); c) (VS < X) / (VS).

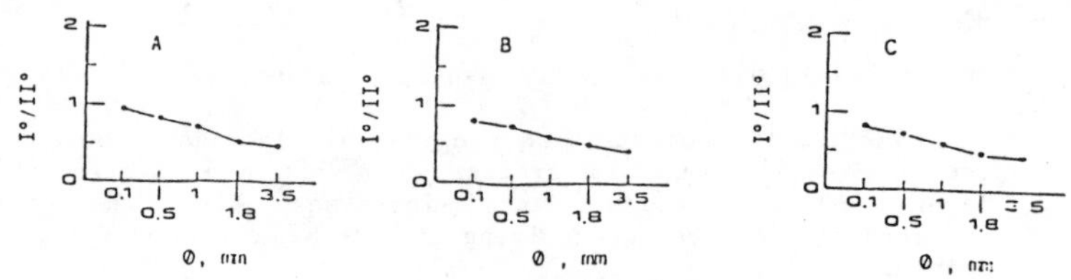

Figure 5 - Fractions of the particulate whose size is less than Ø (abscissa) in summer time (II) and in other seasons(I).
A) (TS < Ø in size)/(TS); B) (VS < Ø in size)/(TS);C)(VS < Ø in size)/(VS).

Table 1 Analyses Plan

	PARAMETERS	ANALYSES X WEEK
Feed	COD TS, VS, TKN, $N\text{-}NH_4$, $N\text{-}NO_3$, $P\text{-}PO_4$, TA, VFA, pH	2 x week
Effluent	COD, TS, VS TKN, $N\text{-}NH_4$, $N\text{-}NO_3$, $P\text{-}PO_4$	6 x week 1-2 x week
Reactor	COD, TS, VS, TKN, $N\text{-}NH_4$, $N\text{-}NO_3$, $P\text{-}PO_4$, TA, VFA, pH, TS, VS, At various reactor positions	2 x week 1 x week
Sludge-Outlet	COD, TS, VS TKN, $N\text{-}NH_4$, $N\text{-}NO_3$, $P\text{-}PO_4$	Always 2 x week
Gas	CO_2 H_2S	6 x week Irregular
Primary Sludge	COD, TS, VS, TRN, $N\text{-}NH_4$, $N\text{-}NO_2$, $P\text{-}O_4$	1 x week
OFMSW	COD, TS, VS, TKN, $N\text{-}NH_4$, $H\text{-}NO_3$, $P\text{-}PO_4$	1 x week
Gas Production		6 x week

Table 2 Controlled Parameters

	PARAMETERS	ANALYSES X WEEK
Digester feed	Flow rate, Temperature OFMSW/Primary Sludge TS	7 Always
Effluent	Flow rate (discontinuous)	7
Sludge Outlet	Flow rate (discontinuous)	7
Reactor	Temperature, Pressure	Continuous

Table 3 Comparison Amongst Data from Treviso City and from N.W.-Italy.

(+) The stuff whose size is 2.5 cm was disregarded; its organic matter percentage is very high.

	Organic	Paper	Plastics	Metals
N. W. - Italy	40%+	23%	7%	3%
Treviso City	48%	21%	7%	3%

Table 4 Primary Sludge Characteristics

Parameters	COD, g/l	TS g/l	VS g/l	TKN mg/l	N-NH_4 mg/l	N-NO_3 mg/l	TA mg/l $CaCO_3$
Mean value ± Standard Deviation	39.0±21.5	41.3±18.1	30.2±12.3	296±149	114±64	0.5	887±436

Table 5 Summary of Selected Data from Digestion Runs at 35 ± 2°C.

Parameters	% OFMSW 16	35	60	90
pH range	6.9 - 7.2	6.9 - 7.2	7.1 - 7.3	7.1 - 7.3
Load				
$KgVS/m^3$ day	1.23 (0.33)	1.32 (0.13)	1.11 (0.25)	1.51 (0.24)
$KgCOD/m^3$ day	2.00 (0.56)	2.30 (0.28)	1.72 (0.42)	2.00 (0.50)
$KgVS/KgX_{VS}$ day	0.033 (0.009)	0.037(0.004)	0.039(0.005)	0.073 (0.012)
$KgCOD/KgX_{VS}$ day	0.052 (0.014)	0.066(0.012)	0.061(0.014)	0.097 (0.025)
H.R.T. day	16.5 (1.1)	16.9 (0.9)	19.5 (2.3)	18 - (3.1)
S.R.T. day	48	43	49	-5
Feed				
gVS/day	4000 (850)	3950 (395)	3260 (780)	4650 (750)
gCOD/day	6650 (1350)	6900 (835)	5050 (1265)	6000 (1500)
Digester				
gTS/1	63.2 (5.3)	62.3 (4.9)	46.3 (2.11)	32.8 (2.3)
gVS/1	38.0 (2.6)	35.5 (2.9)	28.2 (1.6)	21.3 (1.2)
Gas Production Rate 1/D	1810 (145)	2220 (320)	2405 (370)	3560 (365)
Methane Production Rate L/D	1315 (100)	1560 (240)	1615 (240)	2285 (250)
Gas Yield				
1 gas/gVS_{A1}	0.45	0.56	0.74	0.75
1 gas/$gCOD_{A1}$	0.27	0.32	0.48	0.59
1 CH_4/gVS_{A1}	0.33	0.39	0.49	0.45
1 $CH_4/gCOD_{A1}$	0.20	0.23	0.32	0.37

Table 6 Mean Values and Standard Deviations of the Analyzed Parameters

% OFMSW		COD, g/l	VS, g/l	TS, g/l	TKN, mg/gTS	NH_4^+, mg gTS	PO_4^{---}, mg/gTS	TA, mg$CaCO_3$/l
60	FEED	34.5 (6.6)	21.5 (4.0)	27.8 (5.4)	4.6 (1.3)	1.4 (0.7)	6.2 (1.0)	-
	DIGESTER	31.5 (3.5)	28.2 (1.6)	46.3 (2.1)	9.0 (1.0)	5.4 (1.3)	7.7 (0.7)	2920 (180)
	EFFLUENT	13.7 (9.0)	12.0 (5.4)	19.8 (8.7)	20.1 (9.1)	14.8 (7.2)	8.2 (1.1)	-
	SLUDGE-OUTLET	41.1 (8.1)	42.9 (7.3)	87.7 (22.2)	5.7 (1.4)	3.2 (1.3)	8.6 (1.6)	-
90	FEED	36.1 (4.3)	26.2 (2.3)	33.7 (1.9)	4.7 (0.6)	1.2 (0.6)	6.1 (1.2)	-
	DIGESTER	25.0 (5.9)	21.9 (1.8)	33.7 (3.0)	15.5 (1.2)	11.5 (1.1)	9.5 (0.2)	3485 (32)
	EFFLUENT	13.0 (4.7)	11.2 (3.7)	18.6 (5.0)	20.3 (3.4)	14.8 (3.4)	9.3 (1.7)	-
	SLUDGE-OUTLET	25.0 (5.9)	21.9 (1.8)	33.7 (3.0)	15.5 (1.2)	11.5 (1.1)	9.5 (0.2)	

THE TOXICITY EFFECT OF PESTICIDES AND HERBICIDES ON THE ANAEROBIC DIGESTION PROCESS

Fatma A. El-Gohary, Fayza A. Nasr and O.A. Aly
Water Pollution Control Laboratory,
National Research Centre, Egypt

ABSTRACT

The toxicity effect of herbicides and pesticides on the anaerobic digestion process was investigated. For this study, mixtures of cotton stalks, water hyacinth or aquatic weeds were mixed with fresh cow dung.

Results obtained showed that the addition of 5 mg l^{-1} lindane and 5 mg l^{-1} DDT, reduced the total volume of methane gas produced by 34%. An increase in the concentration of each of these compounds up to 10 mg l^{-1} led to a reduction of 45%. Analysis of lindane and DDT in the digested slurry reflected the persistence of these two pesticides.

A concentration of 5 mg l^{-1} and 10 mg l^{-1} of both Gesapax and Gesaprime in the feed reduced methane gas production by 12.5 and 33 percent, respectively. The removal values of Gesapax and Gesaprime were 37% and 40%, at the lower concentration investigated. Raising the concentration added to the feed up to 10 mg l^{-1} for each reduced the removal value of Gesapax by 9 percentage points. On the other hand, the removal of Gesaprime remained unchanged, i.e. 40%.

INTRODUCTION

Adequate knowledge of toxicity phenomena is crucial to the proper design and operation of a biological waste treatment system. This is especially true for the anaerobic waste treatment process because the key group of bacteria in the process, "the methane bacteria," is much more sensitive to environmental conditions than are other groups of bacteria. Although this process has significant advantages over other methods of waste treatment, its use has been retarded because of lack of understanding of toxicity phenomena.

A review of the literature on anaerobic waste treatment indicates considerable variation in the toxicity reported for most substances. The major reason for these variations is the complexity of the toxicity phenomenon.

Heavy metals have been the cause of poor performance of anaerobic waste treatment systems. The variation, however, in reported values is considerable. The high tolerance levels and variability reported are probably caused by the ease with which heavy metals take part in complex-type reactions with the normal constituents of an anaerobic waste treatment unit. The most important complex-type reaction for controlling toxicity in anaerobic waste treatment is the precipitation of heavy metals by sulfides. This was noted by Barth et al. (1965) and Masselli et al. (1961).

Many other groups of substances have been reported to exert a toxic effect on anaerobic waste treatment systems. Long chain fatty acids such as palmitic, stearic, and oleic acids can exert a toxic effect on anaerobic digestion if they are in solution. Highly insoluble calcium salts of these acids can be formed to prevent toxicity (McCarty et al. 1964).

Among other substances reported as toxic to methane bacteria are phenols (Mahr, 1969), methane analogs such as chloroform (Nikitin, 1968), and chlorinated hydrocarbons (Smith, 1973). However, little significant data exists on the toxicity of these substances.

Pesticides and herbicides are now being used on a large scale in Egypt. Pesticides are of different types, e.g. organochlorine, organophosphorus, and carbamates. In spite of the fact that organochlorine compounds are not used now in Egypt except in a few cases, they are still detectable in the environment. Agricultural wastes and aquatic plants are usually contaminated with residues of these compounds (Aly and Badawy, 1984).

Therefore, it was found necessary to study the fate and toxicity effect of these compounds on the anaerobic digestion process.

MATERIALS AND METHODS

Ten batch laboratory-scale digesters, each of 2.5 liters effective capacity, were operated concurrently. Temperature was maintained constant by keeping the digesters in incubators. Mixing of the digesters' contents was accomplished by shaking the fermentation units once each day. Starter seed for anaerobic digestion was prepared by allowing the waste investigated to decompose under anaerobic digestion for two months. The seed was added to all digesters at the rate of 10 percent by volume.

Organic wastes investigated were fresh cow-dung, ground cotton stalks, chopped water hyacinths (*Eichhornia crassipec*), and chopped aquatic weeds (*Echinochloa stagninum*).

The volume and composition of the biogas produced were recorded during digestion. Methane content of the biogas was analyzed by gas chromatography using Varian 2400 GC with a flame ionization detector. Gas chromatographic conditions were:

Column: 6 ft long, 1/8 in ID, stainless steel.

Packing: cromosorb W 80/100

Flow rates: nitrogen 20, Hydrogen 20, and Air 300 ml per minute.

Temperature: detector 150°C, injection 105°C, Column 55°C.

Carrier Gas: nitrogen

Physicochemical characteristics of both the feeding material and the digested slurry were determined according to the American Standard Methods devised by APHA.

Anayltical reference compounds used in this study were DDT, Lindane, Gesapax, and Gesaprime. The first two compounds are insecticides, while Gesapax and Gesaprime are used as herbicides.

Residues of these compounds were identified and quantified by GLC procedure using Varian 3700 GLC fitted with Ni^{63} electron capture detector and a glass column (4 mm I.D. and 2 m length) packed with 3% OV-17 on 80/100 chromosorb W. The column, injector, and detector temperatures were 200°C, 250°C, 300°C respectively. Nitrogen was used as carrier gas at a flow rate of 40 ml/min.

RESULTS AND DISCUSSION

Effect of Lindane and DDT on Anaerobic Digestion of Mixtures of Cotton Stalks and Cow-Dung

As control, various mixtures of cow-dung and cotton stalks containing cotton stalks at a concentration of 20 to 100 percent were subjected to digestion. Initial solids concentration was around 8 percent. The results obtained showed a reduction in the volume of biogas produced from 434.9 l/kg volatile organic matter (VOM) fed to 221.6 l/kg VOM fed by increasing the concentration of cotton stalks from 20 percent to 100 percent (Table 1). Reduction in volatile organic matter followed the same trend. Pretreatment of cotton stalks by soaking in hot water (60°C) for one hour raised the volume of methane produced by an average value of 24 percent, compared to untreated samples (Figure 1-A). Corresponding increase in the reduction of volatile organic matter was 15.3 percent (Table 1).

To study the effect of organochlorine pesticide on the efficiency of the digestion process, three digesters containing 50 percent pretreated cotton stalks were prepared for this purpose. The initial solids content was 8 percent. The first was used as control. To the second digester, 5 mg/l Lindane and 5 mg/l DDT were added. The third one contained 10 mg/l of each of these compounds.

From the results obtained (Table 1), it can be seen that the addition of 5 mg/l Lindane and 5 mg/l DDt reduced the total volume of biogas produced by 24 percent. An increase in the concentration of each of these compounds, up to 10 mg/l, led to a reduction of 34 percent in the total volume of biogas produced (Figure 2-A). Analysis of residual Lindane and DDT in the digested sludge showed that these compounds are not affected by the digestion process (Table 1).

Table 1 Production of Biogas from mixture of cotton stalks and fresh cow-dung.

% Cotton stalks	20		20*		50		50*		50**		50***		60		60*		100		100*	
	Feed	Dig. slurry	Feed	Dig. slurry	Feed	Dig. slurry	Feed	Dig. slurry	Feed	Dig. slurry	Feed	Dig. slurry	Feed	Dig. slurry	Feed	Dig. slurry	Feed	Dig. slurry	Feed	Dig. slurry
T.R. 105 °C gm/L	79.7	49.7	79.7	64.9	80	45.5	80	41.7	80	47.3	80	49.5	79.8	45	79.8	39	80	43.3	80	37
VOM gm/L	49.9	19.9	49.9	35.1	61.4	26.9	61.4	23.1	61.4	28.7	61.4	30.9	66.6	31.8	66.6	25.8	73.12	36.5	73.12	30.1
% VOM	62.6	40.1	62.6	54.1	76.8	59.1	76.8	55.4	76.8	60.7	76.8	62.4	83.5	70.7	83.5	66.2	91.4	84.1	91.4	81.4
% red of VOM		60.2		69.7		56.1		62.3		54.2		49.6		52.3		61.3		50.1		58.9
Temp. °C	22		22		22		22		22		22		22		22		22		22	
D.T. days		40		40		40		40		40		40		40		40		40		40
Total Biogas L/kg VOM fed		434.9		537.1		317.6		389.3		296.4		254.1		262.8		327.3		221.6		284.5
% Methane		46.3		50.5		45.9		63.1		54.5		53.0		59		64.6		46.3		54.0
Methane L/kg VOM red.		334.4		389.2		259.9		394.3		298.1		271.5		296.4		344.9		204.8		260.8
Gesapax ug/L			1.0				4	2					0.8							
Gesaprime ug/L	-	-	2.0	-	-	-	3	2	-	-	-	-	-	-	-	-	-	-	-	-
Lindane ug/L	40.1	33	58.4	58	38.8	30	62.2	60.8	5000	4800	10000	10000	28	18.2	43	43	60.8	50.2	80.2	80
DDT ug/L	5	5	53.2	50	32.1	32.1	28.6	24	5100	5000	10000	9900	36	34	38.2	30.8	46.8	43	15.8	16

* Treated cotton stalks
** Cotton stalks are treated and contaminated with 5 mg/L Lindane and 5 mg/L DDT
*** Cotton stalks are treated and contaminated with 10 mg/L Lindane and 5 mg/L DDT

Effect of Gesapax and Gesaprime on the Anaerobic Digestion of Mixture of Water Hyacinth and Fresh Cow-dung

As control, anaerobic digestion of fresh cow-dung and water hyacinths mixtures containing water hyacinths at concentrations of 0.0 to 100 percent was executed. The results obtained are given in Table 2. It can be seen that the digestion of fresh cow-dung at 8 percent solids concentration and 22°C produced 532.6 l biogas per kg VOM fed, compared to 326.9 l biogas per kg VOM fed when a mixture of 1:1 cow-dung and water hyacinth was digested under the same operating conditions (Figure 1-B). Also the methane content of the biogas produced dropped from 54.3 percent to 45 percent. Corresponding reduction in the removal of volatile organic matter was around 24 percent.

In an attempt to study the effect of the two herbicides on the digestion process, 5 mg/l Gesapax and 5 mg/l Gesaprime were added to the unit containing a mixture of 1:1 cow-dung and water hyacinth. To another unit, a mixture of 10 mg/l of each was added. From the results obtained (Table 2) it can be seen that the reduction in the concentration of the two herbicides investigated ranged from **30-45%.** However, Gesaprime was removed to a greater extent than Gesapax.

The results also showed that the addition of 5 mg/l of both Gasapax and Gesaprime has an inhibitory effect on the digestion process. A reduction of 12 percent in the volume of biogas produced was recorded. Corresponding reduction at the higher concentration investigated (10 mg/l for each) was 27.5 percent (Figure 2-B).

Effect of Gesapax and Gesaprime on Anaerobic Digestion of A **Mixture of** Weeds and Fresh Dow-Dung

The results obtained in the experiment (Table 3 and Figure 2) have shown that the mixing of fresh cow-dung with 10 percent aquatic weeds decreased the volume of biogas produced by 29 percent. The volume of methane gas produced by the sole digestion of aquatic weeds was 111/kg VOM reduced, compared to 529 l/kg. VOM reduced when fresh cow-dung was digested. Corresponding reduction in organic volatile matter was 65 percent and 39 percent for cow-dung and weeds, respectively.

To determine the fate and effect of higher doses of herbicides, 5 mg/l Gasapax and 5 mg/l Gasaprime were added to the digesters containing a feed of 1:1 (W/W) mixture cow-dung and weeds. The results obtained show 37 percent reduction in the concentration of Gesapax. Corresponding removal of Gasaprime was 40 percent. When the concentration of Gesapax was raised to 10 mg/l, the removal value decreased to 29 percent. On the other hand the removal value of Gasaprime was not affected by raising its concentration up to 10 mg/l (Table 3).

Available data (Figure 2-C) showed that the addition of Gesapax and Gesaprime caused a reduction in the volume of biogas produced. While the control digester produced 250 l of biogas/kg VOM fed, the digester fed with a mixture of 5 mg/l of Gesapax and 5 mg/l Gesaprime produced 180 l of biogas/kg VOM fed. This volume was further decreased to 133 l when the concentration of these herbicides was raised to 10 mg/l.

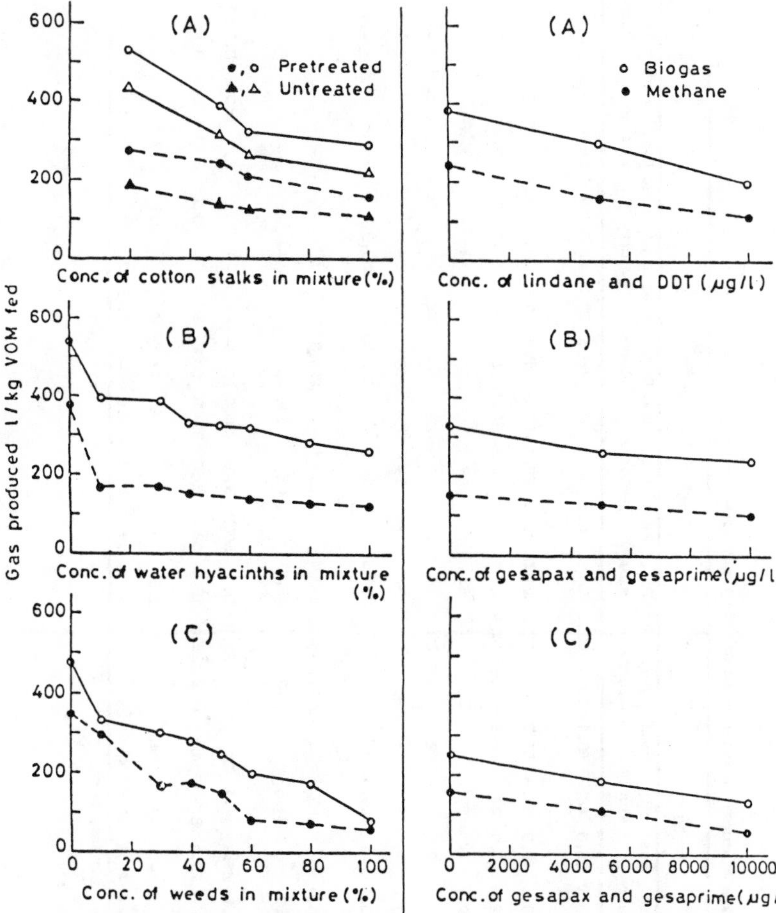

Figure 1 Production of biogas.
(A) Mixtures of cotton stalks and fresh cow-dung.
(B) Mixtures of water hyacinths and fresh cow-dung.
(C) Mixtures of weeds and fresh cow-dung.

Figure 2 Effect of pesticides and herbicides on biogas production.
(A) Lindane and DDT.
(B) Gesapax and gesaprime.
(C) Gesapax and gesaprime.

Table 2 Production of Biogas from mixtures of water hyacinth and fresh cow-dung.

% Water hyacinth	0.0		10		30		40		50		50*		50 **		60		80		100	
	Feed	Dig. slurry	Feed	Dig slurry	Feed	Dig slurry	Feed	Dig. slurry	Feed	Dig. slurry	Feed	Dig. slurry	Feed	Dig. slurry	Feed	Dig. slurry	Feed	Dig. slurry	Feed	Dig. slurry
T.R. 105°C gm/L	78.6	42.6	78.8	48.1	76.6	48.2	80.1	50.6	80.1	52.5	80.1	54.6	30.1	56.6	30.3	58.5	78	56.4	59.6	42.7
VOM gm/L	51.5	15.4	51.2	20.5	49.8	21.4	53.5	24.1	52.0	24.5	52.0	26.5	52.0	28.6	51.9	30.1	54.1	32.5	42.2	26.33
% VOM	65.4	36.2	64.9	42.6	65.0	44.4	66.8	47.6	64.9	46.6	64.9	48.5	64.9	50.5	64.6	51.5	69.4	57.6	70.8	59.3
% red of VOM		70		60		57		55		53		49		45		42		39.9		40
Temp. °C	22		22		22		22		22		22		22		22		22		22	
D.T. days		39		39		39		39		39		39		39		39		39		39
Total Biogas L/kg VOM fed.		532.6		396.4		389.6		336.4		326.9		286.5		236.5		319.9		282.8		262.9
% Methane		413.1		275.5		285.5		277.1		282.4		267.1		221.3		345		296.9		278
Methane L/kg VOM red.		54.3		41.7		41.8		45.3		45.8		45.7		42.1		45.3		42.0		42.3
Gesapax ug/L	16	8.2	206	16.6	18.2	14.1	20	3.2	26.2	16	5000	3000	10000	7220	-	-	8.5	6.1	10.2	8.2
Gesaparime ug/L	12.2	4.6	18.2	9	20.1	8	14.6	11.2	25.4	15.1	5200	2900	10000	6800	2	-	10.2	3	10.8	8.4
Lindane ug/L	2	-	1.2	-	-	-	2.4	-	2.6	2.6	-	-	1.2	1	2	2	3.1	2.8	5	4.2
DDT ug/L	40.8	40	30.2	28	42.6	42.6	30	28	35	34	38	30.2	42.6	42	25.8	22.1	28	28	50	50

* Contaminated with 5 mg/L (Gesapax and Gesaprime)

** Contaminated with 10 mg/L (Gesaprime).

Table 3 Production of Biogas from mixtures of (weeds and fresh cow-dung)

% weeds	0.0		10		30		40		50		50*		50**		60		90		100	
	Feed	Dig. slurry	Feed	Dig. slurry	Feed	Dig. slurry	Feed	Dig. slurry	Feed	Dig. slurry	Feed	Dig. slurry	Feed	Dig. slurry	Feed	Dig. slurry	Feed	Dig. slurry	Feed	Dig. slurry
T.R. 105°C gm/L	80.2	46	79.9	49.9	77.8	49.8	78.6	52	80	56.4	81	56.5	80.1	56.2	79.8	53.6	79.6	52.2	80.1	51.6
VOM gm/L	52.5	18.3	53.5	23.5	53.7	25.7	54.2	27.6	56.0	32.4	60.9	36.4	62.6	38.9	63.9	37.7	65.4	38.5	72.6	44.1
% VOM	65.6	39.8	67	47.1	69	51.6	69	53.1	70	57.4	75.2	64.4	78.1	68.9	80.1	70.3	82.2	73.1	90.6	85.5
% red.of VOM		66.2		56.1		52.3		49		42.1		40.3		38.2		70.3		41.2		39.2
Temp °C	22		22		22		22		22		22		22		22		22		22	
D.T. days		40		40		40		40		40		40		40		40		40		40
Total Biogas L/kg VOM fed		478.1		338.3		301.7		282.2		250		180.6		132.6		198.7		174.3		84.0
% Methane		72.2		67.8		53.7		59.6		59.3		60		47.4		426		45.1		51.9
Methane L/kg VOM red.		529.4		408.9		309.8		343.3		352.3		268.9		154.7		206.6		190.3		111.2
Gesapax ug/L	4.2	2	-	-	1.2	0.8	-	-	6.8	2	5100	3200	10000	7200	-	-	-	-	-	-
Gesaprime ug/L	1.2	-	-	-	0.6	-	1.4	-	-	-	5000	3000	10200	6000	-	-	-	-	-	-
Lindane ug/L	6.1	5	2.6	2.5	-	-	1.8	1	2.2	2.4	3	3	2	2.2	1.4	1.4	0.8	0.8	4	4
DDT ug/L	50.2	50	52.1	48.5	38.7	36	40	40	60.8	60.9	44.2	46.0	28.2	30	36.2	30.4	49.2	42.0	46.0	51.0

* Contaminated with 5 mg/L (Gesapax and Gesaprime)

** Contaminated with 10 mg/L (Gesapax and Gesaprime)

CONCLUSIONS

From the results obtained it may be concluded that agricultural wastes should be subjected to pretreatment before digestion in order to achieve the maximum of biogas production possible. Although these experiments revealed soaking in water for an hour as a pretreatment, other pretreatments, such as acid and alkaline hydroysis of finely divided agriculture residues are practical. Composting may produce materials that might yield higher gas yields when subjected to anaerobic digestion.

REFERENCES

Aly, O.A., M.I. Badawy. 1984. Organochlorine residues in fish from the river Nile, Egypt. Bull. Environ. Contam. Toxicology. 33: 246-252.

Barth, E.F., W.A. Moore, and G.N. McDermott. 1965. Interaction of Heavy Metals on Biological Sewage Treatment Processes. U.S. Department of Health Education and Welfare, May.

Masselli, J.W., N.W. Masselli and G. Burfard. 1961. The Occurence of Copper in Water, Sewage, and Sludge and its Effect on Sludge Digestion. New England Interstate Water Pollution Control Commission, June. Boston, MA

Mahr, I. 1969. Role of lower fatty acids in anaerobic digestion of sewage sludge and their biocenology. Water Res. 3: 507-517.

McCarty, P.L., I.J. Kugelman and A.W. Lawrence. 1964. Ion Effects in Anaerobic Digestion. Tech. Rept. No. 33, Department of Civil Engineering, Stanford University. Stanford, CA

Nikitin, G.A. 1967. Changes in oxidation reduction potential and intensity of methane fermentation during fermentation of different media by a batch culture of methane forming bacteria. Prikl. Biokhim, Mikrobiol. 3: 36 Engl. transl., Appl. biochem. Microbiol (1968) 3,1.

Smith, R.J. 1973. The Anaerobic Digestion of Livestock Wastes and the Prospects for Methane Production. Paper presented at the Midwest Livestock Waste Management Conference, 27-28 November 1973. Iowa State University. Ames, Iowa.

Standard Methods for the Examination of Water and Waste Water. 1975. 14 th ed. APHA, AWWA, and WPCF.

BIOGAS PRODUCTION FROM SOME ORGANIC WASTES

H. Gamal-El-Din, A. El-Bassel, and M. El-Badry
Faculty of Agriculture, Cairo University, Fayum, Egypt

ABSTRACT

Common materials used for biogas generation are animal manures, crop residues, and agro-industrial wastes. In Egypt, most of these materials are used as animal feed or as raw materials for industy. Accordingly, a study was made to investigate biogas production from some other organic wastes. The investigated wastes were geranium flour, akalona, and watermelon residues. Each was digested individually or in different mixtures with cow dung.

The results obtained concerning biogas productivity showed that none of these wastes nor their mixtures with cow dung are superior to cow dung digested separately. However, biogas productivity was found to increase by increasing the percentage of cow dung in the mixture. Explanations for these results and the recommendations to improve the digestion process are presented.

INTRODUCTION

In view of the energy crisis and environmental pollution, biogas technology has recently attracted worldwide attention.

A list of some of the urgent research tasks on biogas was suggested by Sathianathan (1975). This list includes the search for new materials for biogas production. In the present study, a laboratory investigation was undertaken to assess the feasibility of using some organic wastes that were not before investigated as substrates for biogas production. The wastes examined include geranium flour, akalona, and watermelon residues. In Egypt these materials are presently considered useless by-products of geranium oil extraction, the wheat milling process, and watermelon-seed production respectively. Each of the investigated wastes was digested individually or with cow dung at different mixing rations. The cow dung was also digested individually and served as a reference substrate.

MATERIALS AND METHODS

Organic Wastes

Cow Dung Fresh, undiluted dairy cow dung, Brown Swiss (Breed), was collected from the animal Production Farm of Fayum Faculty of Agriculture. The animals were fed a ration that was composed of approximately 50 percent rice straw and 50 percent of: 65 percent cotton seed meal, 20 percent rice bran, nine percent bran, three percent molasses, two percent lime stone, and one percent NaCl. No antibiotics or other additives were incorporated in the animal ration.

Geranium Flour An air dried sample from the residue remaining from geranium plants, *Pelargonium graveolens* Ait, after extraction of the essential oil, was obtained from a geranium oil extraction plant located in Fayum Governerate.

Akalona Akalona is the outer portion (epidermis) of the wheat grains which results from the scouring of the wheat grains during the milling process. A sample of "Akalona" was obtained from the Fayum Mill.

Watermelon Residues *Citrullus Vulgaris* This material was prepared in the laboratory from a watermelon fruit. After separating the seeds, residues of the watermelon fruit (i.e., rind, pulp and juice) were homogenized in a blender prior to use.

Starter

The starter for anaerobic digestion was prepared from the effluent of an actively operating laboratory digester fed with cow dung. The effluent was passed through four layers of cheesecloth to remove the undergraded materials. The starter was added to all digesters at the rate of 10 percent by volume.

Digestion Apparatus

Batch-fed laboratory digesters were constructed as described by Gamal-El-Din (1984). The digester setup consists of a 1.25 liter brown bottle, a gas-measuring cylinder equipped with a leveling bulb, gas sampling port, and a sidearm through which liquids could be withdrawn or added.

Analytical Procedures

Gas Volume The biogas produced was collected and measured by liquid displacement in a calibrated gas cylinder filled with Orsat confining solution: 20 percent sodium sulphate and five percent sulphuric acid in water (Lingle and Hermann, 1975).

Methane Content of the Biogas This was determined by bubbling a known volume of biogas through 20 percent (W/V) potassium hydroxide solution. The loss in volume of biogas was taken as equal to the carbon dioxide. The balance was assumed to be methane (Winter and Cooney, 1980).

Determinations of Total Solids (TS), ash, volatile solids (VS), total volatile acids (TVA), pH, alkalinity, and total nitrogen (TN) were made according to the procedures in "Standard Methods" devised by APHA (1980). Organic carbon (OC) was determined by the rapid titration method of Walkaly and Black (Black et al., 1965).

EXPERIMENTAL

The digesters were fed with cow dung, geranium flour, akalona, watermelon residues, or a mixture of a particular waste and cow dung. The proportions of the cow dung in the mixtures, based on the concentration of the total solids in the feed (five percent TS), were 20 percent, 50 percent, and 80 percent.

The required quantities from each waste or from the waste and the cow dung, to give a total solids concentration of five percent, were mixed with 100 ml of the starter and sufficient amount of tap water to give one liter total volume. Each treatment was replicated three times and a set of digesters was prepared for chemical analysis. The digesters was incubated at 35°C . Mixing of the digester contents was done manually at the time of gas volume measurement.

The volumes of the biogas produced were recorded daily and biogas samples were analyzed periodically. All gas volumes were corrected to 0°C and one atmospheric pressure (STP). The experiment lasted 30 days. Samples from the digester contents were analyzed for TS, VS at the beginning and at the end of the experiment. Determination of pH, TVA, alkalinity was carried out every 10 days.

RESULTS AND DISCUSSION

Feasibility of biogas production from geranium flour (GF), akalona (AK), watermelon residues (WR), and from mixtures of each waste with cow dung (CD) in different proportions was investigated and compared with biogas production from cow dung, which served as a reference substrate.

Prior to the anaerobic digestion, samples from the investigated wastes were analyzed for TS, ash, OC, TN, and pH. The VS and the C/N ratio of each waste were calculated. The data obtained are summarized in Table 1.

Table 1 Characteristics of the Investigated Wastes

Waste	TS %	VS %	VS/TS%	OC %	TN %	C/N	pH in 5% TS slurry
Cow dung	17.63	13.65	77.42	44.01	1.37	32.10	7.10
Geranium flour	92.41	74.32	80.42	40.61	1.35	30.08	4.80
Akalona	91.85	88.32	96.20	51.59	1.03	50.09	6.70
Watermelon residues	6.88	6.12	88.95	49.42	0.36	137.28	5.50

Biogas Production from Geranium Flour (GF)

Geranium oil is one of the major essential oils for export in Egypt. The geranium plant was first grown in Egypt in 1930. Since then, its area has progressively increased to about 14,000 feddans. The geranium cultivated area at Fayum Governerate is about 7,000 feddans.

At present, geranium flour, the residue remaining from geranium plant leaves after extraction of the essential oil, is either left where it is produced or may be used as a source of energy by direct burning.

The results obtained in the present study showed that the geranium flour and its mixtures with cow dung, gave lower biogas volumes than those produced from cow dung. The 100 percent "GF" digesters produced the lowest biogas volumes with low methane content (Figure 1). This result could be explained by the low initial pH value (5.0) of the "GF" digester contents (Table 2). McCarty (1964) reported that methane production proceeds quite well as long as the pH is maintained between 6.6 and 7.6, with an optimum range between 7.0 and 7.2. At pH values below 6.2, acute toxicity occurs.

The analysis of the effluent after the 30 day digestion period (data not presented) indicated that the approximate percentage VS

Table 2 Changes in pH, Alkalinity and Total Volatile Acids (TVA) Concentration During the anaerobic Digestion of Geranium Flour, Cow Dung and Different Mixtures of the Two Wastes at 35°C .

Feed: 5% T S			Digestion Period, Days			
proportion of geranium flour %	cow dung %	Parameter	0	10	20	30
100	00	pH	5.00	4.95	4.85	4.75
		Alkalinity[1]	108.90	376.20	584.10	648.45
		TVA[2]	540.69	597.60	668.74	878.61
80	20	pH	5.75	5.95	5.25	4.95
		Alkalinity	227.70	485.10	940.50	1336.50
		TVA	490.89	1245.00	1814.14	2603.03
50	50	pH	6.30	5.45	5.20	5.25
		Alkalinity	534.60	1485.00	1702.80	1742.40
		TVA	433.97	1173.06	1252.12	2372.62
20	80	pH	6.70	6.65	6.35	6.35
		Alkalinity	772.20	881.10	782.10	1569.15
		TVA	661.63	1010.23	981.77	1216.54
00	100	pH	6.95	6.45	6.55	6.35
		Alkalinity	702.90	861.30	1217.70	1489.95
		TVA	682.97	718.54	704.31	739.89

[1] as mg $CaCo_3$/1. [2] as mg acetic acid/1.

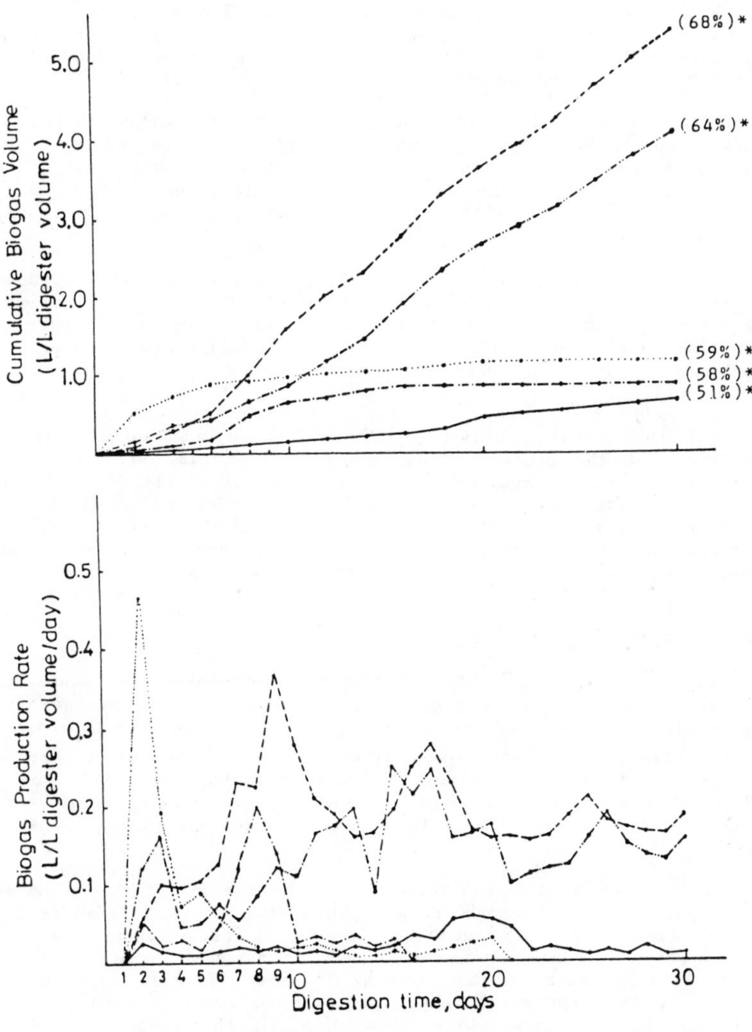

Figure 1 Cumulative biogas volume and biogas production rate (STP) from batch digestion of geranium flour, cow dung and mixtures of the two wastes at 35°C. (geranium flour 100% ——, 80% —·—, 50% ·······, 20% —··—, cow dung 100% ----).

* Average CH_4 percentage.

destruction in the case of geranium flour was 3.9 percent as compared to 36.1 percent for the cow dung. This build-up of solids in the "GF" digesters without a significant increase in TVA concentration suggests that "GF" is resistant to attack by the bacterial population.

In addition to the effect of the low pH, the geranium flour may contain an antimicrobial substance(s). Drabkin and Dumova (1962) studied the phytocidal substances of pelargonium and found that the sap of fresh, crushed leaves and that of autoclaved leaves has an anti-microbial activity which was particularly high in the case of the autoclaved leaves.

The results also showed that the biogas production rate dropped to practically zero in the case of the 80 percent GF and the 50 percent GF digesters before the end of the experimental period (Figure 1). According to Kroeker et al., (1979), digester failure appeared to occur approximately at pH 6.5 when the concentration of TVA was 1650 mg/L as acetic acid.

From the results obtained, it may be concluded that under the conditions of the present study, geranium flour is not a convenient substrate for biogas production, since it is not easily degradable and should be pretreated to be more suitable for bacterial attack. In addition, adjustment of the pH of the waste slurry is needed. However, it seems doubtful whether the costs of pretreatment of the waste and/or chemical addition can be offset by the increase in gas production.

Biogas Production from Akalona (AK)

Akalona represents about 0.5 percent of weight of the wheat grains. At present it is considered as useless residue because it is not usable as animal feed, nor is it profitable for further processing.

The results of the anaerobic digestion of akalona showed that "AK" digesters produced lower biogas volumes than those produced from the "CD" digesters (Figure 2). However, such a result was expected because of the relatively high C/N ratio of the former waste (about 50) (Table 1).

Accumulation of TVA was observed in the akalona digester contents (3282 mg/L as acetic acid/30 days). This caused a drop in pH from 6.6 to 4.6 (Table 3). The combined effect of pH depression and TVA concentration increase may include toxic conditions. It seems that the low alkalinity found in the akalona digester (584 mg/L as $CaCo_3$), at the end of the experiment cannot protect the system. McCarty (1964) indicates that a bicarbonate alkalinity in the range of 2500 to 5000mg/L as $CaCo_3$ provides a safe buffering capacity for anaerobic treatment of wastes.

The performance of the digesters fed with different mixtures of "AK" and "CD" was highly variable, and this makes it impossible to draw general conclusions. However, at about the 20th day of digestion, the cumulative volumes and the methane content of the biogas produced from both the "20 percent AK" and "CD" digesters were approximately equal (Figure 2). This means that akalona, without any pretreatment, can replace about 20 percent of the cow dung total solids for biogas production.

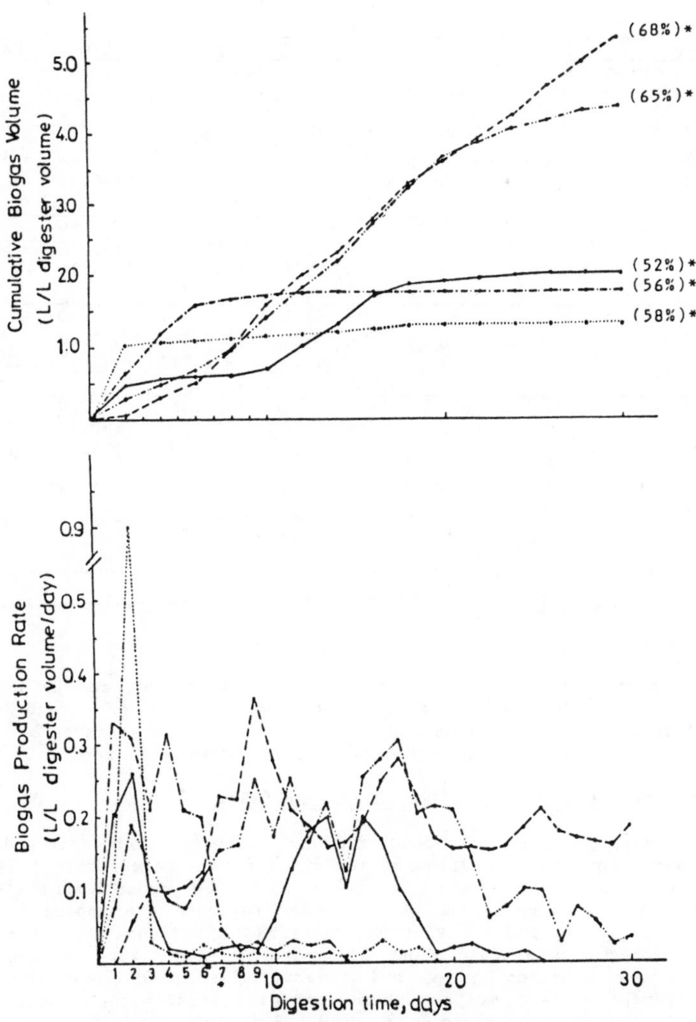

Figure 2 - Cumulative biogas volume and biogas production rate (STP) from batch digestion of akalona, cow dung, and a mixtures of the two wastes at 35°C.
(akalona 100% ——, 80% —·—, 50% ······, 20% —··—, cow dung 100% - - -)
* Average CH_4 percentage.

Table 3 Changes in pH, Alkalinity and Total Volatile Acids (TVA) Concentration During the anaerobic Digestion of Akalona, Cow Dung and Different Mixtures of the Two Wastes at 35°C .

Feed: 5% T S			Digestion Period, Days			
proportion of akalona flour %	cow dung %	Parameter	0	10	20	30
100	00	pH	6.65	4.40	4.60	4.60
		Alkalinity[1]	227.70	326.70	524.70	584.10
		TVA[2]	505.11	1785.69	2276.57	3283.24
80	20	pH	6.65	4.50	4.60	4.60
		Alkalinity	297.00	851.40	445.50	534.60
		TVA	498.00	1316.14	1814.14	2105.83
50	50	pH	6.70	5.05	4.80	5.05
		Alkalinity	514.80	1009.80	871.20	935.55
		TVA	604.71	988.89	1052.91	1241.44
20	80	pH	6.80	6.45	6.20	6.20
		Alkalinity	789.10	762.30	643.50	876.15
		TVA	576.26	811.03	924.86	1070.70
00	100	pH	6.95	6.45	6.55	6.35
		Alkalinity	702.90	861.30	1217.70	1489.95
		TVA	682.97	718.54	704.31	739.89

[1] as mg $CaCo_3$/l. [2] as mg acetic acid/l.

In conclusion, the biogas production from akalona can be improved by nitrogen and/or alkali addition to overcome low pH caused by the rapid formation of the volatile acids. This can be achieved by mixing the akalona with a nitrogen-rich waste such as poultry excreta.

Biogas Production from Watermelon Residue (WR)

In Egypt, more than 3000 feddans of watermelon per year are cultivated for the production of seeds. The watermelons from this area produce about 15,000 tons of juice as a by-product (Khalafallah, 1971). In addition to the juice, other residues are produced, i.e., rind, pulp, and fibers. Khattak et al. (1965) found that charliston watermelon fruits contain about 33.6 percent juice, 43.4 percent rind, and 24 percent seeds, pulp, and fibers. They also found the total carbohydrate content of the watermelon juice (reducing and nonreducing sugars) ranged from 6.26 to 7.28 gm/100 ml juice.

Presently, the watermelon residues are considered a useless by-product of the seed-production industry. Because the watermelon juice is rich in sugars, it can represent a serious pollution problem. Some proposals have been made to recover useful products from the watermelon juice (e.g., Khalafallah, 1971). However, none of the systems so far suggested have yielded a satisfactory result.

Regarding biogas production from watermelon residues, the data obtained in the present study showed that the "WR" digesters produced low biogas volumes of a relatively low methane content; and generally the biogas production and methane content increased by increasing the percentage of cow dung in the mixture (Figure 3). Such a result is not surprising since the C/N ratio (about 137) and the pH value (about 5.5) of the watermelon residues (Table 1) are not favorable for anaerobic digestion. The high C/N ratio and the low pH caused an increase in TVA concentration and a decrease in pH. Both affected the activity of the methane-forming bacteria. Moreover, the marked drop in the pH value of the "WR" digester contents (Table 4) may inhibit the acid-producing bacteria. It was found that the optimum pH for the separate acidogenesis of soluble carbohydrate containing wastewaters is in the range of 5.7-6.0 (Zoetemeyer et al., 1982).

In conclusion, the watermelon residues should be pretreated before they can be successfully treated with the anaerobic digestion process. Pretreatment could include pH adjustment to 7.0 and nutrient addition, particularly nitrogen. Reducing the loading rate may also improve the biogas production. Also two-phase anaerobic digestion could be suitable for treating these residues. Because this type of waste would cause serious pollution, additional work is required to identify the suitable pretreatment.

Table 4 **Changes in pH, Alkalinity and Total Volatile Acids (TVA) Concentration During the Anaerobic Digestion of Watermelon Residues, Cow Dung and Different Mixtures of the Two Wastes at 35°C .**

Feed: 5% T S proportion of water melon residues %	cow dung %	Parameter	Digestion Period, Days 0	10	20	30
100	00	pH	5.60	3.50	3.70	3.60
		Alkalinity[1]	485.10	nd	nd	nd
		TVA[2]	505.11	616.63	988.89	1003.11
80	20	pH	6.00	3.60	3.70	3.65
		Alkalinity	428.70	nd	nd	nd
		TVA	626.06	889.29	1060.02	1113.39
50	50	pH	6.35	4.85	4.85	4.85
		Alkalinity	476.20	564.30	970.20	1074.15
		TVA	640.29	1458.43	2803.03	3834.60
20	80	pH	6.60	5.35	5.20	5.10
		Alkalinity	465.30	1009.80	1366.20	1465.20
		TVA	455.31	832.37	2397.52	2596.72
00	100	pH	6.95	6.45	6.55	6.35
		Alkalinity	702.90	861.30	1217.70	1489.95
		TVA	682.97	718.54	704.31	739.89

[1] as mg $CaCO_3$/l. [2] as mg acetic acid/l.

nd = not determined

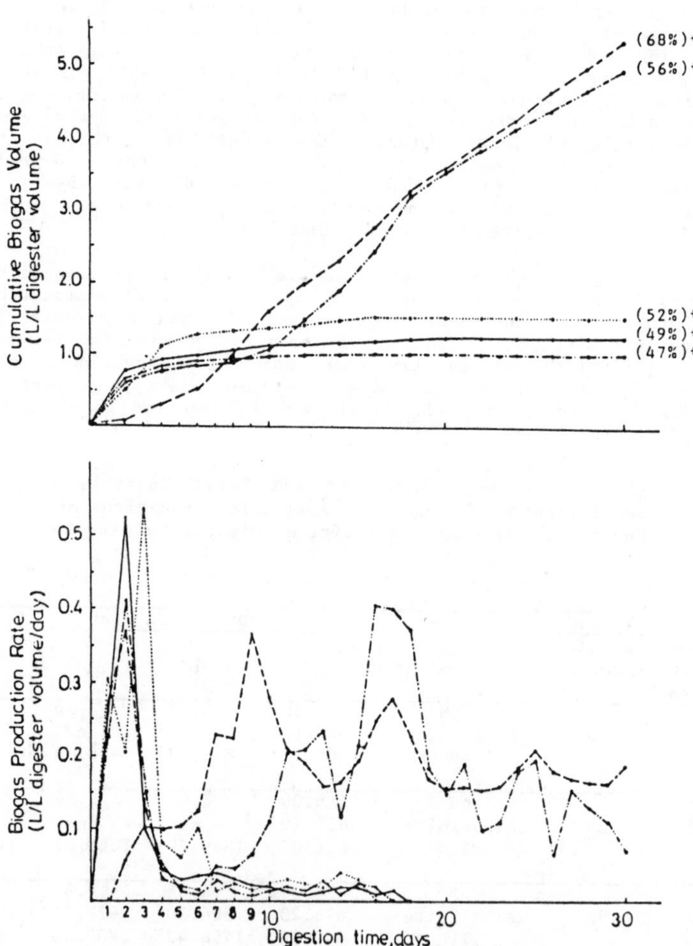

Figure 3 - Cumulative biogas volume and biogas production rate (STP) from batch digestion of water melon residues, cow dung and mixtures of the two wastes at 35°C. (water melon residues 100%——, 80%—·—, 50%·······, 20%—··—, cow dung 100%- - - -)

* Average CH_4 percentage.

REFERENCES

American Public Health Association. 1980. Standard Method for the Examination of Water and Wastewaters. Pub. by APHA-AWWA-WPCF, 15th Edition, Washington, DC.

Black, C.A., D.D. Evans, D.E. Ensminger, J.L. White, F.A. Clark, and R.C. Dirauer. 1965. Methods of Soil Analysis. II. Chemical and Microbiological Properties. Amr. Soc. Agron. Inc., Madison, Wisconsin, USA.

Drabkin, B.S. and A.M. Dumova. 1962. Phytocidal substances of pelargonium. Nauchn. Dokl. Vysshei Shkaly, Bio. Navki. 2:155-159. C.F. Chem. Abs. 57(7), 8907c.

Gamal-El-Din, H. 1984. Biogas from Organic Waste Diluted with Seawater. Paper presented at the International Conference on "State of the Art on Biogas Technology, Transfer and Diffusion." Cairo, November 17th-24th, 1984.

Khalafallah, A.M. 1971. Microbial and Chemical Studies on Watermelon Juice. M.Sc. Thesis. Fac. Agric., Cairo University.

Khattak, J.N., M.K. Hamdy, and J.J. Powers. 1965. Utilization of watermelon juice. I. Alcoholic fermentation. Food Technol., 19:102-104.

Kroeker, E.J., D.D. Schulte, A.B. Sparling, and H.M. Lapp. 1979. Anaerobic treatment process stability. J. Water pollut. Control. Fed., 51:718-727.

Lingle, J.W. and E.R. Hermann. 1975. Mercury in anaerobic sludge digestion. J. Water pollut. Control Fed., 47:466-471.

McCarty, P.L. 1964. Anaerobic waste treatment fundamentals, II. environmental requirements and control. Public Work, 95:123-126.

Sathianathan, M.A. 1975. Biogas: Achievements and Challenges. Association of Voluntary Agencies for Rural Development. New Delhi.

Winter, J.U. and C.L. Cooney. 1980. Fermentation of cellulose and fatty acids with enrichments from sewage sludge. Eur. J. Appl. Microbiol. Biotechnol., 11:60-66.

Zoetemeyer, R.J., J.C. Van den Heuvel, and A. Cohen. 1982. pH influence on acidogenic dissimilation of glucose in an anaerobic digestor. Water Res., 16:303-311.

THE ASSESSMENT OF CELLULOLYTIC ACTIVITIES IN ANAEROBIC DIGESTERS BY THE "TEXTILES COUPON TECHNIQUE"

Mohiy Eldin Abdel-Samie and Mohammed Nabil Mahmoud
Microbial Chemistry Unit, National Research Centre, Egypt

ABSTRACT

A method was developed for the quantitative determination of total cellulolytic activities of anaerobic microflora found in anaerobic digesters. It depends on the determination of loss in weight of a 100 percent cotton textile coupon after direct anaerobic incubation with the sample. The anaerobic Hungate tube was used for this purpose. Triplicates showed a coefficient of variation ranging between six percent and nine percent.

INTRODUCTION

In an actively methanogenic digestion process having a cellulosic material as a substrate, the anaerobic cellulose hydrolysis is the rate-determining step. In anaerobic digestion of cellulosic materials, what concerns us is the assessment of actual cellulolytic activity of a whole collection of cellulose decomposers under the prevailing conditions in the digester, and not the total viable count of the different cellulose decomposers.

Anaerobic, nonsporeforming cellulose decomposers are strict anaerobes for which oxygen is not only inhibitory but severely toxic. Because of their requirement for stringent anaerobic conditions, starting from media preparation before sterilization, and going through all the microbial manipulation techniques, Hungate developed his well known technique (Hungate, 1950). Accordingly, microbial counting of these strict anaerobes will be cumbersome, especially when sample dilutions have to be done in either the roll-tube viable count method or the most probable number method.

In the present work, the samples taken to be representing the actual digester conditions were used directly without any dilution. A pre-weighed coupon of 100 percent cotton textile is dipped into the sample. The tubes were closed under oxygen-free gassing and then incubated for a suitable time. Afterwards, the coupon was removed, washed, oven dried, and weighed. The loss in weight will be taken as a function of cellulolytic activity of the respective sample.

EXPERIMENTAL

Two types of tubes were used for the sake of availability and convenience. The first type was the screw-capped tube, filled with the samples, leaving no air entrapped below the cap. The other type was the serum tube which can receive a rubber stopper and an aluminum cap which can be crimped after filling the tube. In the latter, the tube can be half-filled, provided deoxygenation is performed according to the Hungate technique (Hungate, 1969).

The Use of the Screw-Capped Tubes

Samples (25ml) were taken to fill totally screw-capped tubes (1.8 x 10 cm). The cap's interior contained a black rubber disc for hermetic sealing. A cellulose textile coupon (100-200 mg, 3x5cm) was dipped in each sample. A wrought-iron nail was put in each tube, as it was found that this addition could shorten the lag phase. The tubes were closed tightly and incubated for not more than 10 days. Enough replicates (six tubes for each determination) were made in order to be able to inspect one tube each two to three days, and the rest of the replicates were removed when a considerable portion of the textile coupon was degraded; then each coupon was withdrawn, using a forceps, and placed in a petri dish. A mild stream of tap water was allowed to fall on each petri dish (a manifold was used) for ca. 15 minutes in order to remove all the adhering materials. The petri dishes were drained and then put (with the coupons) in an oven at 105°C . After thorough drying (to constant weight), the weight loss was calculated as a percent based on the original weight of each coupon.

The Crimped-Capped-Serum Tubes

In order to keep the textile coupon stretched and expose the largest surface area possible to the bulk of the sample, the coupon was wrapped-stretched around a short glass rod (ca. 5 cm long) and was fixed at each end with a nylon thread. Serum tubes were partially filled (half-filled) with the sample. The wrapped coupon around the glass rod was dipped into the sample. All these manipulations were done using the Hungate technique (Hungate, 1969). The tubes were stoppered with flanged black-rubber stoppers. Aluminum caps were fitted and crimped according to Balch and Wolfe (1976).

An experiment was done to test the validity of the determinations of cellulolytic activities in actual runs of the semi-continuous operation of digesters. Two digesters, one liter capacity each, were assembled from three-necked, round-bottom flasks as well as the necessary accessories needed for daily feeding and removal of samples and gas collection (Figure 1). Each was charged with diluted cow dung (1:1 with water) to which a 10 percent inoculum from an active digester was added. Each one received 700 ml. Removal of 50 ml sample and addition of the same volume of diluted cow dung was performed daily. This gave a hydraulic retention time of 14 days. Duplicate determinations of cellulolytic activities were made for each sample taken. Screw-capped tubes were used for these determinations.

Another experiment was carried out to test for the variabilities between replicates and the effect of the incubation time on the

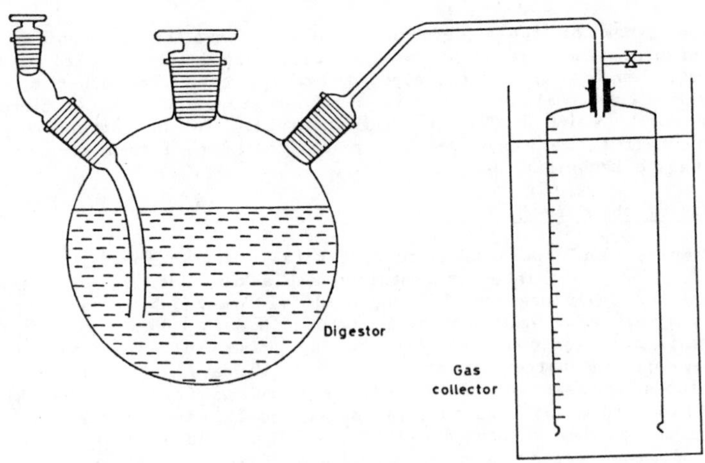

Figure 1 Experimental Assembly

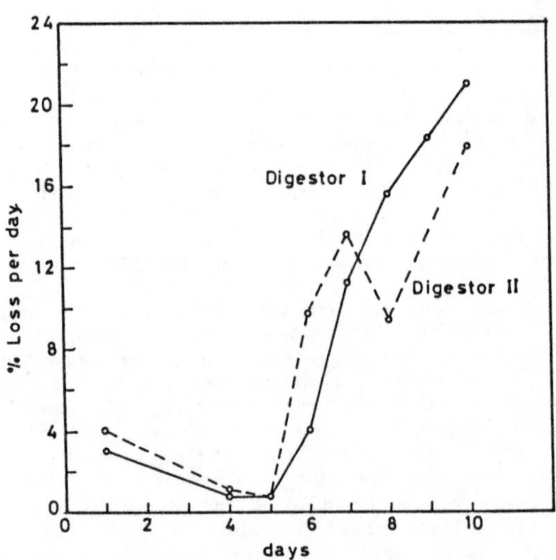

Figure 2 Cellulolytic Activity Assessment in two Semi-Continuous Digesters

consistencies of the estimated cellulolytic activities. The serum tubes and the stretched coupon on the glass rods were used for this experiment in order to satisfy the requirements known for maximal activity determinations.

RESULTS AND DISCUSSION

The results are reported in some detail to show the variabilities between the analytical duplicates and between the two digesters (Tables 1.a., 1.b., and Figure 2). Except for the data of the eighth day, there is a close concordance between the two digesters. However, sometimes the differences between the duplicates were quite considerable and on one occasion the coupon was completely digested (eighth day of digester I). These differences might be due to the inadequate anaerobic conditions which depended on "filling to the top." In addition, daily mixing was not feasible without incorporating

Table 1a. Weight Loss of Cellulose Textile Coupons After Different Incubation Periods. Digester I.

	Weight of Coupon						
Days	before incubation mg	after incubation mg	Weight loss mg	%	Mean loss %	Incubation period in days	Mean loss per day in %
1	156.8	108.6	48.2	30.7	25.5	8	3.2
	168.2	134.2	34.0	20.2			
4	183.4	175.4	8.0	4.4	4.6	6	0.8
	147.8	140.6	7.2	4.9			
5	146.6	140.9	5.7	3.9	4.8	6	0.8
	126.4	119.2	7.2	5.7			
6	172.8	130.7	42.1	24.4	24.7	6	4.1
	109.5	82.2	27.3	24.9			
7	155.1	48.4	106.7	68.8	67.4	6	11.3
	113.0	38.4	74.6	66.0			
8	151.5	20.0	131.5	86.8	93.4	6	15.6
	136.3	00.0	00.0	100.0			
10	163.2	60.0	103.2	63.2	83.4	4	20.9
	128.3	31.4	97.3	75.8			

Table 1b. Weight Loss of Cellulose Textile Coupons after Different Incubation Periods. Digester II.

Days	Weight of Coupon before incubation mg	Weight of Coupon after incubation mg	Weight loss mg	Weight loss %	Mean loss %	Incubation period in days	Mean loss per day in %
1	161.2	126.6	34.6	21.5	32.5	8	4.1
	158.6	89.5	69.1	43.6			
4	159.7	148.7	11.0	6.9	6.3	6	1.1
	145.4	137.2	8.2	5.6			
5	163.0	157.0	6.0	3.7	4.3	6	0.7
	130.0	123.7	6.3	4.8			
6	101.3	52.1	49.2	48.6	58.9	6	9.8
	140.4	43.2	97.2	69.2			
7	97.4	13.8	83.6	85.8	82.0	6	13.7
	141.8	31.0	110.8	78.1			
8	148.9	63.9	85.0	57.1	58.0	6	9.5
	163.1	67.1	69.0	58.9			
10	104.4	34.8	69.6	66.7	71.5	4	17.9
	144.9	34.3	110.6	76.3			

a small solid body in the tube. Also, the textile coupons might have been crimped in some tubes and stretched in others, resulting in unequal exposure to the flora of the sample.

These pitfalls had been remedied in the second experiment where the serum tubes were filled under adequate anaerobic conditions (the Hungate technique) and the coupons were stretched around glass rods which helped to make mixing more adequate by slow reciprocation of the tubes. The data in Table 2 show that the percent coefficient of variation between triplicates ranged between six percent and nine percent, which is quite acceptable. The data show, also that the

Table 2. Quantification of Cellulose Activities by the Textile - Coupon and Glass Rod Technique

Time (Days)	Before Incubation	After Incubation	Loss mg	Loss %	Average ± S.D.	% Loss Per Day
	243.1	229.8	13.3	5.5	6.3±0.5	2.2
3	203.2	190.0	13.2	6.5		
	181.9	169.2	12.7	7.5	(8%)*	
	228.2	201.5	26.7	11.7	13.1±1.2	2.2
6	219.6	191.3	28.3	12.9		
	193.7	165.5	28.2	14.6	(9%)*	
	230.8	193.6	37.2	16.1	17.1±1.2	1.7
10	229.7	192.5	37.2	16.2		
	202.1	163.8	38.3	19.0	(6%)*	
	194.8	144.8	50.0	25.7		
15	201.1	151.5	49.6	24.7	26.5±1.9	1.8
	195.6	138.6	57.0	29.1	(7%)*	

S.D. = Standard Deviation
( )* Coefficient of Variation

calculated daily percent loss ranged from 1.7 percent, for 10 and 15 days incubation to 2.2 percent for three and six day incubation. This might point to the preference for an incubation period of one week.

Work is in progress for the perfection of this technique.

REFERENCES

Balch, W.E. and R.C. Wolfe, 1976. Apple. Environ. Microbiol. 29:540.

Hungate, R.E. 1950. Bacteriol. Rev. 14:1.

Hungate, R.E. 1969. In J.R. Norris and D.W. Ribbons (eds.) Methods in Microbiology 3B, p. 117. Academic Press, New York.

BIOGAS PRODUCTION FROM ANTIBIOTIC-CONTAMINATED COW MANURE

H. Gamal-El-Din
Faculty of Agriculture, Cairo University
Fayum, Egypt

ABSTRACT

Batch laboratory digesters were utilized to study the biogas production from digestion of cow manure contaminated with procaine penicillin (PP), ampicillin (AMP), tetracycline hydrochloride (TC.H), oxytetracycline (OTC), oxytetracycline hydrochloride (OTC.H), and chloramphenicol (CAM). Doses of the antibiotics added to the cow manure were based on the average recommended therapeutic dose. The digesters were fed with cow manure slurry of 5 percent TS and incubated at 35°C for 30 days.

The results showed that all the tested antibiotics, over the range of concentrations used, have an inhibitory effect when expressed as a reduction in the biogas volume produced. However, the degree of inhibition varied according to type of antibiotic, its concentration and the period of digestion.

Results of the effect of a constant dose (12.5 mg antibiotic per liter feed slurry) indicated that, in terms of the reduction in biogas volume produced within 30 days, OTC, AMP, and OTC.H were the most effective antibiotics; PP and CAM were the least effective, while TC.H had no effect. However, the tested antibiotics varied in their effect on the methane content of the biogas. Therefore, in terms of reduction in the volume of methane produced within 30 days, the tested antibiotics could be ranked in a decreasing order as follows: AMP, OTC, OTC.H, PP, CAM, and TC.H. The corresponding reduction percentages were 33, 32, 25, 14, 11, and 10 respectively.

INTRODUCTION

Anaerobic digestion is a complex process, and its efficiency depends on bacterial population, type of feed, and environmental conditions. If environmental or feeding conditions are changed suddenly, or if a toxic material is introduced into the digester, the process may become unstable and the digester may operate at reduced efficiency. The most common toxic materials which could reach the digester are heavy metals, antibiotics, phenols, chlorine compounds, detergents, and sulphur compounds (Stafford, et al., 1980).

Antibiotics have been used routinely for more than 25 years in animal feeds at subtherapeutic levels to promote growth, to increase efficiency of feed utilization, and to prevent diseases; and at therapeutic levels to treat diseases (Langlois et al., 1978). However, a fraction of the ingested antibiotic can be excreted in an active form. For example, Wise et al. (1976) have reported that quantitative bioassays of fresh feedlot manure revealed that when the diet of the cattle was supplemented with chlorotetracycline, approximately 75 percent of the antibiotic was excreted. It has been suggested that excreted antibiotics in feedlot waste may affect waste biodegradation (Clark, 1965; Morrison et al., 1969). Hobson and Show (1976) reported that certain antibiotics used to treat diseases are effective in killing digester bacteria, but little experimental evidence has been obtained. Bryant (1979) also reported that data were lacking on the possible effect of feeding drugs, such as monensin and antibiotics, on methane production from wastes.

This paper describes a study of the effect of antibiotics on the performance of anaerobic digesters and biogas production. In the present study, however, no attempt has been made to assess the ultimate toxicity levels of antibiotics on the anaerobic digestion process.

MATERIALS AND METHODS

Animal Manure: antibiotic-free cow manure was collected from the livestock farm of the faculty of Agriculture at Fayum.

Inoculum: this was prepared from effluent obtained from an active, operating laboratory digester being fed with cow manure.

Antibiotics: six antibiotics were tested in this study, procaine penicillin "PP", Tetracycline hydrochloride "TC.H", Chloramphenicol "CAM" (El-Nasr Pharmaceutical Chemicals Co., Egypt), Ampicillin "AMP" (Vetamplius, Fatro Spa, Italy) Oxytetracycline "OTC" (Terromycin-100, Pfizer) and oxytetracycline hydrochloride "OTC.H" (PAN Terramycin, Pfizer).

The antibiotic concentrations chosen for this study were based on the assumption that 10, 30, or 60% of the administered therapeutic dose could be excreted unchanged in the animal manure. In order to compare the effect of the selected antibiotics at a constant concentration, concentrations close to "12.5 mg/l feed slurry" were rounded up. The tested antibiotic concentrations in milligrams per liter feed slurry were: PP (3.0, 12.5, 18.0; 1 mg=1000 i.u.), AMP (12.5, 37.5, 75.0) TC.H (3.7, 12.5, 125, 22.5), OTS (12.5, 37.5, 75.0), OTC.H (12.5, 33.7, 67.5) and CAM (2.5, 7.5, 12.5).

Digestion Apparatus: the laboratory batch digesters utilized in these experiments were as described by Gamal-El-Din (1984).

Analytical Methods: determinations of total solids (TS), volatile solids (VS), total Kjeldahl nitrogen (TKN), organic carbon (OC), total volatile acids (TVA), and pH were performed according to procedures outlined in Standard Methods (APHA, 1980).

The volume of biogas produced was measured by liquid displacement of a saturated solution of sodium sulphate containing five percent by volume of sulphuric acid.

The carbon dioxide content was determined by bringing a volume of biogas into contact with a saturated solution of potassium hydroxide. The balance was assumed to be methane.

EXPERIMENTAL PROCEDURE

The required quantities of cow manure to give a total solids concentration of five percent were placed in the digestion bottles and mixed with 100 ml of inoculum, and sufficient tap water was added to give a total volume of 950 ml. The digester's contents were then mixed and the pH measured (it was found to be in the range of 7.0-7.1).

To assure uniformity, all the digesters were operated, at 35°C, for one week before the addition of antibiotics. The digesters which gave comparable gas production rates were selected for the experiments.

The antibiotics were prepared as solutions in distilled water. An amount of each antibiotic, sufficient to bring the concentration of the antibiotic in the feed slurry to the desired level, was injected into digester after being made up to 50 ml volume with distilled water. The control digesters received an equivalent amount of distilled water. Hence, the total volume of feed slurry was one liter in each digester. Each treatment was run in duplicate, and the digesters were incubated at 35°C for 30 days.

Biogas volumes were recorded daily and the gas assayed for its methane content at five day intervals. All gas volumes were corrected to 0°C and one atmosphere pressure (STP). At the end of the 30-day experimental period, samples of the digester contents were analyzed to determine VS, pH and TVA concentration.

RESULTS

Prior to the experiments it was found that the manure contained 19 percent total solids. On a dry weight basis it contained 66.3 percent volatile solids, 1.64 percent total Kjeldahl nitrogen, and 37 percent organic carbon.

The process parameters chosen for this study were daily gas production, gas composition, volatile solids destruction, pH, and total volatile acid concentration. From these data, the gas production was calculated on a volumetric (l gas/l digester/day) and total volatile solid fed (l gas/g VS added) bases.

The results on biogas production are summarized in Table 1, and the percentage reduction in biogas volumes resulting from the addition of antibiotics are presented in Table 2. It is apparent from the results that, with the exception of CAM at a 2.5 mg/l conc., all the tested antibiotics, over the range of concentrations used, have an inhibitory effect when expressed as a reduction in the biogas volume produced. However, the degree of inhibition varied according to type of antibiotic, its concentration, and the period of digestion. In some cases the effect was small and temporary (e.g. PP at a 3 mg/l conc.) while in other cases (e.g. OTC.H at a 67.5 mg/l conc.), the effect was severe and continuous. On the other hand, complete inhibition of biogas production did not occur with any antibiotic tested even at the highest concentration.

The results (Table 1) also showed that when the antibiotic concentration in the feed slurry was increased, there was generally a greater decrease in biogas production. However, the effect of AMP at the three concentrations tested was approximately the same during the first few days of digestion.

Irrespective of concentration, the tested antibiotics varied in their inhibitory effect with respect to the time of digestion (Table 2). Some antibiotics (AMP, TC.H, PP and CAM) caused a marked reduction in gas production at the beginning of digestion, but the effect decrease thereafter. In contrast, the inhibitory effect caused by OTC and OTC.H was found to increase during the experiment.

As regards the composition of the biogas (Table 1), no noticeable differences were observed in the average percentage of methane in the biogas produced from control, OTC, OTC.H and CAM treatments. On the other hand, a relatively lower methane percentage, compared with control, was found in the biogas produced from AMP, PP and TC.H treatments.

The VS destruction rate, concentration of TVA and pH were determined in the digested slurry at the end of the 30-day experimental period, and the results are given in Table 3. The results indicate that VS destruction was depressed by the presence of antibiotics. Furthermore, the VS destruction percentage tended to decrease as the antibiotic concentration increased.

The data in Table 3 also show that addition of antibiotics, in general, caused a slight increase in TVA concentration compared with control. However, in some cases the TVA concentration was moderately high (1300-1500 mg/l as acetic acid), and in one case only (OTC.H at a 67.5 mg/l conc.), the TVA concentration was above 2000 mg/l as acetic acid. However, these results (Table 3) indicate that TVA concentration seems to be independent of antibiotic concentration. There was a slight decrease in the pH of the digesters' content (Table 3). This decrease was generally less than one pH unit. The lowest pH value (6.20) was that of the digested slurry with OTC.H treatment at a concentration of 67.5 mg/l, which was related to an increase in TVA concentration up to 2130 mg/l as acetic acid.

In order to facilitate a comparison between all antibiotics under investigation, data on the effect of the constant concentration (12.5 mg antibiotic per liter feed slurry) are compared in Table 4. In terms

Table 1 Effect of Antibiotics on Biogas Production from Batch Digesters Fed with Cow Manure Slurry of Five Percent TS Incubated at 35°C for 30 Days. (Antibiotic concentrations in mg/l feed slurry and all gas volumes corrected to STP)

Parameter	Control	PP*			AMP			TC.H			OTC			OTC.H			CAM		
		3.0	12.5	18.0	12.5	37.5	75.0	3.7	12.5	22.5	12.5	37.5	75.0	12.5	33.7	67.5	2.5	7.5	12.5
Cumulative biogas volume (l/l digester/30 days)	5.3	5.3	5.1	4.7	3.8	3.7	3.3	5.3	5.3	4.3	3.6	3.2	2.7	4.0	2.4	2.0	5.3	5.0	4.7
Biogas production rate (average, l/l digester/day)	0.18	0.18	0.17	0.16	0.13	0.12	0.11	0.18	0.18	0.14	0.12	0.11	0.09	0.13	0.08	0.07	0.18	0.17	0.16
Biogas yield (average, l/g VS added)	0.16	0.16	0.15	0.14	0.12	0.11	0.10	0.16	0.16	0.13	0.11	0.10	0.08	0.12	0.07	0.06	0.16	0.15	0.14
Methane content (average, %)	61	62	55	54	56	54	55	61	55	56	61	61	61	60	61	60	62	60	61

Table 2 Effect of Antibiotics on Biogas Production, expressed as percentage reduction based upon control value, at Different Periods of Digestion in Batch Digesters Fed Cow Manure Slurry of Five Percent TS at 35°C. (Antibiotic concentrations in mg per liter feed slurry).

Digestion Period, Days	PP*			AMP			TC.H			OTC			OTC.H			CAM		
	3.0	12.5	18.0	12.5	37.5	75.0	3.7	12.5	22.5	12.5	37.5	75.0	12.5	33.7	67.5	2.5	7.5	12.5
5	15	23	25	77	75	76	15	25	33	17	31	41	12	37	52	0	6	27
10	6	7	20	70	71	76	11	21	32	25	33	44	19	43	62	0	6	17
15	0	3	13	46	61	61	11	12	32	30	33	41	21	45	62	0	6	13
20	0	4	13	34	47	56	6	13	31	33	32	42	25	50	60	0	8	13
25	0	4	11	31	31	46	5	9	25	31	36	46	25	52	60	0	6	12
30	0	4	11	28	30	39	0	0	19	32	40	49	24	55	63	0	5	12

*1 mg=1000 i.u.

Table 3 Effect of Antibiotics on the Characteristics of Digested Slurry from Batch Digesters Fed with Cow Manure at Five Percent TS Incubated at 35°C for 30 Days. (Antibiotic concentrations in mg per liter)

Parameter	Control	PP*			AMP			TC.H			OTC			OTC.H			CAM		
		3.0	12.5	18.0	12.5	37.5	75.0	3.7	12.5	22.5	12.5	37.5	75.0	12.5	33.7	67.5	2.5	7.5	12.5
VS destruction %	19.7	20.5	16.7	13.4	12.1	11.4	8.4	20.1	18.9	11.7	12.6	10.5	8.4	14.2	10.5	6.7	18.5	17.2	15.8
TVA mg/l as acetic acid	568	446	549	1394	1025	1083	1029	1286	1063	722	617	800	960	756	1543	2130	826	983	754
pH	6.70	6.75	6.60	6.50	6.55	6.55	6.45	6.75	6.60	6.65	6.55	6.65	6.60	6.55	6.40	6.20	6.70	6.75	6.75

Table 4 Effect of Antibiotics on Performance of Batch Anaeroic Digesters Fed with Cow Manure Slurry of Five Percent TS, Incubated at 35°C for 30 Days, All Gas Volumes at STP.

Antibiotic 12.5 mg/l	Biogas productivity				Methane productivity			Analysis of digested Slurry (30 days)		
	1/1/day	1/g VS added	Total vol. % 1/30 days	Reduction %	Content %	Total vol. 1/30 days	Reduction %	VS des. %	TVA mg/1 as acetic	pH
PP*	0.17	0.15	5.1	4	55	2.8	14	16.7	549	6.60
AMP	0.13	0.12	3.8	28	56	2.1	33	12.1	1025	6.55
TC.H	0.18	0.16	5.3	0	55	2.9	10	18.9	1063	6.60
OTC	0.12	0.11	3.6	32	61	2.2	32	12.6	617	6.55
OTC.H	0.13	0.12	4.0	24	60	2.4	25	14.2	756	6.55
CAM	0.16	0.14	4.7	12	60	2.9	11	15.8	754	6.75
Control	0.18	0.16	5.3	--	61	3.2	--	19.7	568	6.70

*1 mg=1000 i.u.

of the reduction in total biogas volume produced within 30 days, the data indicate that OTC, AMP and OTC.H were the most effective inhibitory antibiotics; PP and CAM were the least effective, while TC.H had no effect. However, since the tested antibiotics varied in their effect on the methane content in the produced biogas, they could also be ranked, on the basis of their inhibitory effect on methane production, in a decreasing order, as follows: AMP, OTC, OTC.H, PP, CAM and TC.H. The corresponding percentage reductions in the methane volume produced within 30 days were 33, 32, 25, 14, 11 and 10 respectively.

DISCUSSION

The uncontrolled use of antibiotics in order to prevent diseases in grouped farm animals could be a possible hazard in operating farm-waste digesters (Hobson et al. 1981). However, the results obtained in this study have demonstrated that not all antibiotics have the same inhibitory effects on anaerobic digestion process. This was considered to reflect the differing mode of action and the antimicrobial spectrum of the tested antibiotics. The higher inhibitory effect of AMP, OTC, and OTC.H was expected since they are broad spectrum antibiotics. However, TC.H and CAM which are also broad spectrum antibiotics, had a relatively small effect. There is no obvious explanation for the low inhibition of TC.H. In the case of CAM, the low inhibition could be partially due to presence of acetic acid in the digester contents, since acetic acid assists destruction of the antibiotic by catalytic action (Brander and Pugh, 1977). PP is a narrow spectrum antibiotic, and this may explain its low effect upon biogas production.

The inhibition of biogas production resulting from the addition of antibiotics could be due to inhibition of one or more of the bacterial groups involved in the anaerobic digestion process. The data obtained in the present work could suggest, in general, that the inhibitory effect of antibiotics on biogas production is most likely due to the inhibition of the microorganisms involved in the destruction of volatile solids.

It should be recognized that in the present work, a manure artificially contaminated with antibiotics was used to feed the digesters. However, if other work would be undertaken using a manure excreted from antibiotic treated animals, the results could be somewhat different. This is because antibiotics will affect the chemical, physical, and microbiological properties of the animal manure. For example, Patten and Wolf (1979) found that feces from cattle receiving OTC contained a higher concentration of potassium, more fungi, and fewer volatile solids than did wastes from control animals.

From a practical point of view, the results of the present study, however, suggest that when animals are given antibiotics therapeutically, their manure should not be used for anaerobic digestion. This is particularly important when antibiotics such AMP, OTC, or OTC.H are used. In addition to their high inhibitory effect, these antibiotics could be excreted in an active form in large amounts. For example, the percentage excreted unchanged in urine from animals given AMP or OTC was previously found to be 90 or 70 respectively (Brander and Pugh, 1977).

REFERENCES

American Public Health Association. 1980. Standard Methods for the Examination of Water and Wastewater (15th ed.). APHA. Washington, D.C.

Brander, G.C. and D.M. Pugh. 1977. Veterinary Applied Pharmacology and Therapeutics. (3rd ed.). ELBS and Bailliere Tindall. London.

Bryant, M.P. 1979. Microbial Methane production: Theoretical aspects. J. Animal Sci. 48:193-201.

Clark, C.E. 1965. Hog waste disposal by lagooning. Amer. Soc. Civil Eng. 91:27-30.

Gamal-El-Din, H. 1984. Biogas from organic waste diluted with sea-water. Proc. International Conf. "State of the Art on Biogas Technology, Transfer and Diffusion". Cairo, Egypt, November 17-24, 1984.

Hobson, P.N. and B.G. Show. 1976. Inhibition of methane production by Methanobacterium formicicum. Water Res. 10:849-852.

Hobson, P.N., S. Bousfield, and R. Summers. 1981. Methane Production from Agricultural and Domestic Wastes. Applied Science Publishers, Ltd. London.

Langlois, B.E., G.L. Cromwell and V.W. Hays. 1978. Influence of chlortetracycline in swine feed on reproductive performance and on incidence and persistence of antibiotic resistant enteric bacteria. J. Animal Sci. 46:1369-1382.

Morrison, S.M., D.W. Grant, M.P. Nevins, and G.K. Elmund. 1969. Role of Excreted Antibiotic in Modifying Microbial Decomposition in Feedlot Waste. Proc. Cornell Univ. Agr. Waste Manag. Conf. Syracuse, N.Y.

Patten, D.K. and D.C. Wolf. 1979. Effect of Antibiotics in Beef Feces on Nitrogen and Carbon Mineralization in Soil. Agronomy Abstracts, 1979. Ann. Meeting. Colorado University.

Stafford, D.A., D.L. Howkes, and R. Horton. 1980. Methane Production from Waste Organic Matter. CRC Press, Inc. Boca Raton, Florida.

Wise, D.L., R.L. Wentworth, and R.G. Kispert. 1976. Fuel Gas Production From Selected Biomass Via Anaerobic Fermentation. Presented at the U.S.-Japan Joint Cooperative Seminar Entitled "Biological Solar Energy Conversion." Miami, Florida, November 15-18, 1976.

BIOGAS FROM LIQUID AGROINDUSTRIAL WASTES DERIVED FROM BANANA AND COFFEE PROCESSING

J. F. Calzada, C. Rolz, and M. C. Arriola
Applied Research Division
Central American Research Institute for Industry (ICAITI)
Guatemala

ABSTRACT

Current practices in the industrial utilization of biomass yield large amounts of liquid discharges with high loads of volatile solids. Several high-rate anaerobic digestion technologies have been used to generate biogas from some of these effluents, especially in the industrialized countries.

Laboratory and pilot plant experiences have demonstrated the wider application of these technologies, when properly adapted to local conditions. Results from case studies in Central America will be discussed. They include the utilization of coffee and banana wastes, using the upflow anaerobic sludge bed (UASB) and sponge reactors.

Pretreatment of the materials is usually required in most highly concentrated effluents. A preliminary acidification has been suggested, either as part of the previous process (retting, for example) or as a first phase of anaerobic digestion (two separate reactors). The final discharges from the digesters may also require further purification before being dumped into ravines or water bodies. Physical and chemical methods, sometimes associated with photosynthetic microbial communities, have been assayed for the coffee and banana residues.

INTRODUCTION

Bananas and coffee constitute primary sources of income for most of the countries around the Caribbean Sea. In Tables 1 and 2, the annual production of both commodities in Central America is presented.

The coffee beans of the trade are prepared in the region through the "wet" method (wet beneficio). Fresh fruit berries are squeezed and the pulp is separated and discarded. A fermentation in open tanks follows and the beans from which the pectin mucilage has thus been removed are washed and dried. The last cover, the parchment, is only removed prior to shipment. This traditional method yields at least two major discharges: pulp and wastewaters from the washing.

Table 1 Production of Green Coffee in Central America (10^3t)

Country	1980	1981	1982
Costa Rica	109	120	104
El Salvador	157	165	161
Guatemala	152	163	173
Honduras	47	64	75
Nicaragua	49	59	61

Source: FAO (1983)

Table 2 Production of Bananas in Central America (10^3t)

Country	1980	1981	1982
Costa Rica	1092	1144	1150
Guatemala	650	650	655
Honduras	1401	1425	1240
Nicaragua	150	157	157

Source: FAO (1983)

Alternative methods have been introduced to utilize coffee processing byproducts. The flow diagram for one of them is presented in Figure 1. It can be seen that pressing is a key operation in order to obtain a solid residue for further utilization as feed, fuel, or in the preparation of compost. However, the liquid fraction (approximately 40% of the weight of fresh pulp) is still an environmental problem, given the richness of the effluent in sugars, pectins, and other soluble organic and inorganic compounds.

Bananas are separated from the false stem and usually only the fruit and fruit trunk are removed from the plantation, due to agricultural practices. The rotten stems and leaves are used for soil conservation. The fruit trunk does not have an application at the present. It is discarded into local landfills. It has been observed that a natural retting process takes place where these trunks are discarded into swampy areas.

The study of retting of banana fruit trunks has been undertaken by the Central American Research Institute for Industry (ICAITI) under a grant from the German Office for Foreign Technical Aid (GTZ). As it can be seen from Figure 2, two major liquid discharges result from the process: fresh juice and retting wastewaters. Both should be considered as substrates for methanogenesis.

In this paper, results are presented for the utilization of coffee pulp juice, coffee washing waters, and fresh juice from banana fruit trunks for the production of methane.

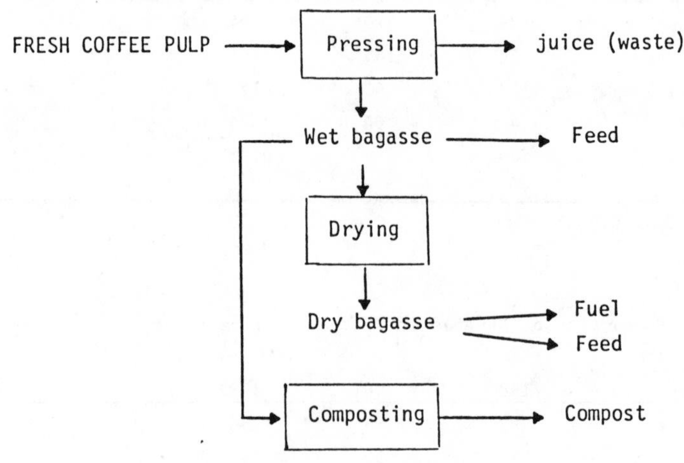

FIGURE 1: UTILIZATION OF COFFEE PULP

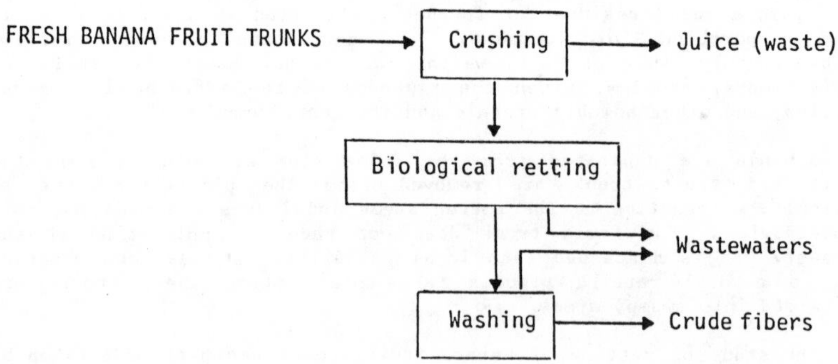

FIGURE 2: UTILIZATION OF BANANA FRUIT TRUNKS

The final discharges from the digesters may also require further purification before being dumped into ravines or water bodies. This goal can be achieved through the use of photosynthesis systems.

RESULTS

In Table 3, it is possible to compare the polluting potential of four different liquid discharges: fresh and acidified pulp juice, concentrated wastewaters from the washing of fermented coffee beans, and juice from the fresh banana fruit trunk (Calzada, 1983; Calzada, et al., 1984).

Table 3 Banana and Coffee Liquid Wastes

Parameter	Coffee Pulp Juice	Coffee Wastewaters	Acidified Coffee Pulp Juice	Banana Fruit Trunk Juice
% Total Solids	3.0- 8.2	2.2- 2.6	3.0- 4.8	2.3- 2.6
% Volatile Solids	2.0- 5.3	1.6- 1.8	1.9- 3.2	0.6- 1.0
Chemical Oxygen Demand, g/l	40.5-55.8	9.4-15.0	-	10.5-14.3
% Nitrogen	0.6- 0.07	0.11	0.06-0.07	0.02-0.04
pH	3.5- 4.5	4.5- 5.0	3.0- 4.3	5.3 -6.1

Three types of high-rate (microbial retention) digesters were used to treat the effluents. The up-flow anaerobic sludge bed reactor (UASB) developed by Lettinga and co-workers (1980) and a reactor packed with unreticulated polyurethane foam (Calzada, et al., 1984b) were used for a second reactor of a two-phase anaerobic system handling coffee pulp juice as substrate. Small laboratory reactors into which the liquid flowed and then passed through reticulated polyurethane foam (0.42) were used for the one-phase digestion of banana fruit trunk juice. The results of the assays with the three digestors appear in Tables 4, 5, and 6.

Table 4 Methanogenic Conversion of Acidified Coffee Pulp Juice Using Unreticulated Foam Reactors

HRT (days)	Load (kg VS/m^3d)	Gas Production (m^3/m^3d)	% CH_4	pH
3	8.5	2.12	81	7.4
2	15.2	2.66	82	7.2
1	16.4	1.94	-	7.3

Table 5 Methanogenic Conversion of Acidified Coffee Pulp Juice Using a UASB Digestor

HRT (days)	Load (kg VS/m^3d)	Gas Production (m/m^3d)	% CH_4	pH
5	8.8	2.18	92	7.6
4	11.0	2.98	95	7.2
3	14.7	1.87	67	7.2
2	22.0	4.58	77	7.2

Table 6 Methanogenic Conversion of Fresh Juice from Banana Fruit Trunk in Anaerobic Digester with Reticulated Polyurethane Foam

HRT (days)	Load (kg VS/m^3d)	Gas Production (n^3/m^3d)	% CH_4	pH
	3.5	1.91	37	7.8
3	3.5	0.96	36.6	7.6

Wastewaters from the washing of coffee beans vary in concentration depending upon the processing practices and the presence of a sedimentation unit as primary treatment. The raw material enters as a one-stage anaerobic process through an up-flow chamber. Results are shown in Table 7.

Table 7 Methanogenic Conversion of Coffee Washing Waters Using An Up-Flow Digester

HRT (days)	Load (kg VS/m^3d)	Gas Production (m^3/m^3d)	% CH_4	pH
5	3.6	0.65	82	7.3
10	1.8	0.60	60	7.0
20	0.9	0.45	70	7.2

Finally, the polishing of the effluent from the methanogenic reactor through two different photosynthetic systems is presented in Table 8.

Table 8 - Polishing of Liquid Effluents from a Methanogenic Reactor Using Acidified Coffee Pulp

System	Time (days)	pH	COD (mg/l)
Water hyacinth pond	0	7.6	813
(*Eicchornia crassipes*)	21	8.1	494
	40	7.7	16
Algae pond	0	6.8	1540
(*Anabaena cylindrica*)	20	8.4	506
	45	9.6	425

Research activities continue at ICAITI, looking for improved efficiencies in both methanogenic and photosynthetic processes for wastewater treatment. However, the impact of the anaerobic methanogenic reaction is evident for any selected treatment.

REFERENCES

Calzada, J. F. 1983. Biogas de subproductos del beneficio humedo de café. Memorias Segundo Simposium Panamericano de Combustibles y Productos Quimicos via Fermentacion. ICAITI-MIRCEN Biotecnologia, Mexico.

Calzada, J. F., M. C. Arriola, H. O. Castaneda, J. E. Godoy, and C. Rolz. 1984b. Methane from coffee pulp juice: experiments using polyurethane foam reactors. Biotechnol Lett 6 (6):385-388.

Calzada, J. F., E. Porres, A. Yurrita, M. C. Arriola, F. Micheo, C. Rolz, and J. F. Menchu. 1984a. Biogas production from coffee pulp juice: one and two-phase systems. Agric Wastes 9 (3):217-230.

Lettinga, G., A. F. M. van Velsen, S. W. Hobma, W. de Zeeww, and A. Klapwijk. 1980. Use of the upflow sludge blanket reacter concept for biological wastewater treatment especially for anaerobic treatment. Biotechnol Bioeng 22 4:669.

FAO. 1983. Monthly Bulletin of Statistics 6 (9).

A SIMPLE, RAPID AND ACCURATE METHOD FOR DETERMINATION OF CARBON DIOXIDE IN BIOGAS

A. Ellegard and H. Egnéus
Bioquest HB
Skârgardsgatan 4, S-424 58 Gôteborg, Sweden

BACKGROUND

CO_2 determination in biogas is a very important and sensitive parameter for monitoring the digestion process.

During the start-up of a digester, the increase and subsequent stabilization in CO_2 content indicates that the digestion process is progressing.

During routine operation, monitoring of the CO_2 content gives a background value for the operation of the specific digester.

When the biogas production decreases, an increase in CO_2 content (from the background value) is a strong indicator that the decrease is due to indigestion, rather than to other causes such as gas leakage.

CO_2 concentration can be measured in a number of ways, e.g. by gas chromatography. However, these methods often imply that gas samples or heavy and expensive equipment have to be transported long distances.

We have developed a simple and rapid method that has proved to be reliable and easy to employ at any place where measurement is needed.

THEORY

CO_2 is easily absorbed in basic solutions of high concentration. When the pH is high the reaction:

$$CO_2 + 2OH^- \longrightarrow CO_3^{2-} + H_2O$$

goes to completion.

If impurities of H_2S are present, they will follow the same route, and thus be measured as CO_2:

$$H_2S + 2OH^- \longrightarrow S^{2-} + 2H_2O$$

The level of H_2S is usually so low that it does not affect the determination appreciably.

MATERIAL

A graduated 5 ml glass tube (Figure 1) as used for fermentation experiments (Saccharometer according to Dr. Einhorn) is used as the measurement device. It is also possible to use a bent glass tube, sealed on one end.

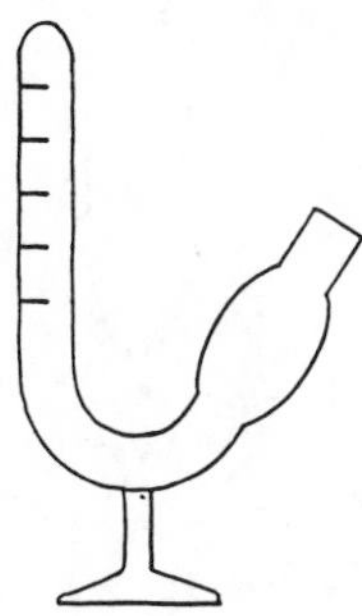

Figure 1 Saccharometer according to Dr. Einhorn.

A standard 5ml medical syringe with a long (5cm) needle is used for withdrawing the samples (Figure 2).

A highly concentrated solution of sodium or potassium hydroxide, , is the reaction medium. To make sure the concentration is high enough, it is easiest to use a saturated solution.
CAUTION: THIS SOLUTION IS CAUSTIC. (Figure 2)

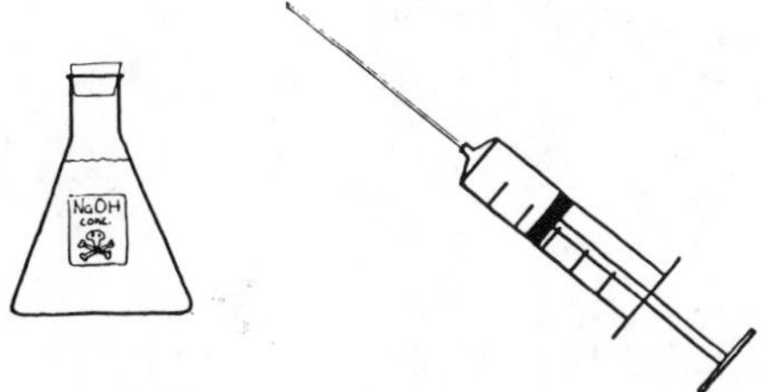

Figure 2 Solution of Na or K hydroxide, 5ml disposable syringe.

PROCEDURE

The gas sample is withdrawn from an open gas valve (pipe, tube). Make sure gas is actually streaming from the opening while sampling. (Figure 3)

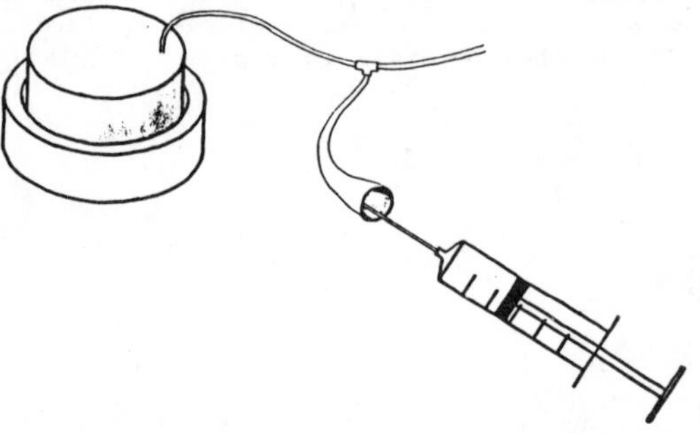

Figure 3 Take gas sample in streaming gas.

Fill the syringe with biogas and empty it 3-5 times to make sure that only biogas will be sampled. When the sample to be measured is withdrawn, the syringe should be overfilled (6ml) and the excess gas should then be slowly expelled. This is to make sure that there is no vacuum in the syringe.

The syringe is then quickly placed in the measuring tube and the gas is slowly injected into the hydroxide solution. (Figure 4)

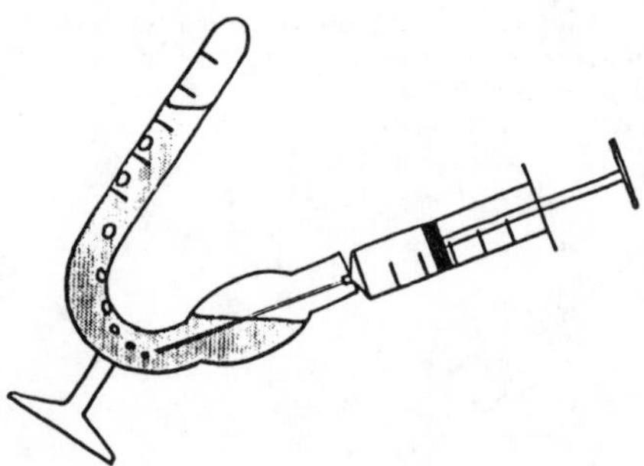

Figure 4 Inject gas sample slowly into tube.

Wait for one minute and then read at the meniscus or as calibrated. The reading represents the volume of biogas minus CO_2 and H_2S, usually pure CH_4. (Figure 5).

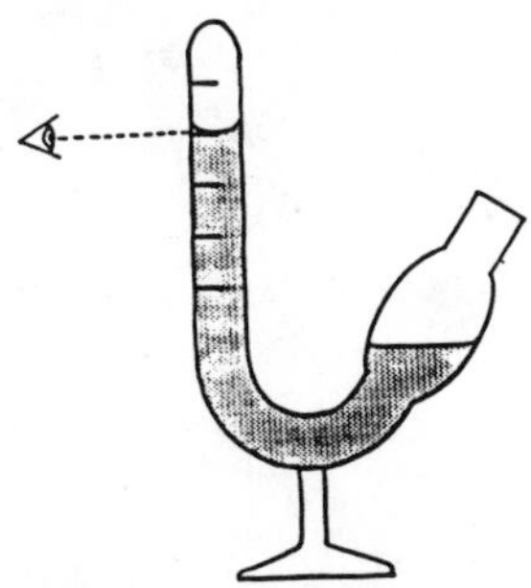

Figure 5 Read at meniscus or as calibrated.

ESTIMATION OF METHANE CONTENT

Using the method for estimating the CH_4 content is also feasible, since in most applications most of the remaining gas is methane when CO_2 and H_2S have been removed. However, before using the method for this purpose, the methane content should be determined at least once by gas chromatography during the routine operation of the digester to determine the standard composition of the biogas. Lacking such faciliites, at least a flame test should be undertaken. During the start up and when the digester gives a bad performance, the method should not be used as for estimating the methane content of the biogas.

VALIDATION

The method as described here has been used by Metangruppen in field observations and laboratory studies for six years. Its accuracy in determining the CO_2 content is $\pm 3\%$ according to gas chromatographic analysis on several occasions.

The method is also used for routine analysis by AC Biotecnics (ANAMET, BIOMET) in all laboratory investigations. Its accuracy in CH_4 determination was determined in a test series and found to be $\pm 2\%$ in that application (Bjorn, Frostell, pers. comm.)

We believe that the method can be very valuable in monitoring rural digesters in developing countries, either as a part of a survey or as a means of operation routines.

USE OF DIGESTED MATERIALS

ASSESSMENT OF ANAEROBICALLY DIGESTED SLURRY AS A FERTILIZER AND SOIL CONDITIONER

H. Moawad, L.I. Zohdy, S.M.S. Badr El-Din, M.A. Khalafallah, and H.K. Abdel-Maksoud
Soil Microbiology Laboratory, National Research Centre, Cairo, Egypt

ABSTRACT

The work reported here constitutes a major part of the endeavors undertaken by the Fertilizer Evaluation Group of the National Research Centre Biogas Project.* It aimed to assess the fertilizer value of the digested materials as well as to develop appropriate handling methods and land use techniques.

The effluents evaluated were produced from a variety of organic substrates in both laboratory and field digesters with cattle dung as the main constituent. Assessment techniques embodied laboratory, greenhouse, and field tests using both sandy and clay soils. Proper physical, chemical, and bacteriological examinations were employed. Treatments included blank runs, farmyard manures, and chemical fertilizers in order to compare them with the digested effluent.

Key results may be summarized as follows:

1. The fertilizer value of the digested effluents obtained from the anaerobic digestion of cow dung was assessed. All the effluents collected either from laboratory digesters operated at 22°C and 37°C or from village digesters contained higher concentrations of available nitrogen (N) and phosphorus (P) than the feed material or traditional farmyard manure. The ammoniacal nitrogen amounted to 40% of the total nitrogen of the slurry, while the available phosphorus concentration reached 25% of the total P content.

2. The direct use of the digested effluents induced retardation of corn and wheat seed germination and plant growth. The fractionation of the effluents to supernatant and sludge clearly indicated that the phytotoxic substances were found as soluble compounds in the supernatant only. This phytotoxic activity of the effluents could easily be removed by their storage under natural field conditions for 5 to 10 days prior to their use as a fertilizer.

*A multidisciplinary project entitled "Development and Application of Biogas Technology for Rural Areas of Egypt," carried out under the auspices of the Academy of Scientific Research and Technology and the U.S. Agency for International Development.

3. Several handling and storage techniques were tested. Although the air drying of the effluents increased organic matter and mineral content of the fertilizer, it resulted in high loss of nitrogen, amounting to 29 percent of its original value. The absorption of the effluent on silt decreased the nitrogen loss to 2 percent, but the product was low in organic matter and nutrients. Postcomposting of the effluent with farm wastes resulted in increasing the total nitrogen content as well as the quantity of the manure.

4. Application of the digester effluents to sandy soils improved the soils' physical and chemical properties and stimulated biological activity, as indicated by the increase in numbers of those microbial groups contributing to the soils' fertility.

5. Field evaluation showed that the effluent was superior as a fertilizer to farmyard manure. Application at 5 tons/acre to wheat gave almost the same yield as 20 tons of farmyard manure/acre. The application of the effluent to clay soil increased significantly the uptake of macronutrients and micronutrients by wheat compared with farmyard manure in the same soil.

INTRODUCTION

Although the heavily subsidized energy sources of many developing countries are a disincentive for the diffusion of biogas technology for fuel production, the organic fertilizer obtained as a by-product from this technology encourages farmers to consider its acceptance as a multipurpose technology.

It is well known that the "green revolution" of the last three decades is based, in particular, on the wide production and application of mineral fertilizers for growing crops. However, the practice of soil fertilization with mineral fertilizers is not without adverse, long-term environmental consequences with regard to increasing the food supply: there is growing concern about the deterioration of soil fertility and environmental quality as a result of using vast amounts of chemicals and inorganic fertilizers to produce more food. Fertilization by organic materials is, therefore, an alternative approach for restoring environmental resources while minimizing the reliance on fertilization by inorganic chemicals and maintaining soil productivity. This emphasizes the need for seeking new organic manures. Effluent produced from biogas plants would be a new source of manure in areas where biogas technology is introduced. The manurial value of the digested effluent discharged from biogas plants has been studied by various investigators (Mishra, 1954; Acharya, 1958; Laura and Idnani, 1971, 1972a, b, c; Idnani and Varadarajan, 1974; Barnett et al., 1978; Zohdy and Badr El-Din, 1983 a, b, c; and Zohdy et al., 1983).

At the National Research Centre, Egypt, bench- and pilot-scale studies have been carried out over the past 7 years. One of the major objectives of these studies has been to evaluate the effluents from biogas plants for their efficacy as fertilizer. During these investigations, various experiments were conducted to study the following:

1. The composition of effluents from biogas plants;
2. Phytotoxic effect of these effluents;
3. Changes in fertilizer value of effluents during handling and storage;
4. Digester effluents as soil conditioners; and
5. Yield response to digester effluent application to soils.

This report presents the results of these various experiments for assessing the usefulness of digester effluent as a fertilizer and soil conditioner.

METHODOLOGY

Effluents Sources

Sources of effluents for the experiments were as follows:

a. Laboratory digesters loaded with 4, 8, and 12 percent of cow dung and operated for 45 days at 22°C and 37°C; and

b. A Chinese digester (6 m^3) fed with 10 percent solids fresh cow dung. For details see Zohdy and Badr El-Din, 1983a, b, c; and Zohdy et al., 1983.

Detection of Phytotoxicity

Maize pot experiments were conducted in sandy soil. Effluent as such, or effluent after fractionation to sludge and supernatant, as well as farmyard manure (FYM) and urea, at the rate of 80 mg N/kg soil were mixed with the soil. Ten seeds of corn were planted in each pot after 0, 5, 10, and 15 days from application of fertilizers. Tap water was pipetted from time to time to keep soil water holding capacity at 60 percent. Treatments were replicated four times. After 20 days the dry weight of seedlings was determined.

Methods of Handling Effluents

Effluents were handled as follows:

a. By air drying under farm conditions;
b. By absorption on silt, 1:1 (w/w)*;
c. By postcomposting for 60 days with tree leaves and soil at a ratio of 2 effluent:1 leaves:1 soil (w/w/w).

Both b and c techniques were supplemented with 2.5 percent calcium carbonate and superphosphate 2.0 percent (w/w). For more details, see Zohdy et al., (1983).

Fertilization Effect of Effluents on Corn (Pot Experiment)

Pots were filled with 2.5 kg of sand. Effluent as such, or effluent after fractionation to sludge and supernatant, as well as FYM and urea, were added at the rate of 80 mg N/kg soil. Two corn seedlings were kept for 60 days in each pot. Dry weight, total nitrogen, and phosphorus were determined.

Fertilization Effect of Effluents on Wheat (Pot Experiments)

Pots were filled with 7 kg of a virgin sandy soil. In organic manure treatments, effluent, air-dried effluent, effluent absorbed on silt, composted effluent, or FYM were applied 7 days before sowing,

---

* Weight by weight.

each at the rate of 20 tons/acre. In the urea treatment, urea and superphosphate were applied to the soil at the rate of 80 kg N/acre and 20 kg P_2O_5/acre, respectively. Urea was added in equal instillations at 20 and 40 days after sowing. Twelve seeds of wheat were sown in each pot and were thinned out to 8 plants on germination. Each treatment, as well as the controls, were replicated 6 times. Soil samples were taken from each treated pot and control after 0, 30, 60, 90, 120, and 150 days from sowing for microbiological and chemical analysis.

Fertilization Effect of Effluents on Field-Grown Wheat in Clay Soil

A field experiment at El-Manawat village was conducted to study the effect of effluent absorbed on silt on the yield of wheat and uptake of nutrients as compared with the effect of ordinary FYM and urea. Nitrogen-fixing *Azospirillum lipoferum* was used as peat inoculum to evaluate whether this organism could be stimulated in its N-fixing capacity by the manuring treatments. The experiment was carried out using the complete randomized plots at the site of the digester. Each plot was 1/800 acre in area (5 m^2), and the treatments were replicated 4 times. The treatments were as follows:

1. Untreated, check plots.
2. Untreated plots + seeds inoculated with *Azospirillum*.
3. 100 kg urea + 100 kg superphosphate + 50 kg potassium sulfate/acre.
4. 100 kg urea + 100 kg superphosphate + 50 kg potassium sulfate/acre + seeds inoculated with *Azospirillum*.
5. 5 tons/acre digester effluent mixed with equal quantity of silty soil (w/w).
6. 5 tons/acre digester effluent mixed with equal quantity of silty soil + seed inoculated with *Azospirillum*.
7. 20 tons/acre digester effluent mixed with equal quantity of silty soil.
8. 20 tons/acre digester effluent mixed with equal quantity of silty soil +seed inoculated with *Azospirillum*.
9. 20 tons/acre ordinary FYM.

Organic manures, superphosphate, and potassium sulfate were applied during the preparation of the soil before sowing. Urea was applied in two equal doses, before the second and the third irrigation (30 and 45 days after planting). Seeds for treatments 2, 4, 6, and 8 were inoculated by N-fixing *Azospirillum*. The inoculation was done by coating the seeds with gum arabic and peat carrier at the rate of 10 g peat inocula (10^8 cells/g)/100 g seeds. At the harvest time, the yield of seeds and straw of every plot was recorded. The nutrient uptake (mg/plot) was determined by analyzing the seed yield and the straw for N, P, Mn, Zn, and Cu.

Microbiological and Chemical Analysis

The total bacteria were enumerated by serial dilution technique on soil extract yeast agar medium (Mahmoud et al., 1964). For microorganisms that decomposed humic acids or fulvic acids, Ladd's (1964) medium supplemented with humic acids or fulvic acids (1 percent) was used. The most probable number of *Azotobacter*, N_2-fixing clostridia, aerobic cellulose decomposers, and anaerobic cellulose decomposers were determined by multiple-tube dilution technique (5 tubes) using the following media, respectively: Ashby's medium, Winogradsky's medium, Dubos medium, and Omeliansky's medium (Allen, 1953). Computations were made using Cochran's table (1950).

Organic carbon, ammoniacal nitrogen, nitrate nitrogen, total nitrogen, available phosphorus, and total phosphorus were determined by the methods given by Jackson (1958). Organic constituents, total soluble substances, cellulose, hemicellulose, and lignin as well as humic and fulvic acids were determined according to Kononova (1960). Total volatile acids were determined by Ducleux method (Taha and Abdel Samie, 1961). Manganese, zinc, and copper contained in plant samples were determined by digestion of the sample with (HNO_3 + $HClO_4$) followed by atomic absorption spectrophotometry.

RESULTS

Composition of Effluents From Biogas Plants

Chemical Analysis of the Effluents From Batch Digesters. The results in Table 1 show the chemical composition of the effluents collected from the batch digesters operated at 22°C and 37°C using three loading rates of cow dung (4, 8, and 12 percent). The anaerobic fermentation

Table 1 Analysis of Effluents From Batch Type Digesters Operated at Different Temperatures and Loading Rates of Cow Dung (on dry basis)

Chemical Analysis As % of Total Solids	% of Cow Dung (w/v) in Feed to Digesters 12	8	4
NH_4-N	0.9 (0.7)	0.8 (0.8)	0.7 (0.7)
NO_3-N	0.0 (0.0)	0.0 (0.0)	0.0 (0.0)
Organic N	0.9 (1.0)	1.0 (0.7)	0.6 (0.7)
Available P	0.2 (0.4)	0.2 (0.3)	0.2 (0.2)
Total P	0.6 (0.7)	0.5 (0.7)	0.5 (0.5)
Organic C	35.8 (36.5)	33.9 (36.0)	38.7 (34.9)
Ash	38.2 (37.1)	41.5 (38.0)	36.8 (39.8)
Hemicellulose	35.4 (35.1)	37.6 (35.4)	40.4 (38.8)
Cellulose	8.4 (7.8)	6.1 (6.9)	5.9 (5.5)
Protein	5.4 (5.9)	5.9 (4.1)	3.5 (4.3)
Lignin	11.6 (17.2)	11.8 (13.2)	12.4 (15.4)
C/N ratio	20.8 (21.9)	18.9 (25.1)	28.3 (25.8)

NOTES: Numbers outside parentheses are % in effluents obtained at 37°C after 30 days of digestion. Numbers in parentheses are % in effluents obtained at 22°C after 45 days.

of 8 percent and 12 percent of cow dung at both temperatures gave effluents with very close chemical composition. The effluents from digesters loaded with 4 percent of cow dung at both temperature regimes had slightly lower values of NH_4-N, organic N, total P, ash, cellulose, and protein.

Effect of Continuous Feeding on Effluent Composition. Table 2 shows the effect of continuous feeding of the Chinese-type digester used on the chemical composition of the effluent. The data show the changes in the slurry composition during the fermentation process. Significant increases were recorded in the effluent content of NH_4-N, organic N,

Table 2 Effect of Continuous Feeding on Chemical Composition of Effluent Obtained From the Chinese-type Digester

Composition as % of Total Solids	Days			
	0	70	100	175
NH_4-N	0.07	0.33	0.63	0.32
NO_3-N	0.00	0.00	0.00	0.00
Organic N	1.15	2.01	1.48	1.22
Available P	0.42	1.05	0.79	0.97
Total P	1.46	1.84	1.79	1.93
Total volatile acids	0.05	0.68	0.61	0.15
Organic C	39.79	34.97	42.61	40.83
Ash	32.34	39.72	27.38	29.77
pH	7.90	7.30	7.20	6.90
C/N ratio	32.62	14.90	19.94	26.34
Humic acid (HA)	2.30	3.98	6.59	4.57
Fulvic acid (FA)	3.18	6.64	6.68	8.88
HA/FA ratio	0.72	0.60	0.99	0.52
Total soluble substances	8.48	15.07	19.42	11.79
Hemicellulose	35.60	27.09	27.80	--
Cellulose	8.10	7.18	8.14	4.54
Lignin	36.15	37.54	36.10	28.41
Protein	7.63	12.62	9.24	28.55

available P, total P, humic acids, fulvic acids, total soluble substances, and total proteins.

Phytotoxic Effect of Digester Effluent. Application of the supernatant separated from the digester effluent to sandy soil retarded growth of maize seedlings as evidenced by reduction of plant dry weight (Table 3). This effect was not observed with the sludge or the effluent application. The prefertilization of soil 5 to 10 days before planting greatly reduced the toxic effect of the supernatant.

Changes in Fertilizer Value of Digester Effluents During Handling and Storage. The two major problems encountered in the utilization of digester effluents are the phytotoxic effect of the effluent, particularly on young plants, resulting from direct application of the effluent on land (Table 3) and the loss of fertilizer elements, especially nitrogen, during storage. These two problems could be overcome by the use of the proper handling techniques. Three methods of handling effluent were studied.

Table 3 Phytotoxic Effect of Biogas Fertilizer in Sandy Soil Using Maize As Test Plant

Fertilizer	Planting Schedule After Fertilizer Application (days)			
	0	5	10	15
	Dry weight mg/plant			
FYM	162	162	174	180
Effluent	164	170	186	190
Sludge	162	167	176	176
Supernatant	127	150	162	178

NOTE: FYM: Farmyard manure.

Table 4 Chemical Changes in Air-dried, Absorbed on Silt and Composted Effluents During Storage (% of dry materials)

Constituent	Air-dried Effluent		Effluent Absorbed on Silt		Composted Effluent	
	0	60	0	60	0	60
Organic carbon	41.70	27.90	10.30	8.20	20.90	15.80
Cellulose	20.30	14.50	18.50	7.60	32.30	17.90
Hemicellulose	29.90	27.90	16.30	14.60	34.30	30.20
Lignin	27.60	-	18.60	17.50	31.30	25.60
Humic acids/ fulvic acids	0.30	0.60	0.20	0.10	0.10	0.30
NH_4-N	0.46	0.02	0.05	0.01	0.10	0.10
NO_3-N	0.00	0.01	0.01	0.04	0.06	0.08
Organic N	1.16	1.02	0.20	0.20	0.11	0.19
Total P	1.57	1.65	0.27	0.30	0.34	0.35
Available P	0.58	1.07	0.06	0.07	0.07	0.08
C/N ratio	25.70	26.40	40.10	32.20	77.60	55.50

a Initial analysis.
b Analysis after 60 days.

The chemical and microbiological changes in the composition of the products derived from the three handling methods were assessed (Tables 4 and 5). The air dried effluent was superior to both effluent absorbed on silt and composted effluent as indicated by the higher concentrations of fertilizer elements, the available P and the narrow C/N ratio. However, the loss of nitrogen during 60 days of storage was extremely high in air-dried effluent compared to the other two treatments.

The low values of chemical constituents in effluent absorbed on silt and composted effluent are mainly due to the dilution of the digester effluent with the carrier material. As the result of organic matter decomposition, decreases were recorded in carbon, cellulose, and lignin content of all three manures. The rate of cellulose degradation was, however, higher than that of lignin.

Table 5 Microbiological Changes in Air-dried Effluent, Silt + Effluent and Compost During Storage (counts/g dry material)

	Air-dried Effluent		Silt + Effluent		Compost	
Microorganisms	0[a]	60[b]	0	60	0	60
Total bacteria ($\times 10^8$)	2.0	13.0	4.0	15.0	6.0	11.0
Azotobacter ($\times 10^4$)	4.0	46.0	0.2	0.2	3.0	126.0
Clostridia ($\times 10^5$)	6.0	1.6	4.0	0.7	22.0	16.0
Aerobic cellulose decomposers ($\times 10^4$)	16.0	160.0	0.4	7.0	1.2	160.0
Anaerobic cellulose decomposers ($\times 10^4$)	22.0	2.0	6.4	0.2	3.2	1.7
Humic acids decomposers ($\times 10^5$)	12.0	49.0	1.0	3.0	24.0	82.0
Fulvic acids decomposers ($\times 10^5$)	20.0	20.0	2.0	6.0	12.0	79.0

[a] Initial analysis.
[b] Analysis after 60 days.

The humification process took place in all manures during the storage period. The humic acid fraction was higher than the fulvic acid fraction in air-dried and composted effluent than in effluent absorbed on silt. This implies that the rate of humic acid decomposition in the last treatment was higher than the rate of its formation.

The major drawback of air drying the effluent as a storage technique is the loss of 29 percent of its nitrogen. In contrast, a gain in nitrogen was recorded in composted effluent. This was correlated to the pronounced increase in *Azotobacter* population (Table 5). The postcomposting of the effluent and the absorption on silt are, therefore, suitable means to conserve the fertilizer value of the biogas effluent until it is used. The variation in the C/N ratio of different biogas fertilizers was further reduced with extended storage to between 18.8 and 26.2 (Table 6).

Table 6 Chemical Analysis of Biogas Fertilizers and Farmyard Manure (% of dry materials)

Constituents	FYM	Effluent	Dried Effluent	Silt + Effluent	Composted Effluent	Sludge
Organic C	3.55	41.32	25.90	8.19	13.60	13.00
NH_4-N	0.04	0.68	0.20	0.01	0.01	0.04
NO_3-N	0.00	0.00	0.02	0.05	0.09	0.00
Total N	0.34	1.91	1.29	0.31	0.72	0.61
Total P	0.10	1.57	1.70	0.26	0.30	1.67
Available P	0.03	0.58	1.08	0.07	0.10	0.68
C/N ratio	10.30	21.60	20.00	26.20	18.80	21.20

Comparison Between the Effluent Manure and Traditional Farmyard Manure. The comparison between the average values of samples collected from digester effluents and FYM showed that the former contained higher concentrations of basic nutrients than the latter (Table 6). Thus, effluents obtained from biogas plants are superior in fertilizer value when compared to the traditional FYM.

Effluents As Soil Conditioner

Digester effluent and FYM were applied to a virgin sandy soil at the rate of 20 tons/acre. The effect of the manure treatments on chemical and biological properties of the soil were studied. The addition of the manures stimulated the growth of bacteria, fungi, aerobic cellulose decomposers, *Azotobacter*, and N-fixing clostridia. Of the organic manures, both effluent and FYM had a stimulative effect (Figures 1 and 2). The various manures enriched the soil with available nitrogen and phosphorus as well as the soil organic matter (Figure 3 and Table 7).

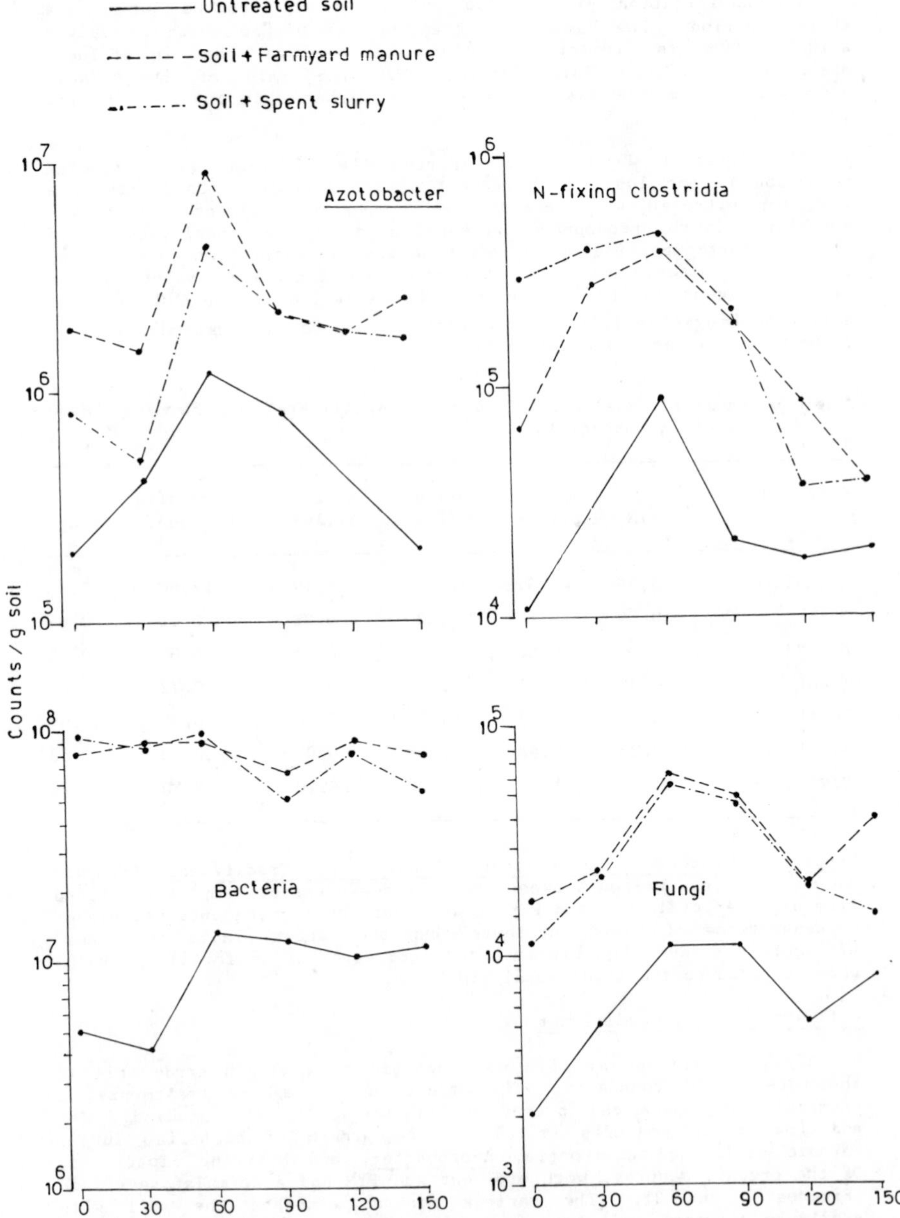

Fig. 1: Changes in densities of microbial groups in sandy soil treated with spent slurry and farmyard manure.

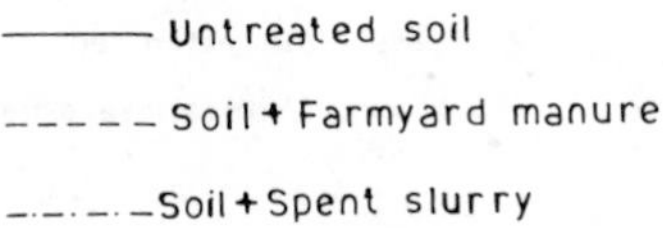

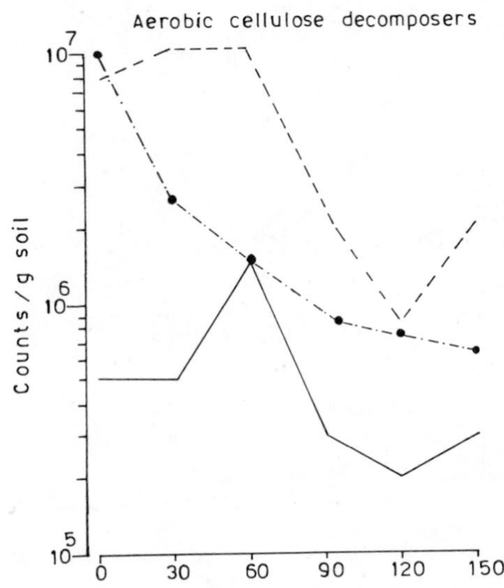

Fig. 2: Changes in densities of aerobic cellulose decomposers in sandy soil treated with spent slurry and farmyard manure.

Fertilizer Value of the Digester Effluents

Several pot and field trials were conducted to assess the nutrient uptake and yield response of wheat and maize to digester effluents application. Different kinds of effluents were included in this study. For comparison, FYM was used as a standard fertilization program.

Effect on Nurtient Uptake. In a pot experiment using different fertilizers, the biogas effluent gave higher maize dry-matter yield and nitrogen and phosphorus uptake than the FYM treatment. However, the mineral fertilization resulted in the highest values of dry-matter and nutrient uptake (Table 8). The sludge treatment came next, in its effect, to the effluent.

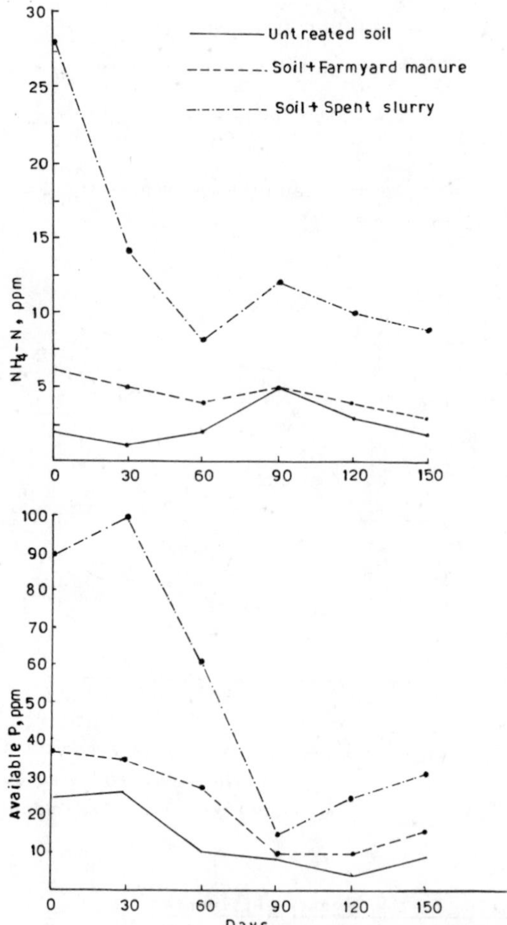

Fig.3: Changes in ammoniacal nitrogen and available phosphorus in sandy soil treated with spent slurry and farmyard manure.

The uptake of N and P by wheat from different effluents in the sandy soil pot experiments is presented in Table 9. The data show that the uptake of these nutrients from the digester effluents were significantly higher than their uptake from the FYM treatment. The fertilization with effluent absorbed on silt resulted in high nitrogen recovery compared to other treatments, while the manuring with

Table 7 Changes in Organic Nitrogen and Carbon in Sandy Soil Treated With Different Fertilizers

Number of Days	Untreated Soil	Soil Treated With		
		Urea	FYM	Effluent
		Organic N, ppm		
0	18	18	80	359
30	17	29	78	353
60	13	31	75	342
90	5	24	71	322
120	9	26	71	324
150	14	28	72	326
		Organic carbon, ppm		
0	260	260	3,760	12,930
30	109	160	1,120	4,950
60	91	120	983	3,402
90	83	101	926	2,993
120	71	86	844	2,869
150	76	76	786	2,817

Table 8 Effect of Fertilization on Nutrient Uptake and Dry-Matter Yield of Maize in Sandy Soil (60-day pot experiment)

Fertilization	Dry Weight (g/pot)	Uptake (mg/pot)	
		N	P
Control	1.3	11.6	3.5
Urea and superphosphate	4.0	64.9	21.1
FYM	1.9	28.1	5.6
Effluent	2.7	41.2	9.2
Sludge	2.1	26.7	7.6
Supernatant	1.2	15.7	3.9

effluent, dried effluent, and effluent absorbed on silt gave significantly higher P uptake than any other treatments.

A field trial in clay soil was conducted to evaluate the N, P, Zn, Mn, and Cu uptake from different fertilizers including the absorbed effluent. Inoculation of wheat seeds with Azospirillum was included to assess possible stimulative effect of organic manures on the growth, proliferation, and N fixation by this bacterium. Table 10 shows that the Azospirillum alone or with the effluent did not induce significant

Table 9 Uptake of Nitrogen and Phosphorous by Wheat and Percentages of Their Recovery

Fertilization Treatments	N Uptake (mg/pot)	N Recovery (%)	P Uptake (mg/pot)	Recovery (%)
Control	25.9	-	6.5	-
Urea/superphosphate	150.2	23.9	33.7	27.2
FYM	41.6	3.3	28.0	21.5
Effluent	64.3	2.4	60.0	53.4
Dried effluent	36.1	2.0	41.6	35.2
Silt + effluent	79.9	18.1	39.9	32.4
Composted effluent	50.5	5.0	25.9	19.4
Sludge	13.9	1.6	20.5	14.0
L.S.D.* (1%)	16.9		(10.6)	
L.S.D. (5%)	12.5		(7.9)	

*Least significant difference.

increase in nitrogen uptake by the plants. The fertilization with 5 tons of effluent/acre resulted in the same macronutrient and micronutrient uptake as 20 tons/acre of FYM. Moreover, the nitrogen uptake recorded in 20-tons effluent treatment was similar to that of urea treatment. The phosphorus uptake was significantly higher in the biogas effluent treatment than in the same rate of FYM

Yield Response to Fertilizer Application. The yield response of wheat to digester effluents in virgin sandy soil and clay soil are presented in Tables 11 and 12, respectively. It is clear that the effect of effluents (except the sludge) on wheat seed yield in sandy soil was superior to the FYM (Table 11). The manure prepared by absorbing the effluent on silt gave significantly higher seed yield than any other effluent treatment and similar yield to the urea fertilizer treatment.

Table 10 Effect of Fertilization on Nutrient Uptake by Wheat in Field Experiment at Manawat

Fertilization Treatments	Uptake[a]				
	N	P	Zn	Mn	Cu
Control	67	6.99	216	44.4	29.0
Azospirillum	72	7.68	207	42.0	29.6
Urea	171	18.89	283	61.4	42.4
Urea + *Azosp.*	177	18.93	283	61.3	44.0
20 t FYM/acre	120	16.10	321	68.0	49.6
5 t effluent/acre	112	12.98	312	72.7	51.2
5 t effluent + *Azosp.*	120	13.18	311	70.9	61.0
20 t effluent/acre	163	24.37	425	89.7	65.8
20 t effluent + *Azosp.*	168	25.42	402	91.7	67.2
L.S.D.* (1%)	13.4	3.18	38	10.0	6.7
L.S.D. (5%)	8.7	2.35	26	6.9	4.6

[a]N and P uptake, g/5 m^2, Zn, Mn and Cu uptake, mg/5 m^2
*Least significant difference.

In the clay soil field experiment (Table 12), the application of effluent absorbed on silt at the rate of 5 tons/acre gave almost the same yield increase obtained using 20 tons of FYM. Significant differences in grain yield were recorded between the 5 and 20 tons of effluent application. On the other hand, no significant differences in yield were found between the fertilization with 20 tons/acre of digester effluent and the regular nitrogen/phosphorus/potassium (NPK) program. The inoculation with *Azospirillum* in both fertilized and unfertilized treatments did not induce significant increases in wheat seed yield.

DISCUSSION

The need for organic matter in agriculture is well recognized. The effluent that results from the anaerobic digestion of animal wastes has high nutrient content and increased nutrient availability (Idnani and Varadarajan, 1974; Sathianathan, 1975; Barnett et al., 1978; and NAS, 1977). In this work the nitrogen and phosphorus content in the digester effluent increased slightly more than the feed material as a result of the decrease in the total solid content. However, the ammoniacal N and available P increased in the effluent to 40 percent and 50 percent of its total content as a result of the mineralization process. Acharya (1958), Hart (1963), and Idnani and Varadarajan (1974) reported that 15 percent to 24 percent of the dung nitrogen was converted into the ammoniacal form.

Table 11 Effect of Organic Manures on Dry-Matter Yield of Wheat Grown in Virgin Sandy Soil (g/pot)

Treatment	Straw	Seeds
Urea	7.70	5.29
FYM	3.00	1.92
Effluent	4.91	5.30
Dried effluent	3.45	4.22
Silt + effluent	5.53	5.91
Composted effluent	4.56	4.35
Sludge	3.40	1.75
L.S.D.* (1%)	1.32	1.29
L.S.D. (5%)	0.97	0.95

*Least significant difference.

Table 12 Yield Response of Field-Grown Wheat to Various Fertilization Treatments in Clay Soil

Treatment	Straw	Seeds
	kg/plot[a]	
Untreated soil	6.15	1.3
Soil + <u>Azospirillum</u>	6.75	1.5
Soil + Urea	13.60	3.2
Soil + Urea + <u>Azosp</u>.	15.00	3.7
Soil + 20t FYM	10.30	2.5
Soil + 5t effluent	10.30	2.3
Soil + 5t effluent + <u>Azosp</u>.	11.40	2.6
Silt + 20t effluent	11.40	2.9
Soil + 20t effluent + <u>Azosp</u>.	11.70	3.2
L.S.D.* (1%)	3.5	0.9
L.S.D. (5%)	2.6	0.6

[a]Area of plot: 5 m^2.
*Least significant difference.

During the fermentation about 26 percent to 43 percent of the organic matter in feed material was digested. The main reduction occurred in the cellulose and hemicellulose constituents of feed materials. However, lignin content was relatively stable in the digested slurry. Similarly, Prasad et al. (1970) and Barnet et al. (1978) showed that lignin was degraded slowly and its percentage increased after fermentation. However, a greater decrease of hemicellulose and cellulose fraction was observed.

The pot and field experiments conducted in this study clearly show that the digester effluents have high manuring value compared to other traditional organic manures. The effluent treatment of clay and sandy soils significantly increased the yield and nutrient uptake of wheat over the farmyard manure, air-dried effluent, and compost treatments. This is in complete agreement with the work of previous authors in this field (Desai and Biswas, 1945; Rosenberg, 1952; Laura and Idnani, 1972c; Barnett et al., 1978). In another work, Mishra (1954) reported that the air-dried effluent gave higher yield than liquid effluent. Although the yield in the air-dried effleunt treatment was inferior to that with the original effluent, it was statistically higher than for FYM. This confirms the results of several workers (Desai and Biswas, 1945; Rosenberg 1952; Mishra, 1954; and Laura and Idnani, 1972 a,b,c). The low recovery of nutrients from the air-dried effluent may be caused by the destruction of colloidal micelles due to desiccation in the sun, which results in "irreversible flocculation" and thus subsequently to slow microbial decomposition of the effluent. Similar findings were reported by Laura and Idnani (1972) and Joffe (1955). The results obtain by Laura and Idnani (1971) also showed that mineralization of nitrogen in liquid effluent was greater than in air-dried effluent. On the other hand, the lowest recovery of nutrients in the composted effluent treatment may be attributed to the immobilization of inorganic nutrients by microbial population following the organic manuring. Alexander (1977) reported that the amount of mineralized nitrogen in soil from the organic material increases if it contains simpler nitrogeneous compounds, and, on the other hand, if it contains easily decomposable carbon there is immobilization of soil nitrogen. The absorption of biogas effluent by silty clay soil, as a handling technique, produced a manure that enhanced the nutrient uptake and yield of wheat as compared with fresh effluent as such. The superiority of silt + effluent manure is likely due to the higher mineralization rate of organic materials as well as to conservation of nutrients through the binding capacity of the silty and colloidal fraction of the soil. This could prevent the nutrients from leaching with water runoff and the loss via volatilization.

The limitation to the use of digester effluents as fertilizer stems from the difficulty associated with the handling of the liquid effluent and its storage until the most suitable application time. Several handling and storage techniques were assessed in this paper. The major disadvantage to the air drying of effluent is the remarkable loss of N through volatilization, which amounted up to 29 percent of its original value. Acharya (1958) pointed out that about 16 percent of the nitrogen in the digested effluent is present as dissolved ammonia that may be easily evaporated on standing, as was shown by Idnani and Varadarajan (1974). Postcomposting of the effluent and absorption on silt are, therefore, suitable means for conserving the fertilizing value of effluent until its usage. The absorption of effluent by silt reduced the loss of nitrogen to a large extent, but organic matter content of the fertilizer was also reduced. This is in agreement with the results reported by McGarry and Stainforth (1978). The fertilizing value of the absorbed effluent can be raised by repeating the operation of soaking and drying. On the other hand, the use of effluent in postcomposting of agricultural wastes was found to increase the nitrogen, possibly due to assimilation of atmospheric nitrogen by microorganisms. The counts of N-fixing bacteria were found to increase during the composting of agricultural wastes (Zohdy et al., 1982).

Another problem associated with the direct use of effluent is the phytotoxic effect, especially on seedlings. Many investigators (Patrick and Koch, 1958; Patrick et al., 1963; and Patrick, 1971) have shown that toxic substances were formed during the anaerobic digestion of the organic residues, and such toxic substances were found to be soluble in water. The depressive effect of the effluent on maize seedlings was not observed if the effluent was added for 5 to 10 days before planting; this may be due to the decomposition of the toxic materials in the effluent by soil microorganisms during storage. Similar results were reported by Boerner, 1961; Patrick, 1971;and Badr El-Din and Gamal El-Din, 1981. Sathianathan (1975) also stated that the digested effluent gave better results after it had been aged in the open air for a week or two.

This study showed that the proper recycling of digester effluents from biogas plants for agricultural purposes provides a good incentive to farmers for accepting biogas technology. The application of a high-quality biofertilizer in farming practices means less dependence on the highly subsidized mineral fertilizers. It also helps to build up the soil fertility, especially in marginal soils. The addition of organic manures to soil was found to enhance the soil biological activity (Mahmoud et al., 1968; and Taha et al., 1968), enrich the soil with available nitrogen and phosphorus (Mulder, 1976; and Sarkhadi, 1979), and increase the soil organic matter content (Taha et al., 1968), thereby improving the overall soil fertility.

REFERENCES

Acharya, C.N. 1958. Preparation of fuel gas and manure by anaerobic fermentation of organic materials. ICAR, Research series No. 15.

Alexander, M. 1977. Introduction to Soil Microbiology. 2nd ed. John Wiley & Sons, New York.

Allen, O.N. 1953. Experiments in Soil Bacteriology. Burges, U.S.A.

Badr El-Din, S.M.S., and Gamal El-Din. 1981. Effect of antimicrobial substances from the residues of certain higher plants on seed germination, seedling growth and rhizosphere microflora of cotton plant. Egypt. J. Microbiol. 16:25.

Barnett, A., I. Pyle, and S.K. Subramanian. 1978. Biogas Technology in the Third World: A Multidisciplinary Review. IDRC (International Development Research Center), Ottawa, Ontario, Canada.

Boerner, H. 1961. Experimental investigations of soil fatigue on the apple. Beitr. Biol. Pflan. 2:36.

Bunt, J.S., and A.D. Rovira. 1955. Microbiological studies of some subantarctic soil. J. Soil Sci. 6:119.

Cochran, W.G. 1950. Estimation of bacterial densities by means of the most probable number. Biometrics. 6:105.

Desai, S.V., and S.C. Biswas. 1945. Manure and gas production by anaerobic fermentation of organic wastes. Indian Farming 6:67.

Hart, S.A. 1963. Digestion tests of livestock wastes. J. Water Poll. Cont. Fed. 35:6.

ICAR. 1976. the Economics of Cowdung Gas Plant. Indian Council of Agricultural Research (ICAR), New Delhi.

Idnani, M.A., and S. Varadarajan. 1974. Fuel Gas and Manure by Anaerobic Fermentation of Organic Materials. Technical Bulletin No 46. ICAR, New Delhi.

Jackson, M.L. 1958. Soil Chemical Analysis. Prentice-Hall, Englewood Cliffs, N.J.

Jenkinson, D.S. 1965. Studies in the decomposition of plant material in soil. I. Losses of carbon from C^{14} labelled rye grass incubated with soil in the field. J. Soil sci. 6:104.

Joffe, J.S. 1955. Green manuring viewed by a pedologist. Advances Agron. 7:142.

Kononova, M.M. 1960. Soil Organic Matter, Its Nature, Properties and Methods of Studying It. 2nd ed. Pergamon Press, London.

Ladd, J.N. 1964. Studies on the metabolism of model compounds related to soil humic acid. I. The decomposition of N- (O-carboxy phenyl) glycin. Australian J. Biol. Sci. 17:153.

Laura, R.D., and M.A. Idnani. 1971. Retarding effect of desiccation on nitrogen mineralization in organic manures. Current Sci. 40:158.

------. 1972a. Mineralization of nitrogen in manures made from spent slurry. Soil Biol. Biochem. 4:239.

------. 1972b. Composting of agricultural waste materials by using spent slurry as a starter for decomposition. Indian J. Agric. Sci. 42:246.

------. 1972c. Effect on wheat yield and nitrogen uptake from manures made from spent-slurry. Plant and Soil 37:283.

Mahmoud, S.A.Z., M. Abou el-Fadl, and M. Kh. El-Mofty. 1964. Studies on the rhizosphere microflora of a desert plant. Folia Microbiologica 9:1.

Mahmoud, S.A.Z., S.M. Taha, A.H. El-Damaty, and M.S.M. Moubarek. 1968. Effect of green manuring on the fertility of sandy soils at the Tahreer Province; U.A.R. (United Arab Republic i.e. Egypt) I. Lysimer experiments. J. Soil Science U.A.R. 8:113.

McCalla, T.M. 1974. Use of animal wastes as a soil amendment. J. Soil Water Conservation. Sept.-Oct., p. 213.

McGarry, M.G., and J. Stainforth. 1978. Compost, Fertilizer, and Biogas Production From Human and Farm Wastes in the Peoples Republic of China. IDRC, Ottawa, Canada.

Mishra, U.P. 1954. Production of Combustible Gas and Manure From Bullock Dung and Other Organic Materials. M.Sc. thesis. Indian Agr. Research Institute, New Delhi.

Mulder, J. 1976. Application of animal manure in Ijssel-meerpolders. Stikstof 82:305.

National Academy of Sciences (NAS) 1977. Methane Generation From Human, Animal and Agricultural Wastes. NAS, Washington, D.C.

Patrick, Z.A. 1971. Phytotoxic substances associated with the decomposition in soil of plant residues. Soil Sci. 3:13.

Patrick, Z.A., and L.W. Koch. 1958. Inhibition of respiration, germination and growth by substances arising during the decomposition of certain plant residues in the soil. Can. J. Botany 36:621.

Patrick, Z.A., T.A. Toussoun, and W.C. Synder. 1963. Phytotoxic substances in arable soils associated with decomposition of plant residues. Phytopathology 53:152.

Prasad, C.R., K.C. Gulati, and M.A. Idnani. 1970. Changes in biochemical constituent of some organic waste materials under anaerobic methane fermentation. Indian J. Agr. Sci. 40:921.

Rosenberg, G. 1952. Methane production from farm wastes as source of tractor fuel. J. Ministry Agr. Fish. 58:487.

Sarkhadi, J., H. Balla, and M. Kramer. 1979. Direct and residual effects of organic and mineral fertilizers applied in long-term experiments. Tagungsbericht, Academie der Landwirtschaftswissen-Schaften der Dutschen Domokratischen Republick 162:91.

Sathianathan, M.A. 1975. Biogas Achievements and Challenges. Association of Voluntary Agencies for Rural Development, New Delhi.

Sinha, M.K. 1972. Organic matter transformation in soil. I. Humification of C^{14} tagged oats roots. Plant and Soil 36:283.

Taha, S.M., and M. Abdel Samie. 1961. Mish spore-forming anaerobes. II. Volatile acids production. Annals Agric. Sci. Fac. of Agric. Ain Shams Univ., Cairo 6:65.

Taha, S.M., A.H. El-Damaty, S.A.Z., Mahmoud, and M.S.M. Moubarek. 1968. Effect of green manuring on the fertility of sandy soil at the Tahreer Province, U.A.R. II. Field experiments on virgin soil. J. Soil Sci. U.A.R. 8:129.

Tunney, H. 1980. Fertilizer value of animal manures. Farm and Food Res. 1:78.

Zohdy, L.I., S.M.S. Badr El-Din, M.A. Khalafallah, and H.K. Abdel Maksoud. 1982. Handling and storage of digested slurry produced from the biogas digester. Egypt. J. Microbiol. (in press)

Zohdy, L.I., and S.M.S. Badr El-Din. 1983a. Evaluation of digested slurry as a fertilizer. I. Its microbial and chemical composition and its phytotoxicity. Egypt. J. Microbiol. 18:69.

------. 1983b. Comparison of biogas slurry treatments. I. On the biological and chemical properties of a virgin sandy soil cultivated with wheat. Egypt. J. Microbiol. 18:115.

------. 1983c. Comparison of biogas slurry treatments. II. On the wheat yield, nitrogen and phosphorus uptake and nitrogen balance in a virgin sandy soil. Egypt. J. Microbiol. 18:127.

Zohdy, L.I., S.M.S. Badr El-Din, and M.A. Khalafallah. 1983. Handling and storage of the digested slurry produced from the biogas digester. Egypt. J. Microbiol. 18:107.

REPEATED APPLICATION OF ANAEROBICALLY DIGESTED SLURRY AND ITS EFFECT ON THE YIELD AND NPK UPTAKE OF WHEAT, TURNIPS AND ONION PLANTS

L.I. Zohdy, R.A. Abd El-Aziz, M.H. Enany,
A.S. Turky, and S.M.A. Radwan
Soil Microbiology Unit, National Research Centre, Cairo, Egypt

ABSTRACT

Fresh biogas slurry*, air dried slurry, silt plus slurry, composted slurry, farmyard manure, and cow dung were added to virgin sandy and clay soils on a dry basis at the rate of 2 percent (20 ton/Feddan**). These manures were applied three times to the same soil for two consecutive years carrying an ordinary crop rotation (i.e., wheat-turnips-onions) in a greehouse-controlled pot experiment. The effect of the manure treatments on the dry matter yield and uptake of nitrogen, phosphorus, and potassium were compared to an inorganic fertilizer treatment and a control treatment.

All manures increased the yield of the three crops planted during the employed rotation cycle. The first application of dried and fresh slurry gave the highest amount of dry matter yield of wheat relative to all other applied manures. The second application dose of these manures induced a moderate dry matter turnips yield. The third application of fresh and dried slurry gave the least increases of the onion dry matter yield. The uptake of nitrogen, phosphorus, and potassium seems to differ with the type of manure applied, the kind of crop and the soil type.

INTRODUCTION

There is no doubt that the whole spectrum of organic material for use in agriculture must be closely investigated with a view to use them to the best advantage, but at the same time avoiding dangers of overmanuring, pollution, and development of toxic substances. Several workers have commented on the fertilizer quality of the digested slurry (Acharya, 1958; Indnani and Varadarjan, 1974; and Zohdy and Badr El-Din, 1983). Also, the effect of biogas slurry on the yield of wheat plants and on the biological and chamical properties of a virgin sandy soil were studied by Zohdy and Badr El-Din 1983a and b.

* 'Slurry' and 'biogas slurry' are used throughout this paper to signify anearobically digested slurry in biogas-producing units.
** A Feddan is about an acre of land (1 Feddan = 4,200 m^2)..

The present work studies the effect of the repeated application of the biogas slurry as it is, and after its treatment for a longer storage period, in comparison with other manures applied to the same soil for two consecutive years carrying an ordinary crop rotation (i.e., wheat-turnips-onions). The effect on the dry matter yields and nitrogen, phosphorus, and potassium (NPK) contents and uptakes were studied.

METHODOLOGY

Materials

Virgin sandy and clay soils collected from Helwan were used. Their physical and chemical properties are shown in Table 1. Fresh biogas slurry used in this experiment was obtained from one of the National Research Centre (NRC) demonstration units (the Diab Indian-type unit at Omar Makram village). Air-dried slurry, silt plus slurry, and composted slurry were prepared as previously described by Zohdy et al., (1983). The ordinary farmyard manure and the fresh cow dung were obtained from the NRC farm at Shalakan, Kaluebia province. The chemical analysis of organic manures used are shown in Table 2.

Table 1. Physical and Chemical Properties of Virgin Sandy and Clay Soils

Constituents		Sandy Soil	Clay Soil
Clay		5.90	34.00
Silt		4.80	49.50
Sand		89.30	26.50
Textures		Sandy	clay loam
$Ca\ CO_3$		1.25	1.80
pH		7.90	7.85
Organic carbon	ppm	320	2540
Total nitrogen	ppm	18	131
$NH_4 - N$	ppm	7	36
$NO_3 - N$	ppm	3	18
C/N ratio		17.778	19.389
Total phosphorus	ppm	160	260
Total potassium	ppm	140	860
Sodium	ppm	48	100
Iron	ppm	163	172
Zinc	ppm	2	18
Copper	ppm	2	31
Manganese	ppm	67	264
Nickel	ppm	4	34

Table 2. Chemical Analysis of Organic Manures Used

Constituents	Cow Dung	Fresh Biogas Slurry	Air-Dried Slurry	Slurry Plus Silt	Composted Slurry	Farmyard Manure
Moisture (%)	77.366	92.688	8.466	3.105	5.148	7.941
Organic matter (%)	54.535	74.375	50.063	7.147	20.242	13.016
Organic carbon (%)	31.633	43.141	29.039	4.146	11.741	7.55
Total nitrogen (%)	2.560	3.008	2.002	0.230	0.559	0.600
NH_4 - N (%)	0.717	0.842	0.159	0.095	0.038	0.086
NO_3 - N (%)	0.0013	0.0095	0.0005	0.0015	0.0010	0.0013
C/N ratio	12.357	14.342	14.505	18.026	21.004	12.583
Total phosphorus(%)	0.916	0.510	0.415	0.196	0.179	0.262
Available phosphorus (%)	0.163	0.273	0.120	0.028	0.013	0.013
Total potassium (ppm)	2209	1368	42042	2580	12121	9774
Sodium (ppm)	240	5434	2240	680	560	240
Iron (ppm)	1636	1126	1606	1634	1614	1623
Zinc (ppm)	32	31	39	26	52	30
Copper (ppm)	54	45	63	54	11	59
Manganese (ppm)	664	93	417	361	409	511
Nickel (ppm)	109	117	103	76	99	114

Experimental

A greenhouse-controlled pot experiment was planned to be repeated in the same pots for two successive years carrying an ordinary crop rotation (i.e., wheat-turnips-onions). Two sets of pots were prepared using virgin sandy and clay soils. Pots were filled with 10 kg of each soil. The manures were applied two weeks before sowing of the crops, at the rate of 2 percent (20 tons/feddan) on a dry basis at the beginning of every cultivation period, and mixed thoroughly with all the soil in each pot. In the inorganic fertilizer treatment (NPK), 200 kg of urea and 100 kg of each of superphosphate and potassium sulphate per feddan were used. Urea was added in two equal doses at the beginning and after 40 days from sowing. A control with no fertilization was also included and all treatments were replicated four times. After sowing, the moisture content was adjusted to about 60 percent of the water-holding capacity of the soils during the cultivation period.

At harvesting times, plants were removed and separated to straw and seed for wheat. The roots and leaves of turnips and onions were quickly washed with distilled water, dried at 70°C, weighed, pluverized to fine powder, and kept for analysis. The yield response of each crop was measured and analysed statistically.

Chemical Analysis

Total nitrogen was determined by the semi-microkjeldahl method (Allen, 1953). Total phosphorus by the colorimeric analysis given by Jackson, 1958. Total potassium was extracted by the wet ashing method given by Jackson, 1958 and analyzing the digested solution by atomic absorption spectrophotometry.

RESULTS AND DISCUSSION

Effect on the Dry Matter Yield

Wheat Plant. Data in Table 3 show that the application of all manures induced increases in the dry matter yield of straw and grains of wheat, and were statistically higher than control treatments in both soil types. The dry matter yield response of the wheat plant to organic manures, in the case of sandy soil, was in the following decreasing order: dried slurry, fresh slurry, cow dung, farmyard manure, compost, silt plus slurry. In the case of the clay soil, the yield response was in the following decreasing order: dried slurry, silt plus slurry, fresh slurry, farmyard manure, compost, cow dung. Also, the variation on the yield as affected by the manure types was not significant in both least significant difference (L.S.D.) levels in clay soil.

Generally, the responses to all fertilizers in the sandy soil were higher than those in the clay soil. Also, the highest dry matter yields of wheat in both soil types were observed in the dried slurry treatments. This result is in agreement with Mishra (1954) who found that the air-dried slurry gave higher yield over liquid slurry.

Table 3. Effect of Organic Manures on Dry matter Yield of Wheat Grown in Virgin Sandy and Clay Soils (g/pot)

Treatment	Sandy Soil			Clay Soil		
	Grains	Straw	Total	Grains	Straw	Total
Control	3.74	3.70	7.44	11.39	13.68	25.07
Inorganic NPK	6.93	8.78	15.71	16.03	20.01	36.04
Fresh slurry	13.25	12.84	26.09	22.05	18.16	40.66
Dried slurry	15.27	13.42	28.69	26.85	26.42	53.27
Silt plus slurry	8.93	6.35	15.28	23.17	17.78	40.95
Composted slurry	11.79	5.44	17.23	21.92	16.69	38.61
Farmyard manure	9.64	10.70	20.34	21.20	18.62	39.82
Cow dung	11.08	12.17	23.25	14.89	21.23	36.12
L.S.D. 5%	1.98	2.41	2.17	2.53	3.37	5.34
L.S.D. 1%	2.69	3.28	2.96	3.89	4.58	7.27

Turnip Plants. Data in Table 4 show that the dry matter yield of whole turnip plants was statistically higher in all manure treatments than the control. But since the turnip plant is a root crop, the dry matter yield of roots is only discussed. Cow dung treatment gave the highest amount of roots yield in both soil types. Fresh slurry and dried slurry treatments were in the second level. The lowest increases in turnip roots were recorded in farmyard manure and silt plus slurry treatments.

Evans et al. (1977) found that the annual application of solid beef manure (at a rate of 224 metric tons/ha wet weight), liquid beef manure, and liquid hog manure (at a rate of 636 metric tons/ha wet weight) for two years was not detrimental to corn production. Also, the two-year residual effect of these manures was adequate to maintain corn yields at an acceptable level. Pietz et al. (1981a) found that sewage sludge application on stripmined land at the Fulton County land reclamation site has not consistently led to significant corn yield increases.

Table 4. Effect of Organic Manures on Dry Matter Yield of Turnips Grown in Virgin Sandy and Clay Soils (g/pot)

Treatment	Sandy Soil			Clay Soil		
	Roots	Leaves	Total	Roots	Leaves	Total
Control	2.63	1.50	4.13	11.06	3.52	14.58
Inorganic NPK	5.36	4.01	9.37	17.22	13.56	30.78
Fresh slurry	11.02	9.07	20.09	24.22	16.99	41.21
Dried slurry	7.17	5.46	12.63	19.37	14.36	33.73
Silt plus slurry	4.40	2.77	7.17	15.76	10.92	26.68
Composted slurry	6.01	4.50	10.51	17.96	11.63	29.59
Farmyard manure	3.50	2.64	6.14	11.90	9.32	21.22
Cow dung	17.59	6.47	24.06	32.47	15.66	48.13
L.S.D. 5%	1.25	1.29	1.85	2.47	3.80	4.39
L.S.D. 1%	2.06	1.76	2.88	3.73	4.89	6.69

Onion Plant. Data in Table 5 indicate that the dry matter yields of onion bulbs, the economic part of the plant, were statistically higher in all manures treatments than the control. Cow dung treatments gave the highest increases in onion bulbs followed by farmyard manure treatments in both soils. Fresh slurry and dried slurry treatments gave the lowest increases in onion bulbs dry yield.

This lowest increase of onion yield from the application of the third additional dose of fresh and dried slurry, in spite of the fact that they contained higher amounts of fertilizer elements such as

Table 5. Effect of Organic Manures on Dry Matter Yield of Onion Grown in Virgin Sandy and Clay Soils (g/pot)

Treatment	Sandy Soil			Clay Soil		
	Bulbs	Leaves	Total	Bulbs	Leaves	Total
Control	2.74	1.87	4.61	4.67	3.77	8.44
Inorganic NPK	6.53	6.06	12.59	10.05	7.15	17.20
Fresh slurry	4.21	3.84	8.05	6.42	5.37	11.79
Dried slurry	3.88	3.17	7.05	6.79	5.04	11.83
Silt plus slurry	5.95	5.83	11.78	8.18	6.76	14.94
Composted slurry	4.71	4.02	8.73	7.46	6.34	13.80
Farmyard manure	7.01	6.93	13.94	11.52	8.59	20.11
Cow dung	8.02	7.71	15.73	13.59	10.39	23.98
L.S.D. 5%	1.29	0.94	1.52	1.02	1.27	1.62
L.S.D. 1%	1.76	1.27	2.07	1.36	1.73	2.21

nitrogen, phosphorus, and potassium, may be explained by the description of McCalla (1972). Excessive nitrogen from any source may affect crop production by disrupting the nutritional balance and causing lower yields and late maturity, increasing the tendency for the crop to lodge, making the crop susceptible to diseases, or causing undesirable plant composition. Powell and Webb (1972) described other possible reasons for yield reductions such as high soluble salt concentrations or some induced nutrient deficiency that did not exhibit any usual symptoms. Wallingford et al. (1974) found that corn forage yield was reduced by excessive quantities of nitrogen by using beef feedlot-lagoon water as organic fertilizer.

Olson (1971) observed that excessive phosphorus in soils can cause crop production problems before the level of saturation of the soil-fixing capacity is reached. The most usual difficulty arises from the interference of phosphorus with the zinc nutrition of crops. Corn is particularly susceptible to zinc deficiency. Hinsely et al. (1974) found that soybean yields have been increased significantly by different levels of sludge annually applied during the first three years of experimentation. But, in the fourth year, the response to maximum sludge treatments was negative. They explained that the plant toxicity symptoms observed, in the fourth year, may be attributed to phosphorus toxicity, a soil-plant salt interaction, or both.

Wilkinson and Stuedemann (1974) found that if high quantities of potassium are present in fertilizer, grasses absorb more potassium than magnesium or calcium.

Effect on Nitrogen, Phosphorus, and Potassium Uptake

Data in Table 6 show that the uptake of nitrogen, phosphorus, and potassium (NPK) differed with the manures applied, crop in the

Table 6. Effect of Organic Manures on Nitrogen, Phosphorus, and Potassium Uptake of Wheat, Turnips, and Onions Grown in Virgin Sandy and Clay Soils (mg/pot)

Treatment	Wheat			Turnips			Onion		
	N	P	K	N	P	K	N	P	K
				Sandy soil					
Control	59.67	10.22	50.27	70.44	5.90	32.10	93.09	15.18	207.15
Inorganic NPK	195.57	26.37	84.36	184.22	25.05	158.50	336.50	45.32	491.52
Fresh slurry	370.21	49.24	507.13	274.79	55.80	745.45	332.64	31.70	595.07
Dried slurry	285.36	59.90	416.35	185.70	36.97	533.05	214.21	28.74	679.16
Silt plus slurry	137.92	23.68	88.61	90.44	20.09	73.06	340.69	45.22	610.32
Composted slurry	157.49	26.21	92.22	228.23	28.64	283.81	249.04	33.25	412.05
Farmyard manure	160.61	33.65	99.85	78.24	16.04	134.15	368.08	53.09	457.98
Cow dung	205.50	37.52	157.70	612.62	74.26	776.34	519.34	64.64	720.33
				Clay soil					
Control	79.45	11.61	62.97	288.08	35.50	600.88	173.33	27.09	192.33
Inorganic NPK	269.32	49.05	270.57	704.71	85.86	1579.16	482.49	64.22	538.46
Fresh slurry	667.37	82.12	399.83	1102.91	121.12	2566.17	452.97	46.58	633.22
Dried slurry	817.47	58.45	224.55	742.73	105.53	1157.19	449.96	55.96	639.58
Silt plus slurry	489.64	50.64	167.42	497.80	75.45	936.20	348.17	56.49	703.36
Composted slurry	267.14	30.45	110.50	672.53	83.78	1335.16	423.02	53.78	433.59
Farmyard manure	261.24	36.16	144.35	424.60	59.90	761.59	442.54	84.00	795.71
Cow dung	323.44	43.33	239.42	1511.51	169.28	3276.09	684.20	102.25	704.94

rotation, and the soil type. In the case of wheat plants, all NPK uptake figures increased over the control with manure treatments in both soil types. The higher increases in NPK uptake were shown in fresh slurry, dried slurry, and cow dung treatments, and were more obvious in sandy soil. On the contrary, the lower increases were observed in silt plus slurry treatment in sandy soil and compost treatment in clay soil. Evans et al. (1977) found that the three organic manure treatments applied increased the NPK level of the corn plant tissue as compared to the control and to the inorganic fertilizer treatment.

In the case of turnips, the second crop in the rotation, all manure treatments gave increases in the NPK uptake more than the control. Cow dung and dried and fresh slurry gave the highest figures in the content and uptake of NPK. On the contrary, farmyard manure treatment induced the least increases in NPK uptake in both soil types (Table 6).

In the case of onions, the third crop in the rotation, the third dose of all manures applied caused obvious increases in the NPK uptake over the control. The highest NPK uptake was observed in cow dung treatments in both soils. Otherwise, NPK uptake in other treatments varied without any remarkable trend (Table 6). Also, there was no relation between the yield and each NPK uptake of the onion plant. This means that the nutrient contents in the plant tissue were not constant and were quite variable due to the third dose of manures. It has been shown that application of various types of manure (Hensler et al., 1970; McIntosh and Varney, 1972) and feedlot-lagoon water (Walling Ford et al., 1974) significantly changed the chemical composition of plants. Other workers have reported significant changes in the chemical composition of various crops following application of sewage sludge (Sabey and Hart, 1975).

Evans et al. (1977) found that solid beef-, liquid beef-, and liquid-hog manures applied for two successive years to a soil cropped for corn, increased the NPK level of the corn plant tissue as compared to the control and to the inorganic fertilizer treatment.

REFERENCES

Acharya, C.N. 1958. Preparation of fuel, gas, and manure by anaerobic fermentation of organic materials. Research Series No. 15. ICAR.

Allen, O.N. 1953. Experiments in Soil Bacteriology. Burges Publishing Company, United States.

Evans, S.D., P.R. Goodrich, R.C. Munter, and R.E. Smith. 1977. Effect of solid and liquid beef manure and liquid hog manure on soil characteristics and on growth, yield, and composition of corn. J. Environ. Qual. 6(4):361-368.

Hensler, R.F., R.J. Olson, and O.J. Attoe. 1970. Effect of soil pH and application rate of dairy cattle manure on yield and recovery of twelve plant nutrients by corn. Agron. J. 62:828-830.

Hinsely, R.D., O.C. Braids, R.I. Dick, R.L. Jones, and J.A.E. Molina. 1974. Agricultural Benefits and Environmental Changes Resulting from the Use of Digested Sludge on Field Crops. Final Rep. USEPA, Grant No. DOI-UI-0080, 375 pp.

Idnani, M.A., and Varadarajan, S. 1974. Fuel gas and manure by anaerobic fermentation of organic materials. Technical Bulletin No. 46. ICAR.

Jackson, M.L. 1958. Soil Chemical Analysis. Constable and Co., London.

McCalla, T.M. 1972. Think of manure as a resource, not a waste. Feedlot Manage. 14(5).

McIntosh, J.L., and Varney, K.E. 1974. Accumulative effects of manure and N on continuous corn and clay soil. I. Growth, yield, and nutrient uptake of corn. Agron. J. 64:374-379.

Mishra, U.P. 1954. Production of Combustible Gas and Manure from Bullock Dung and Other Organic Materials. M.Sc. Thesis, Indian Agr. Research Institute, New Delhi.

Olson, R.A. 1971. Fertilizer technology and use. Soil Sci. Soc, Am., Madison, Wisc. p. 439.

Pietz, R.I., J.R. Peterson, T.D. Hinesly, E.L. Ziegler, K.E. Redborg, and C. Lue-Hing. 1981a. Sewage Sludge Application to Calcareous Stripmine Soil. I. Effect on Soil and Corn N, P, K. Ca, Mg, and Na. Report No. 81-9, Department of Research and Development, the Metropolitan Sanitary District of Greater Chicago.

Powell, R.D., and J.R. Webb. 1972. Effect of high rates of N, P, K fertilizer on corn (Zea Mays L.) grain yields. Agron. J. 64:653-656.

Sabey, B.R., and W.E. Hart. 1975. Land application of sewage sludge: I. Effect on growth and chemical composition of plant. J. Environ. Qual. 4:252-256.

WallingFord, G.W., L.S. Murphy, W.L. Powers, and H.L. Manges. 1974. Effect of beef feedlot-lagoon water on soil chemical properties and growth and composition of corn forage. J. Environ. Qual. 3(1):74-78.

Wilkinson, S.R., and J.A. Stuedennann. 1974. Fertilizer: animal health problems and pasture fertilization with poultry litter. McGraw-Hill Year book on Science and Technology. McGraw-Hill Book Co., New York, N.Y. pp.180-182.

Zohdy, L.I., and S.M.S. Badr El-Din. 1983. Evaluation of digested slurry as a fertilizer. I. Its microbial and chemical composition and its phytotoxicity features. Egypt. J. Microbiol. 18(1-2):69-77.

Zohdy, L.T., and S.M.S. Badr El-Din. 1983. Comparison of biogas slurry treatments. I. On the biological and chemical properties of a virgin sandy soil cultivated with wheat. Egypt. J. Microbiol. 18(1-2):115-125.

Zohdy, L.T., and S.M.S. Badr El-Din. 1983. Comparison of biogas slurry treatments. II. On the wheat yield, nitrogen, and phsophorus uptake and nitrogen balance in a virgin sandy soil. Egypt. J. Microbiol. 18(1-2):127-134.

Zohdy, L.T., S.M.S. Badr El-Din, and M.A. Khalafallah. 1983. Handling and storage of the digested slurry produced from the biogas digester. Egypt. J. Microbiol. 18(1-2):107-114.

BIOGAS MANURE AS A COMPLETE FERTILIZER, FEASIBILITY FOR EGYPTIAN FARMERS

M.N. Alaa El-Din, I.M. Abdel Aziz,
M.H. Mahmoud, and S.A. El-Shimi
Agricultural Research Center, Giza, Egypt

ABSTRACT

The Ministry of Agriculture in collaboration with FAO and USAID has conducted a series of large experiments in farmers' fields to evaluate the manurial value of biogas effluent, commonly named "biogas manure." The experiments were designed to satisfy statistical analysis needs. The direct and residual effects of biogas manure as a complete substitute for NPK and micronutrients on the growth and yields of maize, wheat, rice, cotton, broad bean, spinach and carrots were evaluated during winter and summer seasons of years 1981-1983. The data were published successively as they emerged. In the present paper, only the marketable yields and yield components as affected by the biogas treatments, rates of application, fresh or mixed with the soil, and storage of biogas manures were compared to the optimal common farmers' methods. Economic evaluations of the application of biogas manure as substitute for N, NPK, and/or NPK micronutrients, including labor work involved, were also conducted.

INTRODUCTION

The rapid population growth in Egypt (about 1.2 million per year) presents a major problem. The area cultivated to provide biomass has remained unchanged for many years. Intensive cultivation has become a necessity and has put a heavy load on soil fertility. This situation has created a kind of balance between man, animals, and soil as competitors for agricultural residues, which in other countries may be difficult to dispose. Crop residues are highly needed as animal feed, burning material, and as organic manures to sustain fertility of old land and to improve it in newly reclaimed soils.

As biomass resources are limited, the order of priorities is as follows: animal feed utilization; burning in open fires for cooking and in poor quality ovens to bake bread; and finally to prepare organic manures. The amounts of biomass available for each of the uses are much lower than needed. Thus, traditional and ancient technologies are still in use in Egyptian rural areas and they need to be replaced or even supplemented with simple technologies which can improve the feeding value of residues and energy release as well as increase the amounts of organic manures and their quality.

Egyptian soils, as in most arid or semiarid regions, are relatively rich in the mineral elements required for plant growth, but deficient in organic matter, nitrogen (N), and micronutrients. They are also cultivated twice a year. This intensive cultivation depletes plant nutrients. Unless these nutrients are replaced in the soil, subsequent crops may give poor yields. Nature's way of keeping the soil fertile is by recycling the organic wastes; i.e., by returning them to the soil. The biogas process presents an improved treatment for both crop wastes and animal manures.

The organic matter content in Egyptian soils ranges from 0 to 2 percent, a level which is considered very poor and needs annual ammendment, especially after erecting the Aswan High Dam (in 1970) which deprived the soil of a major part of annually supplied suspended matter. In addition, projects for expanding the cultivated area by reclaiming desert sandy soils present an additional increasing demand for organic manures.

The importance of organic manures to Egyptian agriculture has been repeatedly proved especially in long-term field experiments as those conducted at Bahteem since 1919. As seen in Table 1 organic manuring could replace nitrogen and phosphate fertilizers and increase the yields of cereal, fiber, and forage crops by 16-172 percent over the optimal chemical fertilization treatments. Organic manuring also was favored over chemical fertilizers for its plant nutrient and soil pH values (Table 2).

The spent slurry, as it comes out of the biogas digester, contains about 90-94 percent moisture and 6-10 percent solid materials. This is called effluent or biogas manure. The solid material (sludge) is composed mainly of organic matter (humus) and plant nutrients. It can directly fertilize crops without the adverse effect, observed by fresh organic materials, of serious competition between the growing plant and the soil microorganism. As with solid bio-fertilizers, the liquid

Table 1 Effect of Prolonged Application of Chemical Fertilizers and Organic Manures on Crop Yields (Evaluation of Results 1919-1955, Values are given as % of the control)

		Mineral Fertilizers		
Crop	Control	Nitrogen	Nitrogen + Phosphorous	Organic manuring
Cotton	100	143	188	181
Wheat	100	147	192	264
Maize	100	155	229	245
Clover	100	100	265	437

Source: Alaa El-Din et al., 1983 a.

Table 2 Effect of Prolonged Application of Chemical Fertilizers and Organic Manures on Crop Yields and the Soil Contents of Plant Nutrients (average values of 43 years evaluation of permanent experiments of Bahteem).

	Control	Chemical Fertilizers			Organic manure
		N*	NP**	NPK***	
Crop Yield % (1959-1962)	100	151	165	143	212.5
Organic Matter %	1.08	1.16	1.17	1.17	2.51
Total Nitrogen %	0.072	0.077	0.077	0.076	0.153
NH_4^+N ppm	4.2	5.1	4.2	5.1	9.2
NO_3^-N ppm	16.9	17.2	17.0	17.7	29.4
P_2O_5 ppm	56.1	51.2	104.4	101.6	313.2
pH	8.2	8.4	8.3	8.3	7.6

*N = Nitrogen
**NP = Nitrogen + Phosphorous
***NPK = Nitrogen + Phosphorous + Potassium
SOURCE: Taha et al., 1966

phase carries trace elements such as zinc (Zn), iron (Fe), manganese (Mn) and copper (Cu). (Sathianathan, 1975; Abdel-Aziz et al., 1982).

Fertilization with biogas manure is expected to make supplemental fertilization with nutrient elements unneccessary (Sathianathan, 1975, and Mahmoud et al., 1982). Hornick et al., (1979) found that effleunt manure supplies NPK nutrients as well as considerable levels of the microelements, i.e. Zn, Fe, Mn and Cu in a suitable form for plants.

Van Buren (1979), reported that maize, rape-vegetable, and wheat yields increased by 28, 25 and 16 percent, respectively, when applying effluent in China.

In Egypt, the Ministry of Agriculture (MOA), based on various findings, conducted a large number of experiments in fields owned or operated by farmers with biogas plants through the FAO/MOA biogas project. The experiments used biogas manure as a complete substitute for earth compost, NPK, mineral fertilizers, and trace elements. The layout of the experiments and the statistical, chemical, and biological analyses of plants and soils are included in the original papers cited by each table. The economic return of each biogas manurial treatment was also evaluated and presented whenever available. Biogas manure, like any other organic manure, has two types of effects--direct and residual. The direct effect of biogas manure is usually attributed to its content of quick-acting macro and micro nutrients, vitamins, growth regulators, and the easily decomposable organic materials. The growing plant can benefit from these items immediately. Biogas manure also contains complex organic compounds needing a long time for mineralization. Since the released elements may not benefit the standing crop, it was necessary to evaluate the residual effect of the biogas manure on the next cropping.

COMPOSITION OF BIOGAS MANURE AND TREATMENTS

The mean values of the chemical analysis of biogas manure as compared with the traditionally prepared earth compost are presented in Table 3. The data showed that biogas manure contained higher concentrations of all elements, especially organic matter content. The figures recorded locally were lower than reported by Maramba et al. (1978) for pig manure in the Philippines.

The manure was applied either as fresh liquid (slurry) or after being stored. Mixing with the soil at rations ranging between 1:1 and 1:2 for biogas manure and soil, respectively, was also practiced and evaluated. Biogas manure mixed with soil and stored for some time (one to three months) brings manure very close to the earth compost traditionally used by the Egyptian farmer.

Table 3 Contents of Organic Matter and Plant Nutrients in Both Biogas Manure and Traditional Earth Compost

Component	Biogas Manure				Earth Compost			
	Total		Soluble		Total		Soluble	
	%	kg/m^3	%	kg/m^3	%	kg/m^3	%	kg/m^3
Organic matter	6.3	6.3	--	--	4.0	40	--	--
N	1.3	13.0	0.165	1.65	0.3	3	0.01	0.1
P_2O_5	0.45	4.5	0.0034	0.034	0.2	2	0.0023	0.023
K_2O	0.24	2.4	0.06	0.6	0.1	1	0.03	0.3
	ppm	g/m^3	ppm	g/m^3	ppm	g/m^3	ppm	g/m^3
Zn	3	3	1	1	1	1	0.6	0.6
Mn	6	6	4.7	4.7	3	3	1.4	1.4
Fe	10	10	6.6	6.6	4.6	4.6	2.0	2.0
Cu	0.2	0.2	0.1	0.1	0.05	0.05	0.001	0.001

SOURCE: Alaa El-Din et al., 1983a

AIM AND SCOPE OF WORK

The current work was carried out to evaluate some economical benefits of biogas manure as compared with traditional fertilizers applied to the major cereal crops: maize, wheat, rice, and broad bean;

the most important fiber and oil crop, cotton; and finally to carrots and spinach as representative of horticultural crops.

The residual effect of biogas manure applied to cotton and rice was also evaluated for the winter crops broad bean and wheat.

The data are presented in 14 tables and discussed below.

RESULTS AND DISCUSSION

Short-Term Effect of Biogas Manure

Maize. Maize plants responded significantly to the direct application of biogas manure alone or in combination with urea as compared with the traditional fertilization treatment (140 kgN/ha urea + 48 m^3 earth compost). Table 4 shows that the higher grain yield and income are obtained when biogas manure is used individually at the rate of 105 and 140 kgN/ha. However, the lower income is obtained when urea fertilizer is used alone (where other agricultural treatments are constant).

Fresh biogas manure (effluent) proved to be the best source of fertilizers as compared with urea or stored biogas manure (one part of effluent to one part of soil or two parts of soil).

Results in Table 5 indicate that the application of fresh biogas manure to the soil correspondingly increased income.

The application of biogas effluent, equivalent to 140 and 210 kg N/ha, increased the net income by 373.42 and 681.60 L.E./ha, respectively, over the net income gained by following the traditional farmer's treatment.

Application of stored biogas manure at the above mentioned N-rates increased the net income with only about 140.04 and 319.38 L.E./ha, respectively.

Data also showed that the foliar spray application of the micronutrients to maize plants fertilized with urea increased the yield and income significantly (Table 6). However, the application of these micronutrients in addition to biogas manure decreased the grain yield of maize, and thus the net income. This could be attributed to the presence of micronutrients in sufficient quantities in the biogas manure. Further addition of the elements might therefore have brought the concentration to a level slightly toxic to the plants.

Wheat. Wheat is one of the major problem crops in Egypt since it is a major food. In 1980 the wheat yield dropped three percent to reach 1.8 million tons, but consumption rose seven percent. To help meet its needs, Egypt imported 4.4 million tons of wheat. Any effort to increase wheat yield is therefore highly needed.

The effect of direct application of biogas manure alone or combined with ammonium sulphate with or without zinc sulphate as foliar spray was investigated in a large field experiment at Moshtohour during 1981/1982. The data are summarized in Table 7.

Table 4 Cost Benefit Data of Applying Biogas Manure to *Zea mays L.* (November 1981)

Nitrogen fertilizers applied to maize as kg N/ha			Grain yield ton/ha	Grain yield[b] increase ton/ha	Economical Value, L.E./ha						
					Urea						
Biogas manure	Urea[a]	Earth Compost			Total price	Cost of distribution	Cost of biogas manure transportation	Cost of earth compost transportation	Total cost of fertilizer used	Yield increase	Gain or loss due to treatments
-	140	48 m^3	7.00	-	61.6	4.50	-	20.00	86.10	-	-
70	-	-	8.37	1.37	-	-	10.00	-	10.00	269.38	345.48
105	-	-	9.43	2.43	-	-	20.00	-	20.00	474.34	540.44
140	-	-	9.50	2.50	-	-	25.00	-	25.00	488.00	549.10
-	140	-	7.27	0.27	61.6	4.50	-	-	66.10	52.70	72.70
17.5	140	-	8.90	1.90	61.6	4.50	2.50	-	68.60	370.88	388.38
35.0	140	-	9.17	2.17	61.6	4.50	5.00	-	71.10	423.58	439.58

[a] One ton urea = 240 $ X 0.83 = 199.20 L.E. (international price, 1981).
[b] One ton maize = 136.5 £ X 1.43 = 195.2 L.E. (international price without transportation, September 1981).
SOURCE: Mahmoud et al., 1982

Table 5: Cost benefit data of applying biogas manure to *Zea mays L.* as Compared with Chemical Fertilizers, 1981.

Treatments: Nitrogen fertilizers as kg N/ha					Economical Value L.E. /ha							
					Urea		Biogas manure: Cost of transportation					
Biogas manure	Urea[a]	Earth Compost	Grain yield[b] ton/ha	Grain yield increase ton/ha	Total price	Cost of distribution	Effluent	Stored	Earth Compost cost of transportation	Total cost of fertilizers used	Yield Increase	Gain or loss due to treatments
--	140	48 m^3	7.00	-	61.60	4.5	-	-	20.00	86.10	-	-
--	140	-	7.26	0.26	61.60	4.5	-	-	-	66.10	50.75	70.75
140 (Effluent)	-	-	9.50	2.50	-	-	25.00	-	-	25.00	312.32	373.42
140 (Stored 1:1)	-	-	7.43	0.43	-	-	-	30.00	-	30.00	83.94	140.04
140 (Stored 1:2)	-	-	8.77	1.77	-	-	-	45.00	-	45.00	345.50	286.59
--	210	-	10.90	3.90	92.40	4.5	-	-	-	96.90	716.28	705.48
210 (Effluent)	-	-	10.23	3.23	-	-	35.00	-	-	35.00	630.50	681.60
210 (Stored 1:1)	-	-	8.40	1.40	-	-		40.00	-	40.00	273.28	319.38
210 (Stored 1:2)	-	-	8.20	1.20	-	-		60.00	-	60.00	234.24	210.34

[a]One ton urea = 240 $x0.83 = 199.20 L.E. (international price, 1981).
[b]One ton maize grain = 136.5 £ x 1.43 = 195.2 L.E. (international price without transportation, September, 1981).
SOURCE: Mahmoud et al., 1984.

Table 6 Cost Benefit Data of Applying Biogas Manure to *Zea Mays* L. as Compared with Chemical Fertilizers, 1981

Fertilizer Treatments as kg N/ha	Grain yield[a] ton/ha	Grain yield increase ton/ha	Economical Value L.E./ha: Urea[b] Total Price	Urea[b] Cost of distribution	Biogas manure cost of transportation	Micronutrient[c] Total price	Micronutrient[c] Cost of distribution	Cost of earth compost	Total cost of fertilizers	Yield increase	Gain or loss due to treatment
140 urea + 48 m^3 earth comp	7.00	-	61.6	4.5	-	-	-	20.0	86.10	-	-
140 urea	7.26	0.26	61.6	4.5	-	-	-	-	66.10	50.75	70.75
140+48 m^3 earth comp.+ 0.03% Zn	8.00	1.00	61.6	4.5	-	5.14	3.50	20.0	94.74	195.20	186.56
140+48 m^3 earth comp.+ 0.03% Fe	7.67	0.67	61.6	4.5	-	5.14	3.50	20.0	94.74	130.78	122.14
140+48 m^3 earth comp.+ 0.03% Mn	9.50	2.50	61.6	4.5	-	5.14	3.50	20.0	94.74	488.0	579.36
140 biogas manure	9.50	2.50	-	-	25.0	-	-	-	25.00	488.0	549.10
140+0.03% Zn	9.03	2.03	-	-	25.0	5.14	3.50	-	33.64	396.26	448.72
140+0.03% Fe	9.13	2.13	-	-	25.0	5.14	3.50	-	33.64	415.78	468.24
140+0.03% Mn	8.43	1.43	-	-	25.0	5.14	3.50	-	33.64	279.14	331.60

[a] One ton of maize = 136.5 £ X 1.43 = 195.2 L.E. (without transportation, September 1981).

[b] One ton urea = 240$ X 0.83 = 199.20 L.E. (international price, 1981).

[c] Micronutrients = Zn, Fe, and Mn chelate foliar spray at the rate of 0.03% (1400 L/ha).

SOURCE: Abdel-Aziz et al. 1982.

Table 7 Cost Benefit Data of Applying Biogas Manure to Wheat as Compared with Chemical Fertilizers (1982)

Fertilizer Treatments as kg N/ha			Yield, ton/ha				Economical Value L.E./ha									
			Total		Increase		Amm. sulphate 20.6% N		Cost of Biogas	Zinc Sulphate		Total cost of ferti- lizer	Yield Increase		Total value of yield increase	Gain or loss due to treat ments
Biogas manure (ef- fluent)	Amm. sul- phate[a]	$ZnSO_4$[b] spray (0.4%) 1000 L/ha	Grain[c]	Straw[d]	Grain	Straw	Total price	Cost of distri- bution	trans- porta- tation	Total price	Cost of spray- ing		Grain	Straw		
-	142.5	-	6.858	12.05	-	-	47.08	3.00	-	-	-	50.08	-	-	-	-
-	142.5	0.4%	7.535	13.28	0.677	1.23	47.08	3.00	-	4.00	5.00	59.08	122.88	184.35	307.23	298.18
142.5	-	-	7.715	12.25	0.857	0.20	-	-	2.00	-	-	20.00	155.55	2.96	158.51	188.54
142.5	-	0.4%	6.858	11.41	-	0.64	-	-	20.00	4.00	5.00	29.00	-	96.43	96.43	117.4
-	215.0	-	7.470	12.59	0.612	0.24	71.03	3.00	-	-	-	74.03	111.08	50.42	161.5	137.5
-	215.0	0.4%	8.618	12.57	1.760	0.52	71.03	3.00	-	4.00	5.00	83.03	319.44	78.66	398.1	365.1
215.0	-	-	8.933	12.61	2.075	0.56	-	-	30.00	-	-	30.00	376.61	83.51	460.12	480.15
215.0	-	0.4%	8.833	12.66	1.975	0.61	-	-	30.00	4.00	5.00	39.00	358.46	91.89	450.35	461.38
71.25	71.25	-	7.595	12.08	0.737	0.03	23.54	3.00	10.00	-	-	36.54	133.77	3.78	137.55	151.04
71.25	71.25	0.4%	7.853	13.17	0.995	1.12	23.54	3.00	10.00	4.00	5.00	45.54	180.59	167.33	347.92	352.41
71.25	142.50	-	8.00	12.51	1.142	0.46	47.08	3.00	10.00	-	-	60.03	207.27	68.7	275.97	265.97
71.25	142.50	0.4%	7.563	12.33	0.705	0.28	47.08	3.00	10.00	4.00	5.00	69.03	127.96	42.23	170.19	151.19
107.50	107.50	-	8.350	12.64	1.492	5.31	35.52	3.00	15.00	-	-	53.52	270.80	87.73	358.23	354.74
107.50	107.50	0.4%	7.375	12.58	0.517	0.53	35.52	3.00	15.00	4.00	5.00	62.52	93.84	79.29	173.13	150.64
142.5	71.25	-	9.000	13.37	2.142	1.32	23.54	3.00	20.00	-	-	56.54	388.77	198.4	587.17	580.66
142.5	71.25	0.4%	6.928	11.59	0.070	0.46	23.54	3.00	20.00	4.00	5.00	65.54	12.71	69.01	56.3	40.79

[a]One ton of Amm. sulphate 20.6% N = 62 $ X 0.83 = 199.20 L.E. (international price, 1982).

[b]One Kg of $Zn\ SO_4$ = 1.00 L.E. (local price)

[c]One ton of wheat grains = 126.9 £ X 1.43 = 181.5 L.E. (international price withou: transportation, July 1982).

[d]One ton of wheat straw = 150 L.E. (local price).

SOURCE: Alaa El-Din et al., 1983b

The application of biogas manure at the rate equivalent to 142.5 and 215 kg N/ha resulted in increasing the grain yield by 0.857 and 1.463 ton/ha, respectively, over the treatments receiving the same N-levels as ammonium sulphate. On the other hand, the straw yield also increased by about 0.200 and 0.320 ton/ha over the traditional farmer's treatments. The net income increased after applying the biogas manure by about 188.54 and 342.65 L.E./ha for the above mentioned N-rates, respectively. The highest yield of grain and straw and the best net income, however, were obtained when wheat was fertilized with 215.0 kgN/ha from both biogas manure and ammonium sulphate at the ratio of 142.5:71.25 kgN/ha respectively.

The data also revealed that the application of zinc sulphate as foliar spray to wheat plants fertilized with ammonium sulphate at the rates of 142.5 and 215.0 kg N/ha increased the yield and net income significantly. The income increases due to application of zinc sulphate were 298.18 and 227.6 L.E./ha for the two N-levels, respectively.

However, with the treatment of 71.25 kg N/ha from biogas manure and 71.25 kg N/ha from ammonium sulphate, the application of zinc sulphate to wheat plants fertilized with biogas manure alone or combined with ammonium sulphate decreased the grain and straw yield of wheat and thus the net income as compared with the same treatments without zinc sulphate.

The above-mentioned data indicate that when biogas manure was used, the application of micronutrients decreased the yield of crops under the conditions of this investigation. Further addition of these elements might have therefore brought the concentration to levels slightly toxic to the plants, as was inferred in the case of maize.

Rice. The effect of biogas manure on rice yield (paddy and straw) was evaluated in a large field experiment at Moshtohour during summer 1982. The comparison was made with chemical fertilization with urea at the recommended rate (107kg N/ha). Zinc sulphate applied at the rate of 48 kg/ha to the nursery was also evaluated.

The results, presented in Table 8, showed that the direct application of biogas manure (effluent) at a rate equivalent to 107 kg N/ha (the recommended N-fertilization level for rice) was the best in paddy yield as compared with urea applied at the same level. Lower yields (paddy and straw) were obtained when applying biogas manure combined with urea to satisfy the recommended N-level. The yield, however, remained higher than the urea-fertilized plots.

The highest income was achieved when applying biogas manure; 166.3 L.E./ha higher income was gained as a result of applying 107 kg N/ha as biogas manure when compared with the same level of N applied as urea.

Combinations of biogas manure with urea to satisfy the recommended N-rate achieved a higher income than urea alone. Their income, however, remained lower than the single application of biogas manure.

Application of zinc sulphate in the presence of urea did not show any improvement to the rice yield. The soil of this experiment

Table 8 Cost Benefit Data of Applying Biogas Manure to Rice as Compared with Chemical Fertilizers (1982).

Treatments: Nitrogen fertilizers (kg/N) Biogas manure	Urea[a]	$ZnSO_4$	Yield, ton/ha: Total Grain[b]	Total Straw[c]	Increase Grain	Increase Straw[c]	Economical value L.E./ha: Urea Price	Urea Cost of distribution	Zn SO_4[d] Price	Zn SO_4 Cost of distribution	Cost of biogas manure transportation	Total cost of fertilizer used	Yield increase Grain	Yield increase Straw	Total value of yield increase	Gain or loss due to treatments
-0	107	-	8.52	9.00	-	-	46.3	6.0	0	0	0	52.3	0	0	0	0
107	0	-	9.02	9.28	0.30	0.28	0	6.0	0	0	15	21.0	134.0	7.0	141.0	166.3
36	71	-	8.85	9.50	0.33	0.50	30.8	6.0	0	0	5	41.8	88.44	12.5	100.94	106.44
71	36	-	8.59	9.14	0.07	0.14	15.6	6.0	0	0	10	25.6	18.76	3.5	22.26	48.96
0	107	+	8.28	9.00	-0.24	0.00	46.3	6.0	6.25	2.0	0	54.55	-64.32	0.0	-64.32	-66.57
107	0	+	5.50	7.38	-3.02	-1.62	0	6.0	6.25	2.0	15	29.25	-809.36	-40.5	-849.81	-826.81
36	71	+	8.19	8.64	-0.33	-0.36	30.8	6.0	6.25	2.0	5	50.05	-88.44	-9.0	-97.44	-95.19
71	36	+	7.00	7.38	-1.52	-1.62	15.6	6.0	6.25	2.0	10	39.85	-434.16	-40.5	-474.66	-462.21

[a] One ton urea = 240 $ X 0.83 = 190.2 L.E. international price, 1981.

[b] One ton rice grain = 670 kg White grain = 268 L.E. international price., 1981.

[c] One ton rice straw = 25 L.E. Local price.

[d] One kg zinc sulphate = 1 L.E. Local price (one ha = 6.25 kg Zn SO_4)

Note: One hectar nursery supplies rice seedlings to cultivate an 8 hectare area

SOURCE: Alaa El-Din et al., 1983b

contained 2.6 ppm zinc, a concentration which is considered sufficient for rice plants. Therefore, no response was evidenced. Application of zinc in the presence of biogas manure showed a depressing effect on the yield. Both paddy and straw yields decreased proportionally with the amount of biogas manure applied. Paddy yields decreased from 8.28 ton/ha when applying 107 kg N/ha urea plus zinc to 5.50, 7.00, and 8.19 ton/ha when applying biogas manure equivalent to 107, 71, and 36 kg N/ha in the presence of zinc, respectively. The net income decreased by increasing the quantity of biogas manure reaching the largest loss (760.24 L.E./ha) in the presence of the highest amount of biogas manure. The toxic effect of biogas manure in the presence of zinc fertilization is simply due to the fact previously discussed in the cases of maize and wheat. Namely, the zinc available in biogas manure together with the zinc sulphate applied brought the zinc level in the soil to a concentration toxic to rice plants while application of biogas manure alone made the zinc content in the soil ideal for both paddy yield and net income.

Broad bean. The direct application of biogas manure to the most important seed legume crop in Egypt, broad bean (*Vicia faba*), and its economic return was studied as compared to nitrogen and phosphorous fertilization. The recommended level of nitrogen (24 kg N/ha) was applied either as biogas manure or as urea. Superphosphate was applied at the rate of 37 kg P_2O_5/ha to urea-treated plants, while biogas treatments were considered as substitutions for nitrogen and phosphate fertilization. The data are given in Table 9.

Applications of biogas manure at rates equivalent to 24 or 12 kg N/ha were superior in their seed and straw yields as compared to the best chemical fertilization treatment. The lower level of biogas manure increased the seed and straw yields by 10.7 and 11.5 percent over the higher level of N and P chemical fertilizers. No differences, however, were recorded between the two levels of biogas manure. The positive effect of biogas manure is therefore considered to be attributed to its content of major and minor elements rather than to the nitrogen. This is expected since seed legumes, especially broad bean, can meet their N demand through the biological fixation of atmospheric nitrogen.

The net income increased by about 235 L.E./ha when applying the low rate of biogas manure as compared to the net income of the urea-phosphate fertilization treatment.

Cotton. Cotton was chosen as the most important fibre and oil crop in Egypt, as well as the major export crop. Biogas manure was evaluated for its effect on both yields of seed cotton and stalks. It was added on the basis of its nitrogen content to satisfy the levels of 107 and 161 kg N/ha. However, it was considered as a replacement for both nitrogen and phosphate fertilizers. The treatments and results are presented in Table 10.

Application of fresh biogas manure equivalent to 107 kg N/ha gave the highest seed cotton yield (3.97 ton/ha or 10.6 Quantar/feddan) being 72 percent higher than the treatment receiving 107 kg N/ha as ammonium sulphate and 37 kg P_2O_5/ha. Biogas manure mixed with the soil (1:1) and stored for two months gave a lower yield of seed (3.25

Table 9 Cost Benefit Data of Applying Biogas Manure to Broad Bean as Compared -with Chemical Fertilizers (1983)

Fertilizer Treatments			Yield ton/ha				Economical Value..E./ha									
			Total		Increase		Urea[c]		Super Phospate[d]							
Biogas manure	Urea kgN/ha	Super phosphate (kg P_5O_5/ha)	Seeds[a]	Straw[b]	Seeds	Straw	Price	Cost of distri-bution	Prie	Cost of distri-bution	Cost of biogas manure transpor-tation	Total cost of fertilizer used	Yield Increase Seeds	Yield Increase Straw	Total Value of yield increase	Cost or loss due to treat-ments
-	-	-	4.96	3.36	-	-	-	-	-	-	-	-	-	-	-	-
-	24	37.0	5.73	4.76	0.77	1.40	10.41	4.0	7.4	3.0	-	24.21	893.55	70.0	963.55	938.74
12	-	-	6.12	5.48	1.16	0.72	-	-	-	-	2.0	2.00	1019.35	158.0	1177.35	1175.35
24	-	-	6.10	5.53	1.14	0.05	-	-	-	-	4.0	4.0	1012.90	157.0	1169.90	1165.90

[a] One ton brad bean seeds = 322.58 (local price)
[b] One ton braod bean straw = 50.00 (local price)
[c] One ton urea = 240 \$ x 0.83 = 190.2 L.E. (international price)
[d] One ton superphosphate 15.5% P_2O_5 = 31 (international prices)
SOURCE: Alaa El-Din et al. 1983b.

Table 10 Cost Benefit Data of Applying Bioga Manure to Cotton as Compared with Chemical Fertilizers (1982)

Treatments: Nitrogen fertilizers (kg N/ha) Biogas Manure	Amm. Sulphate	Super Phosphate kg P_2O_5/ha	Yield, ton/ha: Total Seed Cotton[a]	Total Stalk	Increase Seed Cotton	Increase Stalk	Economic Value L.E./ha: Amm.sulphate[b] 20.6% N Price	Amm.sulphate Cost of distribution	Super phosphate[c] (15.5 P_2O_5) Price	Super phosphate Cost of distribution	Cost of biogas manure transportation	Total cost of fertilizer used	Yield Increase Seed cotton	Yield Increase Stalk[d]	Total value of yield increase	Gain or loss due to treatments
-	107	37	2.31	10.23	-	-	42.59	7.15	7.4	3.0	-	60.14	0	0	0	0
107 (Fresh)	-	-	3.97	11.43	1.66	1.20	0	0	0	0	15.00	15.00	1054.1	12.0	1066.1	1111.24
107 (Stored 1:1)	-	-	3.25	12.15	0.94	1.92	0	0	0	0	30.00[e]	30.00	596.9	19.2	616.1	642.24
-	161	37	3.13	11.26	0.82	1.03	64.09	7.15	7.4	3.0	0	81.64	520.7	10.3	531.0	509.50
161 (Fresh)	-	-	3.42	11.43	1.11	1.20	0	0	0	0	22.50	22.50	704.9	12.0	716.9	753.54
107 (Fresh)	54	-	3.38	12.30	1.07	2.03	21.36	7.15	0	0	15.0	43.51	679.5	20.3	699.8	716.43
161 (Stored 1:1)	-	-	3.28	11.49	0.97	1.26	0	0	0	0	42.00[e]	42.00	616.0	12.60	628.60	646.74
54 (Fresh)	107	37	3.17	12.34	0.86	2.11	42.29	7.15	7.4	3.0	15.0	75.14	546.1	21.10	567.20	552.20

[a] One ton seed cotton yield = 635 L.E. (international price, 1981).
[b] One ton of Amm.-sulphate = 82 $ X 0.83 = 68.06 L.E. (international price, 1982).
[c] One ton super phosphate 15.5% P_2 0_2 = 31.0 L.E. (international price, 1982).
[d] One ton of cotton stalk = 10.0 L. E. (local price)
[e] Cost of stored biogas manure contained the cost of transportation and distribution.

Source: Alaa El-Din et al., 1983b.

ton/ha or 8.7 Quantar/feddan), but higher than the chemical fertilizer treatment and even higher than the treatment receiving 161 kg N/ha as ammonium sulphate. Application of higher levels of ammonium sulphate increased the seed yield from 2.31 ton/ha to 3.13 ton/ha. It remained, however, lower than the biogas manure applied to satisfy the lower level of nitrogen.

The increase in the net income of cotton from the application of fresh biogas manure equivalent to 107 kg N/ha reached 1,111 L.E./ha as compared with those receiving the same nitrogen level as ammonium sulphate in addition to superphosphate. Storing biogas manure reduced the net gain to reach only 642.24 L.E./ha because of the lower efficiency in increasing yield and the higher processing cost as compared with fresh biogas manure. Application of a higher level of biogas manure either alone or in combination with chemical fertilizers gave lower net gain as compared with the lower rate of biogas manure.

Spinach. The results dealing with the effect of ammonium sulphate and biogas manure and their combinations, with or without iron chelate as foliar spray, on the yield and income of spinach are presented in Table 11.

The results show that the application of biogas manure alone at the level of 120 kg N/ha gave the highest yield of spinach as compared with ammonium sulphate at rates of 120 and 180 kg N/ha, with or without using iron chelate as foliar spray. On the other hand, the application of iron chelate with ammonium sulphate significantly increased the yield of spinach, while biogas manure together with iron decreased spinach yield.

The highest income was obtained when biogas manure was applied alone at a level equivalent to 120 kg N/ha. The net income increase was 188.15 L.E./ha over that of the traditional farmer's treatment (120 kg N as ammonium sulphate/ha).

Carrots. The data on carrots yield, as well as the gain or loss due to the application of biogas manure as compared with traditional treatment, are shown in Table 12.

The data indicate that application of biogas manure alone or combined with ammonium sulphate gave the best yields of carrots and the highest income as compared with the mineral nitrogen fertilization using ammonium sulphate.

The application of biogas manure at the rate of 105 kg N/ha increased the net income by 160.321 L.E./ha as compared with the same nitrogen level as ammonium sulphate, while application of a higher level of biogas manure (140 kg N-equivalent) increased the net income by 176.834 L.E./ha as compared with 140 kg N/ha ammonium sulphate. The highest yield and the highest net income were obtained, however, when carrots were fertilized with 140 kg N/ha from biogas manure and ammonium sulphate at the ratio of 105:35 kg N/ha, respectively.

The Residual Effect of Biogas Manure

The residual effect of biogas manure was evaluated using a cereal

Table 11 Cost Benefit Date of Applying Biogas Manure to Spinach as Compared with Chemical Fertilizers (1982)

Fertilizer Treatments			Vegetative	Vegetative	Economical Value L. E./ha							
			yield	yield	Amm. Sulphate[b] 20.6%N		Cost	Fe-Chelate[c]		Total Cost		
Biogas manure	Amm. sulphate (kg N/ha)	Fe Spray	ton/ha	increase ton/ha	Total price	Cost of distribution	Biogas manure transport-ation	Total price	Price of Distribution	fertilizer used	yield increase	Gain or loss due treat-ments
-	120	-	16.05	-	39.65	3.00	-	-	-	42.65	-	-
-	180	-	16.48	0.40	59.47	3.00	-	-	-	62.47	20.00	0.180
-	120	+	19.05	3.00	39.65	3.00	-	5.143	3.50	51.293	150.00	141.357
30	180	+	17.13	1.08	59.47	3.00	-	5.143	3.50	71.113	54.00	25.537
60	120	-	17.63	1.58	39.65	3.00	10.00	-	-	52.65	79.00	69.00
120	-	-	19.36	3.31	-	-	20.00	-	-	20.00	165.50	188.15
120	60	-	18.09	2.04	19.82	3.00	20.00	-	-	42.82	102.00	101.83
120	-	+	14.29	-1.76	-	-	20.00	5.143	3.50	28.643	-88.00	-73.99

[a] One ton Vegetative Spinach = 50.00 L.E. (local price).
[b] One ton Amm. Sulphate = 20.6 %N = 82 $ X 0.83 = 68.06 L.E. (international price, 1982).
[c] One kg Fe-Chelate 14% = 12.00 L.E. (local price).
SOURCE: Alaa El-Din et al, 1983b.

Table 12 Cost Benefit Data of Applying Biogas Manure to Carrots as Compared with Chemical Fertilizers (1982)

Nitrogen Fertilizers kg N/ha as				Economical Value L. E./ha					
Biogas manure	Amm. Sulphate	Vegetative growth yield[a]	Yield increase ton/ha	Amm. Sulphate[b] Total Price ton/ha	Cost of distribu-tion	Cost of biogas manure	Total cost transportation	Vegetative yield increase of fertilizer	Gain or loss due to treatments
-	105	36.145	-	34.691	3.0	-	37.691	-	-
-	140	40.464	4.319	46.254	3.0	-	49.254	107.975	96.412
35	105	44.414	8.269	34.691	3.0	5.00	42.691	206.725	201.725
70	70	45.755	9.610	23.127	3.0	10.00	36.127	229.000	230.564
105	-	41.650	5.505	-	-	20.00	20.000	137.625	155.316
105	35	46.174	10.029	11.564	3.0	20.00	34.564	250.725	253.852
140	-	46.367	10.222	-	-	25.00	25.000	255.550	268.241

[a] One ton vegetative carrots = 25.00 L. E. (local price, 1982).
[b] One ton amm. sulphate 20.6% N = 82 $ X 0.83 = 68.06 L.E. (international price, 1982).
SOURCE: Alaa El Din et al., 1983b

crop (wheat) following rice which has been fertilized with biogas manure or chemical fertilizers in summer 1982. Wheat was cultivated during winter 1982/1983. Broad bean was chosen as representative of winter grain legumes to follow cotton receiving biogas manure during summer 1982. Broad bean was also cultivated in winter season 1982/1983.

The fertilization treatments for wheat and broad bean were the same as for rice and cotton, respectively, as they were cultivated on the same plots.

Residual Effect of Biogas Manure on Wheat. Wheat was cultivated on the same plots of rice and thus had the same fertilizer treatments. Each plot was divided into four equal parts. The first part was left without any treatment to evaluate the combined residual effect of biogas manure while the second was fertilized with 37 kg P_2O_5/ha only, the third part was fertilized with 107 kg N/ha as urea, and the fourth part of the plot received both nitrogen and phosphate fertilization. The data are given in Table 13.

The residual effect of biogas manure applied to rice cultivated in summer 1982 on the following wheat crop was clear when compared with the same effect of urea containing the same level of nitrogen (107 kg N/ha). The effect increased by increasing the percentage of nitrogen fertilizer applied as biogas manure. Grain yield increased by 0.34, 0.61, and 0.91 ton/ha as a result of applying 36, 71, and 107 kg N/ha as biogas manure, respectively. Straw yield showed even higher increases than grain yield, being 0.30, 0.98, and 1.43 ton /ha for the residual effect of the same treatment, respectively. These figures were obtained in comparison with the residual effect of applying 107 kg N/ha as urea to the previous crop.

The net income increased also as a result of the residual effect. Application of one-third of the nitrogen as biogas manure to the previous crop (rice) increased the net income of the following wheat by 107 L.E./ha over that of the treatment which had received the total nitrogen fertilizer dose (107 kg N/ha) as urea. Increasing the ratio of nitrogen applied as biogas manure to cover two-thirds the quantity, increased the net income to reach 258 L.E./ha. Application of the full N-dose as biogas manure to rice gave the highest residual effect on grain and straw yields and consequently exhibited the largest net income (380 L.E./ha).

Application of phosphate fertilizers (37 kg P_2O_5/ha) to the present wheat crop did not reflect any response on the yield of grain or straw, indicating that the phosphates contained in the biogas manure were quite sufficient to meet the demand of both rice and wheat crops. Additional treatment was made on the border that did not receive biogas manure during the former season. Application of P_2O_5 did not show notable differences in respect to the residual effect.

To evaluate the nitrogen fertilizer component in the residual effect of biogas manure, part of the plots received fertilizer with 107 kg N/ha to support the wheat. Additional treatment of the border that did not receive biogas manure in the former season was fertilized with 178.5 kg N/ha. The data presented in Tables 11 and 12 showed much higher yields for all treatments than those not receiving N-fertilizer

Table 13 Cost Benefit Data of Residual Effect of Biogas Manure and Chemical Fertilizers Applied to Rice (summer, 1982) on Wheat Cultivated After Rice (winter 1983)

Treatments: Residual nitrogen fertilizers applied to rice, 1982 kg N/ha		Treatments: Present fertilizer treatments applied to Wheat (82/83)		Yield, ton/ha: Total		Yield, ton/ha: Increase		Total cost of fertilizer used	Economic Value L.E./ha	
biogas manure	Urea	Urea [a] kg N/ha	Super [b] phosphate kg P_2O_5/ha	Grain [c]	Straw [d]	Grain	Straw		Yield increase	Gain or loss due to treatments
-	107	-	-	3.05	5.12	-	-	-	-	-
36	71	-	-	3.39	5.42	0.34	0.30	-	106.71	106.71
71	36	-	-	3.71	6.10	0.61	0.98	-	257.72	257.72
107	-	-	-	3.96	6.55	0.91	1.43	-	379.67	379.67
-	107	-	37	3.17	5.16	-	-	10.4	-	-
36	71	-	37	3.22	5.33	0.05	0.17	10.4	34.58	34.58
71	36	-	37	3.47	5.40	0.30	0.24	10.4	90.45	90.45
107	-	-	37	3.88	5.65	0.71	0.49	10.4	202.37	202.37
-	-	-	37	2.96	4.86	-0.21	-0.30	10.4	-83.12	-83.12
-	107	107	-	5.26	9.90	-	-	50.3	-	-
36	71	107	-	5.39	10.33	0.13	0.43	50.3	68.10	68.10
71	36	107	-	5.64	11.00	0.38	1.10	50.3	233.97	233.97
107	-	107	-	5.82	11.66	0.49	1.76	50.3	352.94	352.94
-	-	178.5	-	5.82	12.00	0.56	2.10	75.6	416.64	391.34
-	107	107	37	5.36	10.02	-	-	60.7	-	-
36	71	107	37	5.55	11.12	0.19	1.10	60.7	199.49	199.49
71	36	107	37	5.69	11.45	0.33	1.43	60.7	274.40	274.40
107	-	107	37	5.98	12.06	0.62	2.04	60.7	418.53	418.53
-	-	178.5	37	6.41	11.38	1.05	1.76	86.0	454.58	439.28

[a] One ton urea = $240 x 0.83 = 199.2 L.E. (international price, 1981).
[b] One ton supherphosphate 15.5% P_2O_5 = 31.0 L.E. (international price, 1981).
[c] One ton of wheat grains = 126.9 L. x. 1.43 = 181.5 L.E. (international price, 1982).
[d] One ton of wheat strau 150 L.E. (local price).

SOURCE: Alaa El-Din et al., 1983b.

in the present season. This indicates that the residual nitrogen from both urea and biogas manure treatments from the previous crop (rice) was not sufficient to give high grain or straw. The yields of those treatments not receiving any fertilizers for wheat ranged between 3.05-3.96 ton/ha and 5.12-6.55 ton/ha of grains and straw, respectively. Those receiving N-fertilizer in addition to the residual effect gave grain and straw yields ranging between 5.26-5.82 and 9.90-11.66 ton/ha, respectively. Application of 178.5 kg N/ha as urea without the residual effect of biogas manure gave more or less the same yields of grain and straw as the treatment receiving 107 kg N/ha in the present season. Additionally, the residual effect of biogas manure applied to rice at the rate of 107 kg N/ha was about equivalent to 71.5 kg N/ha and 37 kg P_2O_5/ha. Equivalent fertilizer quantities could therefore be saved when fertilizing wheat crops cultivated after rice fertilized with biogas manure.

Application of both nitrogen and phosphate fertilizers in the present season did not show a considerable increase in yield as compared with the application of nitrogen fertilizer only. This seems to verify the findings mentioned previously that biogas manure can satisfy the phosphate needs for the present and the following crops.

The net income has also increased as a result of the residual effect of biogas manure under the application of nitrogen and phosphate fertilizers to the present wheat crop. Values of 200, 274, and 419 L.E./ha were recorded for the net income increase after applying biogas manure to satisfy one-third, two-thirds, and full nitrogen needs, respectively.

<u>Residual Effect of Biogas Manure on Broad Bean</u> The residual effect of biogas manure applied to cotton (Summer 1982) was evaluated using the most common food legume grain, namely broad bean.

Broad bean was cultivated (1982/1983) on the same plots of cotton, and thus had the same fertilizer treatments. The treatments and results are given in Table 14.

The residual effect of biogas manure applied to cotton on the broad bean yield was higher than the same effect of ammonium sulphate and superphosphate applied to the previous crop. Biogas manure showed 16.2 percent and 11.1 percent grain and straw yields higher than ammonium sulphate and superphosphate applied to cotton. Stored biogas manure showed here also had lower positive residual effect indicating the losses in nutrients and/or in growth-promoting substances, which might have happened during the storage of biogas manure. Fertilization of broad bean with 24 kg N and 37 kg $P_2\ O_5$/ha as ammonium sulphate and superphosphate, respectively, gave more or less the same yield as revealed from the residual effect of fresh biogas manure applied to the previous crop (cotton) at a rate equivalent to 107 kg N/ha.

Slight increases in grain and straw yields were obtained for the residual effect of higher levels of fertilizers (chemical and biogas manure) applied to the previous crop. The net gain due to the residual effect of biogas manure was 260 L.E./ha higher than the residual effect of applying the same level of N-chemical fertilizer combined with P-fertilizer. The former was slightly higher than the net income of

Table 14 Cost Benefit Data of Residual Effect of Biogas Manure and Chemical Fertilizers Applied to Cotton (summer 1982) on Broad Bean Cultivated After Cotton (winter 1983)

Treatments: Fertilizers applied to cotton (1982)			Yield, ton/ha				Economical Value L.E./ha		
Biogas manure kg N/ha	Amm. sulphate kgN/ha	Super phosphate kg P_2O_5/ha	Total Seeds [a]	Total Straw [b]	Increase Seeds	Increase Straw	Total cost of Fertilizer used	Yield increase	Gain or loss due to treatments
A)	Residual effect after cotton								
-	107	37	4.50	4.42	-	-	-	-	-
107 (Fresh)	-	-	5.23	4.91	0.73	0.49	-	260.0	260.0
107 (Stored 1:1)	-	-	4.97	4.76	0.47	0.34	-	168.6	168.6
B)	Immediate effect								
-	24	37	5.26	5.04	0.57	0.41	24.81	276.2	251.39
A)	Residual effect after cotton								
-	161	37	4.69	4.63	-	-	-	-	-
161 (Fresh)	-	-	5.38	5.00	0.69	0.37	-	241.1	241.1
161 (Stored 1:1)	-	-	5.75	5.11	1.06	0.48	-	365.9	365.9
107 (Fresh)	54	-	5.29	4.90	0.60	0.27	-	207.0	207.0
54 (Fresh)	107	37	5.00	4.66	0.31	0.03	-	101.5	101.5
B)	Immediate effect								
-	24	37	5.26	5.04	0.57	0.41	24.81	204.37	179.56

SOURCE: Alaa El-Din et al., 1983b

[a] One ton fo broad bean seeds = 322.58 L.E. (local price).
[b] One ton of broad bean stau = 50,00 L.E. (local price).

fields fertilized with 24 kg N and 37 kg P_2O_5/ha in the present season. The net income due to the residual effect of higher fertilizer levels (equivalent to 161 kg N/ha) was not much different from that of the lower level. It did increase, however, by increasing the ratio of biogas manure in the fertilizer applied.

The yield results were, however, lower than those revealed from the fresh application of biogas manure at rates much lower than applied to the previous crop. Broad bean grain yields as much as 6.12 ton/ha were recorded when applying biogas manure to broad bean at a rate equivalent to 12 kg N/ha (see Table 9). The difference between grain yields could not be attributed to the different supply of nitrogen or phosphate between the fresh and residual effect of biogas manure, as the first one was due to 12 kg N/ha while the latter was the residual effect of 107 and 161 kg N/ha.

The residual of 107 kg N/ha as biogas manure applied to rice was found to benefit the following wheat with about 71.5 kg N/ha and 37 kg $P_2\ O_5$/ha. In view of the fact that cotton is a longer lasting crop (eight months) than rice (three months after transplanting), the residual effect of biogas manure after cotton evaluated as N-fertilizer would certainly exceed the level of 12 kg N/ha. In other words, the lower yield recorded for the residual effect of biogas manure as compared with the fresh application of very low amounts of nitrogen as biogas manure could not be attributed to the smaller supply of nutrients from the former treatment. The presence of growth-promoting substances repeatedly reported for biogas manure could be a logical explanation for the higher yields recorded from the freshly applied biogas manure.

It is of great interest and importance to simulate several crop rotation options to include the direct and residual effects of biogas manure and to evaluate the net gain of income from biogas application. Taking into consideration the size of the biogas unit and the area of land fertilized with biogas and other fertilizers, a total picture of economic balance for households adopting the use of biogas manure could be drawn. However, berseem clover in this case should be included since this forage crop occupies about half of the area cultivated for winter crops or about 25 percent of the total cultivated area in Egypt. Economic balance models could not be designed without evaluating the effect of the direct and residual effects of biogas manure on berseem clover.

Additional field experiments were therefore planned for the coming winter season 1983/1984 for evaluation of the direct effect of biogas manure, and in winter 1984/1985 to evaluate the residual effect on berseem clover. During these two years of activity, evaluation of the direct and residual effect of biogas manure on productivity of citrus, grapes, and banana farms can possibly be carried out.

ACKNOWLEDGMENTS

The authors wish to thank the Ministry of Agriculture of Egypt, the Agricultural Research Center, and the Soils and Water Research Institute for encouragement and providing facilities. The financial assistance provided by FAO and USAID is highly appreciated.

REFERENCES

Abdel-Aziz, I.M., M.N. Alaa El-Din, M.H. Mahmoud, and S.A. El-Shimi. 1982. Fertilizer value of biogas manure II. Micronutrients for maize. The first OAU/STRC. Inter-African Conference on Bio-fertilizers. Cairo, Egypt, 22-26 March.

Alaa El-Din, M.N., S.A. El-Shimi, M.H. Mahmoud, and I.M. Abdel Aziz. 1983a. Biogas for Rural Egypt: Energy, Manure, Fodder. The Egyptian Experience. (in Arabic) Giza, Egypt. March 1983.

Alaa El-Din, M.N., S.A. El-shimi, M.H. Mahmoud, and I.M. Abdel Aziz. 1983b. The analysis of the impact of biogas technology on Egyptian agriculture. MOA/USAID.

Alaa El-Din, M.N., S.A. El-shimi, M.H. Mahmoud, and I.M. Abdel Aziz. 1984. Effect of storage and moisture content on biogas manure. Second conference, ARC, Giza, 9-11 April 1984

Hornick, S.B., J.J. Murray, G.B. Willson, and C.F. Tester. 1979. Use of sewage sludge compost for soil improvement and plant growth. U.S. Department of Agriculture, Science and Education Administration. Agriculture. Reviews and Manuals. ARM-NE.6.

Mahmoud, M.H., S.A. El-shimi, I.M. Abdel Aziz, and M.N. Alaa El-Din. 1982. Fertilizer value of biogas manure. I. Nitrogen for maize. The first OAU/STRC Inter-African Conference on Bio-fertilizers. Cairo, Egypt. 22-26 March.

Mahmoud, M.H., S.A. El-Shimi, I.M. Abdel-Aziz and M.N. Alaa El-Din. 1984. Effect of storage on biogas manure. An evaluation of biogas manure as a fertilizer for maize. Second Conference, ARC Giza, 9-11 April.

Maramba, Sr., F.D. 1978. Biogas and waste recycling, the Philippine experience. Maya Farms Division, Liberty Flour Mills, Inc., Metro Manila, Philippines.

Sathianathan, M.A. 1975. Bio-gas, Achievement and Challenges. Association of Voluntary Agencies for Rural Development. New Delhi, pp.26-53.

Taha, S.M., S.Z. Mahmoud, A. El-Damaty, and A.N. Ibrahim. 1966. Effect of prolonged use of fertilizers on chemical and microbiological properties of soil. J. Microbiol. UAR. Vol. 1, No. 2.

van Buren A. 1979. A Chinese biogas manual. Popularizing technology in the countryside. Office of the Leading Group for the Propagation of Marshgas, Sichuan (Szechuan) Province, People's Republic of China.

CONTROL OF PATHOGENS

HEALTH RISKS ASSOCIATED WITH THE USE OF BIOGAS SLURRY: AN INTRODUCTORY NOTE

M. Adel A. Tawfik
National Research Centre, Egypt

Where sludge is to be applied to the soil, particularly on land used for growing edible crops that are eaten raw, disinfection of the sludge is important.

Anaerobic digestion of sewage and manure is effective in reducing the number of pathogens, but may not eliminate them completely. The ova of parasitic worms, the cysts of protozoa, and different microbial agents can survive for long periods of time. Eggs of *Ascaris*, *Toxocara*, and *Trichuris* are intestinal nematodes that belong to a group referred to as the soil-transmitted parasites. In a sense, soil serves as an intermediate host for these eggs as they undergo development in any external environment before becoming infective for the next host. The length of this period varies depending upon the worm species and environmental conditions. The eggs of these worms are resistant to a wide range of physical and chemical conditions and are capable of surviving for several years in the soil.

The eggs of the tapeworms *Taenia sp.*, *Echinococcus granuloses*, and *Moniezia sp.* are infective for their intermediate hosts at the time they are discharged from the segments of the worm. Consequently, unlike the eggs of the soil-transmitted nematodes, when these tapeworms eggs are present in raw sewage or manure, they are already infective for the next host.

The oocysts of *Entamoeba histolytica*, *Toxoplasma*, and *Eimeria*, spp. are not infective when passed in the hosts' faeces, but require a few days to reach infectivity in the soil. These oocysts are relatively resistant to many environmental conditions.

Bacterial pathogens of potential risk to humans and animals are known to be found in municipal sludge. *Salmonella* can be reduced 90 to 99 percent by anaerobic digestion. If the sludge is not dried or dewatered, no further treatment reductions will occur prior to land application. *Mycobacteria* are acidfast rods that are responsible for tuberculosis and leprosy in humans and tuberculosis in animals. Anaerobic digestion inactivates about 90 percent of the *mycobacterium* population in sludge. *Mycobacteria* are relatively stable in drying sludge. Once released to the environment, the organisms are quite stable and would likely persist for long periods. Clostridial organisms, mainly Clostridium tetani, Clostridium botulism, and Clostridium perfringens, are of human and animal origin. As anaerobic micro-organisms they can resist anaerobic digestion and adverse conditions, including complete dryness.

From a public health standpoint, the potential for human or animal disease caused by pathogens transmitted from sludge is most important to consider. However, the evaluation of health risks associated with the land application of municipal sludge is very difficult since many factors must be considered. The factors include the types and number of viable stages of pathogens, the manner of application, soil characteristics, land topography, climatic conditions including rainfall and temperature, and the subsequent use of the land. The possibility of disease transmission is greater if fresh vegetables are eaten from, or if grazing occurs on land where sludge has been applied. Therefore, sludge applied to the land surface or incorporated into the soil must receive pretreatment to protect public health.

Natural evironmental conditions will reduce the density of microorganisms. A number of factors such as sunlight, soil moisture, and temperature affect the persistence of the organisms and their ability to survive. Bacteria, protozoa, and viruses generally are inactivated in a few days to a few months, but helminth takes several years. Air drying of digested sludge reduces the number of pathogens but cannot ensure complete safety. Heat drying of either digested or fresh sludge is virtually the only process that ensures complete disinfection. However, the cost of the process is so high that its use is seldom justified. Chlorine can be used to disinfect sludge, but the dose must be high, and the sludge must be well digested or homogenized to permit access of the chlorine to all the pathogens.

INCIDENCE, PERSISTENCE, AND CONTROL OF PARASITIC EGGS AND CYSTS IN ANAEROBICALLY DIGESTED WASTES

M. A. Tawfik, M. A. Hassanain, Nabila Sh. Deghidy
Parasitology and Animal Diseases Laboratory,
National Research Centre, Egypt

ABSTRACT

The effect of anaerobic digestion on the viability and infectivity of *Toxocara vitulorum* eggs and *Eimeria* sp. oocysts was investigated under field as well as laboratory-controlled conditions. Results indicated that *T. vitulorum* could survive the anaerobic treatment whereas *Eimeria* sp. oocysts could not.

INTRODUCTION

The use of anaerobically digested sludge on land to grow crops for human or animal consumption presents a serious risk of parasitic disease transmission . Among parasitic organisms, there are certain ova and cysts that can survive for long periods of time. Different ascarid eggs, trichurid eggs, *Taenia* eggs, and oocysts of both *Entamoeba* and *Eimeria* spp. are highly resistant to a wide range of adverse conditions for a long time.

Hays (1977) and the National Academy of Sciences (1977) reported the parasitic organisms present in the sludge of human origin as: *Entamoeba histolytica*, *Ascaris lumbricoides*, *Trichuris trichura*, *Hymenolepis* sp., *Taenia* sp. *Entrobius vermicularis*, *Ancylostoma*, and *Necator*.

Roberts (1935), Northington et al (1970), and a report of the National Centre for Disease Control (1970) indicated that the applications of sewage effluents and sludge to pastures increased parasitic diseases, especially beef tapeworm (*Taenia saginata*), among cattle.

The following work emphasizes the action of anaerobic digestion of sludge on the viability and infectivity of *Toxocara vitulorum* eggs (representing parasite eggs) and *Eimeria* spp. oocysts (representing cysts of protozoa).

METHODOLOGY

I. Incidence of Ascaris Eggs and Eimeria Oocysts in Different Village Digesters

Ninty six sludge samples, each of 200 ml, were obtained from the different operating digesters, 52 from Omar Makram, 32 from El-Manawat, and 12 samples from Shubra Kass villages. The samples represented the inlet, outlet, stored basin, and stored heap locations. Each sample was collected in a plastic container with its recording date, site of collection, and name of the village. The samples were examined on the collection day by the concentration flotation technique using a saturated common-salt solution.

II. Laboratory-controlled experiments

The experiments were conducted in laboratory-scale digester units, each a 2 l capacity glass bottle. Each unit was closed with double-hole stoppers on the top. One hole was closed permanently after sample collection, while the other was connected by a rubber hose to a sealed container filled with water. The anaerobic digestion for the contained sludge was conducted in an incubator at a constant temperature for a survival time of 45 days. The digesters temperatures were between 16°C and 28°C, representing Egyptian climatic conditions during winter and summer. The pH of the sludge was adjusted between 4 and 9. The normal pH of sludge (i.e., 6.9) was represented in the control bottle digester. Another form of treatment involved adding 10 percent of calcium carbonate to the sludge.

Toxocara vitulorum eggs and *Eimeria* sp. oocysts, of cattle and sheep origin, were inoculated at a rate of 100 of each per ml of the sludge consisting of fresh manure and wastes. The investigated samples were taken every other day during the first week, then every week until the end of the 45-day period.

Aeration of the sludge after 45 days. The anaerobically digested sludge of each treatment was transferred to a suitable container exposed to air at 30°C for 24 hours to partially evaporate the contained wastes. This was followed by the collection of *Ascaris* eggs and *Eimeria* oocysts incubated at 28°C in 1.5 percent formalin for 30 days and in 2.5 percent potassium dichromate for 22 days respectively.

RESULTS AND DISCUSSION

Incidence of Ascaris Eggs and Eimeria Oocysts in Different Village Digesters: As displayed in Table 1, the number of both *Ascaris* eggs and *Eimeria* oocysts decreased gradually from the inlet to the outlet in both digesters of Omar Makram and El-Manawat. A sharp decrease in the number of *Ascaris* eggs is observed. The samples of Omar Makram, included complete disappearance from the stored heap, contrary to findings for El-Manawat. Therefore, it could be concluded that the operating village digesters of Omar Makram are more efficient in destroying *Ascaris* eggs than those of El-Manawat. The process of anaerobic digestion, used in both village digesters, minimized the number of *Eimeria* oocysts with variable results, but the El-Manawat digester is preferred to those of Omar Makaram in this respect.

Table 1 Occurrence of Ascaris sp. Eggs and Eimeria sp. Ooycysts in the Different Village Digesters

Site of samples	Omar Makram						El-Manawat						Shubra Kass					
	Ascaris eggs			Eimeria Oocysts			Ascaris eggs			Eimeria Oocysts			Ascaris eggs			Eimeria Oocysts		
	No.	positive	%	No.	positive	%	No.	positive	%	No.	positive	%	No.	positive	%	No.	positive	%
Inlet	15	5	33.33	15	14	83.33	8	4	50.0	8	5	62.5	6	0	0	6	3	50
Outlet	15	2	13.33	15	12	80.00	8	4	50.0	8	4	50.0	6	0	0	6	3	50
Stored basin	14	1	7.14	14	6	42.85	8	2	25.0	8	1	12.5	-	-	-	-	-	-
Stored heap	8	0	0	8	2	25.00	8	3	37.5	8	1	12.5	-	-	-	-	-	-
Total	52	8	15.38	52	34	85.38	32	13	40.62	32	11	34.37	12	0	0	12	6	50

Sludge samples obtained from Shubra Kass were free from Ascaris sp. eggs, while the occurrence of Eimeria oocysts in both the inlet and the outlet was equal.

It is important to mention that Ascaris eggs taken from the inlet, outlet, and stored basin were identical and of normal shape and size, while those collected from the stored heap were degenerated. Eimeria sp. oocysts detected in the outlet, stored basin, and stored heap at the different digesters were nonsporulated.

II. Laboratory-Controlled Experiments

The results obtained from the laboratory-controlled experiments done to investigate the effect of anaerobic digestion on the viability of T. vitulorum eggs and Eimeria sp. oocysts are shown in Figures 1 and 2. Different temperatures and hydrogen ion concentrations and the addition of calcium carbonate had a lethal effect on the viability of Eimeria oocysts, with an ovistatic action on T. vitulorum eggs. The lethal effect persisted even after exposing the oocysts to the optimum conditions of their development, while 13-19 percent of the eggs developed to the larval stage.

Oxygen starvation, pH, and temperature together affected the destruction of Ascaris eggs. These results are supported by Fairbairn (1957, 1960) and Arthur et al. (1981). On the other hand, Foster and Engibrecht (1973) found that Ascaris lumbricoides eggs survived after 14 days of anaerobic fermentation. Moreover, Reyes et al (1963) stated that the destruction rate of A. lumbricoides var. suum eggs, in all anaerobic digesters, grew with increases in temperature, while Gram (1943) suggested that 10 percent of these eggs survived six months anaerobic digestion at 20°C.

Concerning Eimeria sp. oocysts, the present study as well as that of Soulsby (1968) indicated that oxygen is necessary for the development of oocysts. Similar results were confirmed by Cram (1943) and Fitzgerald and Ashley (1973) with Entamoeba histolytica cysts. They stated that the cysts were completely destroyed in anaerobically digested sludge after more than 12 days at 20°C and after 10 days at 30°C.

CONCLUSION

The results of the present investigation indicated (1) the eggs of T. vitulorum were capable of surviving the anaerobic digestion of sludge for 45 days as indicated by their development to the larval stages, and (2) Eimeria sp. oocysts were unable to survive the anaerobic conditions.

REFERENCES

Arthur, R. G., P. R. Fitzgerald, and J. C. Fox. 1981. Parasitic ova in anaerobically digested sludge. Journal WPCF, 53, 8.

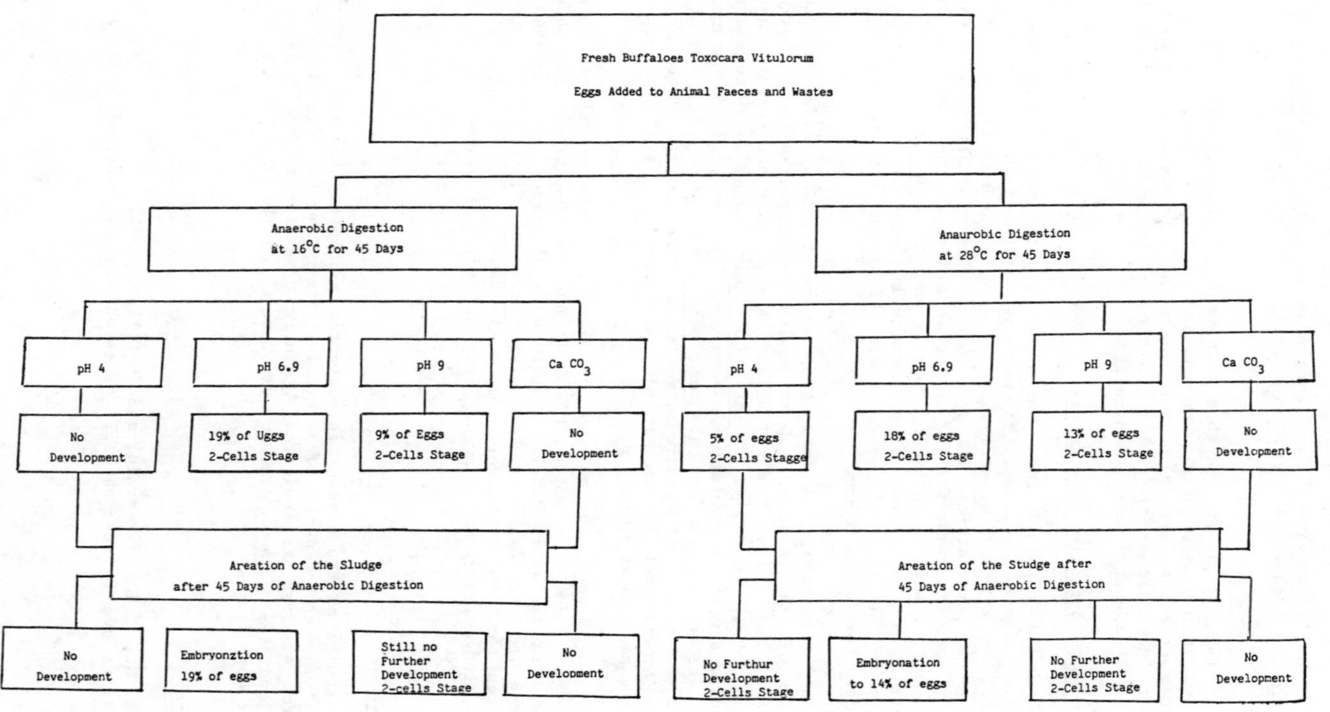

Figure 1 Schematic Diagram of Work Done on Toxocara Vitulorum Eggs Under Anaerobic Digestion of Animal Faeces and Wastes.

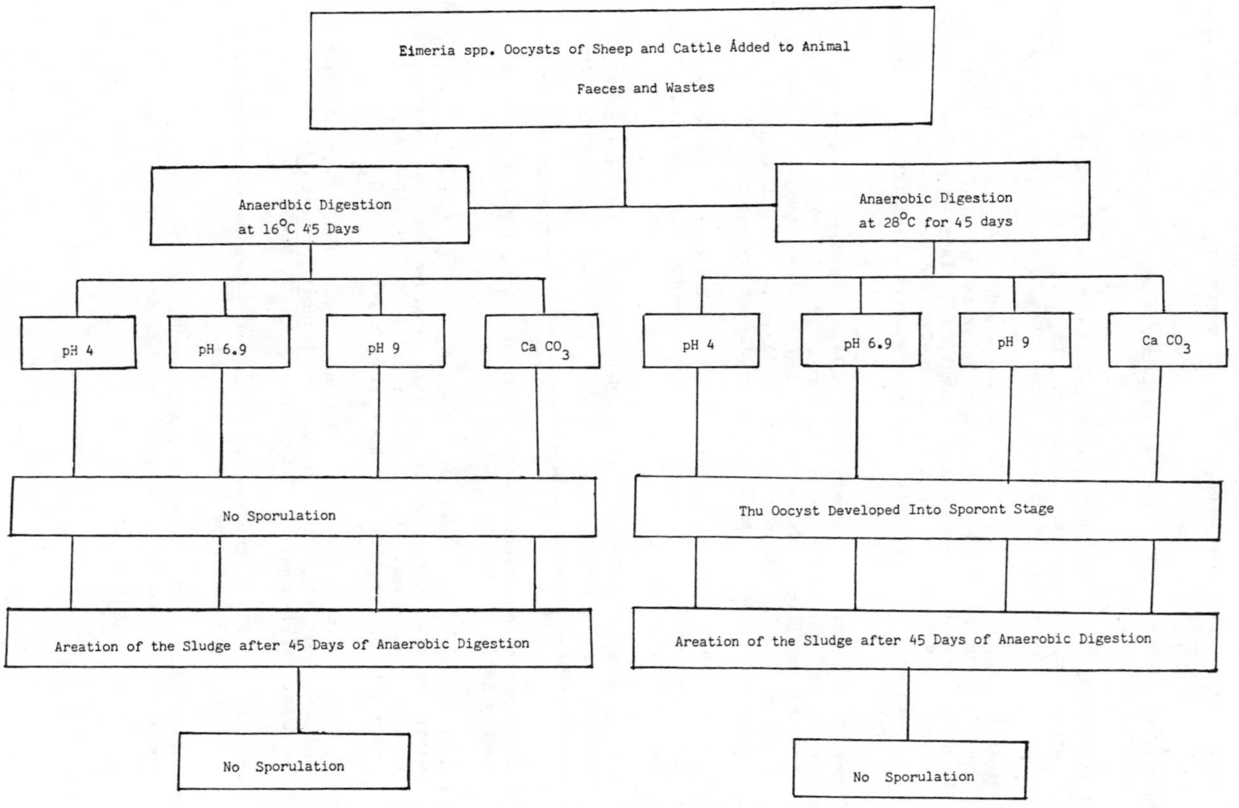

Figure 2 Schematic Diagram of Work done on Eimeria spp. Under Anaerobic Digestion of Animal Faeces and Wastes.

Gram, E. B. 1943. The effect of various treatment processes on the survival of helminth ova and protozoa cysts in sewage. Sewage Wks. J., 15:1119-1138

Fairbairn, Donald. 1957. The biochemistry of Ascaris. Exp. Parasitol., 6:291

Fairbairn, Donald. 1960. Physiological aspects of egg hatching and larval exsheathment in nematodes. In Host Influence of Parasite Physiology. Rutgers University Press, New Brunswick, New Jersey.

Fitzgerald, P., R. F. Ashley. 1973. Experimental studies on parasite survival in digested sludge. In Parasitological Study of Sludge and the Effect of some Human, Animal, and Plant Parameters. Final Report, 1971-1973. Urbana, Illinois: University of Illinois (submitted to the Metropolitan Sanitary District of Greater Chicago).

Foster, D. H., and R. S. Engelbrecht. 1973. Microbial hazards in dispersing of waste water on soil, pp. 247-270. In Recycling Treated Municipal Waste Water and Sludge Through Forest and Cropland. W.E. Stopper and L. T. Kardos, eds. University Park, Pennsylvania: Pennsylvania State University Press.

Hays, B. D. 1977. Potential for parasitic disease transmission with land application of sewage plant effluents and sludges. Water Research 11:583-595.

National Academy of Sciences. 1977. Methane Generation from Human, Animal, and Agricultural Wastes. National Academy Press: Washington, D.C.

National Centre for Disease Control. 1970. Morbidity and mortality. Weekly Rep. 19:100.

Northington, C.W., S.L. Chang, L. J. McCabe. 1970. Health aspects of waste water reuse. Water quality improvement by physical and chemical processes. E. F. Gloyna and W. W. Eckenfelder, Jr. eds. University of Texas Press: Houston pp 619-6565.

Reyes, W. L., C. W. Kruse. M. S. Batson. 1963. The effect of aerobic and anaerobic digestion on eggs of Ascaris lumbricoides var. suum in night soil. Amer. J. Trop. Med. Hyg., 12, 46-55.

Roberts, F. C. 1935. Experiences with sewage forming in southwest. U.S. Amer. J. Public Health 25:122-125.

Soulsby, E.J.I. 1968. Helminths, Arthropods, and Protozoa of Domesticated Animals. Bailliere, Tindall, and Gassell:London

INCIDENCE, PERSISTENCE, AND CONTROL OF SOME PATHOGENS DURING ANAEROBIC DIGESTION OF ORGANIC WASTES

M.A. Tawfik, A.G. Hegazi, H. Soufy and Laila Ali
Laboratory of Parasitology and Animal Diseases
National Research Centre, Egypt

ABSTRACT

The Incidence of Salmonella, Shigella, atypical Mycrobacteria, and Clostridia was studied in three village digesters using organic wastes as a biogas source.

The isolated Salmonella strains were identified as S. pullorum, S. gallinarum, S. entritidis, S. typhi, S. typhimurium and S. dublin. The atypical Mycobacteria were rapid growers and belonging to scotochromogenic and non-chromogenic types. Four types of pathogenic anaerobic clostridia were identified as Cl. welchii, Cl. oedematiens, Cl. tetani, and Cl. septicum. Shigella was not detected in the samples.

The persistence and control of Salmonella, Shigella, atypical Mycobacteria, and Clostridia were also studied under laboratory conditions during anaerobic digestion of the sludge. This study was done under different pH values (i.e., 4,6.9, and 9) and with the addition of 10 percent calcium carbonate at temperature of 16°C and 28°C. These three pH values were chosen to simulate an acid-phase digester, a normal one, and a digester suffering from alkalinity build up, respectively.

INTRODUCTION

The use of sludge obtained from the anaerobic digestion of human and animal wastes as a fertilizer may cause some public health hazards. Many diseases can be transmitted through the improper handling of diseased human and animal excrements. Among the bacterial diseases are those caused by Salmonella, Shigella, atypical Mycobacteria and Clostridia strains (Prissick and Masson, 1956; Brandes, 1960; Clancey, 1964; Paterson, 1965; Smith Holdeman, 1968; Krebs and Kappler, 1971; McColle and Elliot, 1971, and National Academy of Sciences, 1977). Some of these organisms are widely distributed in nature (i.e., soil, dust, air, water, and sewage).

Therefore, this study's goal was to examine the incidence, persistence, and control of Salmonella, Shigella, atypical Mycobacteria, and Clostridia organisms under various conditions prevailing during the anaerobic digestion of sludge.

METHODOLOGY

The investigation consisted of two parts: (1) the isolation and identification of Salmonella, Shigella, atypical Mycobacteria, and Clostridia in 384 sludge samples obtained from different village digesters under operation in El-Manawat, Omar Makram, and Shubra Kass from April 1983 to August 1984; and (2) the determination of the persistence and control of previously mentioned pathogens during anaerobic digestion of the sludge under laboratory conditions.

The experiments were conducted in laboratory-scale digester units, each a 2 capacity glass bottle. Each unit was closed with double-hole stoppers on the top. One hole was closed permanently after the sample collection, while the other was connected by a rubber hose to a water-sealed gas collector.

Three groups of digesters were used, each consisting of four digesters. Using either NaOH or HCl solutions, the pH of each group was adjusted to be 4, 6.9 (normal pH of the sludge) and 9. These pH values were chosen to represent the acid-phase digesters in a two-stage digestion process, a normal one, and one suffering from alkalinity build up. Also 10% calcium carbonate was added as a cheap buffering. The temperature of the digesters was adjusted to 16°C and 28°C for the first and second group of digesters respectively, representing Egyptian climatic conditions during winter and summer. All previous treatments were performed with the addition of the following microorganisms: Salmonella dublin, Shigella flaxinary, two strains of atypical Mycobacteria identified as rapid growers, non-chromogenic and scotochromogenic acid-fast bacilli, and Cl. oedematiens type A. The time of exposure was 45 days during which samples were taken for bacteriological reisolation of the inoculated strains at 0, 3, 7, 14, 21, 28, 35, and 45 days. The inoculated pathogenic strains were previously isolated and identified from the anaerobic digested sludge.

Isolation and Identification of the Pathogens

Isolation and identification of Salmonella and Shigella were done according to Cruickshank and Duguid (1968).

The isolation of mycobacterial strains was performed according to the Beerwerth method (1971) and a preliminary identification of the isolated strains was based on acid fastness, rate of growth, and pigment production according to the Runyon classification of Mycobacteria (1959).

The isolation and identification of Cl. oedematiens were adopted according to Nishida and Nakagawara (1964) and Willis (1977). Plating was done by spreading a 0.1 ml portion of appropriate dilution on blood agar plates. The growing colonies were counted after 48 hours of anaerobic incubation (Garcia and McKay, 1969).

RESULTS

Isolation of Pathogens in Samples Obtained from Different Operating Village Digesters

From 384 samples examined, the positive samples numbered with 19 Salmonella, 12 with atypical Mycobacteria, and 52 with Clostridia containing samples (Table 1). The isolated Salmonella strains were identified as S. pullorum, S. typhi, S. entritidis, S. typhimurium, S. gallinrum, and S. dublin. While, the atypical Mycobacteria are rapid growers and belonging to scotochromogenic and nonchromogenic types. 4 types of pathogenic anaerobic Clostrida were identified as Cl. welchii, Cl. oedematiens, Cl. tetani, and Cl. septicum. Shigella was not detected from any examined samples.

Persistence and Control of Pathogens during Anaerobic Digestion of Sludge under Laboratory Conditions

At temperature 16°C and pH 6.9 Salmonella, Shigella, atypical Mycobacteria were persistent for 14. 7, and 28 days respectively. At a pH value of 4 Salmonella, Shigella, and atypical Mycobacteria were viable until 7, 3, and 14 days post inoculation respectively. At pH 9 all pathogens were not detected from the start. Salmonella only persisted for 3 days post inoculation in 10% calcium carbonate (Table 2).

At temperature 28°C and pH 6.9, Salmonella, Shigella, and atypical Mycobacteria were viable until 21, 7, and 14 days respectively. Salmonella, Shigella, and atypical Mycobacteria persisted for 7, 3, and 7 days at pH 4, while no organisms were reisolated at pH 9. Salmonella and Shigella persisted for 7 and 3 days with the addition of 10 percent calcium carbonate, while atypical Mycobacteria were not detected (Table 2).

The viable count of Cl. oedematiens type A (Table 3) increased from the third day until the seventh day at different temperature and pHs. At 16°C and 28°C the viable count was decreased by 20 percent and 34 percent at pH 4, 20 percent and 30 percent at pH 9 respectively from the original viable count. At pH 6.9 no changes were observed in viable count at 16°C, while 10 percent decreases were manifested at 28°C. The addition of 10 percent calcium carbonate induced a reduction of the viable count after the third day until the end of the experiment. These reductions were 50 percent at 16°C and 60 percent at 28°C.

DISCUSSION

The occurrence percentage of the pathogens in the organic waste material may differ according to the area of grazing, type and nature of foodstuffs, and human and animal health states.

From Table 1 it was observed that Salmonella was isolated from inlet, outlet, and basin at the villages of El-Manawat and Omar Makram digesters, while it was detected in the inlet and outlet only at the Shubra Kass village digester (Table 1). This finding was similar to those reported by Jones at al. (1980).

Table 1a Incidence of <u>Salmonella</u>, <u>Shigella</u>, Atypical <u>Mycobacteria</u>, and <u>Clostridia</u> in Samples Obtained from Different Operating Village Digesters

Source of Isolation		El-Manawat Village Digester												Omar Makram Village Digester												Shubra Kass Village Digester											
		<u>Salmonella</u>			<u>Shigella</u>			Atypical <u>Mycobactria</u>			<u>Clostridia</u>			<u>Salmonella</u>			<u>Shigella</u>			Atypical <u>Mycobacteria</u>			<u>Clostridia</u>			<u>Salmonella</u>			<u>Shigella</u>			Atypical <u>Mycobactria</u>			<u>Clostridia</u>		
		No.	+ve	%	No.	+ve	%	No.	+ve	%	No.	+ve	%	No.	+ve	%	No.	+ve	%	No.	+ve	%	No.	+ve	%	No.	+ve	%	No.	+ve	%	No.	+ve	%	No.	+ve	%
Inlet	116	8	5	66.5	8	0	0	8	2	25	8	5	62.5	15	3	20	15	0	0	15	5	33.3	15	11	73.3	6	1	16.6	6	0	0	6	2	33.3	6	3	50.0
Outlet	116	8	4	50.0	8	0	0	8	0	0	8	5	62.5	15	2	13	15	0	0	15	0	0	15	9	60.0	6	1	16.6	6	0	0	6	0	0	6	3	50.0
Basin	88	8	1	12.5	8	0	0	8	1	12.5	8	3	37.5	14	2	14	14	0	0	14	0	0	14	6	42.8	0	0	0	0	0	0	0	0	0	0	0	0
Heap (Compost)	64	8	0	0	8	0	0	8	1	12.5	8	2	25.5	8	0	0	8	0	0	8	1	12.5	8	5	62.5	0	0	0	0	0	0	0	0	0	0	0	0
Total	384	32	10	31.25	32	0	0	32	4	12.5	32	15	47.0	52	7	13.4	52	0	0	52	6	11.5	52	31	60.0	12	2	16.16	12	0	0	12	2	16.6	12	6	50.0

Table 1b Types of Isolated Pathogens from Different Operating Village Digesters

	El-Manawat			Omar Makram			Shubra Kass		
	Salmonella	Atypical Mycobacteria	Clostridia	Salmonella	Atypical Mycobacteria	Clostridia	Salmonella	Atypical Mycobacteria	Clostridis
Inlet	pullorum gallinarum entritidis typhi typhimrium	scotochoromogenic nonchromogenic	walchii oedematiens	typhi dublin entritidis	nonchromogenic	welchii oedematiens tetani septicum	gallinarum		welchii
Outlet	pullorum gallinarum entritidis typhi	--	welchii oedematiens	typhi entritidis	--	welchii oedematiens septicum tetani	gallinarum	--	welchi
Basin	entritidis	nonchromogenic	welchii oedematiens	typi entritidis	--	welchii oedematiens tetani septicum	--	--	--
Heap (Compost)	--	nonchromogenic	welchii oedematiens	--	sootochromogenic nonchromogenic	welchii oedematiens tetani septicum	--	--	--

Table 2 Persistance of <u>Salmonella</u>, <u>Shigella</u>, and Atypical <u>Mycobacteria</u> During Anaerobic Digestion of the Sludge Under Laboratory Conditions

Days Post Inoculation of Sludge with Pathogens	16°C												28°C											
	pH 4			pH 6.9			pH 9			Calcium carbonate 10 percent			pH 4			pH 6.9			pH 9			Calcium carbonate 10 percent		
	S	Sh	AM	S	Sh	AM	S	Sh	AM	S	Sh	AM	S	Sh	AM	S	Sh	AM	S	Sh	AM	S	Sh	AM
0	+	+	+	+	+	+	+	+	+	+	+	+	+	+	+	+	+	+	+	+	+	+	+	+
3	+	+	+	+	+	+	-	-	-	+	-	-	+	+	+	+	+	+	-	-	-	+	+	-
7	+	-	+	+	+	+	-	-	-	-	-	-	+	-	+	+	+	+	-	-	-	+	-	-
14	-	-	+	+	-	+	-	-	-	-	-	-	-	-	-	+	-	+	-	-	-	+	-	-
21	-	-	-	-	-	+	-	-	-	-	-	-	-	-	-	+	-	-	-	-	-	-	-	-
28	-	-	-	-	-	+	-	-	-	-	-	-	-	-	-	-	-	-	-	-	-	-	-	-
35	-	-	-	-	-	-	-	-	-	-	-	-	-	-	-	-	-	-	-	-	-	-	-	-
45	-	-	-	-	-	-	-	-	-	-	-	-	-	-	-	-	-	-	-	-	-	-	-	-

S = <u>Salmonella</u>   Sh = Shegilla   AM = Atypical mycobacteria

Table 3 Survival of Cl. oedematiens type A During Anaerobic Digestion of Sludge Under Laboratory Conditions

Days Post Inoculation of Sludge with Cl. Oedematiens type A	16°C				28°C			
	pH			$CaCO_3$	pH			$CaCO_3$
	4	6.9	9	10 percent	4	6.9	9	10 percent
	(Number of viable Clostridia (X 10^{-5})/ml)							
0	450	450	450	450	450	450	450	450
3	460	480	470	450	490	490	490	400
7	520	520	470	390	430	560	495	380
14	390	470	430	300	380	500	480	300
21	370	450	400	280	370	460	360	270
28	360	450	400	250	350	430	440	220
35	360	450	380	250	330	405	310	210
45	360	450	360	225	300	405	290	180

Anaerobic digestion of sludge causes death for all atypical Mycobacteria present, so no organisms were isolated from sludge obtained from the outlet (Table 1). The obtained data agree with that of Briscoe (1976) who reported that anaerobic digestion causes death for Mycobacterium tuberculosis. The isolated strains from the stored (compost) heap may be referred to the soil added to the sludge in the field. The results agree with Constallat et al. (1977) and Hammam (1981) who isolated many strains of atypical Mycobacteria from soil samples.

According to the results, Cl. welchii was the most affected microorganism in the different stages of anaerobic digestion (40 percent, inlet, 7 percent in the stored heap). Cl. oedematiens was relatively resistant as its percentage decreased from 33 percent in the inlet to 13 percent in the stored heap, while Cl. tetani and Cl. septicum were highly resistant to anaerobic digestion. These results agree with Abd El Karim (1968) who isolated the same types of Clostridia from Egyptian animal dwellings.

In laboratory-controlled experiments, it was found that 10 percent calcium carbonate and pH 9 were more effective than pH 4 and 6.9 on the persistence of Salmonella, Shigella, and atypical Mycobacteria (Table 2). These results coincided with the findings of the National Academy of Sciences (1977), Farrah and Gaborile (1983) and Corrington *et al.* 1982).

In the case of Cl. oedematins, 10 percent calcium carbonate was a more effective treatment than different pHs. The viable count was decreased 60 percent and 50 percent at 16°C and 28°C respectively, from the original population (Table 3). On the other hand, the viable count decreased 20 percent and 34 percent at pH 4 and 20 percent and 30 percent at pH 9 at 16°C and 28°C, respectively, from the original population. At pH 6.9 no changes were detected in the viable count at 16°C, while 10 percent reduction was observed at 28°C. The obtained results may be due to the favorable anaerobic conditions and the presence of organic matter. The data agreed with those of Garico and McKay (1969).

REFERENCES

Abd El Karim, M. 1968. The Incidence of Animal Infections in the Soil of Some Animal Dwellings. M.D. Vet. Thesis (Hygiene), Faculty of Veterinary Medicine, Cairo University.

Beerwerth, W. 1971. Mycobacteria in the environment of our domestic animals. J.G. Weiszfliler. Atypical Mycobacteria. Publishing House of the Hungarian Academy of Science. pp. 247-252.

Brandes, H. 1960. Tuberculoid changes in mesentric lymph nodes of pigs. Arch. Lebensittelhyg. 11: 193-198.

Briscoe, John. 1976. Public Health in Rural India: The Case of Excreta Disposal. Research paper No. 12. Doctoral dissertation, Harvard University. Center for Population on Studies: Cambridge, Massachusetts.

Clancey, J.K. 1964. Mycobacterial skin ulcer in Uganda: Description of a new Mycobacteria (Mycobacterium buruli). J. Path. Bact. 88: 175-187.

Corrington, E.G. et al. 1982. Inactivation of Salmonella Duesseldorf during anaerobic digestion of sewage sludge. J. Appl. Bact. 53. (3): 331-334.

Constallat, L.F., A.F. Pestanas, A.C. Rodriguas, and F.M. Rodrigues. 1977. Examination of soil in the campinas rural area for microorganisms of Mycobacterium avium-intracellulare scrofulaceum complex. Aust. Vet. J. 53: 349-350.

Cruickshank, R. and L. Duguid. 1968. Medical Microbiology. A guide to the laboratory diagnosis and control of infection eleventh, Livingstone Limited. Edinburgh and London.

Farrah, Samuel. R. and Gabril Bitten. 1983. Bacterial survival and association with sludge floes during anaerobic and aerobic digestion of wastes, water sludge under laboratory conditions. App. Environ. Microbial. 45 (1): 174-181.

Garcia, M.M. and K.A. McKay. 1969. The growth and survival of Clostridium septicum in soil. J. Appl. Bact. 32: 362-370.

Hammam, H.M. 1981. Isolation and characterization of Mycobacteria in Egyptian environment. M.V. Sc. Thesis. Faculty of Veterinary Medicine, Cairo University.

Jones, P.W., M. Lynne, N.H. Renninson, and Redhead. 1980. The occurence and significance to animal health of Salmonella in sewage and sewage sludges. J. Hyg. 84(1): 47-62.

Krebs, A. and W. Kappler. 1971. Contribution to the Epidemiology of Pulmonary Disease Caused by Atypical Mycobacteria. J.G. Weiszfeiler. Atypical Mycobacteria. Publishing House of the Hungarian Academy of Science. pp. 201-205.

McColla, T.M. and L.F. Elliott. 1971. The role of microorganisms in the management of animal wastes in the beef cattle feedlots pp. 132-134. In Livestock Waste Management and Pollution Abatement: The Proceedings of the International Symposium on Livestock Wastes, April 19-22. The Ohio State University, Columbus, Ohio, State Joseph Michigan Society of Agricultural Engineers.

National Academy of Sciences. 1977. Methane Generation from Human, Animal, and Agricultural Wastes. National Academy of Sciences, Washington, D.C. U.S.A.

Nisbide, S. and G. Nakagawara. 1964. Isolation of toxogenic strains of _Cl. novyi_ from soil. J. Bact. 88: 1636-1640.

Paterson, K.J. 1965. _M. fortuitum_ as a cause of bovine mastitis. Tuberculin sensitivity following experimental infection. J.Am. Vet. Med. Ass. 147: 1600-1607.

Prissick, F.H. and A.M. Masson. 1956. Cervical lymphadenitis in children caused by chromogenic _Mycobacteria_. Cand. Med. As. J. 75: 798-803.

Runyon, E.H. 1959. Anonymous _Mycobacteria_ in human diseases. Med. Clin. North America 43:273-280.

Smith, L.D.S. and L. Holdeman. 1968. The Pathogenic Anaerobic Bacteria. Charles Thomas Publisher: U.S.A. pp. 325.

Willis, A.T. 1977. Anaerobic Bacteriology: Clinical and Laboratory Practice. Third Edition, Butterworths. London and Boston.

SURVIVAL OF PATHOGENS AND PARASITES DURING THE ANAEROBIC DIGESTION OF ORGANIC WASTES

A. El-Bassel, H. Gamal-El-Din, I.M. Ghazi and O. Soodi
Faculty of Agriculture, Cairo University, Fayum, Egypt

ABSTRACT

The survival of organisms of the coliform group, *Salmonella* and *Shigella* was investigated during anaerobic digestion of various organic wastes. Preliminary work was also done on the viability of Bilharzia eggs.

The data showed a rapid decrease in the numbers of coliform, *Salmonella*, and *Shigella* groups of bacteria. Their counts decreased by over 90% after 15-30 days in the anaerobic digesters.

INTRODUCTION

The anaerobic digestion of organic wastes and sewage for the production of biogas is a well-known process. With the onset of the energy crisis, many studies were carried out in different countries to improve the process in various directions. These covered increasing the efficiency of biogas production and assessing the effect of variable waste materials and the related hygienic aspects.

In 1959, Harold et al., reported that the possibility for the survival of pathogenic bacteria was a major reason for the slow acceptance and extension of this method for the treatment of sewage sludge.

The direct application of animal manure or human excreta as organic fertilizer can heavily pollute soil and growing plants with pathogenic bacteria (Best, 1969; Hori et al., 1970; Young and Burbank, 1973; and El-Hawaary et al., 1981). The treatment of these wastes by the anaerobic digestion in biogas producing plants before their use as fertilizer decreases the counts of pathogenic bacteria with consequently low possibilities of disease transmission.

The aim of the present work is to follow the effect of anaerobic digestion on the counts of coliforms, *Salmonella* and *Shigella* as well as the viability of parasites during the digestion of different organic wastes, in relation to the efficiency of biogas production.

METHODOLOGY

The following materials and methods were used:

Laboratory digester. It was constructed of a 1 l fermenter (brown bottle), a gas-measuring cylinder equipped with a levelling bulb, a gas sampling part, and a side arm for the withdrawal or addition of liquids (Figure 1).

Inoculum. It was prepared from the effluent of the actively operating laboratory digester fed with cattle manure. The slurry was filtered through cheese cloth to remove undegraded materials.

Organic wastes. Night soil, cow dung, and pulverized garbage were tested at the concentrations of 2% and 5% total solids.

Fermentation experiments. A set of three digesters was prepared for each treatment. The required quantities of each waste material to give the desired concentrations were individually mixed with 100 ml inoculum and adjusted to 1 l with tap water. Control sets containing minerals with 0.1% and 0.2% acetic acid were included. The initial pH was adjusted to 7.0 in all digesters and anaerobically incubated at 35°C. Mixing was done manually before every measurement of the gas volume produced. The experiment was maintained for 100 days.

Biogas analysis. The gas volume was measured daily by the displacement of saturated sodium sulphate solution acidified with 5%

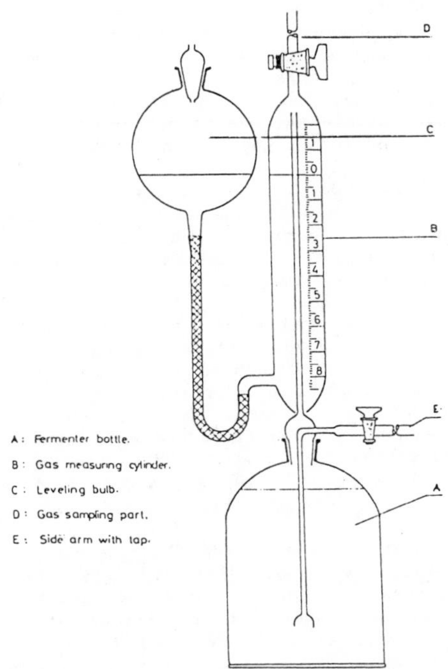

Figure 1 Schematic Diagram of the Digestion Apparatus.

sulfuric acid. The methane and CO_2 content were estimated weekly by bubbling a known volume of biogas in a saturated solution of potassium hydroxide, the loss in gas volume was taken to represent the CO_2 portion, and the balance was assumed as methane.

Chemical analysis. Total solids, total nitrogen, and organic carbon were determined according to APHA (1980).

Microbiological determinations. After mixing, 10 ml samples were withdrawn from each digester and serial dilutions were prepared. The coliforms were counted by the MPN technique using MacConkey broth. The plate count on S.S. and brilliant green agar media was used for counting the *Salmonella* and *Shigella*. The total bacterial counts were determined on tryptone agar medium. The incubation period and temperature were as recommended in APHA. Media were prepared according to Difco manual, 1977.

RESULTS AND DISCUSSION

The data of biogas production and microbial changes during the anaerobic digestion of acetic acid, cow dung, garbage, and night soil are presented in Tables 1, 2, 3, and 4 respectively.

The results showed that the cumulative volume of biogas was markedly higher from the digesters fed with higher concentrations of acid or organic wastes. The acetic acid gave the lowest volume of gas with the relatively highest methane content, while the garbage gave the lowest gas volume and methane content. The largest biogas volume was obtained from the digesters fed with 5% cow dung followed by those containing 2% cow dung, and 5% and 2% night soil respectively. Thus, it is clear that, within the used concentration range, the increase in total solids resulted in a higher biogas production.

The total counts of the facultative anaerobes tend to decrease sharply after 15 days of fermentation, except in the treatments of acetic acid and the 2% garbage. The counts increased after 15 days, then markedly decreased at the same rate as other treatments. Similar results were mentioned by Antoun et al. (1982).

With regard to the counts of coliforms, the general trend was a rapid decrease after 15 days. The decreasing rate ranged from 97.4% to 99.8% in the case of the acetic acid, cow dung, and garbage fed digesters. In the 2% night soil treatment, the counts decreased by 65% only, while the counts decreased by 94% in the digesters fed with 5% night soil. A nearly similar decreasing count of coliforms was mentioned by Nasr (1980), Zehender (1982), and Mohamed (1983). On the contrary, El-Hawaary et al. (1981) reported that anaerobic digestion has no effect on the counts of coliforms.

The survival of *Salmonella* and *Shigella* was strongly affected by the conditions in the anaerobic digesters. In case of the acetic acid, the counts decreased after 15 days by 89.3% and 97.7% at the concentrations of 0.1% and 0.2% respectively. The disappearance of these pathogens continued until it reached only 0.05% of the initial counts after 100 days. The digesters fed with cow dung showed the same trend, the counts decreased by 99.6% and 99.93% at 2% and 5% total solids respectively.

Table 1: Biogas Production, Microbial Changes and Survival of Salmonella and Shigella in Batch Digester Fed With Acetic Acid

Concentration	Time in days	Biogas volume ml/L	Biogas CH_4 %	**T.C. X 10^7	T. coliforms X 10^4	§Redn. %	***S.S. X 10^4	§Redn. %	pH
Acetic acid 0.1%	0	-	-	52	23000	-	58	-	7.0
	15	410	71	67	170	99.26	6.2	89.3	7.5
	30	440	72	2	3.7	99.98	0.42	99.3	7.3
	50	very low gas		1.2	1.6	99.99	0.26	99.6	7.3
	75	very low gas		0.7	0.6	99.99	0.12	99.8	7.2
	100	very low gas		0.1	0.17	99.99	0.03	99.95	7.2
Acetic acid 0.2%	0	-	-	59	23000	-	20	-	7.0
	15	400	74	81	49	99.8	0.46	97.7	7.2
	30	465	76	8	2.3	99.99	0.35	98.25	6.9
	50	very low gas		2.6	1.1	99.99	0.17	99.15	7.0
	75	very low gas		0.9	0.23	99.99	0.06	99.7	6.9
	100	very low gas		0.2	0.11	99.99	0.01	99.95	6.9

Legend
**T.C. = Total counts of aerobes and facultative anaerobes.
***S.S. = Counts of Salmonella and Shigella.
§Redn. % = Percentage of reduction in relation to initial zero-time counts.

Table 2: Biogas Production, Microbial Changes and Survival of Salmonella and Shigella in Batch Digester Fed With Cow Dung (C/N = 22.6)

*T.S.	Time in days	Biogas volume ml/L	Biogas CH_4 %	**T.C. X 10^7	T. coliforms X 10^4	§Redn. %	***S.S. X 10^4	§Redn. %	pH
2%	0			110	230		3.2		7.0
	15	1550	62	16	4.9	97.88	0.35	89.00	6.7
	30	2460	64	0.9	0.4	99.8	0.05	98.4	6.5
	50	2910	63	0.5	0.2	99.92	0.03	99.1	6.5
	75	very low gas		0.1	0.1	99.96	0.02	99.4	6.7
	100	very low gas		0.04	0.1	99.96	0.013	99.6	6.6
5%	0	-	-	110	490	-	6.0	-	7.0
	15	3400	59	23	3.3	99.3	0.26	95.6	6.6
	30	6700	62	1.2	0.23	99.95	0.017	99.7	6.5
	50	7630	62	0.8	0.18	99.97	0.01	99.84	6.5
	75	8025	63	0.25	0.11	99.98	0.009	99.85	6.5
	100	very low gas		0.072	0.11	99.98	0.004	99.93	6.6

Legend
*T.S. = Total solids.
**T.C. = Total counts of aerobes and facultative anaerobes.
***S.S. = Counts of Salmonella and Shigella.
§Redn. % = Percentage of reduction in relation to initial zero-time counts.

Table 3: Biogas Production, Microbial Changes and Survival of Salmonella and Shigella in Batch Digester Fed With Garbage (C/N = 37.7)

*T.S.	Time in days	Biogas volume ml/L	Biogas CH_4 %	**T.C. X 10^7	T. coli-forms X 10^4	§Redn. %	***S.S. X 10^4	§Redn. %	pH
2%	0			45	490	-	0.17	-	7.0
	15	610	37	54	13	97.4	0.00	100	3.7
	30	670	41	0.47	4.9	99.0	-	-	3.7
	50	very low gas		0.26	1.7	99.66	-	-	3.4
	75	very low gas		0.12	0.23	99.95	-	-	3.4
	100	very low gas		0.08	0.11	99.98	-	-	3.2
	0	-	-	100	1300	-	0.37	-	7.0
5%	15	1570	36	6.2	4.9	99.6	0.00	100	3.6
	30	1690	42	0.45	0.07	99.99	-	-	3.6
	50	very low gas		0.21	0.02	99.99	-	-	3.6
	75	very low gas		0.09	0.00	100	-	-	3.4
	100	very low gas		0.04	0.00	100	-	-	3.2

Legend
*T.S. = Total solids.
**T.C. = Total counts of aerobes and facultative anaerobes.
***S.S. = Counts of Salmonella and Shigella.
§Redn. % = Percentage of reduction in relation to initial zero-time counts.

Table 4: Biogas Production, Microbial Changes and Survival of Salmonella and Shigella in Batch Digester Fed With Night soil (C/N = 4.8)

*T.S.	Time in days	Biogas volume ml/L	Biogas CH_4 %	**T.C. X 10^7	T. coli-forms X 10^4	§Redn. %	***S.S. X 10^4	§Redn. %	pH
	0	-	-	90	1.4	-	0.39	-	7.0
2%	15	970	63	3.9	0.49	65	0.20	49	7.1
	30	1180	62	0.78	0.11	92	0.02	95	6.9
	50	2060	63	0.37	0.09	93.6	0.01	97.5	6.9
	75	very low gas		0.16	0.049	96.5	0.01	97.5	6.8
	100	very low gas		0.047	0.049	96.5	0.006	98.5	6.6
	0	-	-	110	3.3	-	0.99	-	7.0
5%	15	1980	49	5.0	0.2	94.0	0.08	92.0	7.3
	30	2810	54	0.35	0.07	98.0	0.02	98.0	7.2
	50	4370	55	0.16	0.06	98.2	0.01	99.0	7.2
	75	5690	59	0.10	0.04	99.0	0.007	99.3	7.1
	100	5970	57	0.09	0.04	99.0	0.004	99.6	7.1

Legend
*T.S. = Total solids.
**T.C. = Total counts of aerobes and facultative anaerobes.
***S.S. = Counts of Salmonella and Shigella.
§Redn. % = Percentage of reduction in relation to initial zero-time counts.

With regard to the night soil treatments, the counts of pathogens also decreased, but at a lower rate. After 15 days the counts had decreased by 49% and 92% with respect to the 2% and 5% total solids. At the end of the fermentation period, the percentage of survivals was 1.5% and 0.4%. The relative high survivals in these digesters may be due to the presence of sufficient soluble nutrients that enabled more cells to survive for a longer period. Harold et al. (1959) mentioned that the natural flora compete with pathogens for essential food in the sludge, and they are more successful than typhoid bacteria.

The nonlactose fermenting bacteria disappeared completely from the garbage treatments after less than 15 days. The pH dropped sharply from 7.0 to 3.7 which affected the survival of Salmonella and Shigella. Jones (1976) found that there is apparently a relationship between the decline in pH and the survival of Salmonella as a reflection of toxicity of acid compounds produced. The high acidity that occurred in the garbage fed digesters may be due to its content of easily digested carbohydrates like bread and food wastes.

With regard to the survival of parasites, a preliminary experiment was carried out on the viability of Bilharzia eggs during the anaerobic digestion of 2% night soil. Eggs were introduced in a 15 days working laboratory digester and well mixed with the slurry. Samples were taken at one-hour intervals and microscopically examined to differentiate between the living and dead embryos. It was found that all embryos died after 15-20 hours. This point of research is still under investigation and more information will be published in the near future.

According to these results, it can be concluded that:

- The decrease in the counts of coliforms and pathogens was over 90% after 15-30 days in the anaerobic digesters.
- The presence of sufficient nutrients (cow dung and night soil) produces a higher percentage of survivals, which indicates that the composition of the waste has a role in the death of pathogens.

The increase of total solids from 2% to 5% enhanced the rapid drop in the counts of coliforms and pathogens during anaerobic digestion. Similar results were obtained by Mohamed (1983). In contrast, Jones (1976) mentioned that survival of Salmonella was longer in slurries with solids content of 5% or more.

- Acidity has a direct effect on the survival of pathogens, the acids produced depending on the constituents of the waste.
- The survival of parasites in anaerobic digesters seems to be much shorter than that of the pathogenic bacteria.

REFERENCES

American Public Health Association. 1980. Standard Methods for Examination of Water and Wastewater. 15th ed., APHA, N.Y. Washington.

Antoun, G.G., S.A. El-Shemi, and M.N. Alaa El-Din. 1982. Survival of certain pathogenic organisms during biomethanation of organic residues. 1st Inter-African Conference on Bio-fertilizers. Cario.

Best, E. 1969. Survival and destruction of Salmonella in liquid manure from cattle and calves. The Veterinary Bulletin 40(3777):614.

Difco Manual. 1977. Dehydrated culture media and reagents for micro-biological and clinical laboratory procedures. 9th ed. Difco Laboratories, Detroit, Michigan, USA.

El-Hawaary, S., M.M. El-Abagy, and N.A. Neweigy. 1981. Effect of biogas on the survival of feacal indicators and Salmonella typhimurium. Res. Bull. No. 1604, Ain Shams University, Faculty of Agriculture, Cairo, Egypt.

Harold, E.L., E. McKinney Rose, and Henry Campbell. 1959. Survival of Salmonella typhosa during anaerobic digestion. II: The mechanism of survival. Sewage and Industrial Wastes 31:23-32.

Hori, D.H., N.C. Burbank, R.E.F. Young, L.S. Lan, and H.W. Klemmer. 1970. Migration of poliovirus type II in percolating water through selected Oahu soils. Advances in Water Pollution Research 2:HA 11/1.

Jones, P.W. 1976. The effect of temperature, solids content and pH on the survival of Salmonellas in cattle slurry. Br. Vet. J. 132:284.

Nasr, A. Fayza. 1980. Production of methane from biomass. Ph.D. Thesis, Faculty of Engineering, Cairo University.

Mohamed, M.E. 1983. Fermentation of city refuses under anaerobic conditions. Ph.D. Thesis, Faculty of Agriculture, Ain Shams University.

Young, R.H.F., and N.C. Burbank. 1973. Virus removal in Huwavan soils. J.AWNA 65:398. (Cited from El-Hawaary, 1981).

Zehender, A.J.B. 1982. Sanitary aspects of anaerobic digestion, survival of pathogens in anaerobic habitats. Regional training course in prospectives of technologies and techniques of applied microbiology and waste recycling. Cairo, Egypt, 6-25 March. pp. 1-5.

ACKNOWLEDGEMENT

The authors thank Prof. Dr. K.H. Knoll, director of Hygiene Institute, Phillips University, Marburg, W. Germany for his cooperation and continuing interest in this research field.

We also thank the Volkswagen Foundation for support and equipment needed for this work.

SECTION EIGHT: COUNTRY PROGRAMS AND PROJECTS

DEVELOPMENT AND APPLICATION OF BIOGAS TECHNOLOGY FOR RURAL AREAS OF EGYPT THE NATIONAL RESEARCH CENTRE PROJECT

M.M. El-Halwagi, A.M. Abdel Dayem, and M.A. Hamad
National Research Centre, Egypt

ABSTRACT

The National Research Centre (NRC) biogas project, from its inception, is outlined. After a short description of its background, objectives, organization, initial development, and research activities, the field demonstration stage with its underlying engineering endeavors is detailed. Future plans and major conclusions evolving from this work are also highlighted. Finally, the key project personnel and the principal project publications are listed.

BACKGROUND AND OBJECTIVE

Since the mid-seventies, the National Research Centre (NRC) initiated laboratory activities in the field of anaerobic digestion of organic wastes. Concurrently, a full-fledged multidisciplinary program was being carefully planned. Near the end of 1978, these endeavors culminated in the formulation and the formal inauguration of a national R&D and demonstration project for the development and application of village-type biogas technology (BGT). The project is managed by the NRC under the auspices of the Egyptian Academy of Scientific Research and Technology (ASRT) and is financially supported by USAID.

The overall objective of the project is to demonstrate, on the basis of an integrated R&D and technology transfer program, that the application of biogas technology is technically feasible, socially acceptable, and economically viable in rural areas of Egypt; and to lay the foundation for subsequent widespread implementation on the national scale. The project is executed by an interdisciplinary team of scientists, engineers, and sociologists.

THE PRELIMINARY FACT-FINDING PHASE

The project started with a short fact-finding in which the BGT state of the art was assessed. An Asian study tour was conducted in China, India, and Thailand. Socio-economic surveys of typical Egyptian villages were undertaken. Two villages representing traditional and new-planned types have been selected for field demonstration. Al-Manawat village represents the traditional, whereas Omar Makram village represents the new-planned type. A third village named Shubra Kass has been selected to demonstrate large-scale biogas system servicing a poultry rearing operation. The latter was carefully chosen because more than 800 poultry houses exist there.

OUTLINE OF THE R&D ACTIVITIES

The second phase of the project encompassed extensive R&D endeavors. Considerable digestibility research work was done to determine optimum conditions conducive to greater efficiency, highest pathogens destruction rate, and diminishing of the effect of toxic and inhibitory materials (particularly those pesticides and herbicides that may contaminate the agricultural feedstocks). Various substrates, including cow dung, night soil, and agricultural wastes (weeds, water, hyacinth, maize, and cotton stalks) were investigated at different mixing ratios, organic loadings, and temperatures. Certain pretreatments, including pre-composting were also examined. Laboratory research was as well conducted on two relevant problems, namely the selective inhibition of hydrogen sulphide and the destruction of ova and embryos of Ascaris. A sizable portion of work was done on the evaluation of digested products such as fertilizer and soil conditioner, as well as on their handling, storage, and application.

The engineering and development work was directed toward the design, construction, testing and operation of different prototype digesters, as well as toward the development of local appropriate digesters constructed, tested, and operated at the NRC extension area. Performance characteristics of these digesters were determined in terms of the gas production rate, gas composition, relation to the effect of pressure and temperature inside the digester, and in terms of the effect of ambient temperature. The rate of destruction of parasites and pathogens was also followed. Internal mixing patterns were determined by measuring the residence time distribution.

As a result, designs suitable for local village conditions could be proposed for field demonstration. These included some modifications of the conventional Chinese and Indian designs, together with development of new designs that may alleviate the major shortcomings of the available technologies under local conditions. These developments encompass:

1. Dry-type fermenter capable of digesting cellulosic wastes. This will solve the problem for families having a limited number of cattle.

2. Horizontal tunnel or tubular-type digester that can solve the problem of the high water table in many localities.

3. Solar-heated digester, either through heating the feed water, or by creating the green-house effect in order to solve the problem of low gas productivity due to low temperature, particularly during winter.

Extensive work was also directed toward the development of appropriate gas-use devices. These include household burners and stoves (both simple and modern), gas lamps, space heaters, refrigeration cooling, and conversion of petrol engines to work on biogas. Performance of these devices was also tested.

THE DEMONSTRATION PHASE

The third phase is the village demonstration phase, which is the central focus of the whole project. The purpose here is to introduce the BGT to the peasant on a concrete level through full-scale demonstration. It was planned to conduct the demonstration in two types of villages: traditional and new planned. These demonstrations are to be followed up and evaluated to assess their viability, social and environmental impacts, and for identification of the optimum course of action for possible implementation on a national scale.

Prior to village demonstration, some family-sized prototype digesters were installed at the demonstration area of the NRC. Three prototypes have been constructed. The first (Figure 1) is a fixed roof 10 m^3 rectangular digester of the water-pressure Chinese design. The second (Figure 2) is a 6 m^3 domed-roof cylindrical Chinese digester. The third prototype unit (Figure 3) is a new adapted design combining the features of both the plug-flow and the Indian movable cap type. Provisions for solar heating of feed water, composting of the digester effluent on the top of the plug flow part, and attachment of a latrine and animal shed were incorporated in this last unit.

Implementation of the village demonstration program at Al-Manawat began in March 1980. Two family-sized digesters, 10m^3 each, were installed first. The first (Figure 4) is a modified Indian type and is attached directly to the animal shed and the latrine of the farm house. The second (Figure 5) is an improved version of the cylindrical wide and shallow Chinese design, installed within the premises of the village collective services unit. Both are still properly functioning. Technical, social, and economic features are monitored on a continuous basis.

The continued manifestation of the field demonstration success in Al-Manawat has motivated a number of village residents to adopt the technology and request constructing digesters for their own use while bearing a part (25%) of its investment cost. Three new units have been recently installed, all of which are either completely novel or have novel design features. The first is a 10 m^3 BORDA-type digester (Figure 6), the second is a 6 m^3 improved Chinese-type digester in which novel construction techniques and internal flow patters are introduced (Figure 7), and the third is a 13 m^3 new type of horizontal plug flow digester with a pyramid-shape roof and floating gasholder (Figure 8).

Omar Makram village demonstration started in the early part of 1981. Three family scale digesters were installed for the use of four families (Figure 9-11). The first two are of the Indian type, but one of them is shared by two households as a pilot experiment of communal sharing facilities. The third unit is the Chinese type, but built by almost purely native Egyptian technology, particularly with regard to the dome section of the unit. All three units have been operating very well and their performance is being followed on a regular basis. However, the pilot communal unit has encountered some difficulties of a social nature.

Controlled comparative field demonstrations of various fertilizers and combinations, including inorganic fertilizers, farmyard manure, and digester effluent manifested the supremacy of the latter.

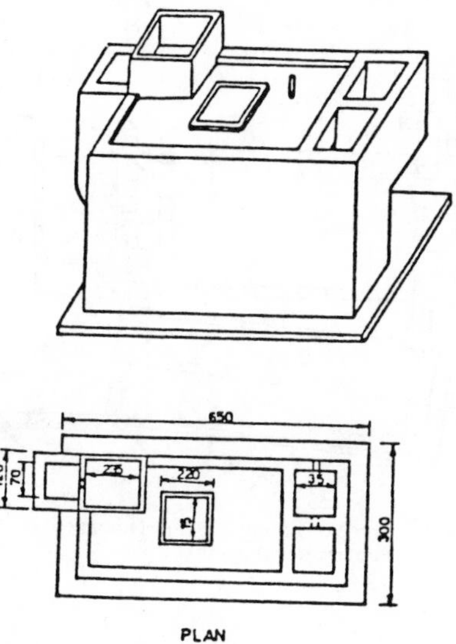

Figure 1 Rectangular Type Chinese Digester

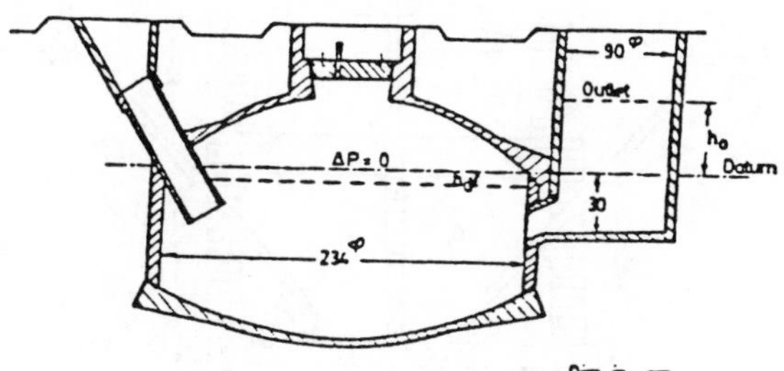

Figure 2 The Chinese Type Digester (5 m^3)

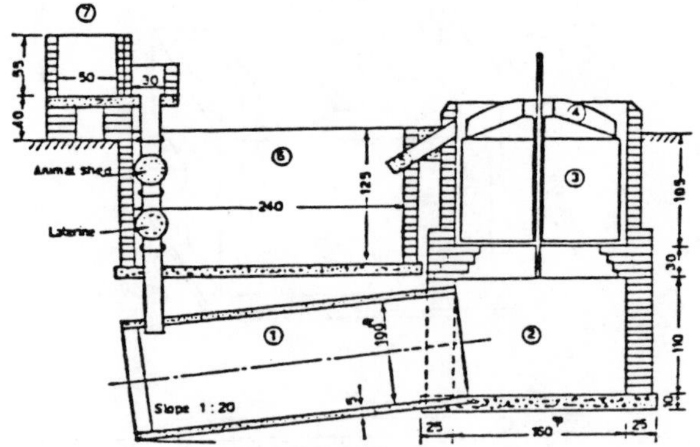

Figure 3 The Modified Indian (horizontal-Vertical) Prototype Digester.

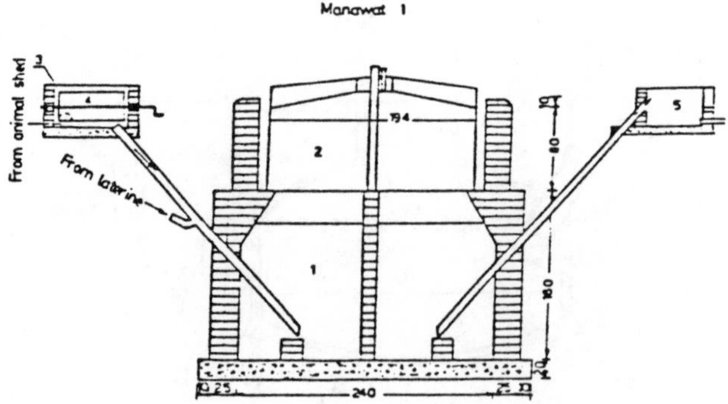

Figure 4 The Modified Indian Type Digester

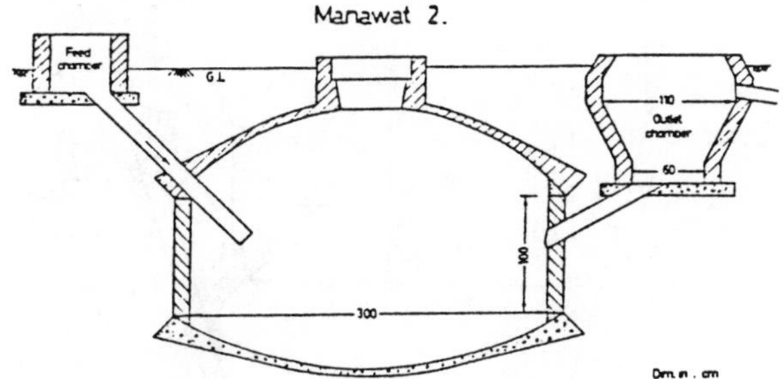

Figure 5 Manawat Modified Chinese Type digester (10 m^3).

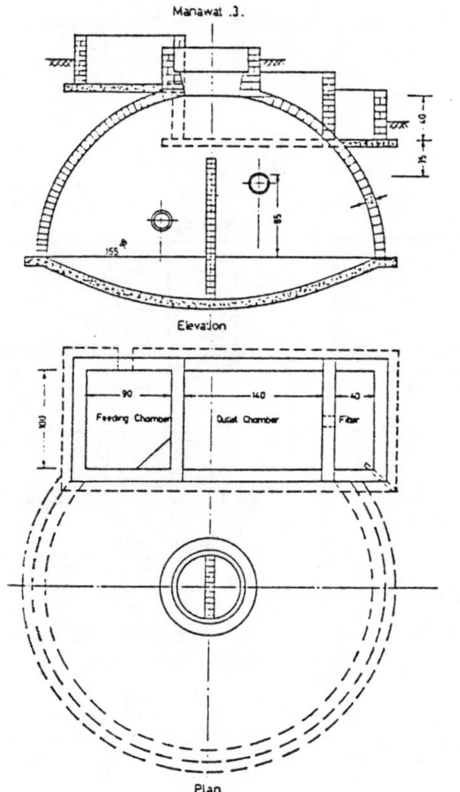

Figure 6 Egyptian-Chinese

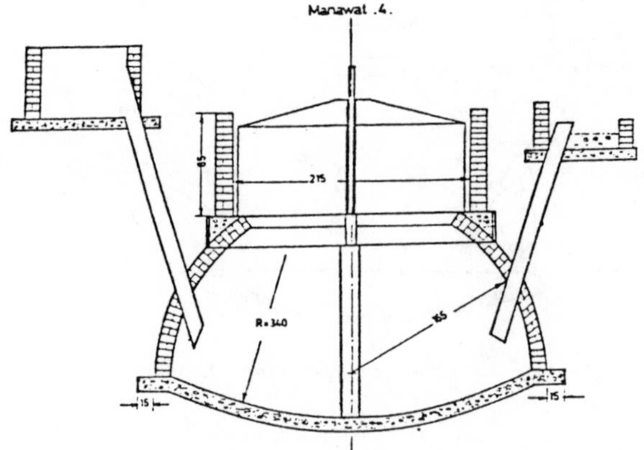

Figure 7 Modified Two-Chamber Borda Type Digester (Maroof-Meet Kados).

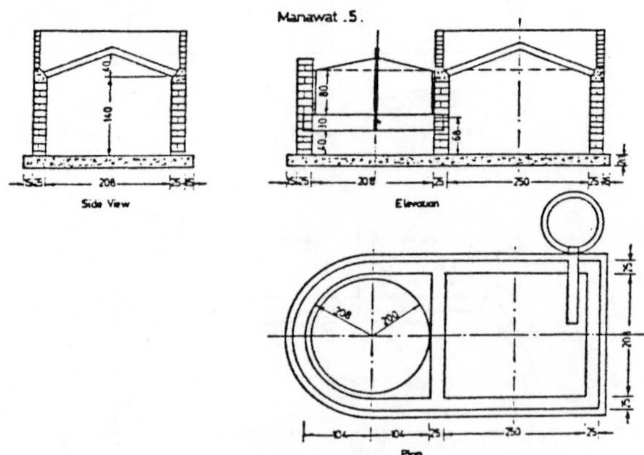

Figure 8 Two-Chamber Tunnel-Type Digester (Rahman-Meet Kados).

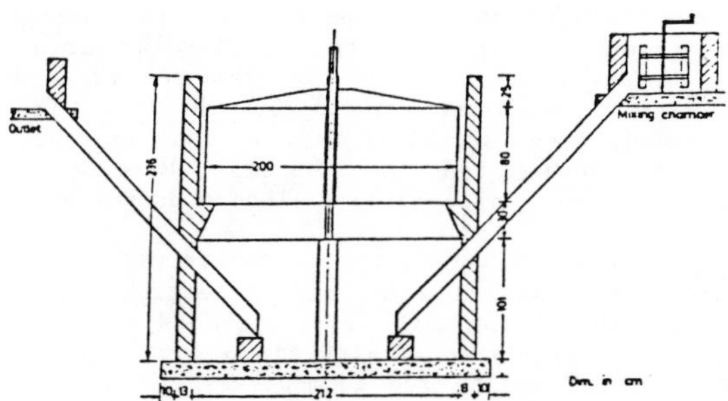

Figure 9 Omar Makram Modified Indian Type Digester (5 m^3)

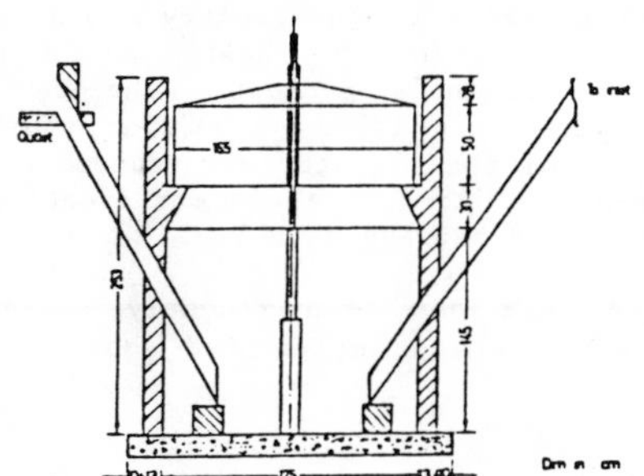

Figure 10 Omar Makram Pilot Communal Unit of the Modified Inidan Type.

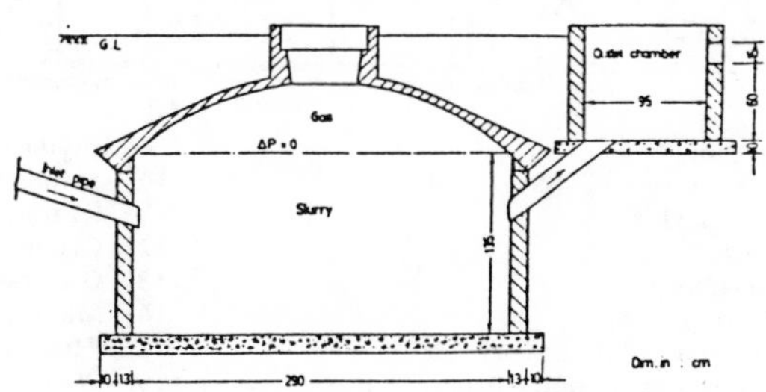

Figure 11 Omar Makram Modified Chinese Type Digester (9 m^3)

A large-scale biogas plant (Figure 12) was installed in May 1982 at a third village called Shubra Kass, where a 50 m^3 biogas plant is attached to a poultry farm. The objective here is to demonstrate and assess the viability of locally introducing a wholly designed and implemented Egyptian biogas system to a poultry-rearing operation. The demonstration site has been carefully selected, since in that particular area, more than 800 similar poultry houses exist. The biogas produced is used to warm up the chicks on the farm. The plant can supply the farm with a considerable portion of its energy needs and overcome the troubles that frequently exist due to the non-available butane gas bottles. Moreover, it recycles the poultry droppings and hence alleviates pollution and disease transfer.

A specifically designed tunnel-type digester with arched roof has been constructed with a separate floating gas collection system. The digester was provided with mechanical mixing rods which did not prove effective, and subsequently gas and slurry-recirculation systems were added to provide more efficient means of mixing and scum-breaking. Heating is provided by circulating hot water through a tube bundle immersed in the digester. In the initial plant the water was to be heated by burning 10-15 m^3 of biogas per day in a gas water heater. At a later stage, a 4 kVA Honda gasoline engine-generator was modified to operate on biogas. The power is to be used for illumination in the poultry house and for other electrical equipment at the site. Water is heated in a specially-designed heat exchanger around the engine's exhaust. The device proved to be very effective and the hot water is now used for slurry mixing and digester heating. Heat loss from the digester was reduced by enclosing the digester in a plastic greenhouse. Through these modifications, the productivity of the unit is now estimated between 1 to 1.2 m^3 biogas/m^3/day.

A large biogas unit has been recently designed for the Aluminium Company of Egypt and will be implemented shortly. The feed to the unit

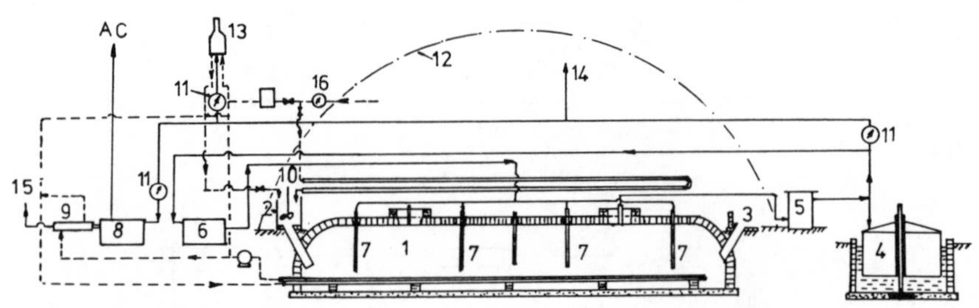

1- Tunnel digester
2- Feed chamber
3- Effluent outlet
4- Gasholder
5- Gas Scrubber
6- Gas Compressor
7- Gas mixing distributors
8- Electrical power generator
9- Engine waste-heat recycle
10- Solar water heater
11- Gas meters
12- Greenhouse
13- Gas heater
14- Main biogas line
15- Flue gases
16- Water meter

Figure 12 The 50 m^3 Tunnel Type Digester (Shubrakass Village)

will be mainly the cow dung collected from about 500 cows and calves of one of the animal sheds owned by the company. The gas will be supplied to a nearby housing complex belonging to the company, while the effluent will be used for desert land reclamation. The unit will be of the plug-flow type with a separate floating gasholder. Its size is approximately 320 m^3, and is estimated to be capable of producing gas sufficient for about 200 households. The unit will be highly mechanized. It will use slurry pumps for feeding and heating by both passive solar means and biogas combustion for controlling the digestions temperature.

FUTURE PLANS

The National Research Centre in its capacity as the largest national R&D organization and sole multidisciplinary institution in Egypt, is planning to continue with target-oriented development endeavors relevant to BGT within the wider context of a full-fledged biomass energy, resource recovery and recycling program.

Over the short-term of a two-year span, emphasis will be on:

1. Appraisal of the designs already implemented by the NRC and perfection of those most promising under local conditions.

2. Development work on two new digesters that have good prospects in coping with the two major constraints of lack of space and insufficient number of animals. The first is a solar-heated above-ground, small-sized digester; and the second is a version of what is called a "dry-fermenter" that would be fed principally with agricultural residues.

3. Erection and performance assessment of the large-scale biogas plant to be attached to Misr Aluminium Company animal rearing operations.

4. Completion of the NRC site for the "Integrated Biomass Utilization Technologies Program." This site will cover development work, demonstration, training, and extension activities. The program encompasses:

o Integrated biogas system embodying: several types of digesters, an animal shed, a fish pond, a green-house, and a specialized laboratory for developing appropriate biogas-use devices.
o Development of improved biomass-burning stoves and ovens. Three designs have already been developed.
o Development of portable biomass gasification units.
o Municipal solid waste composting systems. A pilot plant for mechanized composting is already in operation as a result of joint cooperation with a British firm (DAD).

CONCLUSIONS

1. Field demonstrations manifested the technical feasibility of appropriately designed and installed rural biogas systems. Implementation and operation could be managed with local resources, most of which can be accessible within the village setting itself.

2. There are positive indicators of the special acceptability of the rural biogas systems if they are properly demonstrated. Benefits of the

system are readily visible and perceptible, particularly when it comes to the improvements in the general sanitation level and quality of life.

3. Despite the relatively high initial investment cost, the system can become economically viable if applied to the appropriate situations and if all the tangible benefits are optimally realized. These benefits should include those accrued from the fuel gas, organic fertilizer, and sanitary disposal of wastes.

4. Presently available designs for household units mandate for their viability the fulfillment of two prerequisites:

a. Sufficient space within the house premises for installing the biogas plant including ancillary treatments.
b. Ownership of at least four large animals.

5. Large units attached to animal rearing operations should be relatively mechanized and preferably heated.

6. Widespread propagation of BGT in the Egyptian rural areas necessitates development of modified and new designs that provide for:

a. Large-scale prefabrication of whole units or parts of, to reduce costs and alleviate leakage and other construction problems resulting from poor workmanship that can exist under village conditions.
b. Relatively shallow underground structure in order to avoid the problem of the prevailing highwater table.
c. An appropriate heating system to raise gas productivity, particularly during winter.
d. Possibility of using agricultural residues in case of insufficient manure substrate.
e. Minimum space requirement.

7. On the basis of the presently accessible technologies, it is estimated that there are good prospects for installing approximately 400,000 units all over rural Egypt, provided that proper institutional infrastructure is developed and serious endeavors at all levels are undertaken. We believe that we should not go biogas unless we are sure and extremely serious.

8. BGT prospects can be realized if novel designs can be developed in such a way that the major constraints of lack of space and insufficient animals are somewhat alleviated. The NRC program emphasizes this important concept in its future engineering development work.

9. Government support, especially in terms of financial incentive, grants, and soft loans, is indispensable for the propagation of BGT, particularly under the existing highly-subsidized fuel pricing structure.

BIOGAS PRODUCTION FROM KITCHEN REFUSES OF ARMY CAMPS OF EGYPT USING A TWO STAGE BIOGAS DIGESTER

M.N. Alaa El-Din, H.A. Gomaa, S.A. El-Shimi and B.E. Ali

ABSTRACT

Kitchen refuse and other garbage sources in the army camps present a disposal problem. They require fuel to burn and cause pollution. Cooking in some army camps is done using diesel fuel in low efficiency stoves that cause large energy losses as well as health hazards to cooking personnel. Biogas technology may provide a solution to all these problems.

The aim of the present study is to evaluate a special two-stage biogas digester designed to match the needs of one of the army camps in Egypt through actual operation, and consequently extract the technical knowledge needed to help optimize the plant design for future population.

A two-stage biogas plant of 190 m^3 total capacity (150 m^3 digesting volume) was designed. The plant included a fixed film compartment (20 m^3).

The plant was operated for 422 days. During this period, 203 tons of camp refuse was fed to the digester at the average rate of 480.9 kg/day fresh garbage, containing 248.5 kg/day total solids. The dry material consisted of 85.5 percent dry bread pieces, 9.8 percent kitchen refuses, and 4.7 percent spoiled cooked food. The plant produced 84668 m^3 biogas with the average of 200.6 m^3/day or 1.337 m^3 biogas/m^3 digesting slurry/day. The rate of biomethanation was 1.009 m^3/kg VS added, which is very high. However, considering the fact that most of the materials used were very easy to decompose (baked bread and spoiled cooked food) this efficiency falls within the expected range.

Investment costs amounted to L.E. 20,000. Based on international fuel prices, preliminary economic evaluation indicates a pay-back period of less than six months. Such positive indicators encouraged the Ministry of Defense to finance an extended program for the popularization of biogas technology in the army camps of Egypt.

INTRODUCTION

Army camps provide three meals every day that are usually composed of one cold meal and two hot meals. Cooking food and boiling tea are estimated to consume about 150 gm/person/day diesel fuel or kerosene. Often, cooking is carried out in low efficiency stoves that over-consume fuel and are also hazardous to the cooking personnel.

It is also not uncommon that food catches the smell of fuel. Kitchen refuses accumulating from the preparation of vegetables, meat and fruits present about 82 g/person/day fresh material. Residues after saving food present a sizable source, namely, about 65 g/person/ day bread pieces and about 53 g/person/day cooked food (Alaa El-Din et al., 1983).

These residues are a continuous headache to camp operators. They are transported outside the camp, spread with diesel fuel, and burned every day to avoid attraction of flies, mice, and odor.

Biogas generation from such refuse, partially hydrolized residues, was not a difficult task to accomplish. The task was to choose the proper design and to operate the plant under optimal conditions. The aim of the present experiment was to develop a biogas digester suitable for this condition, to evaluate its production, to optimize the operation of the plant and the utilization of the gas, and to determine the suitable directions for popularization among other camps.

MATERIALS AND METHODS

Materials Fed to the Digester

A mixture of bread pieces, vegetable and fruit residues, and cooked food was fed to the digester at a variable rate ranging between 235 kg and 1194 kg fresh material/day. The added dry matter ranged between 91-866 kg/day. During 422 days of operation, 203 tons of fresh materials were fed (481 kg/day) containing 104.9 ton dry material (248.5 kg/day). Dry bread constituted 85.5 percent and kitchen refuses 9.8 percent, while cooked food totaled only 4.7 percent. An analysis of the materials is given in Table 1.

Plant Description

The plant is composed of the following (Figure 1).

Inlet tank: (6 m^3) which receives one-day residues mixed with water and kept for one day to develop organic acids (Table 2) and represents the acid phase of the plant.

Table 1 Composition and Analysis of Some of the Materials Fed to the Two-Stage Biogas Plant at one of the Army Camps in Egypt.

Material	Total solids TS, %	Volatile Solids VS, %	Total Carbon C, %	Total N, %	C : N ratio
Dry bread pieces	95.60	97.00	56.26	1.37	41.1
Eggplant residues	8.96	76.91	44.61	1.75	25.5
Ochra "	19.06	69.88	40.53	3.62	11.2
Green beans "	19.65	83.35	48.34	2.20	22.0
Tomato "	5.60	94.40	54.75	2.25	24.3

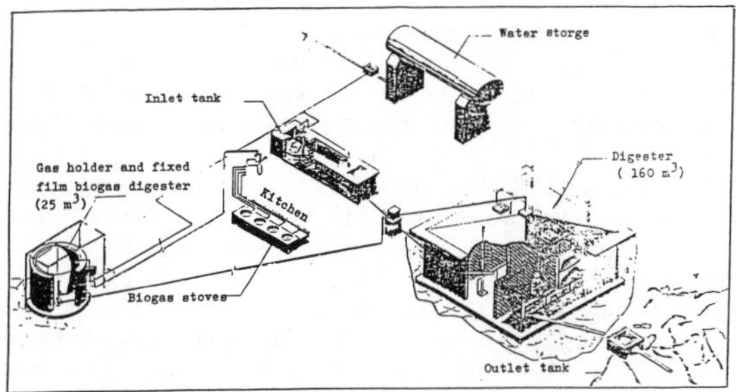

Figure 1 Biogas Unit at One of Army Camps

Main digester: Total volume 160 m^3 containing four chambers with incomplete partition walls as described in Figure 2. The digesting volume is 130 m^3 only. The digester is fitted with a heat exchange system connected to two boilers run on diesel fuel or biogas and a circulation pump for heating water 0.375 kWh. A circulation pump for digesting slurry to the system (1.5 kWh) is connected. An air compressor is connected to the gas phase and is operated to push biogas to the bottom of each of the chambers and to the gasholder for mixing the digesting slurry. (2 kWh).

Gasholder: The gasholder is of the low pressure floating type. The tank has 30 m^3 capacity. Originally, it stored gas over water but later it was converted to a fixed-film digester (20 m^3) capacity by constructing four chambers in the tanks and fixing an iron skeleton on which polyurethane sheets of 1 cm thickness and reinforced with polyurethane threads were stretched. Another group of polyurethane sheets were fixed between the stiffening arms of the mild steel gasholder. The fixed film was developed using active sewage sludge from the nearly sewage treatment plant.

Operation of the System

1. Materials are fed into the acid phase tank (inlet). Vegetable residues are grouped by a specially designed grinder before feeding. Five to 5.3 m^3 of water (later from the effluent of the fixed film digester) are then added and the material is mixed (hand mixers).

2. Feeding is done after one day (to allow formation of acids) by opening the valve leading to the main digester.

Table 2 pH Values of Digesting Slurry in the Different Digester Stages

Digesting State	pH Range
Inlet-acid phase	3.70-4.30
Main digester	7.10-7.35
Outlet of main digester	7.15-7.45
Fixed film digester fed with outlet slurry of main digester	7.35-7.60

3. The acid slurry goes to the first chamber. Mixing in the digester is done three times daily by circulating the slurry from the 4th chamber to the 1st, 2nd, 3rd, and 4th chamber for five minutes each. Another mixing with compressed biogas is done once daily for five minutes in each of the four chambers of the main digester and in the fixed film one.

4. The digesting material lasts for a hydraulic retention time of 21.7 days and leaves the main digester for the fixed film digester (30 m^3) to stay there for a hydraulic retention time of 3.33 days.

5. Bioenergy feeding with 6 m^3 fresh slurry containing 1.5-14.43 percent total solids; the same amount leaves the 4th chamber of the main digester and goes to the fixed film digester to push 6 m^3 of the exhausted slurry to be used for moistening the fresh feed.

6. The biogas goes to the kitchen to be used in burners designed to match the purpose and is used for cooking meals in large containers (100 liters each); no special treatment of the gas is made.

7. The excess gas is stored in the gasholder which rises up and the fixed film in it leaves the digesting slurry. While the stored gas is consumed, the gasholder and its fixed film sheets go down. Consequently, the gas production of this unit increases again, representing a type of self-regulating system.

Experimental

Experiments were carried out to investigate the effect of the following factors on gas yield (m^3/day), gas productivity of the plant (V/V/day), and the efficiency of generating biogas from materials fed (m^3/kg VS added):

1. Concentration of feeding materials
2. Heating of the digester
3. Converting of the gasholder to a fixed film biogas digester.

The gas amounts were measured daily. The methane and CO_2 contents of the biogas were also measured. Total solids (TS), volatile solids (VS) and pH, and total and ammoniacal nitrogen were also measured.

Methods of Analysis

Methane content in the biogas: Gas yield was determined daily by a wet gasometer GA type MPS 6. Methane content in the biogas generated was estimated using the following technique as described by Wujcik and Jewell (1980). Gas samples were withdrawn into a 50 ml syringe, and 0.5 ml gas samples were injected into GOW MAC gas chromatograph model 750P, fitted with a 120 cm long and 0.2 cm diameter stainless steel column filled with 5 percent OV 101 on CHROM-PAW 80-100 mesh and with dual flame ionization detector. Carrier gas was nitrogen at a flow rate of 28 ml/min, hydrogen by attached hydrogen generator and provided at the rate of 30 ml/min. Air at the rate of 300 ml/min was applied for the flame. The operation temperatures were 75°C for column oven, 100°C for injection port, and 150°C for detector. Standard curves were prepared using pure methane gas (Messer Gressheing GMBH, Frankfurt, FAG) and were used as a reference for calculating methane concentration in the biogas produced.

<u>CO_2 Content</u>: Carbon dioxide was measured by Orsat's apparatus using caustic potash solution for CO_2 absorption (American Public Health Association, 1976). The results were used as a check for methane content.

Total solids (TS) were determined according to the methods as recommended by the American Public Health Association (1976).

Volatile solids (VS) determined by glowing the dried samples at 650°C to constant weight (American Public Health Association, 1976).

Organic carbon was determined according to Black et al. (1965).

Ammoniacal nitrogen content was determined according to the methods described by Piper (1950).

Total nitrogen in dried samples was performed by the Kjeldahl method as recommended by Jackson (1973).

RESULTS AND DISCUSSION

During a long period of operation (422 days), the two-stage biogas digester was closely monitored. The results are presented in Table 3 and illustrated in Figures 2, 3, 4, and 5.

The plant was operated for 47 days during the summer August 1 to September 16, 1983, without heating. Biogas production averaged 130.9 m^3/day and the rate m^3 biogas/m^3 digesting slurry/day was 1.01 for the same period. The efficiency to convert volatile solids in the digesting slurry into biogas (m^3/kg VS added) was fairly high, being 0.896 on the average. Heating the digester to reach 32°C-34°C did not bring any improvement. During a period of 86 days from October 2 to December 26, 1983, the average gas production was 125.4 m^3/day (slightly lower rate of 0.97 V/V/day), and the efficiency was 0.88 (m^3/kg VS added). The net gain of biogas was, however, much lower than in the first period; as in the second period 20 m^3 biogas/day were consumed to operate the boiler and an additional 4-5 m^3/day to generate electricity enough to operate the 0.375 kWh hot water circulation pump for 24 hours. Heating in the summer period was therefore stopped completely, and a partial heating (one boiler) was instituted during October, November, March, and April; full heating (two boilers) was operated during December, January, and February. The consumption of biogas to heat the unit, to circulate the slurry and to press the biogas in the digesting slurry for mixing could be summarized as follows:

1. <u>Heating</u>: Two boilers are operated for full heating and consume 20 m^3 biogas/day each. The hot water circulation pump (0.375 kWh) is operated for 24 hours, and consumes 4 kW/day equivalent to 4.5 m^3 biogas/day.

2. <u>Mixing</u>: a. Slurry circulation pump (1.5 kWh) is operated for 25 minutes 3 times/day and consumes 1.8 kW/day equivalent to 0.9 m^3 biogas/day. b. Agitation by compressing biogas into the slurry. A compressor of 1.5 kWh capacity is operated for 25 minutes once every day and consumes 0.83 kW/day equivalent to 0.4 m^3 biogas/day.

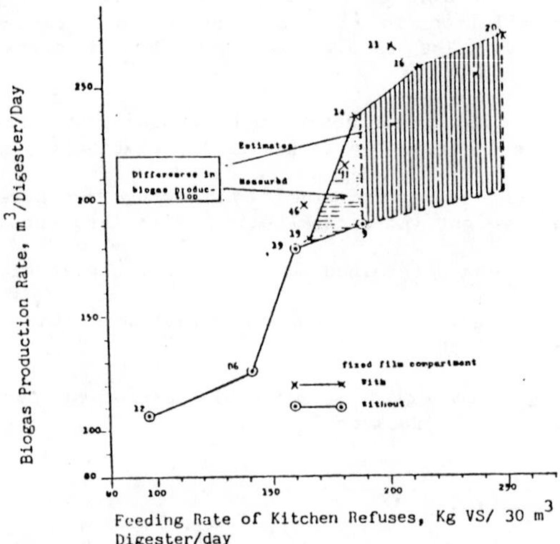

Figure 2 Efficiency of Biogas Generation from Ketchen Refuses of Army Camp in Egypt as Olle Ctdd by Increasing Rates and Converting the Gas Holder to a Fixed Film Extended Digester, Numbers Indicate the Evaluation Periods in Days.

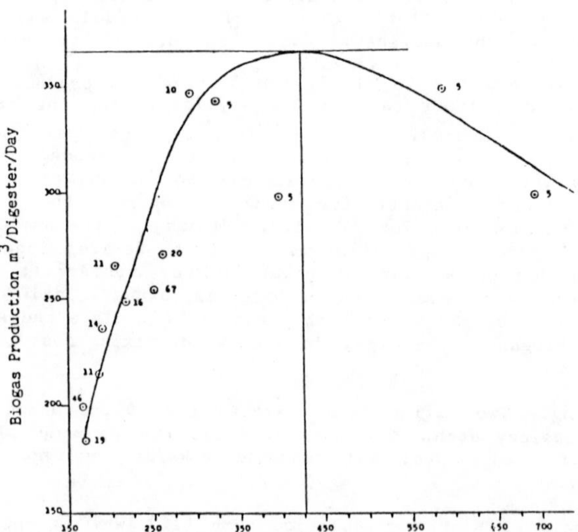

Feeding Rate of Kitchen Refuses, Kg/ VS 130 m^3 Digester/Day

Figure 3 Biogas Production of Army Camp Biogas Plant (Fitted with a small Fixed Film Digestion Oompartment) as Affected with Loading Rate of Kitchen Refuses, Numbers Indicate the Evaluation Periods in Days.

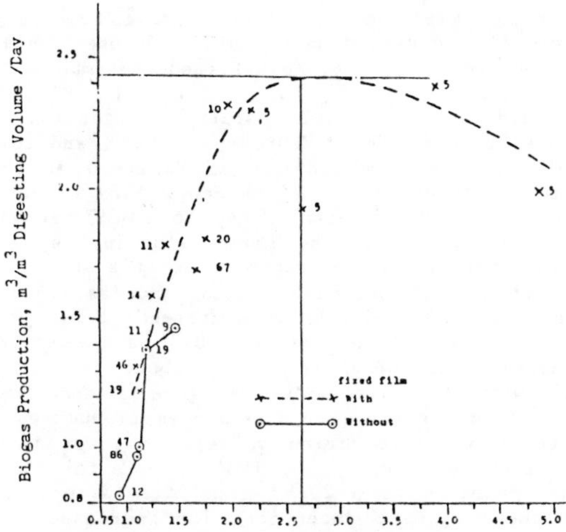

Loading Rate of Kitchen Refuses, Kg VS/m^3
Figure 4 Productivity of Army Camp Biogas Digester as Affected by Loading Rate of Kitchen Refuses and Extending the Digester (130 m^3 Slurry) with 20 m^3 Fixed Film Digester in the Gas Holder, Numbers Indicate Evaluation Periods in Days.

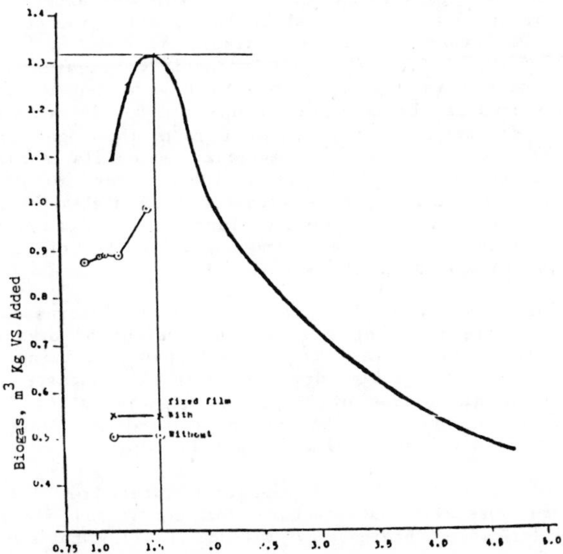

Loading Rate of Kitchen Refuses, Kg VS / m^3 Digesting Slurry/day
Figure 5 Efficiency of the Army Camp Digester in Generating Biogas from Biomass as Affected with the Rate of Loading and the Presence of Fixed Film in the Gas Holder.

Biogas consumption will be 46 m^3/day during December, January, and February; 26 m^3/day during March, April, October, and November, and only 2 m^3/day during June, July, August, and September.

Mixing by circulation of digesting slurry and gas circulation was stopped for 10 days November 27 to February 2, 1984, and this led to a decrease of the efficiency from 0.891 m^3/kg VS added in the previous period to 0.810 m^3/kg VS or about 8.2 percent. The effect of mixing is usually higher than 8.2; figures of up to an 80 percent increase were reported (van Buren, 1979). The lower value in the present study could be due to the shorter time of exposure to this stress. Hashimoto (1982) found, however, that continuous mixing increased gas production by only 8.11 percent as compared with intermittent mixing. In the case of our fermenter, the continuous mixing would have cost about 36 m^3 biogas/day to increase the productivity by about 20 m^3. Increasing the feeding rate with volatile solids (Figure 2 and Table 3) was accompanied with a clear increase in the biogas production rate. The level of the daily biogas production values was lower in the period between August 1, 1983 to February 19, 1984 than in the following when the gasholder was converted to a 20 m^3 fixed film digester. This was expected because the hydraulic retention time increased from 21.7 to 25.4 days, and the enriched active methane bacteria structure enhanced the biomethanation of the volatile solids available.

When comparing two periods similar in the feeding rate, mixing, and heating, one before February 6-14, 1984 and one after April 24-May 9, 1984 the effective action of the fixed film, it becomes clear that the gas productivity of the unit has increased from 1.462 to 1.650 V/V/day. The biomethanation efficiency of the plant (m^3 biogas/kg VS) has increased from 0.993 to 1.145 or about 15 percent. The contribution of the fixed film compartment is accordingly 32.9 m^3/ m^3/day, produced from 20 m^3 of the outlet slurry. Productivity figures around 1.645 m^3/m^3/ day under this loading rate level were repeatedly recorded. This is considered fairly high for an unheated but intermittently mixed small biogas digester. The production rate of biogas increased by increasing the loading rate of digesting materials showing maximal production of about 368 m^3/day at a loading rate of about 390 kg VS/day (Figure 3). Loading this digester with higher rates than the optimal one (2.6 kg VS/m^3 digesting slurry/day) led to a decrease in both total production (m^3/day) and productivity of the digester (m^3/m^3/day) as indicated in Figure 4.

Increasing the rate of loading usually increases the gas generation, but the yield of gas per unit weight of organic material decreases. Shelat and Karia (1977) showed that increasing the loading rate from 1.17 to 5.29 kg dry solids/m^3 of digester volume/day increased the total volumes of biogas of organic matter from 0.22 to 0.47 m^3/m^3/day but reduced the volume of gas per unit weight of organic matter from 0.22 to 0.09 m^3/kg solids/day.

Miner and Smith (1975) indicated that rates from 1.6 to 5.9 kg VS/m^3/day are generally satisfactory for continuous flow digesters. Similar results were obtained by Velsen (1977), who showed that a stable digestion could be maintained up to a space load of 5.4 kg VS/m^3/day and the maximum space load at which a satisfactory reduction of objectionable manure odor was attained appeared to be about 3.6 kg VS/m^3/day. The optimal loading rate as kg dry solids/m^3 of digester volume/day is also reported by Bene et al. (1978) to be 0.8 kg/m^3/day.

Table 3 Biogas productivity and efficiency of biomethanation of kitchen refuses in a two stage fermentation biogas plant at one of the army camps in Egypt.

Observation periods dates	Days	Treatment			Feeding rates, Kg/day			Biogas production m^3		
		Mixing pump	gas	Heating	Fresh	T.S./ ferm-	V.S./ m^3	Total/ day	/m3 dig. slurry/	/Kg V.S. added
A Digesting volume 130 m^3										
1/8-16/9/83	47	x	x	-	431.2	182.6	1.124	130.9	1.010	0.896
2/10-26/12/83	86	x	x	x	379.0	177.8	1.095	125.4	0.970	0.881
27/12/83-7/1/84	12	x	x	x	428.8	121.2	0.932	106.3	0.820	0.877
8/1-26/184	19	x	x	x	486.1	201.6	1.241	179.7	1.380	0.891
27/1-5/2/84	19	x	x	x	456.5	198.2	1.220	128.5	0.989	0.810
6/2-14/2/84	9	x	x	x	471.1	239.0	1.472	190.0	1.462	0.993
15/2-19/2/84	5	x	x	x	235.0	91.0	0.560	134.0	1.031	1.841
Average values without fixed film	188				406.2	179.5	1.381	134.6	1.035	0.937
B-Gasholder converted to fixed film digester 20 m , Total digesting vol. 150 m										
20/2/9/3/84	19	x	x	x	413.2	210.5	1.121	183.7	1.225	1.091
10/3-14/3/84	5	x	x	x	632.0	496.8	2.649	290.0	1.933	0.730
15/3-25/3/84	11	x	x	x	459.1	230.5	1.229	215.5	1.437	1.169
26/3-30/3/84	5	x	x	x	694.0	406.0	2.166	344.0	2.293	1.059
31/3/13/4/84	14	x	x	x	427.5	236.6	1.262	237.2	1.581	1.253
14/4-23/4/84	10	x	x	x	599.0	369.4	1.970	348.0	2.320	1.178
24/4-9/5/84	16	x	x	x	418.0	270.6	1.443	247.8	1.650	1.145
10/5-15/7/84	67	x	x	x	584.0	312.7	1.668	254.1	1.694	1.058
16/7-20/7/84	5	x	x	-	1194.0	866.0	4.619	299.0	1.993	0.432
21/7-9/8/84	20	x	x	-	583.0	326.2	1.739	271.5	1.810	1.040
10/8-14/8/84	5	x	x	-	1060.0	732.4	3.906	350.0	2.393	0.613
15/8-25/8/84	11	x	x	-	455.5	256.4	1.368	266.4	1.776	1.299
25/8-9/11/84	46	x	x	-	463.9	208.2	1.110	197.6	1.317	1.186
Average	234				543.9	304.0	1.621	253.8	1.692	1.044
Total figures	422				202,960.0	104885		84668		
Average					480.9	248.5		200.6		1.009

The efficiency of the system to produce biogas from the volatile solids added to it (Figure 5 and Table 3) increased by increasing the loading rate unit reaching the maximal efficiency value of 1.32 m^3 biogas/kg VS added at a loading rate of 1.6 kg VS/m^3 digesting slurry/day. Higher loading rates decreased the efficiency sharply.

It is therefore suggested that this type of unit, used for optimal treatment of kitchen refuse as a way of getting rid of them, should be loaded up to about 2.6 kg VS/m^3 digesting slurry/day. Further attention should be given to optimize the acid fermentation phase and to study the destruction rates of VS, N-balances, salt accumulation, and the survival of pathogens.

The plant was operated for 422 days. During this period 203 tons of camp refuse were fed to the digester at the average rate of 480.9 kg/day fresh garbage containing 248.5 kg/day total solids. The dry material was composed of 85.5 percent dry bread pieces, 9.8 percent kitchen refuse, and 4.7 percent spoiled cooked food. During this period, the plant produced 84668 m^3 biogas with the average of 200.6 m^3/day or 1.337 m^3 biogas/m^3 digesting slurry/day. The rate of biomethanation was by 1.009 m^3/kg VS added very high. However, considering the fact that most of the materials were very costly to decompose (baked bread and spoiled food), this efficiency falls within range.

The amount of biogas produced was equivalent to 51.164 liters of kerosene (LE 12784 at world market prices). This amount, however, saved the total fuel consumed previously for cooking in this camp, taking into account the very low efficiency of the stoves. The export value of this fuel amounts to 44.6 x 10^3 US $/year (49.9 x 10^3 LE) while the cost of construction and modification did not exceed 20 x 10^3 LE. Payback within less than six months has encouraged the army authorities to finance a strengthened program for popularization of biogas technologies in the army camps in Egypt.

ACKNOWLEDGMENT

The authors wish to thank the Ministry of Defense, the Engineering authority, and the Engineering Department for their interest and support. They wish also to thank the Ministry of Agriculture of Egypt, the Agricultural Research Center, Soils and Water Research Institute for providing laboratory and technical facilities.

REFERENCES

Alaa El-Din; M.N.; Y.H. Hussein; H.A. Gomaa; S.A. El-Shimi and B.E. Ali. 1983. Biogas from Camp Residues, Source of Energy, Manure and for Pollution Control. 95 pp (in Arabic).

American Public Health Association. 1976. Standard Methods for the Examination of Water and Wastewater. Washington, D.C. 14th edition.

Black, C.A.; D.O. Evans; L.E. Ensminger; J.L. White; F.E. Clarck and R.C. Dineuer. Methods of Soil Analysis, II. Chemical and Microbiological Properties. American Soc. Agron. Inc. Madison, Wisconsin, USA.

Hashimoto, A.G. 1982. Effect of mixing duration and vacuum on methane production rate from beef cattle waste. Biotechnol. Bioeng. 24(1):9-24.

Jackson, M.L. 1973. Soil Chemical Analysis. Prentice-Hall of Indian Private, New Delhi.

Miner, J.R. and R.J. Smith. 1975. Livestock waste management with pollution control. North Central Reg. Res. Pub. 222 Midesuest Plan Service Handbook, MWPS, 19.

Shalat, R.M. and G.I. Kari. 1977. Utilization of Biogas in South Gujarate Region. Dept. Civil Engineering, S.V. Regional College and Engineering and Technology. Gujarate, India.

van Buren A. 1979. A Chinese Biogas Manual, Popularizing Technology in Countryside. Intermediate Technology Publications Ltd., London WC 2E 8HN.

van Velsen, A.F.M. 1977. Anaerobic digestion of piggery waste. 1. The influence of retention time and manure concentration. Neth. J. Agric. Sci. 25:151-169.

Wujcik, W.J. and W.J. Jewell. 1980. Dry Anaerobic Fermentation. Biotechnology and Bioengineering. Symp. No. 10, p. 43-65. John Wiley & Sons, Inc. New York.

AN INTEGRATED RENEWABLE ENERGY SYSTEM

PROJECT OVERVIEW

Engineering Group	Agricultural Group
A. Fathy	A. El-Bassel (PI)
A. Serag El Din	H. Gamal-El-Din
G. B. Salem	I. Ghazi

Cairo University, Egypt

American Counterpart: E. Lumsdaine, Dean of Engineering
University of Michigan, Dearborn

ABSTRACT

An integrated renewable energy system is being implemented (near Fayum) for research and demonstration purposes. It encompasses wind, solar energy, and biogas. The system concept, design bases, components, basic technical data, and anticipated scope of operation are outlined.

PROJECT OBJECTIVES

The aim of the project is the development of a renewable energy hybrid system that is economically competitive with other conventional energy sources. The system should be compatible with technical and socio-economic conditions prevailing in Egypt. The technology for the production of energy from renewable sources is already well-developed. Wind machine, solar collector, photovoltaic cell and biogas digester designs have, in general, been standardized. Unfortunately, taken individually, each of these systems has serious economic drawbacks, which have so far prevented their widespread application.

We thus propose to design and construct a power plant depending solely on renewable energy sources and demonstrate that it can be economically viable if taken as an integrated package and if maximum use is made of all its by-products.

DESIGN CONSIDERATIONS

In order to achieve the economic goal of the project, the following considerations had to be factored into the design of the plant.

1. Use renewable energy systems that offer the lowest cost per unit energy produced. In this context we use solar energy for low temperature heating only and eliminate the use of photovoltaic cells as part of the plant. The P.V. cells would only be used to run our scientific measuring and monotoring equipment, which is not an integral part of the plant.

2. Eliminate costly energy storage systems for wind energy and solar energy and rely on biogas when the wind is not blowing for the production of energy.

3. Maximize the efficiency of the biogas digesters by using solar heaters to heat their contents and a Savonius type wind rotor to stir their contents.

4. Obtain maximum use of the by-products resulting during anaerobic digestion. This is achieved by using the slurry resulting from the digesters and the CO_2 (after its separation from the biogas) to grow algae and raise fish in algae ponds. In addition, the sludge remaining in the digesters and settling in the algae pond can be used as a high quality fertilizer of considerable economic value.

SYSTEM COMPONENTS

Attached is a conceptual diagram of the proposed system.
The main components are:

- Three digesters for anaerobic digestion and the production of biogas.
- Two gas holders for the storage of the produced biogas.
- Two gasoline engines converted to run on biogas to produce electricity if needed.
- An algae pond for the growth of algae and the raising of fish.
- A main wind machine for the production of electricity.

The auxiliary components include:

- A gas purification system. This includes a filter to absorb CO_2 from the biogas. The CO_2 is subsequently released to algae pond by heating the lime water using solar energy.
- Solar collectors to heat the content of the digesters.
- Stirring mechanism to stir the contents of the digesters. Different stirring concepts will be tested, among them the use of a Savonius rotor.
- Monitoring and recording equipment.

OPERATION CONCEPT

The production of electricity from the wind machine when the wind is blowing and from biogas through the converted engine when the wind is not blowing. Any excess gas produced can be stored in the gas holders (the cheapest way of storing energy).

The CO_2 contained in the biogas and the slurry produced by the digesters are used to grow algae in the algae pond. Such algae can be

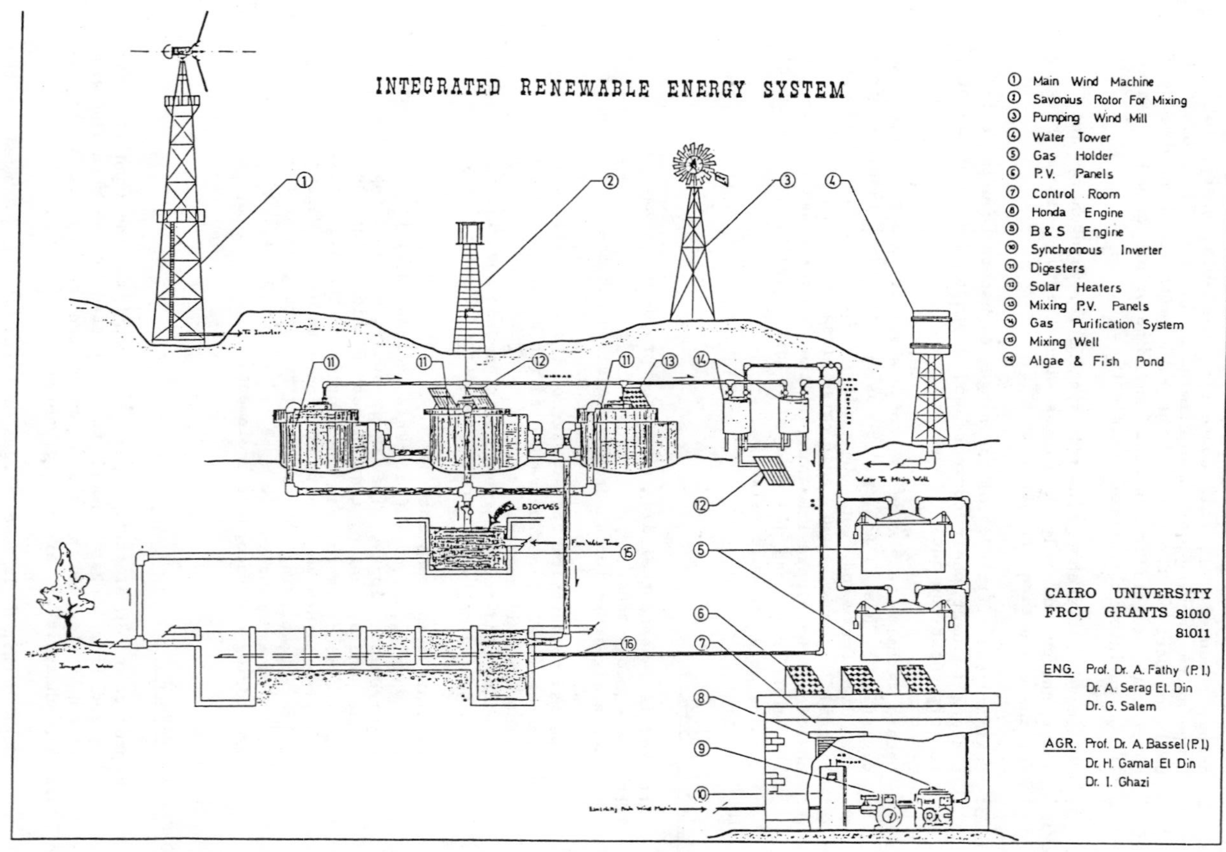

Conceptual Diagram of the Proposed System

used as feed for animals, poultry or fish and has considerable economic value at virtually no extra cost.

The sludge remaining in the digesters and settling in the algae pond is a high-qualtiy fertilizer that also has considerable economic value.

Finally such a plant can have considerable economic impact on rural development, since it helps the farmer solve his sanitation problem in a pollution-free manner. Moreover, it could easily be integrated, either within human settlements, or within cattle or poultry raising farms. In the latter, it can result in a considerable decrease in the cost of the produced meat or poultry.

TECHNICAL DATA

Number of digesters: 3
Digester working volume: 15 m^3
Daily slurry input: 0.5 m^3/ digester (8 percent total solids)
Retention time: 30 days.
Anticipated gas production rate: 0.5 m^3 gas/m^3digester/day.
Total anticipated gas production rate: 22.5 m^3 biogas/day.
Number of gas holders: 2
Gas holder capacity: 10 m^3
Sedimentation tanks: one 9 x 2 x 1.25 m^3
Algae ponds: six 4 x 2 x 0.5 m^3
Fish ponds: two 4 x 2 x 1.25 m^3
Engines: one Honda Engine rated at 2 kW
one Briggs & Stratton Gasoline Engine Rated at 1 kW
Anticipated electric power output from the biogas: 28 KWh/day
Main Wind Machine:
Type: APS - 78-2 with a 24" Zephyr type alternator
Cut-in Speed: 5 mph
Rated Speed: 20 mph
Rated Output: 2 kW
Maximum output: 2.5 kW

BIOGAS TECHNOLOGY FOR WATER PUMPING IN BOTSWANA*

M. Khatibu
Rural Industries Innovation Centre, Botswana

Biogas technology has good potential in Botswana. The Rural Industries Innovation Centre (RIIC) has built and tested three types of digesters: the horizontal, Chinese, and Indian in 10 m^3 size. The Indian type has proven to be the most advantageous technically. A diesel engine was modified and operated on dual fuel and exhibited good results. Encouraged by these developments, a 75 m^3 floating-type digester has been built for the purpose of pumping water. Over a period of 12 months of tests, the plant with its dual-fuel engine was a successful operation. Future plans encompass introduction and popularization of the technology in Botswana. A 120 m^3 digester is under construction. Three other plants will follow for different uses and in different parts of the country.

Rural Industries Innovation Centre (RIIC) is an affiliate of Rural Industries Promotions which is a non-profit association and parastatal organization under the Ministry of Commerce and Industries. RIIC obtains its funds from the Government of Botswana and donor agencies. In the case of the Biogas Project, it was initially funded by the Friedrich Ebert Foundation (FEF), West Germany, for research and development (R&D) work and the International Foundation for Science (IFS), Sweden, for the Diphawana Community Biogas plant used for water pumping.

Rural Industries Promotions has set RIIC as an Appropriate Technology Centre to carry out some R&D work with an objective of improving living standards of the rural poor by developing and innovating technologies which are viable socially and economically.

---

* Only a summary of the original paper is included here

REPORT OF BURKINA FASO TO CONFERENCE

Traore S. Alfred
Intitut Superieur Polytechnique
B.P. 7021 Universite de Ouagadougou, Burkina Faso

INTRODUCTION

Burkina Faso is a sahelian non oil-producing country. It consumes a small amount of commercial energy and now imports only 150,000 - 200,000 tons of oil or derivatives per year.

Except for a few food industries, Burkina Faso had not yet developed big industry. So the main power consumers are families. They get power from non-commercial sources (wood, charcoal, straws) representing up to 91 - 94% in total energy used all over the country. Wood is the common energy source in both cities and rural areas. Its utilization is 1.7 kg per day and per inhabitant. The consequences are drastic for environmental safety due to:

- rapid depletion of trees;
- irreversible advance of desert; and
- shortage of rains causing a decrease in food production.

With respect to this situation, two solutions must be adopted:

1) Reforestation programs, using tree species with high rate of growth; but this is very expensive (U.S. $200-250/acre); and

2) New and renewable energies (biogas/solar energy). In 1976, in Burkina Faso, a research and development program began on waste valorization through biogas fermentation; these waste materials include plant biomass and industrial and animal refuses.

BIOGAS PRODUCTION IN BURKINA FASO

Potentialities of biomass in Burkina Faso are important. One acre produces about 0.52 tons of dry agricultural material. About 80 - 90% of this biomass is abandoned in fields and/or lost on firing. To reduce the loss of the energy, information is given to people so that they can use this biomass to produce biogas. Now, we can enumerate up to 20 pilot sites where digesters of small volumes (4 - 10m^3) are functioning. In respect to the nature of available substrates, we are using a discontinuous system for a 30 - 50 day incubation time. The yields are rather low: 100 - 110 liters of biogas per kilogram of dry material. In addition, daily productivities are low because no fermentation parameter (temperature, pH, inoculum, ...) is controlled: 1 - 2m^3 of biogas per 6m^3 of loading volume.

Until now, biogas has not been popularized in Burkina Faso because two kinds of constraints hinder its development.

1) Economic contraints:

The building cost of 1m^3 is U.S. $80 - 100. The low income of people in Burkina Faso does not allow many families to have biogas digesters for their own use. Now we are trying to use local building materials so that we can decrease the costs of digesters.

2) Scientific constraints:

Growth conditions must be improved so that the daily biogas productivities can be economically interesting. To achieve this goal the following points must be taken into consideration:

- inocula, very active in polymers hydrolysing, must be found in some specific biotopes (hot streams, termites, miches, etc.);
- all of the parameters influencing biological activities must be controlled; and
- it could be interesting to study the continuous fermentation system which produces the biogas at a higher rate in comparison with discontinuous system.

PROSPECTING OF NEW SOURCES OF BIOMASS FOR BIOGAS FERMENTATION IN BURKINA FASO

A national program has been undertaken on biogas technology, sponsored by the Minister of High Education and Scientific Research. It focuses on:

- screening of substrates of fermentation;
- lowering building fees of digesters; and
- adoption of new designs of digesters capable of improving biological yields.

So we used Calotropis Procera as a new source of biological energy. It is a latex plant, broadly found in sahelian countries, and especially in Burkina Faso. It contains some cardenolides, very toxic for animals. For this reason, they do not use this plant as feed. Its biochemical analysis results are gathered on the following table:

C/N Ratio	17.20 ± 1.20
Ashes (%)	9.70 ± 1.40
Volatile solids (%)	80.50 ± 0.90
Ether extract (%)	7.40 ± 0.90
Crude protein (%)	16.40 ± 1.20
Crude fiber (%)	56.80 ± 2.30

Fermentation experiments have been conducted at 30°C and at varying pH values. The load rate was 1%. Inocula of beef belly have been used. Maximum productivity rates have been obtained at pH 7.0 and 7.5, corresponding 3.6 l biogas/kg per day and per liter. The biogas thus obtained included 65% in methane and about 35% in CO_2. Therefore, Calotropis Procera constitutes a prospective source of biomass for biogas production.

CONCLUSION

Biogas technology is still in the early stages of development in Burkin Faso. The biogas produced in our country is, to date, used for the satisfaction of several needs:

a) energy for lighting;

b) pharmaceutical product conservation in rural hospitals;

c) cooking in some schools and training centers; and

d) water drawing with motorpump functioning on biogas as energy. In this case, energy need is 450 - 500 l biogas per hour and per horse power.

BIOGAS PLANT IN IVORY COAST

Reinhard Henning

OEKOTOP*, West Germany

SUMMARY

This project was undertaken by the GTZ, West Germany (German Agency for Technical Cooperation). OEKOTOP, Berlin, West Germany, acted as consultant.

The paper describes a larger-sized biogas plant for a commercial unit, in this case a combined feedlot and slaughterhouse. At the present stage, it is still a pilot project, but serves as the first phase for the full implementation later on. The main objective is to generate electricity for the slaughterhouse plant, which is generated at present using diesel generators. The project has operated successfully for the past two years.

Two significant innovations have to be specifically mentioned:

- The pit is dug into the soil without any coating or plastering. The loss of liquid has become marginal after a short time of biological self-sealing. This method provides a sizable reduction in the cost of the plant.
- The cover of the pit is made of an expandable synthetic rubber sheeting, which acts as a variable size gas holder. The material is still in good condition after two years of operation.

* For full report, reference can be made to either:
- GTZ, P.O. Box 5180, 6236 Eschborn,

or

- OEKOTOP GmbH, Paul-Lincke-Ufer 41, 1000 berlin 36, F.R.G.

BIOGAS FROM BIOMASS FOR A KENYA FARM SERVICE CENTER

Donald L. Day, Agricultural Engineering Department,
University of Illinois, U.S.A.

ABSTRACT

The objective is to provide fuel, feed, and fertilizer by utilizing the wastes and by-products from a Kenya farm service center. Each farm service center is planned to operate a four-hectare farm: one hectare of corn, one hectare of wheat, and two hectares of a legume/grass mixture. There would also be 10 dairy cows, 35 pigs, and 300 chickens. Wastes and byproducts from these crops and livestock can be used to produce methane and possibly alcohol to supply locally needed fuels and help operate a food processing plant.

We are collecting background information on the quantity and characteristics of the wastes and byproducts. The next step will be laboratory research on mixtures of the wastes and by-products to match biogas production with energy requirements. Finally, the residues from biogas production will be used to provide feed and/or fertilizer for the farm service center or the local community.

Also included is a brief report of the anaerobic digester on a swine farm at the University of Illinois. The digester is part of a project of an integrated farm-fuel system.

INTRODUCTION

Fuel, feed and fertilizer are critical resources that must often be imported at great expense. The high cost of these resources, along with high interest rates, are major economic problems for agriculturalists, especially in developing countries. Conversion of locally available byproducts (biomass) to fuel, feed and fertilier can greatly improve community development and welfare. Waste materials and byproducts have at least four advantages as sources of biomass: (1) they are available nearby, (2) they are continuously available, (3) they generally have negative value and may be a source of pollutants, and (4) they are amenable to biological upgrading to useful products.

This project is an engineering technology element of a USAID Title XII Strengthening Grant to the Universiy of Illinois. Our project is entitled Microbial Conversion of Biomass into Fuels, Feed, and Fertilizer in Developing Countries. We are presently working with Egypt and Kenya. The Egyptian phase has to do with the fungal fermentation of crop residues to enhance them as a livestock feed. We have done some work using a white-rot fungi, pleutotus ostreatus, to break down lignin for better fermentation and improve the protein content. The Kenya phase involves anaerobic digestion of livestock manure to provide needed fuel. This paper is on the Kenya phase of our project.

Kenya has an agricultural-based economy, accounting for about 29 percent of the total gross national product of the whole country. Kenya also has a free enterprise system and large colonial farms are breaking up into small farms owned by Kenyans. The average size of a small farm in Kenya is about 2-3 hectares. The government has placed a high priority on developing the small farm and agri-related businesses which sustain small farming operations.

Nakuru is located in Kenya's richest agricultural land and has been selected as the agricultural basis for this study. The study is for energy needs and alternative sources by utilizing wastes and byproducts from agricultural production and processing.

BIOGAS PRODUCTION

The study is for a farm and processing plant such as is being promoted by International Farming Systems (IFS, 1980). This is a system of agricultural services for subsistence farmers and is designed to be the "hub" of a community supplying training, marketing, and food processing in fertile but under-productive areas.

An example farm service center is planned to operate a four-hectare farm: one hectare of corn, one hectare of wheat, and two hectares of a legume/grass mixture. There would be 10 dairy cows, 35 pigs, and 300 chickens. However, only about half of the manure from the dairy cows is expected to be collectable. There would also be a food processing plant (cannery) to handle two to five tons of raw product per day. This makes for an integrated farming system.

Table 1 gives the amounts and characteristics of manure as excreted (feces and urine but no bedding) for livestock on rations to give high production of meat, milk and eggs (ASAE, 1983 and MWPS-18). Table 2 gives amounts of biogas expected from the volatile solids of these manures (MWPS-19, 1975). A literature review of the anaerobic production of biogas is given by Day (1983). Biogas is about 60 percent methane.

The biogas expected from the manure of 10 dairy cows (only half collected), 353 hogs, and 300 chickens is given in Table 3 based on data of Tables 1 and 2. The total biogas of $13m^3$/day has an energy equivalent of 8 L/day of diesel fuel. If the biogas were used in an engine generator at an overall efficiency of 20 percent, it would produce 16 kWh/day or 0.7 kW. Table 3 gives a guide to the amount of biogas needed for cooking, lighting and running engines (Eggerling et al., 1981). At the rate of $0.25m^3$ of biogas per person per day for cooking, this gas would be sufficient for about 52 people.

Table 1. Manure production and characteristics per 454 kg (1000 lb) live weight.* (ASAE, 1983 and MWPS - 18, 1979)

Item	Units	Dairy		Beef		Swine			Poultry		
		Cow	Heifer	Yearling 182-318 kg (400-700 lb)	Feeder >318 kg (>700 lb)	Feeder	Breeder	Sheep	Layer	Broiler	Horse
Raw Waste (RW)	kg/day	37.2	38.6	40.8	27.2	29.5	22.7	18.1	24.0	32.2	20.4
	lb/day	82.0	85.0	90.0	60.0	65.0	50.0	40.0	53.0	71.0	45.0
Feces/Urine Ratio		2.2	1.2	1.8	2.4	1.2		1.0			4.0
Density	kg/m^3	1005.0	1005.0	1010.0	1010.0	1010.0	1010.0		1050.0	1050.0	
	lb/cu ft	62.7	62.7	63.0	63.0	63.0	63.0		65.5	65.5	
Total Solids (TS)	kg/day	4.7	4.2	5.2	3.1	2.7	1.9	4.5	6.1	7.7	4.3
	lb/day	10.4	9.2	11.5	6.9	6.0	4.3	10.0	13.4	17.1	9.4
	% of RW	12.7	10.8	12.8	11.6	9.2	8.6	25.0	25.2	25.2	20.5
Volatile Solids	kg/day	3.8			2.7	2.2	1.4	3.8	4.3	5.4	3.4
	lb/day	8.6			5.9	4.8	3.2	8.5	9.4	12.0	7.5
	% of TS	82.5			85.0	80.0	75.0	85.0	70.0	70.0	80.0
BOD_5†	% of TS	16.5			23.0	33.0	30.0	9.0	27.0		
COD‡	% of TS	88.1			95.0	95.0	90.0	118.0	90.0		
TKN§	% of TS	3.9	3.4	3.5	4.9	7.5		4.5	5.4	6.8	2.9
P‖	% of TS	0.7	3.9		1.6	2.5		0.66	2.1	1.5	0.49
K#	% of TS	2.6			3.6	4.9		3.2	2.3	2.1	1.8

* Numerical values for kg/day/1000 kg live weight are the same as those for lb/day/1000 lb live weight.

† 5-day biochemical oxygen demand.

‡ Chemical oxygen demand. § Total Kjeldahl nitrogen. ‖ Phosphorus as P. # Potassium as K.

Table 2. Guidelines for anaerobic digesters.

Parameter	Swine growing-finishing	Dairy	Beef	Poultry, layer	Poultry, broiler
Estimated fraction[a] of N as total ammonia-N in digested effluent	0.5	0.2	0.25	0.75	0.75
Dilution as manure/(manure + water)	1:2.9[a]	undiluted	1:1.32[b]	1:8.3[a]	1:10.2[a]
Estimated dilution water required lb H_2O/1000 lb animal	123	0	18.6	387	653
Hydraulic detention time (day)	12.5	17.5	12.5	10	10
Loading rate					
lb VS/day ft^3	0.13	0.37	0.37	0.13	0.1
kg VS/day m^3	2	5.9	5.9	2.1	1.6
Digester volume per animal unit					
ft^3/1000 lb animal	38	23.1	16	72	120
m^3/1000 kg animal	2.38	1.44	1	4.5	7.5
Estimated fraction of VS destroyed at this detention time	0.5	0.48	0.45	0.6	0.6
Ratio COD/VS	1.19	1.06	1.12	1.28	1.28[c]
Estimated total gas production (20 °C) per unit of VS removed					
ft^3/lb VS	11.9	10.6	11.2	12.8	12.8
m^3/kg VS	0.745	0.663	0.7	0.8	0.8
Estimated gas production (20°C) per animal unit if digester is loaded according to this table					
ft^3/day 1000 lb animal	28.6	43.7	29.8	72.3	92.1
m^3/day 1000 kg animal	1.79	2.73	1.86	4.52	5.76

[a] Ammonia toxicity criteria.

[b] Volatile fatty acids criteria, compared with dairy for want of a better criterion.

[c] Taken to be the same as that for layers.

(MWPS - 19. 1975)

FERTILIZER PRODUCTION

Sludge from the digester would contain essentially all the plant nutrients of the manure entering the digester. Table 1 also gives typical values of nutrients in manure as excreted. There are, however, losses of nutrients depending upon the methods of collection, handling,

Table 3. Expected biogas production from manure.

$$1/2\ (5\ \text{cows}) \times 544.3\ \text{kg} \times 2.7\ \frac{m^3}{(\text{day})(1000\ \text{kg})} = 7.35\ m^3/\text{day}$$

$$3\ 35\ \text{hogs} \times 57\ \text{kg} \times 1.8\ \frac{m^3}{(\text{day})(1000\ \text{kg})} = 3.56\ m^3/\text{day}$$

$$3\ 300\ \text{poultry} \times 1.4\ \text{kg} \times 5.1\ \frac{m^3}{(\text{day})(1000\ \text{kg})} = 2.14\ m^3/\text{day}$$

$$\text{Total} = 13\ m^3/\text{day}\ (461\ ft^3/\text{day})$$

$$\text{Gross energy content} = 13\ \frac{m^3}{\text{day}} \times 22.4\ \frac{MJ}{m^3} = 291\ \text{MJ/day}\ (277{,}000\ \text{btu/day})$$

Table 4. Guide values for gas consumption. (Eggerling et al., 1981)

Cooking:	0.25 m^3 (8 cu ft) per person per day
Lighting:	0.12 - 0.15 m^3 (4-5 cu ft) per hour per lamp
Driving engines:	0.45 m^3 (15 cu ft) per HP per hour

and storage prior to entering the digester. Up to 50 percent of the nitrogen can be listed by normal methods of collection and storage. This is because about half of the nitrogen in excreta is in the ammonia form and can volatize away in a few weeks time. Therefore, a waste management system that conserves nitrogen is desirable such as daily collection and loading into the digester.

Assuming 50 percent of the nitrogen is lost in collection and storage, sludge from the digester for the farm service center would be expected to contain 423 kg/yr N, 224 kg/yr P and 527 kg/yr K. This would furnish the required fertilizer nutrients to grow 3 hectares of corn yielding 7.5 m^3/ha of grain. The sludge would be sufficient for up to twice this much corn land if nitrogen losses could be reduced or eliminated.

PLANS

We are collecting background information on the quantity and characteristics of the wastes and byproducts from a Kenya farm service center. The next step will be laboratory research on mixtures of the wastes and byproducts to match biogas production with energy requirements. Finally, the residues from energy production will be studied as a source of feed or fuel for use on the farm service center or in the local community.

DIGESTER AT THE UNIVERSITY OF ILLINOIS

A digester has been built on the University of Illinois new swine farm at Urbana-Champaign, Illinois. The farm has a capacity equivalent of a farrow-to-finish facility marketing 3,000 pigs per year. It is a totally enclosed, modern confinement facility with partially slotted floors. The manure produced from the animals in some of the buildings is automatically scraped from under the slats several times each day to a central sump and then is pumped to the anaerobic digester.

The main tank of the anaerobic digester unit is composed of four compartments: gas storage, gas processing, the main reactor, and sludge storage (see Figure 1). The four connected compartments are cylindrical in shape and are divided by reinforced concrete walls. The total unit is laid horizontally in the ground. A separate tank is also provided for mixing, processing, and preheating the incoming manure. The tanks were fabricated on-site from galvanized steel. The interiors and exteriors of the tanks were insulated with spray-on polyurethane

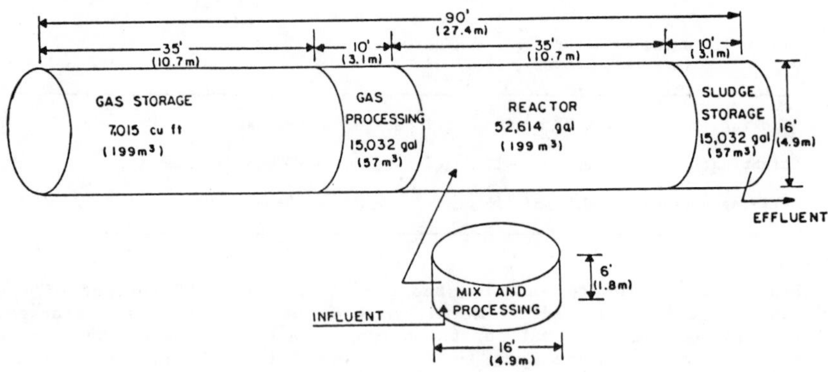

Figure 1 The anaerobic digester on the Swine Research Farm at the University of Illinois (constructed by Energy Resource Systems, Dewey, Illinois).

and then sealed on the inside with spray-on rubber lining. The remainder of the unit not buried in the ground was mounded over with earth to provide additional insulation. The unit is further explained by Fedler (1983).

The digester is part of a research project of an integrated farm fuel system as depicted in Figure 2. Briefly, it works as follows. Fuel for the tractor, combine and truck can be obtained by making alcohol from 10 percent of an Illinois corn crop. Stillage residue from the alcohol plant is a good feed protein and can be used in a nearby feedlot instead of being dried. Manure from the feedlot can be used in a digester to produce methane to operate the alcohol plant. Finally, sludge from the digester can go to the crop land to furnish part of the needed fertilizer, thus completing a beneficial energy and ecological cycle. This project is further explained by Day et al. (1984) and by Rodda and Steinberg (1981).

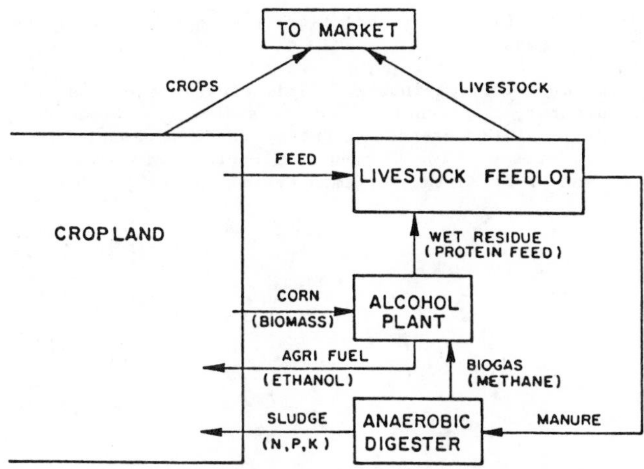

Figure 2 Flow scheme of the proposed integrated farm fuel system.

REFERENCES

ASAE. 1983. Manure production and characteristics data D-384. Yearbook, American Society of Agricultural Engineers. St. Joseph, MI. p. 436.

Day, D. L. 1983. Anaerobic treatment and production of biogas, Chapter 12 in: Livestock Manure Management, Text and Reference. Agricultural Engineering Dept., University of Illinois. Urbana, IL p.12 1-12.30

Day, D. L., M. P. Steinberg, and E. D. Rodda. 1984. Integrated alcohol-methane production systems for midwest farms. Proc., Symposium on Energy from Biomass and Wastes VIII, held Jan. 30-Feb. 3 at Lake Buena Vista, FL. Mtg. sponsored by the Institute of Gas Technology. Chicago, Il.

Eggerling, G., R. Guldager, H. Guldager, G. Hilliges, L. Sasse, C. Tietgin, and U. Werner. 1981. Biogas plants building instructions. German Appropriate Technology Exchange. Eschborn, Federal Republic of Germany.

Fedler, C. B. 1984. The anaerobic digester on the swine research farm at the University of Illinois. Proceedings: Livestock Waste Management Conf., held March 12 at Champaign, IL. Agri. Engr. Dept., Univ. Il. at Urbana-Champaign, IL. pp. 42-47.

IFS. 1980. International Farming Systems, Inc. Wheaton, IL 8 pp.

MWPS-19. 1975. Livestock waste management with pollution control. Midwest Plan Service. Ames, IA. p.51

MWPS-18. 1979. Livestock waste facilities handbook. Midwest Plan Service. Ames, IA.

Rodda, E.D. and M. P. Steinberg. 1981. Energy analysis of an agricultural, alcohol fuel system. Proceedings: Third International Conference on Energy Use Management, Oct. 26-30. W. Berlin, Germany. In: Beyond the Energy Crisis, Ed. by Fazzolore, R. A. and C. B. Smith, Pergamon Press. Pp. 1807-1814.

MECTAT'S EXPERIENCE IN THE TRANSFER OF BIOGAS TECHNOLOGY

Boghos Ghougassian, MECTAT Coordinator,
MECTAT (Middle East Centre for the Transfer of Appropriate Technology) *
Beirut, Lebanon

ABSTRACT

This article summarizes the relatively recent experience of MECTAT in the field of biogas technology. The first two small-scale demonstrations in Lebanon and Syria are described. Preliminary views on the prospects of biogas technology in some Arab countries are also presented.

INTRODUCTION

This brief essay concentrates on the recent field experience MECTAT* has acquired during the past two years, and, an overview of biogas technology applications in the rural areas of some countries of the Arab World.

The transfer of biogas technology is just one of MECTAT's activities in the field of local renewable energy sources development, which range from solar food cooking ovens to efficient woodburning cook stoves.

FIELD EXPERIENCE IN ESTABLISHING BIOGAS DIGESTERS

Thus far, MECTAT's experience includes the establishment of two small scale biogas plants (1984), for demonstration and research purposes.

The first prototype plant was established on the balcony of MECTAT's office in Beirut. It is a fixed dome type. Two 40 liter plastic containers were used as digesters. Three-quarters of each was charged with slurry (dry chicken droppings mixed with water), and the remaining quarter used as biogas storage chambers. The feeding mode is of batch type. No seeding sludge is inserted in the digesters.

---

*The Middle East Center for the Transfer of Appropriate Technology (MECTAT) is a non-profit organization. MECTAT is a division of Middle East Engineers and Architects (MEEA), based in Beirut. Our overall goals are the transfer of Appropriate Technologies to rural and poor areas in the fields of:

- renewable sources of energy;
- water resources development;
- sanitation;
- food production, processing and storage;
- income generating activities; and
- other related subjects.

Gas production started after fifteen days. At first, gas was noticed leaking from the pipe-connecting joints. With the problem checked, gas pressure registered over 40 cm of water column daily. The gas is being burnt regularly with our engineers monitoring its performance and replacing the slurry with a new batch of waste matter once the gas production stops.

The produced gas burns; its flame is as blue as that of any ordinary gas stove.

For pipings, 11 mm (internal diameter) transparent plastic pipes are being used. A discarded conventional gas stove is utilized as burner, after enlarging the outlet jet hole to 1.5 mm.

Figure 1 presents the layout of the fixed dome, batch type, biogas digester unit.

Our investigations confirmed that this small prototype biogas production plant was, and still is, the first biogas producing unit ever installed in Lebanon.

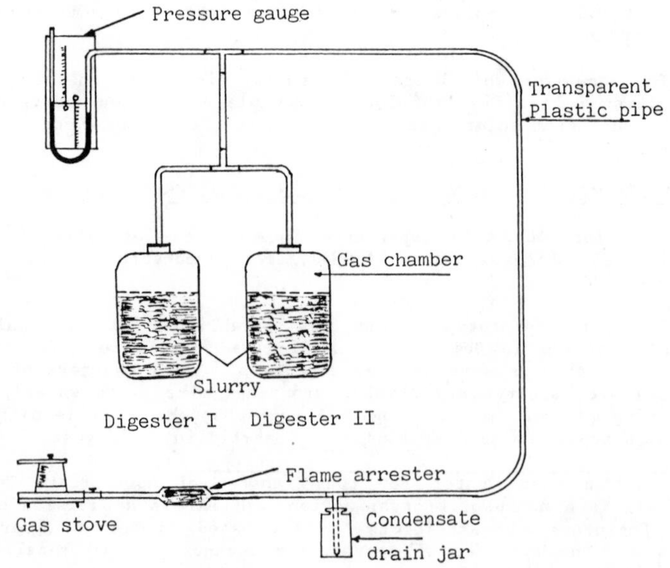

Figure 1 The Layout of a Small Scale (Fixed dome and batch type) Biogas digester (Lebanon)

The second biogas digester plant was established on the premises of the Rural Development Center (RDC) of Salkhad town in Syria. RDC Salkhad is one of seven RDCs presently existing in the Republic of Syria, overseen by the Syrian Ministry of Social Affairs and Labor.

A MECTAT expert ran a training session at RDC Salkhad on this topic. A group of thirteen young men with various backgrounds including agricultural engineers, health workers, technicians, and RDC staff members, participated in this ten-day workshop, and were familiar with many aspects of biogas technology. With their active participation, the prototype biogas plant was established.

This project was exectuted for the UNICEF Regional Office of the Middle East and North Africa (MENA). UNICEF/MENA has launched (in 1983) a regional appropriate technology (AT) program for Child Survival and Development.

All the construction materials were prepared before the arrival of the MECTAT expert. Four empty metallic barrels were used in the construction of the biogas unit. Two barrels (200l capacity each) served as fermentation chambers of continous feeding type; the other two served as a floating type gasholder. The inlets and outlets of the digesters were made out of 3" galvanized iron pipes.

Twenty-two mm (diameter) plastic pipes were used for carrying the produced gas. Another kind of discarded gas stove was used as burner. A pressure gauge was made out of transparent (10 mm diameter) plastic pipe. For details, refer to Figure 2.

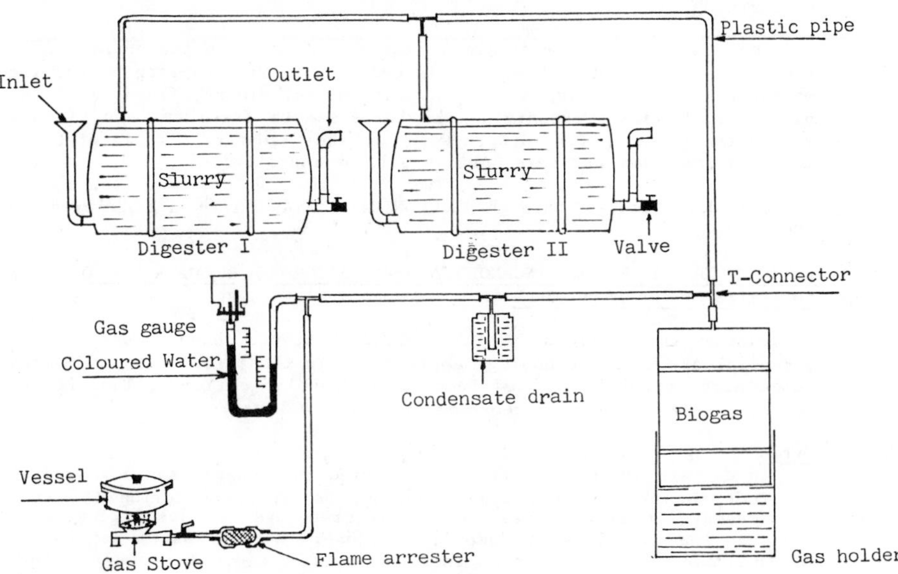

Figure 2 The Layout of a Small Scale above Ground Horizontal Type Biogas Plant with Floating Type Gas Holder (Syria).

The two barrels were charged with domestic animal wastes (fresh cow dung 20%, chicken drippings 60%, and sheep droppings 20%). Water was added in the proportion of eight to one. All were well mixed in a container and then charged into the two barrels. Sludge from an active septic tank was mixed, too, as seeding material.

Given that the location is in a cold region, the digesters were put inside a greenhouse. This will increase the digester temperature by at least 10°C relative to the ambient temperature.

The officials claimed that this prototype unit was the first digester plant in the country.

One of the participants was assigned to take care of the system and keep records on the daily fed material, as well as the gas production rates.

Two months later, a MECTAT expert visited RDC Salkhad for running another AT training session for UNICEF/MENA, where he was informed that the digester is not producing biogas. After an investigation it was found that:

- The digesters (empty barrels) had previously contained insecticides;
- The chicken droppings contained a high percentage of sawdust (containing lignin) and medications, such as antibiotics.

Undoubtedly these two factors have killed the gas producing bacteria.

The plant now is under investigation. The local technician, upon the advise of the MECTAT expert, is undertaking step-by-step correcting measures. These include: emptying the initial slurry; flushing of the barrels with water; recharging the barrels with fresh cow dung (mixed with water); adding lime or ash-water in the digester. If all of these measures fail, new and safe barrels will be installed.

The developments through the UNICEF/SYRIA office are relayed to MECTAT.

AN OVERVIEW OF BIOGAS TECHNOLOGY APPLICATION IN RURAL AREAS OF SOME COUNTRIES OF THE ARAB WORLD

During their field visits, MECTAT experts have observed the potential of biogas technology application in the rural areas of Syria, Yemen Arab Republic, Peoples Democratic Republic of Yemen, Tunisia and Lebanon. Their findings are summarized here below:

Syria

In certain areas the cow dung is dried into dung cakes and used as fuel for cooking and space heating, particularly during the cold season. In other areas the animal droppings are just accumulated near the dwellings or dumped as refuse. Animal wastes are rarely utilized as fertilizers. This, in part, is due to the fact that the villagers are unaware of its benefits.

The application of biogas technology is relatively easy in areas where piped water exists. However, government support is needed, particularly for providing the building materials.

Soils of the country, in general, lack organic matter content. The effluents of biogas plants would greatly contribute to the food production.

Yemen Arab Republic (YAR)

In the mountainous areas of YAR, villagers mix the animal wastes with water and keep them in a pit for fermentation. After a period of 15-30 days, they take the mixture out of the pit, make dung cakes, and dry them under the sun for the purpose of producing household fuel for winter use.

Actually they are at the half-way mark in producing the methane gas. They just need air and water-tight digesters for collecting the produced gas.

In the Thihama coastal zone of the country, the animal wastes are piled next to dwellings. These are not used in agriculture. In this zone the daily average temperature is always 25°C . Ground water is available in most of the villages and towns. These two factors are very good preconditions for the application of biogas technology.

Biogas could be utilized for cooking, lighting, and water pumping if diesel motors are adapted for this job.

People's Democratic Republic of Yemen (PDRY)

In all the rural areas of the PDRY, the animal droppings are just piled next to dwelling areas. These are never composted for use as organic fertilizers.

In agricultural areas, where water is abundant, biogas plants could bring great benefits. The gas could be used for water pumping, cooking, and lighting purposes. The effluent of the biogas plants can boost agricultural production.

All of the fossil fuels are imported. However, there are good prospects that oil exists in the country.

Tunisia

In the northern province of Tunisia, an experimental biogas plant already exists.

In most of the rural areas, the animal wastes are under-utilized and the soils lack organic matter.

Whenever water exists, biogas technology can boost the agricultural production and the gas could be used for cooking, lighting, and water pumping purposes.

Lebanon

Although during the last decade, the domestic animal population has greatly declined in the country, there is still a potential for installing biogas plants on poultry and cattle farms.

Biogas could be used for space heating, cooking, and electric power generation.

Traditionally, the animal wastes, after composting, are utilized in agriculture.

CONCLUSION

If biogas projects in a given country are backed by the active support of its government, and that of international organizations, the biogas technical capabilities of the country could be developed within a decade, and the proliferation of biogas plants could subsequently be achieved.

In all the above-mentioned countries, oil production is either limited or non-existent. Fuel wood is scarce, and the soils lack organic matter. Hence, the biogas technology might bring in the future, many real benefits such as increased food production, improvement of environmental sanitation and personal health, check deforestation, etc. (particularly when we think of scarcer oil in the future).

THE PRODUCTION AND USE OF BIOGAS: AT THE "LABORATOIRE D'ENERGIE SOLAIRE" (LESO) IN BAMAKO, MALI

Yava Sidibe
Laboratoire d'Energie Solaire
BP 134, Bamako, Mali
October 1984

ABSTRACT

A short outline on Mali's R&D endeavors in the field of biogas technology is presented. Future programs, diffusion efforts, and major socio-economic constraints are also highlighted.

INTRODUCTION

Mali, like many West African countries, is faced with a very difficult energy situation. In the modern energy sector, it is completely dependent on outside sources for petroleum supplies. The traditional energy sector, which in Mali represents over 90% of the energy consumption, is also problem-plagued. The over-exploitation of forest reserves has caused a serious firewood crisis.

Resort to renewable energy resources is therefore quite necessary. Biogas presents one prospective option.

BIOGAS RESEARCH AND DEVELOPMENT PROGRAM AT LESO

As part of the Renewable Energy Project funded by USAID, LESO has been researching and developing biogas production and utilization technologies since 1980. It has conducted studies on the discontinuous anaerobic fermentation of cow manure, using 200 liter oil-drum digesters. Tests in a larger Indian-type digester were conducted to confirm the results of the small-scale study for digesters of greater volume. Gasoline pump engines and kerosene refrigerators have also been converted for use with biogas and tested accordingly.

FUTURE RESEARCH PROGRAM:

The following further studies are planned:

- the continuation of tests on the anaerobic fermentation of animal wastes,

- the continuation of tests on modified pump motors, taking into consideration the problems encountered in the first trials,
- the continuation of tests on converted kerosene refrigerators,
- the production of biogas by the fermentation of cow manure and crop wastes,
- the improvement of biogas purification systems using local materials (e.g., laterite).

DIFFUSION EFFORTS

In addition to laboratory testing, LESO has been studying the adaptibility and social impact of appropriate technologies for several years. Studies in four zones of Mali have attempted to determine the following:

- the availability of organic matter and water,
- the available building materials,
- the energy needs,
- the existence of social structures capable of maintaining biogas systems,
- the possibilities for the use of digester waste.

A digester was installed in Keleva, a village 70 km from Bamako. The digester is situated near a maternity unit and the biogas produced was used to run a refrigerator to store medicines, two lamps for the school, and a central kitchen for use by the maternity patients' families. The effluent produced by the digester was used on the school field. After the first follow-up evaluation, it was found that the digester was no longer being filled with manure. It seems that the villagers became discouraged for several reasons. There were technical problems with the digester, the cows are not kept near the village all year, and it seemed the central kitchen was monopolized by the head mid-wife. LESO is currently trying to resolve this problem with the help of the local authorities. The installation of four additional biogas digesters at other sites is planned.

CONCLUSION

The experiments done at LESO on biogas have given the engineers a certain mastery of the technology, both of the production of biogas and of the utilization of the gas for pumping water, refrigeration, cooking, and lighting. Although the results obtained so far are very encouraging, important work remains to be done in the areas of improving the utilization devices, reducing the costs of installing digesters, and increasing the range of materials used as digester feed, etc. Through these efforts, it is hoped that biogas technology can be adapted to village conditions in Mali. The socio-economic studies done by LESO point to several constraints of the village world:

- the poor purchasing power of villagers,
- the lack of easily available cow manure during certain times of the year due to herd migration,
- the scarcity of water in most villages.

These constraints must be taken into consideration when planning research activities, for the goal of such reasearch is to develop technologies which satisfy needs of the country's population. As these needs are pressing, it is important to avoid the loss of time in "reinventing the wheel." It is for this reason that exchanges between researchers of different countries must be encouraged and supported by, for example, conferences such as the one we are attending. We hope that our contribution will further this goal.

BIOGAS IN MOROCCO

Abdelhaq Fakihani
Director-General of the "Centre de Developpement des Energies Renouvelables" (CDER)

ABSTRACT

This paper summarizes the relevant aspects of biogas technology in Morocco. The biomass resource base and biogas potential is first outlined. Projects implemented and prospects are next addressed. Finally, pertinent conclusions and policy directives are highlighted.

THE POTENTIAL OF BIOGAS IN MOROCCO

The valorization of organic wastes through methanic fermentation is only one possible means of utilizing biomass, and the CDER is interested in all sources of renewable energies. Therefore it seems to us indispensable to present first a short evaluation of the biomass potential in Morocco.

First it should be noted that statistical sources in this field very often are lacking and that the figures presented here are usually the result of preliminary studies collected by the CDER. They need to be refined.

Biomass sources in Morocco consist essentially of:

- Forestry biomass, already extremely exploited, but with important losses.
- Agricultural biomass, consisting of residues or byproducts of agricultural or agro-industrial activities.
- Industrial or urban wastes, such as municipal wastes or waste waters.
- Wastes or byproducts resulting from cattle-raising, such as manure or residues from slaughterhouses.

The forestry biomass available for an energetic valorization, which we will consider here, consists of residues of existing activities (sawdust, industrial wood, charcoal production, etc...). We will not take into account the wood already exploited as firewood or for charcoal production, which should be decreased rather than increased for ecological reasons.

The gas wasted during charcoal production is estimated to be 51000 TEP (ton oil equivalent) and the wastes of forestry industries to be 12000 TEP.

A CDER study dating from 1983 has evaluated the total agricultural wastes in Morocco to be 1.6 million tons.

Finally, the resources in alfa-alfa are estimated to be 200,000 tons annually. A first project of a power plant of 3 MW using alfa-alfa is being studied.

Neither are other sources of biomass (industrial and urban wastes) known with precision. One recent CDER study, however, gives an estimate of 265,000 TEP (or 11 percent of the national electric consumption) to the energetic potential of municipal wastes in Morocco, which are presently used partly as compost only.

We are going to evaluate in detail the potential of biogas production. CDER has completed a certain number of sectorial studies which allowed us to reach the following tentative results.

The sources of methane-producing wastes will come from:

- animal excrements recoverable in the cattle-raising sector, including the slaughterhouses.
- certain wastes of the agro-industries.
- urban wastes to a lesser extent (waste waters or municipal wastes).

I will talk here only of the first ones, as information on the others is insufficient. In any event, the valorization techniques perfected today involve (most of the time) only the animal wastes.
All quantified data come from the Ministry of Agriculture and Land Reform and cover the years 1982 and 1983.

Species	Number	Wastes recoverable T (DM)*/year	Biogas Potential 103 m3
Bovines	3×10^6	500,000	97
Ovines	12×10^6	200,000	40
Caprines	7×10^6	-	-
Porcines	negligible	-	-
Poultry	6.3×10^6	135,000	48
Equines	1.7×10^6	-	-
Camels	0.13×10^6	-	-
Slaughterhouses	213,000 T	3,900	1

(*DM = Dry Matter)

The theoretical potential of the recoverable fraction of animal excrements or wastes, coming from cattle-raising, is therefore estimated to be 186 x $10^6 m^3$ of biogas per year corresponding to 92,500 TEP.

The table here below summarizes the figures quoted previously. Let us remember that we are considering here only the wastes or by-products of already exising activities.

Source	Equivalent 10 TEP
Residues from forestry exploitations	12
Lost gas during charcoal production	51
Vegetation wastes	256
Municipal wastes	265
Animal wastes	62
Total	676

The primary energy consumption, outside wood and charcoal, was 470 x 10^3 TEP in 1982, the energetic biomass potential, represents therefore about 14 percent of this comsumption.

Obviously, potential does not mean that all this resource is exploitable, but these figures are enough to show the interest for Morocco in investing in this field.

IMPLEMENTATIONS AND PROSPECTS

Objectives

At the national level, the present objectives of the development policy concerning anaerobic digesters are:

- Assess clearly the constraints impeding the diffusion of this technology and define a potential market by sector of activities and by region.
- Develop reliable digesters of moderate cost, responding to the constraints identified above.
- Train the personnel necessary in diffusion of digesters at CDER, as well as interested institutions, such as the Offices of Agricultural Development.
- Find and promote original formulae of financing, for example with the Agricultural Credit Bank (Caisse de Credit Agricole) which funds the development of this sector.

Roles of CDER

The role that CDER will have in the dissemination of this technology is primarily that of coordinator and starter, since CDER has national authority in the field of renewable energies.

We can detail these activities under the following items:

- coordination of efforts expended in this field
- dissemination of results of the work done in Morocco or abroad
- research necessary to the support of work aimed at developing digesters adapted to local conditions (nature of wastes, climate, size of installations, available raw materials)
- evaluation of experiments and current projects

- definition of a policy of dissemination of biodigesters
- assistance to the search and availability of funds necessary to the support of these activities
- on the other hand, CDER does not have the role of installing thousands of digesters and will leave it up to the institution involved in rural areas.

As for promotion and the dissemination of biogas systems, CDER is interested in two large categories of applications:

Those concerning rural areas with individual digesters or digesters adapted to small agricultural farms, and other environments such as agro-industries, municipalities.

CDER until now was involved more in the first domain than in the second, essentially because the request for advice and studies happens to be stronger in that sector than in the other.

Means of Intervention

In order to reach the objectives established, CDER proposes to use the following:

- Implementation of pilot or demonstration project, possible with collaboration of other institutions
- Organizations of seminars, such as the one we have held in February 1984, and which has marked the beginning of the dissemination of digests at the national level. More generally, interventions with mass media to disseminate information on projects already completed.
- Assistance to research activities of other institutions and proposal of topics of research in this field (for example a team of the Faculté des Sciences in Marrakech works with on the scientific monitoring of one of our projects.
- Research in the biogas laboratory of CDER (which will be in operation by the beginning of next year) to assist experiments in the field and to conduct some internal to CDER
- Make available to private enterprises funds necessary to cover part of the financing of a project.

Projects

I thought it would be interesting to present, briefly, the projects in this field already implemented or in the stage of final study.

In operation:

- an Indian-type digester designed by the CDER and built under its supervision in 1983, operating well for over a year. Its characteristics are as follows:

Volume:	$6m^3$
Production:	0.6 to 1.8 m^3 biogas/day
Digested matter:	dilute manure (150 l/day)
Cost:	About $1000

A plastic-material digester has just been installed side by side with this digester. Its characteristics are similar to the first one, and thus we can perform a technico-economic comparison of these two installations.

- Several digesters of the Chinese-type, built in collaboration with the Offices of Agricultural Development. The oldest has been operating for almost a year, and a report comparing the performance of various types of digesters will be soon published by the CDER.

In the stage of construction or in final study:

- Several individual digesters, in collaboration with the Offices of Agricultural Development
- Two or three digesters projected for cattle-raising farms with stable cattle
- A digester for a poultry-raising farm
- A digester for the slaughterhouse of Marrakech
- Valorization of municipal wastes and waste waters

On the other hand, the CDER will undertake (during the forthcoming year) several studies on:

- the adaptation of internal combustion engines to biogas
- the discontinuous digesters requiring less water
- the applications of agro-industries

CONCLUSIONS

The expedience with which digesters were adapted in Morocco, following a very modest start by the CDER, is encouraging us to intensify our efforts in this field and to assist any individual or institution willing to build a digester.

It is obvious that this action cannot continue today in the same manner, as the CDER is not in a position to respond alone to the numerous requests coming from potential users.

This is why the CDER, from the start, had to rely on institutions related to rural areas, which will, with our assistance, disseminate digesters in their sector.

On the other hand, we are establishing contacts with municipalities, collectivities, and soon with industries, in order to study the possibility of building large digesters.

We heard very often from travelling experts that the experience of individual digesters in the world does not spell "success", except in rare cases. However, we have decided before accepting this claim, to run our own experiment. The CDER will continue simultaneously to develop both individual and industrial digesters. The demands and problems in rural areas are, in fact, such that a dissemination of technology, even on a small scale, would contribute efficiently to the resolution of the problem of energy supply to rural areas.

We also think that the development of renewable energies should evolve on the basis of a pragmatic approach. It is possible that anaerobic digesters are well adapted and accepted in one area of Morocco and not in another, for socio-cultural reasons. This is why the cooperation with Offices of Agriculture Development, which work on a regional level, are of importance, as they know best the rural areas and are best qualified to judge and undertake the actions leading to the dissemination of a technology.

Therefore, after only 18 months experience on anaerobic digestion, it is difficult and premature to project with accuracy the future of biogas in Morocco.

Like any emerging technology in a country, it is necessary for a large scale dissemination of biogas technology to show and prove its technological maturity; its investment and operating costs must be well known and mastered, and information about this technology must be well spread by a competent and credible agency.

All this implies important and long-term action. The CDER has undertaken this action but has just begun. However the rapid reactions triggered by our action, the interest shown by several individuals and institutions, and the effective collaboration with Offices of Agricultural Development, lead us to think that biogas systems represent a solution to some energy problems in Morocco and are destined for great dissemination.

The efforts of the CDER in this domain will bear on the development of models technically and economically adapted to local conditions, on the training of competent technicians, on the technico-socio-economic evaluation of existing projects, and on the dissemination of information to potential users.

BIOGAS ACTIVITIES IN SUDAN

Yaman Fadlalla (Ms.)
National Council for Research
Energy Research Council
Renewable Energy Research Institute

ABSTRACT

Sudan has great potentialities for biogas production. Biogas technology has been recently introduced in the country, mostly on the research and demonstration levels. It uses various available substrates, particulary cow dung, sewage, and water hyacinth. These activities are briefly reviewed in this paper.

INTRODUCTION: ENERGY SITUATION IN SUDAN

Sudan is a non-oil producing country. It depends mainly on imported oil to meet its needs for commercial energy. This imposes a heavy burden on the country's balance of payments. The ever-increasing worldwide oil prices impose serious problems on oil availability, while the problem of transportation to remote areas increases cost and makes availability unreliable.

The largest consumer of energy is the domestic sector where it is consumed mainly as direct heat. The main source of energy for this sector is firewood and charcoal which are supplied by natural forests and the scattered trees and bushes of the Savanah regions. In cultivated areas, the agricultural residues are sometimes used to produce domestic heat.

The continuous removal of these forests for mechanized agriculture, energy needs, and over grazing -- these factors in addition to the even growing desert add to the problem of energy in Sudan.

BIOGAS POTENTIALITIES

Biogas technology is a simple technology generally utilizing waste organic materials to yield a useful fuel, methane gas. It is an indispensable source of energy in regions where the stock materials are present, and where the other conventional sources of energy are not readily available. In this respect, Sudan can be counted as one of those countries which have abundant sources for biogas production and which can have great contribution in solving her energy problems locally and indigenously.

Sudan has an estimated livestock population of over 56 million (1981); over 20 million of these are cattle. The manure of these animals can be used for biogas generation. A total of about 250 million tons of animal manure is produced annually but only 10 percent of it is available for biogas production, and this is due to the nomadic nature of the animal owners.

Water hyacinth, an aquatic weed naturally infesting the White Nile and its tributaries (about 3250 Km along the river), can be anaerobically digested to yield biogas. Also papyrus, another aquatic weed infesting most of the swamps area in the Southern region, can be used as a substrate for biogas production. These two weeds produce annually about 9 million tons of dry matter [1].

Sudan is an agricultural country and there are abundant amounts of agricultural residues, e.g. cotton stalks, different cereals residues, ground-nut shells, rice husks, sesame stalks etc. These residues produced annually in large quantities reach about 20-30 million tons. These residues can be used in certain proportions, in a biogas producing system, raw or pretreated.

The fast growing sugar industry results in production of large quantities of molasses amounting to 450 thousand tons/year and also 1.13 million tons of bagasse [2]. Molasses, beside its use for alcohol and yeast production, can also be used for biogas production.

It is obvious that Sudan has rich sources for biogas production. These sources, if properly invested, will definitely substitute for a great part of the energy required.

RESEARCH AND DEVELOPMENT IN BIOGAS PRODUCTION

Water Hyacinth

Since late 1950's, water hyacinth has infested the White Nile and its tributories creating great problems regarding navigation, transportation, fishing, irrigation, and public health. In 1972 a joint Sudanese-German Project was initiated to deal with the chemical and mechanical control of this weed. Part of this project is its utilization as energy source. Further, some trials with small biogas digesters, 200 litre, were done in 1976. These were found to produce 450 litre of biogas/kg dry matter during a retention time of four months [3].

Research on biogas generation from water hyacinth was carried on and a number of digesters were constructed at Taweela village (17 Km north Kosti). These are:

Two Dome-Type digesters (Chinese design) having total volumes of 12 and 18 m^3;

One Bell-Type digester (Indian design) having a total volume of 8.0 m^3; and

Two Tube-Type (butyl-rubber) digesters (England) each having a total volume of 2.5 m^3.

A Total gas volume of 0.75 m^3/kg dry matter was obtained using 5 percent total solid at a temperature of 37°C and pH about 7.0. Methane concentration was found to be about 60-80 percent in the mixture.

The gas produced by one of the Chinese digesters was used to operate a modified gas refrigerator, gas ovens and burners, and a number of lamps at the village health center. The gas supply from the other unit was used to provide the needs of an elementary school. The yield of the other three digesters, the Indian and the butyl-rubber, was conveyed to adjoining household to fulfill their energy needs. Unfortunately the work on these digesters was terminated by the end of the project in 1980. But recently, in April 1984, a rehabilitation program sponsored by the Renewable Energy Research Institute was realized. In this program, the Indian and one of the Chinese digesters were cleaned, repaired, and recharged with water hyacinth. Now they are working perfectly and the neccessary data is being recorded [4].

A thorough research on biogas production from water hyacinth has been done at the Faculty of Agriculture, University of Khartoum[5]. In the work, the parameters governing the process have been studied and water hyacinth has been recommended as a rich source for biogas production.

Cow dung

In 1979 the Norwegian Church Aid-Sudan Project constructed a small continuous biogas digester utilizing cow dung, at Torit in the Eastern Equatorial Province Southern region. It consists of three sealed drums of total volume of 0.594 m^3 and fermentation space of 0.446 m^3. Another plant was constructed by the Young Science Society at Jube University. It has been operating since 1980 to supply gas to Bunsen burners in the science laboratories[1].

Since 1982 the Biomass Department of the Renewable Energy Research Institute is carrying out intensive research on biogas technology. In regard to cow dung as substrate, an intensive laboratory research work has been carried out using small-sized (4 liter) digesters. That work has revealed valuable results concerning the different parameters governing the process of biogas generation [6]. Also two other units have been constructed (one and three-sealed-barrels units). In the first one-barrel unit, the barrel serves as a digester, as well as a gas-holder, while the other unit, with three sealed barrels, has a separate gasholder. These two units are used for outdoor demonstration.

Molasses

Since January 1983, the Biomass department is executing a project on biogas generation from molasses and alcoholic distillery refuse. This project is sponsored by the National Council for Research. Until now valuable positive and promising results have been achieved. The laboratory research work (the first phase of the project) is supposed to be finished during the coming few months, then the second phase can be started. In the second phase, a medium-size biogas digester will be constructed at one of the sugar factories. The operation of such pit will be based on the results obtained from the first phase, and subsequent work regarding optimization of operation under field conditions will be carried out. The work on this project is supposed to be finished by June 1985 [7].

Biogas from Sewage Treatment Plant

In Khartoum, the national capital, there are three sewage treatment plants. Some suggestions have recently been presented by KUP, an engineering consultant company in West Germany, to construct covers to the sludge basins and to do necessary modifications to convert the plant into a biogas producing unit. The local authorities have not taken a decision yet. The Renewable Energy Research Institute sees that a decision in this matter can not be drawn without actual study of the case [8]. A team of researchers from the institute is involved in this work. Our one-barrel digester has been used in the preliminary test of the sludge as a precursor for methane. The first trials proved successful and now intensive laboratory research work is being carried out regarding analysis of sludge and the detailed study of the mechanism.

Other Sources

Biogas can be obtained through bacterial anaerobic fermentation of any organic matter, provided that certain requirements have been fulfilled. A wide range of suitable substrates is found in Sudan. The biomass department has prepared a future program in which a thorough survey will be done regarding identification of suitable substrates, their biogas potentialities, and their different probable combinations.

CONCLUSION

It is clear that Sudan has great potentialities for biogas production. Biogas as a cheap and simple source of energy can help solve the energy problem, especially in rural areas of the country, where the feed-stock materials are abundant and the commercial sources of energy -- oil, electricity, butagas etc. -- are absent, rare, or very difficult to obtain.

Biogas technology has been recently introduced into the country and most, if not all, of the work done in this field is for research and demonstration only. The dissemination of such technology among the rural and suburban communities in the country needs an intensive awareness program to convince the inhabitants of the feasibility of the technology. Generally, and with some effort, biogas technology will have a promising future in Sudan.

REFERENCES

1. Renewable Energy Assessment for the Sudan. 1982. National Energy Administration. Khartoum.

2. Sugar and Distillery Corporation. April 1983. Report on Sugar Factories Bioproduct Productivity.

3. Ottmar, P., K. Werner, and K. Heinz. 1983. Utilization and Control of Water Hyacinth in Sudan. The German Agency for Technical Cooperation (GTZ).

4. Hassan, B. and F. Yaman. April 1984. Report on Rehabilitation of Biogas Units at Taweela Village.

5. Hassan, B., 1983. M.Sc. Thesis. Faculty of Agriculture, University of Khartoum.

6. Biomass Report on Biogas Production from Cow Dung. April 1984.

7. Yaman, F. 1984. Report on Anaerobic Fermentation of Molasses and Waste Water from Molasses Fermentation for Production of Biogas.

8. Shomo, S,. March 1984. Proposal for Production and Utilization of Biogas at Khartoum Sewage Treatment Plant.

BIOMASS AND BIOGAS IN TUNISIA
A SUMMARY PRESENTATION

M'hamed Brini
National Agronomic Institute of Tunisia, Tunisia

BIOGAS PROGRAM

Organizations Involved

- National Center of Scientific and Technical Research (Bordj Cedria)
- National Agronomic Institute of Tunisia (Tunis)
- National School of Engineering (Tunis)
- Office of Central Tunisia Development (Kasserine)
- Office Sylvo-Pastoral of Sedjenane (Sedjenane)

Achievements and Projects

- Study of methanation of poultry manure in laboratory-scale digesters.
- Full-scale farm digester of $10m^3$/size, with floating roof. The biogas produced from cattle waste is used for lighting and cooking (Sedjanane) (Project with West Germany)
- Full-scale farm digester (Kasserine). Project with U.S. Agency for International Development.
- Development and Diffusion of Biogas Technology (Project with United Kingdom).
- Research and Development of Biogas (Project PNR with EEC)

UGANDA BIOGAS PROJECT

Jo Ilukor, Department of Physics, Makerere University,
Kampala, Uganda

ABSTRACT

Until 1979, the biogas-related activities in Uganda were mostly confined to basic laboratory-scale studies. After that, however, clearly applied objectives, including extension to rural areas, have been set and plans have been initiated for implementation. Such early endeavors are outlined in this presentation.

INTRODUCTION

Interest in the renewable sources of energy in Uganda dates back about thirty years. This concern has included geothermal energy, biogas, fuelwood, and wind energy. The achievements then lay in the basic studies of the potentials presented by each energy source, whereas little was seen in terms of extension services. Indeed, a large-installation biogas plant can be seen in the heart of the capital city, Kampala, built about 1968. Since this project was not commissioned by 1971, when Uganda's development stagnated for a decade, it never captured much attention and until now few people knew about it.

The current Research and Development Biogas Program took root soon after the oil crisis precipitated in 1973. Even then, it was initially confined to basic studies, until after 1979, when clear objectives, including extension to rural areas were drawn. These objectives in the Program may be summarized as:

1- Development and construction of biogas digesters which are best suited to Uganda, considering the potentials in raw materials, climatic conditions, and cultural habits of the people;

2- Development of biogas appliances with optimal combustion efficiency;

3- Development of testing and evaluation facilities, which can be used for monitoring the performance of biogas digesters and appliances in the country (these include precautions for hazards and safety in utilizing the gas);

4- Extensions of applications of biogas as a fuel to rural areas, in particular, bearing in mind the social and cultural life-style of various areas in the country.

RESEARCH AND DEVELOPMENT PROJECTS

Digesters

Biogas research and development has involved, from the very early stages, design, construction and operation of different types of digesters. The continous-type or floating dome and semi-batch types have been emphasized. This is because the removal of the effluent is relatively easier for these types of digesters, compared to that from the batch-type. In addition, the two types singled out are most likely to be acceptable in various communities in the country.

The performance of each type of digester, operating under different conditions of temperature, alkalinity of slurry, and digester pressure, has been evaluated for three substrates in particular. The first of these substrates is cow dung; the second is straw from maize stalk and elephant grass; and the third a mixture of straw and cow dung.

Biogas Composition

Literature reports quite widely that biogas is composed mainly of methane and carbon dioxide, plus traces of carbon monoxide, nitrogen, water, hydrogen and hydrogen sulphide. This is not to mention others that may be produced by either malfunctioning or peculiar substrate.

In this respect, carbon dioxide content is measured as x% and used as a basis for the determination of the percentage of methane (100-x)%, assuming that the rest of the gases are negligible.

The Gas Composition Analysis Project, however, has involved using different types of gas analyzers in order to determine percentage composition of biogas constituents. The analysis is carried out at various stages of gas production from the initial loading of the digester, through the stabilized operation of the digester to the storage systems.

Biogas Appliances

Appliances utilizing biogas are being fabricated and so far include stoves for cooking, lamps, space heaters and dryers. Testing and evaluation of these appliances is performed in relation to the determined composition of the biogas. In addition, the internal combustion engine, forming part of this project, is expected to utilize biogas which is free of impurities like hydrogen sulfide.

Fertilizer

The effluent from the digester has been considered a fertilizer. The proportionate increase in the nitrogen content of the effluent, compared to the inlet substrate, elevates the biogas plant to a dual purpose reactor.

Storage and distribution of the effluent is a study of its own, while field trials on a variety of crops have indicated positive results, both in terms of increased yields and improvement of the soil. Experiments have been run on cassava as a root crop, maize as a grain, soya bean as a protein source, and a number of vegetable crops. In essence, recycling of agricultural waste is a logical consequence in the extension work of the project.

OPERATING FACILTIES

The main equipment facilities include Orsat-gas-analyzers, pH meters, gas meters, and plastic bags for auxillary storage of biogas.

DIFFUSION EFFORTS

With respect to research and development, a number of experiments are assigned to undergraduate students so that they become familiar with methods of production, and utilization and evaluation of the gas. Seminars and mass media coverage have been organized from time to time. Simplified literature has been handed out to some organizations which are interested in the development of energy resources and share concern for environmental pollution. A small book entitled "Introduction to Biogas Systems" has been produced, and it is a scientific manual based on the results of the project.

Extension of the Project into rural population is planned through the up-country schools, and the already established research centres. Such places will have biogas plants of their own, operating to supply gas in the laboratories, kitchens etc., and they will also serve as demonstration centres. On-the-spot evaluation of the acceptability of the new technology in the rural area is being pursued by a number of research assistants, living within the village communities and using biogas alternatively with fuelwood and kerosene.

The attention of the government of the Republic has been drawn to the applicability and potential of the new and renewable sources of energy including biogas. The government has already included these energy sources into its overall national energy policy.

THE EXPERIENCE OF THE DEVELOPMENT AND RESEARCH OF BIOGAS TECHNOLOGY IN THE RURAL AREAS OF CHINA

Li Jiazao and Deng Heli
Chengdu Biogas Research Institute, Ministry of Agriculture
Animal Husbandry and Fisheries, People's Republic of China

ABSTRACT

Expanded use of biogas in China requires that the conditions of biogas production and use match the needs and resources of the Chinese countryside. It is important that the digesters are not beyond the means of the farmers, that a wide variety of substrates can be used for fermentation, and that biogas can be produced throughout the year. Chinese research on the efficiency and economy of various designs and operating methods for biogas units is described.

INTRODUCTION

China is well known for coordinating rural biogas development with rural conditions. Since China's rural population makes up 80 percent of its total population, the development of rural biogas is of great importance. Chinese peasants are both users and producers of biogas, but because China is a developing country, neither individuals nor the state can invest much in the development of biogas. These factors in China's biogas development determine the characteristics of Chinese rural biogas digesters. They are small in size, low in cost, of semi-continuous mode of operation, and ambient temperature fermentation units using stalk materials as the principal substrate.

Recent research by Chinese scientists and biogas workers has further improved the Chinese hydraulic digester. Studies on the fermentation conditions and gas production potential of fermentation materials have provided the basis for formulating better fermentation techniques and these make it possible for the peasant households to get more biogas and the advantage of steady gas supply. The research and application of sealing paint have basically solved the problem of gas leakage caused by poor-quality building materials.

Progress in China's rural biogas development and research includes:

I. Research and extension of biogas fermentation techniques

1. Research on the application of the fermentation techniques

(1) Fermentation techniques of the Chinese rural domestic hydraulic digester at different environmental temperatures.

Results of the fermentation at different environmental temperatures (Table 1) indicates that whether in the north or south areas of China (provided the operation is done according to the fermentation technique), the Chinese rural hydraulic digesters can be operated year round and supply biogas steadily.

(2) Fermentation techniques of layer and full charge of raw materials

The main characteristic of this technique is that it can charge 2/3 more stalks than the conventional techniques, increase the utilization of stalks in the digesters and make up for the shortage of animal manure. At present, Sichuan province has about 30,000 hydraulic digesters adopting this fermentation technique. The gas production rate is 0.10 - 0.15 m^3/m^3day or 0.20 - 0.25 m^3/kg TS. The changes in gas yields before and after the use of this technique in 34 digesters is indicated in Table 2. About 25% more gas is produced than with conventional fermentation.

(3) Research on the dry fermentation

Since some parts of China are short of water for agriculture, households in these areas use solid manure. This high concentration of solids or dry fermentation technique has been studied by many biogas researchers (see Table 3). Results indicate that the dry fermentation of 20 - 35% solid matter is suitable for hydraulic digesters and is accepted by the peasant households. At the same time, it can greatly increase the gas production rate per volume of digester.

2. Research on the techniques for increasing the gas production rate

The data listed in Table 4 are the research results of new techniques developed by Chinese biogas workers. These new techniques give much higher gas production rates than the present techniques, thus providing the basis for further utilization of the Chinese rural biogas fermentation techniques. Popularization of these techniques remains for future research.

3. Research on fermentation substrates

Research on the abundant fermentation raw materials in China has been discussed in a review entitled ""Study on fermentation of the Chinese hydraulic biogas digesters" by Wu Jinpeng (published on No.3, Journal of China Biogas, 1983). Since China is a cotton producing country, research has been done on cotton stalks as fermentation raw materials. Han Tianxi (1983) has conducted research on this topic. Results are shown in Table 5. Although the gas production potential of cotton stalks is lower than that of pig dung and rice straw, cotton stalks have the advantage of high initial gas production. Therefore, cotton stalks may also be used for biogas production.

II. Improvement of Chinese digesters

In order to bring the strong points of the Chinese biogas units into full play and overcome their weak points, Chinese biogas workers have done research on the following aspects:

Table 1 The Application of the Fermentation Technique for the Chinese Rural Domestic Hydraulic Digester at Different Temperatures [1]

Site	Season										Average
	Summer - Autumn					Winter - Spring					
	Ambient Temp. °C	Soil Temp. °C	Digester Temp. °C	Rate of gas production (m^3/kg TS)	Rate of gas production (m^3/m^3day)	Ambient Temp. °C	Soil Temp. °C	Digester Temp. °C	Rate of gas production (m^3/kg TS)	Rate of gas production (m^3/m^3day)	(m^3/m^3day)*
Wangkui** (Heilong-jiang)	17.5	10.4	19.0	0.212	0.172	-16.9	8.0	12.3	0.111	0.075	0.125
Dandong** (Liaoning)	22.0	15.1	16.0	0.199	0.158	-5.3	9.2	7.4	0.132	0.106	0.130
Wujin (Jiangsu)	25.2	20.3	24.0	0.234	0.199	3.0	9.5	11.3	0.132	0.112	0.149
Dayi (Sichuan)	23.4	20.0	21.3	0.274	0.233	9.6	18.4	19.1	0.289	0.232	0.232
Yichang (Hubei)	26.7	23.7	24.0	0.281	0.239	5.0	11.1	11.8	0.215	0.194	0.216
Wuzhou (Guangxi)	-	-	-	-	-	6-7	16.0	14.0	0.234	0.187	-
Chongqing (Sichuan)	26.3	22.8	24.7	0.270	0.229	10.5	13.0	14.1	0.218	0.175	0.191

* The temperature of 1.2 m depth soil.

** In Wangkui a greenhouse provides warmth and in Dandong a rice straw layer 1.5 m thick provides insulation.

Table 2 Comparison of the Experimental Results of the Layer Full Charging Technique and the Conventional Technique 2*

Year / Month, Gas yield (m^3)	1	2	3	4	5	6	7	8	9	10	11	12	Total
1st year conventional technique	366	360	216	545	674	657	858	770	703	1038	681	438	7306
2nd year conventional technique	441	443	593	896	411	782	696	773	752	572	445	435	7239
3rd year layer full charging technique	219	508	387	804	809	386	928	1049	1081	1026	849	572	9067

* The technique has been described in the Teaching Materials for the International Biogas Training Course jointly held by FAO - UNDP - China.

Table 3 Research on Dry Fermentation with Various Materials

Fermentation concentration (%)	Fermentation raw material	Fermentation time (day)	Digester volume (m^3)	Building material	Digester temperature (°C)	Rate of gas production (m^3/m^3day)	Digester cost (Yuan/m^3)
20.0[3]	Mixture of Cattle, Horse, Pig and Chicken dungs	120	2.0	Iron and Steel	30-40	1.630	125
20.0[3]	Horse dung + Corn stalks	120	2.0	Iron and Steel	30-40	0.960	-
30.0[3]	Horse dung + Corn stalks	120	2.0	Iron and Steel	30-40	0.870	-
27.5[4]	Wheat straw + Rice straw + Pig dung	130	2.4	Semi-red mud plastics	26.4	0.236	-
25.0[5]	Horse dung + Corn stalks	120	6-7	Concrete	-	0.236*	116
10.0[5]	Horse dung + Corn stalks	120	6-7	Concrete	-	0.141*	-
35.0[6]	Pig dung + Corn stalks	60	1.0	Concrete	ambient	0.560	-
10.0[6]	Pig dung + Corn stalks	60	1.0	Concrete	ambient	0.290	-

* T test of the gas production rate of 25% and 10% fermentation concentrations show that the difference was very significant (P 0.01).

Table 4 Research Results Using New Fermentation Techniques

Fermentation technique	Fermentation raw material	Test time (day)	Rate of gas production (ml/g TS)	Fermentation temperature (°C)	Rate of gas production (m^3/m^3 day)
Anaerobic filter[7]		15	-	-	0.957
Semi-two step fermentation[8]	Corn stalks + Pig dung	60	-	35.0	1.580
Two step fermentation[9]	Pig dung	87	154	22-25	1.590
Two step fermentation[10]	Waste liquid after processing of sisal hemp leaves*	300	-	37.0	1.026

8 The data are quoted from the material (1983) by Zhou Mengjing. The characteristic of this technique is that two plastic solid acidized-beds are added on the hydraulic digester and acid is flushed by water circulation into the hydraulic digester.
* See "Tropical Plant Journal." 3:77-86 (1983)

Table 5 Research on the Gas production Potential of the Cotton Stalks [11]

Raw material	Gas production rate		Gas production rates during successive fermentation day (ml/g TS day)				
	ml/g TS	ml/g VS	0-20	21-40	41-60	61-80	81-125
Pig dung	359.3	440.4	4.07	4.31	4.65	2.21	1.21
Cotton stalks	301.2	312.1	8.02	5.36	1.68	-	-
Rice straw	381.7	444.3	5.98	7.30	5.23	0.57	-

1. New digester building materials

(1) The semi-plastic digester (Figure 1)

In the semi-plastic digester the fixed dome of the hydraulic digester is replaced by a plastic film dome, which can not only absorb solar energy to increase the digester temperature, but is also convenient for the discharge of fermentation materials. The experimental data listed in Table 6 indicates that this kind of digester has a high gas production rate (0.23 - 0.38 m^3/m^3 day). This digester is greatly influenced by ambient temperature. The quality of plastic film needs to be improved also.

(2) The red-mud plastic gas holder digester (Figure 2)

Fang Gouyuan et al. (1983) of Chengdu Biogas Research Institute conducted research on using the red-mud plastic film gas holder. The

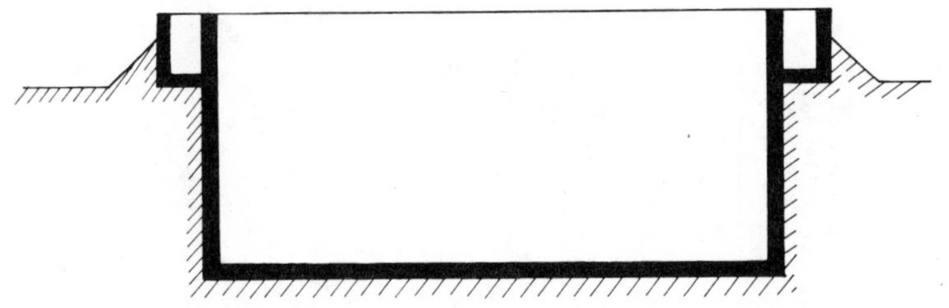

Figure 1 Sectional drawing of semi-plastic digester

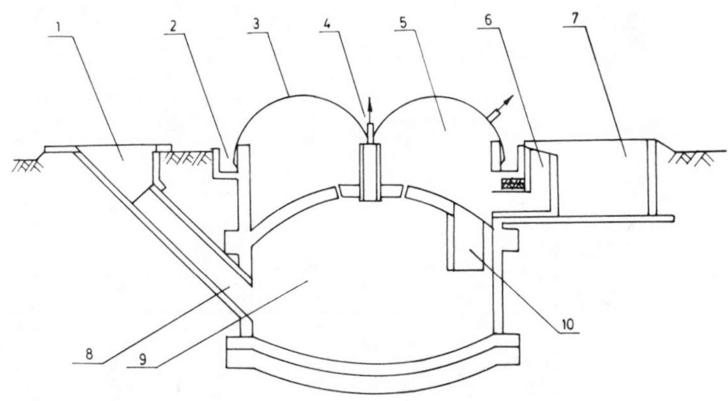

Figure 2 Schematic diagram of test digester

1. Inlet
2. Water-sealing groove
3. Red mud plastic gas holder
4. Gas guiding pipe, removable cover
5. Second fermentation chamber
6. Overflow
7. Manure storage chamber
8. Inlet tube
9. First Fermentation chamber
10. Water return tube

red-mud plastic chamber has the multi-functions of water pressure, gas storage, fermentation and heat absorption. It can not only absorb the heat of solar energy, but also can keep the fermentation material moist and settled to avoid scum formation, thus increasing the gas production rate by an average of 12.2% over the common hydraulic digester (Table 7).

(3) The iron-made domestic digester

This type of digester is designed for using biogas for half a year in the cold areas of China (annual average temperature is 2 - 5°C; monthly average temperature below 0°C lasts 5 months per year; the

Table 6 Research on Semi-Plastic Digesters

Season	Fermentation time (day)	Digester volume (m^3)	Fermentation material	Fermentation concentration (TS %)	Total gas yield (m^3)	Gas rate of digester volume (m^3/m^3)	Gas rate of volatile solid (m^3/kg VS)
Sept.-Oct.[12]	60	3	Sheep dung	20.0	42.1	0.24	0.143
Sept.-Oct.[12]	60	3	Sheep dung	20.0	40.5	0.23	0.132
June-Aug.[13]	70	3	Pig dung + Horse dung +	20.0	71.8	0.34	0.218
June-Aug.[13]	90	3	Pig dung + Horse dung + Rice straw	20.0	102.7	0.38	0.312
June-Aug.[13]	90	3	Pig dung + Horse dung + Rice straw	20.0	100.4	0.37	0.305

Table 7 Comparison of Gas Production Rates Between the Red-Mud Plastic Film Gasholder Hydraulic Digester and the Ordinary Hydraulic Digeser [14]

Experimental treatment		Digester volume (m^3)	Fermentation time (Month)	Fermentation material	Digester temperature (°C)	Rate of gas production (m^3/m^3day)
1	Test digester	6	7-9	Wheat chaff + Human waste + Cow dung	25.3	0.146
	Conventional digester	6	7-9		23.3	0.124
2	Test digester	6	7-9	Wheat chaff + Human waste + Cow dung	24.8	0.140
	Conventional digester	6	7-9		23.7	0.136
3	Test digester	6	7-9	Wheat chaff + Human waste + Cow dung	26.2	0.132
	Conventional digester	6	7-9		25.7	0.117
4	Test digester	6	7-9	Wheat chaff + Human waste + Cow dung	26.1	0.150
	Conventional digester	6	7-9		25.1	0.130

depth of frozen soil is 1.5 m - 1.7 m). This digester has a small volume, a high gas production rate (see Table 8) and is gas leakage-proof. A 2 m^3 digester of this type can supply enough biogas for a family of four for cooking and lighting. Although the cost of this type of digester is as high as 100 yuan, it is worth further research and popularization in accord with the development of the steel industry in China.

2. Use of solar energy with biogas digesters

(1) The combination of hydraulic digester with solar energy heat-collector (Figure 3)

Cai Lianhai et al. (1983) installed a solar energy heat-collector on the hydraulic chamber of a digester as a means of heating the fluids

Table 8 Gas Production Effect of the Iron-Made Digester[15]

Digester type	Digester volume (m^3)	Fermentation concentration (%)	Fermentation time (month)	Fermentation material	Ambient temp. (°C)	Digester temp. (°C)	Rate of gas production (m^3/m^3 day)
Floating-holder digester	0.5	14.0	8-9	Horse dung + Corn stalks	19.7	29.9	0.63
Jar-type digester	2.5	20.0	8-9	Horse dung + Corn stalks	-	-	0.68

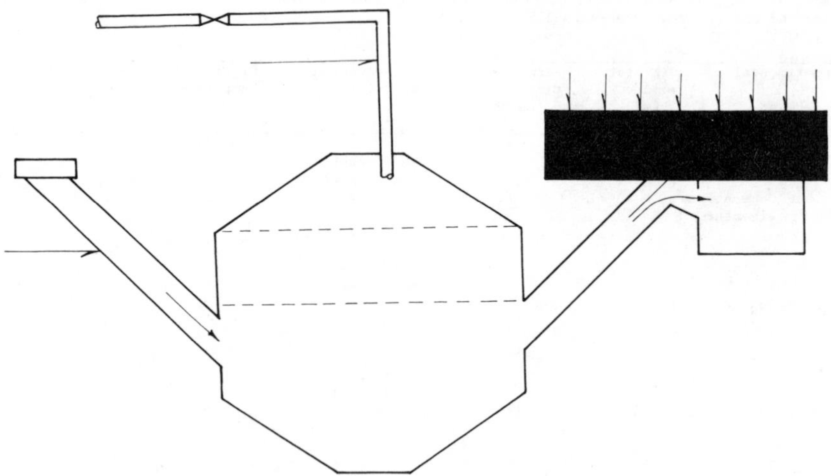

Figure 3 Schematic diagram of the combination of a biogas digester with a solar energy heat collector

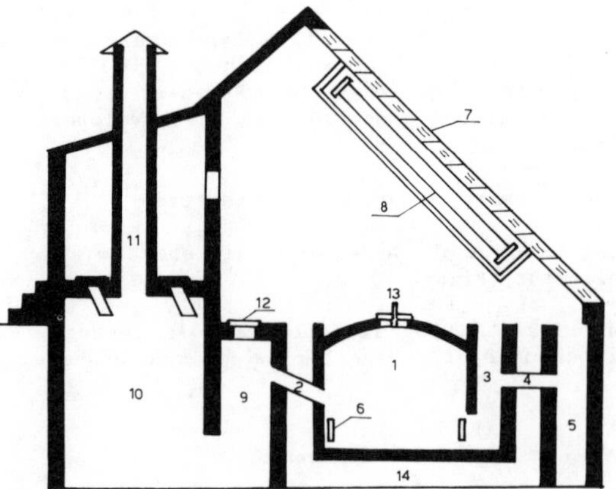

Figure 4 Sectional drawing of a solar energy feedback heated digester

1. Fermentation chamber
2. Inlet tube
3. Outlet chamber
4. Overflow
5. Manure-used pit
6. Radiating fin
7. Glass greenhouse
8. Solar energy hotwater device
9. Pretreatment chamber
10. Manure storage chamber
11. Odor exhaust hole
12. Removable cover of pretreatment chamber
13. Removable cover of fermentation chamber
14. Insulating layer

in the main body of the digester. The maximum water temperature in the heat-collector can reach 60°C, with a mean water temperature of 40°C - 50°C. As a result, the digester temperature can increase by 3 - 5°C and the gas production rate can increase by 30% (see Table 9). This system of digester heating does not require a great investment, and can be adopted in areas with abundant sunshine.

(2) The solar energy feedback heated digester (Figure 4)

This type of digester was designed and built by the Jilin Provincial Sanitation and Epidemic Prevention Station. It consists of a hydraulic tank (with an insulating layer around it), a solar energy greenhouse and a solar energy hot-water device. Two year's practical operation has indicated that the digester temperature can reach 22°C in the winter and 33°C in the summer. In summer, the gas production rate of a 11.7 m^3 digester is 0.318 m^3/m^3 day. It is 0.082 m^3/m^3 day in winter. This type of digester is suitable for low temperature areas.

(3) The solar energy hot-water digester installed with heat-exchanger inside (Figure 5)[18]

This type of digester can carry hot water from the solar energy hot-water device into a coiled pipe installed in the fermentation tank of the digester. A 1 kW water pump is used to circulate the hot water through the pipe and increase the temperature of the fermentation fluid. The 27 m^3 collection area can provide 7,500 Kcal/day (temperature increase stage) and 4,000 Kcal/day (constant temperature stage) to a digester of 15 m^3 volume. The average digester temperature is 32.8°C; gas production rate is 0.54 m^3/kg TS (pig dung) and 0.40 m^3/m^3 day. Therefore, this is an effective design to lengthen the days of gas production in low temperature areas.

Table 9 The Test of the Hydraulic Digester with the Solar Energy Heat-Collector[16]

Experimental treatment	Season	Ambient temp. (°C)	Tank temp. (°C)	Water temp. of heat-collector (°C)	Rate of gas production (m^3/m^3day)	Increased rate of gas production (%)
Test digester	Summer	21.0-30.0	25.3	40.0	0.21	31.3
Conventional digester	Summer	21.0-30.0	20.5	-	0.16	-
Test digester	Summer	21.0-30.0	25.3	40.0	0.28	100.0
Conventional digester	Summer	21.0-30.0	20.5	-	0.14	-
Test digester	Winter	11.0	-	-	0.20	25.0
Conventional digester	Winter	11.0	-	-	0.16	-

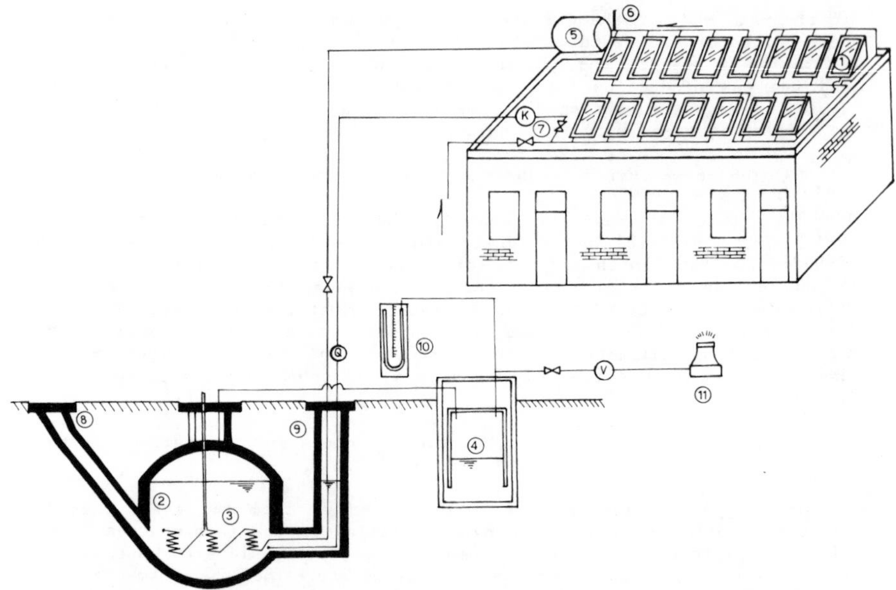

Figure 5 Schematic drawing of the combination of a solar energy heat-collector with a biogas digester

1.	Plate hot-water device	8.	Inlet
2.	Biogas digester	9.	Outlet
3.	Coiled tube heat exchanger	10.	Biogas barometer
4.	Gas storage tank	11.	Bas burner
5.	Water tank	Q	Flowmeter
6.	Thermometer	K	Water pump
7.	Valve	V	Gas meter

3. Avoiding the adverse effects of low soil temperature on fermentation

(1) Above ground digesters

In the Heilongjiang province of China, the soil temperature at 1.6 m is above 8°C for less than 4 months per year and the ambient temperature is above 8°C for 6 months per year. In order to investigate the effects of soil temperature on the fermentation in summer months, the Heilongjiang Provincial Academy of Agricultural Science conducted a comparative fermentation test of digesters above and under the ground. The experimental hydraulic digesters are 4 m^3 in volume. Horse dung is used as the raw material at a concentration of 12.4%. The test results are given in Table 10.

Jilin province has popularized the small-size plastic floating-holder digester built above the ground or half above the ground, small spherical digesters, single-opening hydraulic digesters and modified and strengthened plastic digesters. The gas production rate for these units is 0.12 - $0.47m^3/m^3$day [23] This level of gas production can satisfy the summer gas consumption in the low temperature areas.

(2) Insulating devices for the hydraulic digester

Insulation may be added above or around the hydraulic digester to avoid heat loss caused by low soil temperatures. The Jilin Provincial Academy of Agricultural Science accomplished this by setting up a roof-type shed one meter high above the digester and digging a thermal insulation ditch around the digester. The part of shed exposed to the sun is made of plastic for absorbing the solar energy, and the opposite part of the shed is made of stalks and mud with a thickness of 3 - 4 cm for temperature maintenance. The thermal insulation ditch is filled with ensilage from corn stalks to increase the temperature. The volumes of the test and control digesters are both 6 m^3, the fermentation concentration in both is 30%, and horse dung and corn stalks are used as raw material. The results in Table 11 show that the effect of insulation is very significant. The digester with insulation can produce biogas even in the winter, and conversely biogas can not be produced if no insulation layer is added.

III. Analysis of direct economic benefit of Chinese rural biogas development

The indirect benefits of biogas fermentation are great, especially in increasing of biomass-energy use, in recycling waste material from rural agriculture and industry, and in establishing and maintaining a good agricultural ecological system. This section provides a brief analysis of the direct economic benefits of biogas production.

1. Macroscopic economic benefit evaluation[24]

By the end of 1980, Sichuan province had constructed 3.15 million biogas digesters with material cost per digester of 36 Yuan and 0.8 million digesters with a material cost per digester of 100 Yuan, for a total cost of 190 million Yuan. From 1973 to 1980, the government invested 50 million Yuan in biogas development in this province. So the investment totaled 240 million Yuan. From 1973 to 1980, 2.1 billion m^3 of gas has been produced with a calorific value equivalent to 2.1 million tons of raw coal. The heat efficiency of biogas is 60% and that of coal is 15%. One cubic meter of biogas can be used instead of 4 kg of coal and 2.1 billion m^3 biogas equates to 0.84 million tons of coal, which cost 340 million Yuan (40 yuan/ton coal). Thus it can be seen that the economic benefit of the China's investment in the biogas development is quite significant.

2. Analysis by county units

Table 12 shows the evaluation of the economic benefits from biogas production in Haian county, Jiangsu province and Chongqing county, Sichuan province. The results indicate that the benefit is significant. The investment recovery periods of those two counties are 4 years and 2.6 years respectively, and the profit rates of return are 24.76% and 38.74% respectively.

Table 10 Fermentation Comparison Test of Aboveground and Underground Digesters[17]

Month	Ambient temp. (°C)	Soil temperature (°C)			Digester temperature (°C)		Gas yield (m^3)		Rate of gas production (m^3/m^3 day)	
		0.8m	1.6m	3.2m	Under-ground digester	Above-ground digester	Under-ground digester	Above-ground digester	Under-ground digester	Above-ground digester
6	20.0	10.6	4.4	2.5	9.4	20.6	1.70	22.65	0.01	0.19
7	22.8	15.7	9.5	4.5	12.5	25.9	9.29	35.54	0.07	0.29
8	21.1	17.1	12.4	7.1	15.8	25.8	12.94	29.46	0.10	0.24
9	14.4	15.1	12.8	8.9	15.6	16.9	8.36	4.02	0.07	0.03
Total	-	-	-	-	-	-	32.29	91.67	-	-
Average	-	-	-	-	-	-	-	-	0.06	0.19

Table 11 Fermentation Comparison Test of a Hydraulic Digester with Insulation and a Digester without Insulation[20]

Month	Ambient temperature (°C)		Soil temperature (°C)		Insulation layer temperature (°C)		Digester temperature (°C)		Rate of gas production (m^3/m^3day)	
	Max.	Min.	Inside the shed	Outside the shed	Without insulation layer	With insulation layer	Without insulation layer	With insulation layer	Without insulation layer	With insulation layer
11	- 2.2	-11.1	9.8	9.3	-	18.7	9.5	16.2	-	0.098
12	- 3.8	-14.2	4.5	6.4	-	12.1	7.1	11.6	-	0.068
1	- 8.0	-19.2	4.0	2.6	-	20.2	5.1	12.2	-	0.048
2	- 1.8	-13.4	4.0	1.1	-	14.8	3.4	13.7	-	0.070
3	4.0	- 7.6	1.5	0.3	-	9.1	3.0	11.2	-	0.077
4	27.0	- 3.7	6.8	0.6	-	11.6	3.4	15.5	-	0.047

Table 12 Evaluation of the direct economic benefits of biogas production in Haian and Chongqing counties[21,22]

Evaluation item			Item No.	Place	
				Haian county	Chongqing county
Total digesters			1	2583	58803[1)]
Expenditures	Investment	Unit price (Yuan/each)	2	112.73	56.08
		Total cost (ten thousand yuan)	3	29.12	412.21
	Annual investment (Yuan/each Year)	Dismantlement fund[2)]	4	7.52	3.74
		Repair fund	5	2.00	2.96
		Fund for operation and management	6	9.60	9.60
		Total (Yuan)	7	19.12	16.30
	Annual investment of the whole county (Ten thousand/Year)		8	4.94	95.85
Revenue	Gas yield (m^3/Year)	Average gas yield of per digester	9	213.5	164.0
		Total of the whole county	10	551470.0	9643692.0
	Direct economic benefit per digester (Yuan/each Year)	For cooking	11	40.14	30.83[3)]
		For lighting	12	1.36	4.80
		As fertilizer	13	5.53[4)]	7.62
		Total (Yuan)	14	47.03	43.25
Benefit analysis	Biogas cost (Yuan/m^3)		15(= 7 : 9)	0.090	0.099
	Annual direct profit of the whole county (Ten thousand Yuan/Year)		16(= 1 x 14)	12.15	254.32
	Annual net profit of the whole county (Ten thousand yuan)		17(=16 - 8)	7.21	158.47
	Investment recovery period (Year)		18(=3 : 17)	4.0	2.6
	Fund profit rate (%)		19(=17 : 3)	24.76	38.44

NOTE

1) The real number of digester is 73504, but only 80% of them can produce biogas, thus is 58803.

2) The dismantlement expenses are cost - 15 years.

3) 1 kg stalks = 3500 Kcal x 0.1 = 350 Kcal; 1 m^3 biogas = 5,500 Kcal x 0.6 - 330 Kcal; 1 m^3 9.4 kg stalks; 1 kg stalks cost 0.02 yuan; 164 m^3biogas cost 30.83 yuan.

4) 100 kg of biogas fertilizer equal to 0.28 kg NH_4HCO_3, 10968 kg biogas fertilizer 30.71 kg NH_4HCO_3. 1 kg NH_4HCO_3 cost 0.18 Yuan thus, the utilization of 10968 kg biogas fertilizer can save 5.53 yuan.

Table 13 Analysis of Economic Benefits of Biogas Development in Several Production Teams[22,25]

Place	Number of Digesters (N)	Investment		Annual revenue			Benefit analysis			
		Investment $1 m^3$ digester (Yuan/m^3)	Feeding of $1 m^3$ digester (kg TS/m^3)	Annual gas yield (m^3/Year)	Annual output value (Yuan/Year)	Annual operation expenses (Yuan/m^3 Year)	Biogas production cost (Yuan/m^3)	Annual net profit (Yuan/m^3 Year)	Investment recovery period (Year)	Investment effect coefficient*
Hengqi 3rd team (Quxian, Sichuan)	44	10.98	50.7	37.0	2.96	1.46	0.039	1.50	7.3	0.14
Beigang 2ne team Wuhan, Hubei)	60	39.35	11.4	65.0	5.20	3.63	0.056	1.57	25.1	0.04
Qunfeng 7th team Huangbo, Hubei)	43	22.22	50.8	75.0	6.00	2.94	0.039	3.06	7.3	0.14
Lousheng tai 7th team Mianyang, Hubei	54	14.97	49.8	52.2	4.20	2.30	0.044	1.90	7.9	0.13

* Investment effect coefficient = 1/investment recovery period

3. Analysis by production team units

Table 13 shows the economic benefits of biogas production in 4 production teams (respectively subordinate to 4 counties in Sichuan and Hubei provinces). Although the benefits are dependent on the level of construction and management of digesters in different areas, undoubtedly economic benefits have been brought to the peasant households.

Results in Table 14 are from 34 digesters operated by the Hongqi No. 3 production team in Quxian county, Sichuan province. Through linear regression analysis of the related main indexes of economic benefit interrelation have been developed. The results are as follows:

Table 14 Linear Regression Data for Various Economic Indexes of the Biogas Production[22]

Regression item		Related Coefficient
Gas production rate	vs. Investment per m^3 digester	0.4690*
Gas production rate	vs. Annual production cost per m^3 digester	0.4869*
Gas production rate	vs. Costs of biogas production per m^3	-0.3352
Annual net profit	vs. Investment per m^3 digester	-0.2931
Annual net profit	vs. Annual production cost per m^3 digester	-0.3037
Annual net profit	vs. Gas production rate	0.6195*
Annual net profit	vs. Costs of biogas production per m^3	-0.9179*
Investment effect coefficient	vs. Investment per m^3 digester	-0.4876*
Investment effect coefficient	vs. Gas production rate	0.3724**
Investment effect coefficient	vs. Annual production cost per m^3 digester	-0.5089*

* N=34; $r_{(32),0.01}=0.4365$; $P<0.01$;
** N=34; $r_{(32),0.05}=0.3395$; $0.05>P>0.01$.

Table 15 Comparison of the Effective Calorific Value from the Direct Burning and Biogas by Fermentation of Bio-matter[26]

Bio-matter	Direct burning			Biogas production			Increase of effective joule by biogas fermentation
	Heat value (KJ/kg) A	Heat efficiency (%) B	Effective joule value (KJ) C=AxB	Heat value (KJ/kg) D	Total efficiency of energy conversion (%) E	Effective joule value (KJ) F=DxE	(%) G=(F-C)/C
Corn stalk	17354	10	1735	18746	19.67	3689	112.7
Wheat stalk	16483	10	1648	17817	18.35	3268	98.4
Rice straw	155595	10	1559	16774	19.21	3222	106.7
Cow dung	14698	10	1470	15803	15.48	2447	66.2
Pig dung	15382	10	1538	16549	19.68	3256	111.7

(1) The gas production rate, annual net profit and investment effect coefficient are significantly correlated with most economic indexes, namely operation effect of digesters have achieved the desired goal of the design.

(2) The gas production rate is not correlated with the biogas production cost. That is, the increase of gas production rate does not reduce the biogas production cost, hence there are some problems in the management of biogas production.

(3) The annual net profit is not correlated with the investment of per cubic meter digester and the annual production cost of per cubic meter digester. Thus in order to increase the annual net profit, it is necessary to reduce the investment in the digester construction and annual production management expenses as much as possible.

4. Comparison of economic benefits from methods of stalk utilization

Data listed in Table 15 compare the heat values obtained from the fermentation of stalks from three kinds of crops and manures of two kinds of animals with that obtained from direct burning. Fermentation is much more economical than combustion.

To sum up, Chinese biogas workers have done a great deal of work in applied research on biogas fermentation appropriate for conditions in the countryside. At the same time, Chinese biogas workers have made some theoretical preparations for further development of rural biogas production to make the biogas less expensive, more effective and more acceptable and practical for rural households.

REFERENCES

1. Haung Kui. 1984. Biogas Fermentation Technique of Domestic Hydraulic Biogas Digesters in Rural Areas of China. (Unpublished ed Technical Report).

2. Xu Yizhong. 1984. Full-charging Biogas Fermentation Process. (Appraised Technical Report).

3. Biogas Research Group of Soil and Fertilizer Institute, Agricultural Academy of Jilin Province. 1983. Study of High-Efficiency (dry) Biogas Fermentation Process and Device in Rural Areas. Tech. Rep. of Annual Conference of Biogas Association of China.

4. Biogas Institute of Jiangsu Province. 1983. Study of the Process of Biogas Fermentation with Compost as Raw Materials. Tech. Rep. of National Biogas Research Conference.

5. Sandong energy Institute. 1983. Solid Biogas Fermentation. Tech. Rep. of National Biogas Research Conference.

6. Zhou Mongjin, Wang Yuzhu, and Yang Xousan. 1983. Study of Dry Methane Fermentation. Tech. Rep. of National Biogas Research Conference.

7. Shao Xihao, Tu Zhuxin, Hong Yiwu, and Zen Jiang. 1983. Study of fermentation technique to convert plant into biogas using anaerobic filter. Tech. Rep.of Biogas Newsletter No. 18.

8. Zhou Menjin, and Ynag Xousan. 1983. Solid Bed Acidification Half Two-Phase Fermentation. Tech. Rep. of Annual Conference of Biogas Association of China.

9. Xu Jequan, Ciao Doqun, Yang Chunfu, and Shao Qixue. 1983. Two-phase Biogas Fermentation of Pig Dung-preliminary Lab-scale Study at Ambient Temperature. Tech. Rep. of Annual Conference of Biogas Association of China.

10. He Yingbo, fang Riming, and Si Tuhau. 1983. Study of Two-Phase Biogas Fermentation from the Juice of Sisal Hemp Leaves. Res. Rep. of Annual Conference of Biogas Association of China.

11. Han Tianxi. 1983. Study on Feasibility of Cotton Stalks Used as Raw Material of Biogas Fermentation. Res. Rep. of National Biogas Research Conference.

12. Fu Jianhui. 1983. Study of High Concentration Fermentation in Half-Plastic Biogas Digester. Res. Rep. of Annual Conference of Biogas Association of China.

13. Shao Zhaoyi. 1983. Study of Red-Mud Plastic and its Application. Tech. Rep. of Annual Conference of Biogas Associations of China.

14. Fang Gouyuan. 1983. Experimental Report on New Type of Biogas Digester Fit up with Red-Mud Plastic Gasholder. Res. Rep. of Annual Conference of Chinese Biogas Association.

15. Biogas Research Group of Soil and Fertilizer Institute, Academy of Agriculture of Jilin Province. 1983. Preliminary Study of Household Iron-Made Biogas Digester and its Fermentation Process in Rural Areas. Tech. Rep. of National Biogas Research Conference.

16. Cai Nianhai. 1983. Study of Combination of Biogas Digester and Solar Energy Collector for Raising Gas Yield. Tech. Rep. of National Biogas Research Conference.

17. Sanitation and Epidemic Prevention Station of Jilin Province and Huaide County. 1983. Temperature Rising by the Use of Solar Biogas Digester in Northern China. Tech. Rep. of Annual Conference of Biogas Association of China.

18. Science and Technology Committee of Xian. 1983. Application Study of Combination of Biogas and Solar Energy. Tech. Rep. of Annual Conference of Biogas Association of China.

19. Soil and Fertilizer Institute, Academy of Agriculture of Heilongjiang Province. 1983. Experimental Report of the Fermentation Technology and Regular Pattern of Gas Production of Digesters Built on the Ground.

20. Wei Jishan, Miu Zhexei, and Zhao Hale. 1983. Study of Rural Household Biogas Digesters in Cold Regions. Tech. Rep. of National Biogas Research Conference.

21. Rural Biogas Benefit Investigation Group of Biogas Association of China. 1983. Analytical Report on the State of Art, Benefits and Prospect of Biogas in Haian County. Report on Plenary Session of Annual Conference of Biogas Association of China.

22. Gao Yutian, and Tang Chejiang. 1984. Evaluation on biogas economic benefits. Editorial Office of Biogas Journal of China.

23. Zhang Qiyu, and Han Yuzhen. 1983. New Designs of Biogas Digesters Utilizable to Cold Regions. Tech. Rep. of Annual Conference of Biogas Association of China.

24. Yao Yongfu. 1983. Prospect of Biogas Development in China in View of its Economic Benefits. Rep. of National Biogas Research Conference.

25. Biogas Extension Office of Hubei Province. 1983. Investigation and Analysis of Cost and Benefits of Biogas Extension in Rural Areas. Res. Rep. of Annual Conference of Biogas Association of China.

26. Shao Xihao, Xu Nansun, and Chai Lixiong. 1984. Study of energy conversion rate of biogas fermentation. China Biogas 3:40-43.

THE BIOGAS PROGRAM IN INDIA

T.K. Moulik, Indian Institute of Management, Ahmedabad, India
J.B. Singh, Action for Food Production (AFPRO), New Delhi, India
and
S.K. Vyas, Punjab Agricultural University, Ludhiana, India

ABSTRACT

In this presentation, the historical developments of India's biogas program are outlined. Throughout, the program strategy and the organizational framework are emphasized. Finally, key problems and future directions are highlighted.

HISTORICAL BACKGROUND

Historically, India's biogas development and promotion program can be broadly classified into four stages:

1930-1950's	-	Period of indigenous technology development
1960's	-	Research and limited field promotion
1970's	-	National mass promotion and coordinated research
1980's	-	Accelerated promotion and infrastructure/ institutional development.

Although initial attempt to develop an indigenous biogas technology in India was started by a group of scientists in the 1930's, it was only in 1951 that J.J. Patel could evolve a workable biogas floating-dome biogas design called Gramalaxine. By 1965, Mr. Patel further improved this design and standardized it. It was interesting to note that initially India's interest in biogas technology had been to produce inexpensive and richer manure. By the 1950's, the interest shifted to gas production and energy.

By 1955, about 500 floating-dome biogas plants were installed in India, with full government support. However, all these plants eventually failed due to design problems. It was only in 1960 that the government encouraged KVIC (the Khadi Village Industries Corporation) to take up biogas as one of its rural development programs, with some provision of subsidy and incentive. In fact, KVIC, in spite of slow progress, kept the biogas technology alive in India. By 1975, KVIC could install only about 12-15 thousand biogas plants, and most of them were small family-sized plants.

Apart from KVIC's official program, supported by the government, the period of the 1960's has been crucial for India's biogas program. It was during this period that a large number of national research institutes/organizations started research work on biogas technology, mainly in relation to design parameters and microbial processes. An exclusive biogas research station was created at Ajitwal, UP.

RECENT DEVELOPMENTS

The 1970's energy crisis brought the biogas program in India into national focus. Biogas promotion became the responsibility of the Agriculture Ministry of the Government of India. Given the energy crisis and the continuing need for energy, biogas has been promoted as an alternative cooking fuel. The government's commitment in terms of manpower, finances, and other infrastructure has increased. Eventually, an offical national biogas development and promotion program was initiated in 1977, which ultimately became an important element of the Prime Minister's 20-point development program.

Meanwhile, the Department of Science and Technology initiated an all-India-coordinated biogas research program. A large number of scientists, research institutes, and organizations have been brought into the fold to undertake an organized and coordinated research program. On the other hand, there has been a major shift in strategy and organizational thrust in India's biogas program. Concerned about the high cost of the floating-dome design, India has been in search of a cheaper biogas technology. Through its adaptive research, a modified, Chinese-type, fixed-dome design was developed, named Janata, which has been officially promoted with similar financial incentives and other supports. Similarly, with its largely enhanced target and accelerated program, it was found to be necessary to decentralize and diversify the supplementing agencies. KVIC alone was found to be not adequate to implement the program. A large number of NGOs as well as government organizations, were officially brought into the implementation program. Thus, India's multi-model and multi-agency strategy in biogas programs was implemented, and still continues to be the key element in the program. By 1980, the total number of biogas plants installed in India was approximately 80,000.

Apart from scale economy and the potentiality of multiple-use of biogas as other than cooking fuel, India has been facing the problem of limited market segment of potential biogas customers. There are only about 15-20 million households having more than 3-4 cattle, who are the potential customers of biogas technology. A large majority of 85 million rural households, particularly cattleless households, are not reachable by biogas technology. In response to this problem, an organized government-sponsored, research-cum-field experimentation program on large-size village-community and institutional plants was launched in 1980. About 150 large plants were installed in all parts of India, of which 85 were institutional plants. Based on the experiences of these experimentations, a target of 1,000 large-sized community/institutional plants was set up for the 6th plan period (1980-85).

A significant organizational thrust was given to the biogas program in India in 1982. With the expanded program and India's search for alternative energy sources, it became necessary to create an exclusive organization to coordinate the program. Accordingly, a separate department, the Department of Non-Conventional Energy Sources (DNES), was created at the center to coordinate and implement the program.

Similarly, at the State level, a special nodal agency for implementing the program was created. It has clearly been realized that the success of the biogas program is largely dependent on creating capacity at the decentralized levels. A massive manpower development program through training is organized through various government agencies and NGOs in which AFPRO (Action for Food Production) played a most significant role. Simultaneously, a conscious and organized publicity has been mounted using all possible media. A recent documentary film made by India's Film Division on biogas energy, for example, has recently been awarded an international prize. The impact of these organized efforts is conspicuous in the sense that biogas technology has become widely known to the Indian public.

Future Plans

By 1983, the total number of biogas plants in India rose to 0.28 million. Encouraged by the progress and the infrastructure development , India started a crash program of 0.15 million biogas plants to be installed in 1984-85, the last year of the 6th plan. In the 7th plan period, a target of six million plants has been proposed.

What is most interesting and encouraging is the fact that 40 percent of the new customers of biogas plants in India are not dependent on subsidy or loan. Similarly, the interest in using biogas as propulsion energy in pumping, power-generation, and threshing has been increasing steadily to the extent that a number of reported engineering industries have entered the market in manufacturing dual-fuel engines and generating sets.

MAIN CONSTRAINTS, RESEARCH ALLOCATIONS, AND FUTURE DIRECTIVES

The foregoing brief account perhaps gives a very optimistic picture. This does not mean that there are no problems. What is important to note is the fact that the biogas program has reached a stage in India where one feels confident it can solve the problems. There are problems of reduction in gas production in winter, slurry disposal, microbial efficiency, use of diversified feedstock, rich-poor differential in biogas beneficiaries, cost of biogas plant and operating failures of installed plants. Many of these problems need to be tackled through continuous research, which India had been consciously promoting. In fact, as high as 20 percent of the budget allocation for the biogas program has been earmarked for research. While the research has been a continuing commitment, India's biogas program envisages the following future directions:

1. With a massive effort in manpower training, the operating failure rate in installed plants in India has been considerably reduced in recent years (not more than 10 percent). An organized rectification program of the non-operating plants has been initiated at the government expense of Rs 1,000 per plant.

2. A massive national demonstration-cum-popularization program on the manurial value of biogas slurry has been taken up in order to emphasize this immediately-perceptible benefit, which has so far not been adequately promoted.
3. Apart from small family-size plants, a clear program of large size village-community and/or institutional plants is planned with similar emphasis and support. It is in this direction that the large-sized plants are conceived as the nucleus of India's initiated program of an Integrated Rural Energy Centre, which is planned to meet the total energy demand of a given rural area.

SUMMARY OF THE NEPAL BIOGAS PROGRAM

H.G. Gorkhali
Biogas and Agricultural Equipment Development Co. Pvt. Ltd.
Kathmandu, Nepal

ABSTRACT

Biogas technology has good prospects in Nepal. At present, more than 1,600 family-size units and 24 community plants are in operation. This presentation which encompasses historical background, present situation, research and development, and future plans, outlines the key points relevant to Nepal's biogas program.

COUNTRY BACKGROUND

Nepal is a small, landlocked, mountainous country situated between Tibet, the autonomous region of the People's Republic of China, to the North, and India to the East, West, and South. The country has a land area of 141,000 sq. km. and its altitude varies from 100m to 8848m above sea level. Nepal has at present a population of 16 million persons, which is increasing at the rate of 2.66% per annum. It has a heterogeneous population with 17 major language groups and many smaller groups. Culturally, the country is a combination of a large number of caste and ethnic groups. Average per capita income in Nepal is estimated at about US $140 per annum and the literacy rate is approximately 24 %.

Geographically, the country can be divided into three regions: mountains, hill, and terai. Mountains and hills comprise about 75% of the total area, and the remaining 25% falls in the terai plain. Road and transport facilities have been rapidly expanding in recent years, but a large number of hill and mountain areas remain without a modern road link.

Nepal's economy is predominantly rural and its mainstay is agriculture. Agriculture contributes over 66% of the gross domestic product and provides more than 80% of the export earnings. Nepal's industry is mainly agro-based. Approximately 94% of the population live in rural areas. More than 90% of the population is dependent upon agriculture for its livelihood. Approximately 75% of the population are poor, disadvantaged small farmers who own less than one hectare of land, or are landless, and work as agricultural laborers, or are engaged in traditional local crafts.

APPROPRIATENESS OF BIOGAS PROGRAM IN NEPAL

More than 90% of the population of Nepal is engaged in traditional subsistence agriculture, and 66% of the gross domestic product is derived from agriculture. The bulk of the energy needs for economic production is supplied by manual labor and draft animals. The remaining domestic energy demand is for cooking fuel (chiefly firewood). Firewood is rapidly becoming scarce and expensive. A study of energy use in Nepal shows that per capita fuel wood consumption for the hills and terai are respectively 556 kg and 439 kg per year. This demands substantially more fuelwood than the present rate of renewable growth, depleting forest cover, and causing severe soil erosion affecting both fertility and crop production.

Nepal has great potential to develop hydropower in order to meet the energy needs of the population, but hydropower cannot be developed very quickly, and it is expensive to use. Fossil fuels such as coal and petroleum are not present in Nepal, and imported fuels are expensive. The use of solar and wind energy has limitations, and appropriate technology has to be developed to suit socio-economic conditions.

The livestock population of Nepal is estimated to consist of 3.4 million cattle, 3.6 million oxen, and 4.2 million buffalo. The total of the three classes averages 0.75 livestock units per capita. Since there is no current census of livestock, by district, this average has been used to determine livestock available in the terai. The resulting figure is 4.77 million units. The daily average production of manure is 10 kg per animal, however field losses and other uses eliminate 40% of this from use as fuel. The net annual production is thus 2.2 tons per animal for a total of 10.5 million tons. Theoretically this is sufficient manure to supply 525,000 family-sized biogas units.

Both in the hills and terai, the rural families burn fuelwood or dry dung cakes as fuel for cooking purposes. In this way the rural population destroys valuable natural resources which otherwise could be used for productive purposes. The time involved in collecting fuelwood or making dung cakes is enormous, and if farm families were relieved of this operation they could utilize the spare time for livestock production, preparation of quality milk products, cottage industries, protection and improvement of exisiting forest and grazing land, and other labor-intensive activities, such as horticulture and vegetable cultivation. To achieve this, it would be necessary to provide fuel for cooking and lighting to each rural household at an inexpensive rate.

Regarding the fertilizer, it has been estimated that every ton of wet dung, when passed through a biogas plant, undergoes fermentation, and the resulting spent slurry, after drying, gives an additional quantity of plant nutrient at the rate of 4.735 kg of nitrogen, 4.735 kg of phosphoric acid, and 2.5 kg of potash. Thus theoretically if all the available 10.5 million tons of wet dung was passed through biogas plants, it would yield an additional quantity of 49,717 tons of nitrogen, 49,717 tons of phosphoric acid, and 26,250 tons of potash.

The fuel requirement, per person, per day, both for cooking and lighting is estimated at 10 to 12 cft of biogas. One adult animal (partly stable bound) on an average gives 6 kg of wet dung, which is enough to generate 6 cft of biogas. In other words, two adult cattle would meet the fuel requirements of one person per day. Thus theoretically, 4.77 million livestock will meet the fuel requirements of 2.38 million people.

It will thus be seen that biogas technology has enormous potential to meet at least the partial fuel requirement of the population. This could considerably reduce the importation of petroleum fuels. The biogas technology also assures wide-ranging socio-economic benefits for the prosperity of rural households. In this context it seems that the biogas program in Nepal must be of two types: extension and promotion programs to make the present technology available to those who can afford it, and research and development programs to develop biogas systems which are cheaper to build, require less maintenance, and are more appropriate to Nepal.

HISTORY OF BIOGAS PROGRAM IN NEPAL

The first biogas plant in Nepal was built in around 1955, and later one or two private owners built their own plants. It was not until 1974-75 that real interest in biogas was expressed in Nepal. The government designated 1975-76 as Agricultural Year, and a plan was developed to build 250 plants. During the year, 196 plants were built by various contractors using interest-free loans from the Agricultural Development Bank of Nepal. During 1976 a joint project, "Study of Energy Needs in the Food System," which included a strong emphasis on biogas, was set up between the Department of Agriculture and Peace Corps, with funds from USAID. To meet the demand of the rural population, the Biogas and Agricultural Equipment Development Company Pvt. Ltd. was established in 1977 as a specialized institution with joint investment of the Agricultural Development Bank, Nepal Fuel Corporation, and the United Mission to Nepal. The company aims to set up research offices, fabrication units, and sales and service centers at strategic places in the country to build and service plants.

Today, more than 1,600 household-size plants and 24 community-type plants are in operation, out of which 906 units are of drum design and the remaining units are the drumless design. In the beginning, the Indian drum-type of gas plant was the only widely used design available. It has some well known problems, but the design was accepted. After much research and testing, the Chinese type of gas plant, which is constructed entirely using cement masonry and works on the displacement principle, was adopted. This plant, at present, is the type most frequently built. The drum-type plants are now used only for larger plants, which are constructed for running engines for milling purposes. The reasons for switching from drum-type to the drumless design are lower plant costs, less repair and maintenance, higher gas yields, easier operation, and less material transportation.

ORGANIZATIONS INVOLVED IN BIOGAS ACTIVITIES

There are three main bodies in Nepal interested in biogas at present: the Soil Science Section of the Department of Agriculture, the Research Center for Applied Science and Technology; and the Biogas and Agricultural Equipment Development Company Pvt. Ltd. The Biogas Company is the primary organization constructing and developing biogas plants in Nepal. The company attempts to cover all the areas feasible for biogas plant utilization from two regional offices, 10 branch offices, one research unit, and one workshop with a total staff of 140. The biogas plants built by the company include a seven year operation and maintenance guarantee. Due to this guarantee, more than 95% of the plants are still in operation. Since this is a private company, without government or other aid, all the costs of promotion, extension, training, R&D, etc. have to be passed on to the farmer, resulting in expensive plants. The high cost of the plants has discouraged sales, preventing the widespread use of this technology.

RESEARCH AND DEVELOPMENT

The Soil Science Section of the Department of Agriculture is studying the use of digester effluent as a manure. The Development and Consulting Services of the United Mission to Nepal is working to improve the effectiveness of the spent slurry as a manure. The Biogas Company is involved in various activities for biogas utlization, uses of alternative feedstocks for biogas production, use of compost for heat generation to heat plants, maximization of gas production in the winter, low cost biogas plant design and application, social and economic impacting of community owned biogas plants, running diesel engines from biogas for grain milling, irrigation water pumping, and cottage industries.

FUTURE BIOGAS PROGRAMS

As a result of the usefulness of the biogas technology, and due to the increasing demand from the rural people, the Biogas Company is planning to expand its activities to construct 2,400 plants of various sizes over the next three years. The executing agency for this expanded program will be the Agricultural Development Bank of Nepal (ADBN). It will extend loan facilities to individual borrowers through its already established network of branch offices extending throughout the country. The ADBN will also extend loans to the Biogas Company in order to expand fabrication facilities and to procure the needed raw materials.

BIOGAS TECHNOLOGY DEVELOPMENT AND DIFFUSION: THE PHILIPPINE EXPERIENCE

Norberto A. Orcullo, Jr.
Ministry of Energy, Republic of the Philippines

ABSTRACT

An overview of the Philippine biogas program and experience, particularly over the past seven years, is presented. The national energy scene, biogas potential, R&D activities, commercialization endeavors, system economies; and its change with size, available incentives, constraints, and possibilities, for future plans are covered. Presently, there are around 600 operational installations throughout the country producing approximately 26,000 m^3 biogas per day. Economic feasibility increases with size and extent of product utilization.

INTRODUCTION

There is no question that energy, or the lack of it, will be a critical factor in determining the shape of the Philippine future. The impact of the energy crunch, which started in 1973, has been felt and will continue to affect the economy. The government, therefore, has embarked on an energy development program essentially meant to substitute imported energy for indigenous resources and to attain a certain level of sufficiency to propel its development. The program set out to explore local resources of conventional fuels such as oil, hydro, and geothermal steam. To rationalize the whole effort, nonconventional energy resources were also given a priority. Thus, in January 1977, Presidential Decree No. 1068 was promulgated, which was essentially aimed at accelerating the research, development, demonstration, and utilization of the so-called nonconventional energy resources. This Decree triggered the birth of the National Nonconventional Energy Resources Development Program (NERDP), being currently implemented by the Bureau of Energy Development (BED). To support the program and to effectively carry out its mandate, several Letters of Instructions were further issued by the President, specifically identifying government agencies and instrumentalities to be involved in the process.

The initial effort under the NERDP was centered on biogas technology. This was a logical choice as the resources are readily available and the technology is already well proven in other countries. Moreover, a number of biogas systems had also existed even before the government institutionalized the energy industry with the creation of the Ministry of Energy. The immediate undertaking of NERDP was the construction/dispersal of demonstration and pilot models in all regions of the country, coupled with promotional activities such as the conduct of workshops/seminars and the distribution of "how to" brochures pertaining to the biogas plant. We are fortunate to have individuals, private firms, educational establishments, and other institutions who supported our efforts.

The new sources for biogas technology abound due to the number of livestock raised nationwide. As shown in Table 1, about 1.28 billion cubic meters per year of biogas is potentially available from livestock, distillery wastes, and aquatic plants such as water hyacinth. Besides deriving energy from polluting wastes, biogasification also brings about a healthy environment and provides organic fertilizer, which, potentially, can displace imported commercial fertilizer.

Table 1 Biogas Potentials of the Philippines

Feedstocks	Biogas Potentials (in '000 m^3 per year)	% Share
1. Hog manure	412,698	32.18
2. Carabao manure	487,184	35.66
3. Cattle manure	200,424	15.63
4. Chicken dung	144,911	11.30
5. Duck's manure	12,471	0.98
6. Distillery wastes	48,423	3.77
7. Water hyacinth*	6,084	0.48
	1,282,195	100.00%

*Estimate for Laguna Lake only.

THE FIRST SEVEN YEARS

For the period 1977-1984, the Ministry of Energy (MOE), funded over one-hundred projects related to renewable energy technologies development and promotion (see Table 2). The funding came in the form of grants to research institutions and other appropriate organizations. It is to be noted that while the program (NERDP) has funded about ₱77 million (US$3.86 million), expenditures on biogas technology accounted for roughly 2.95% as compared with other technology areas. The 10 projects on biogas meant an expense of ₱2.27 million (US$199,921), or an average of 0.227 million per project. Although a lesser amount was infused for biogas technology development under our program, the technology has taken a long stride in terms of diffusion to various areas (whether urban or rural, public or private sector). As shown in Figure 1, practically all of the provinces in the

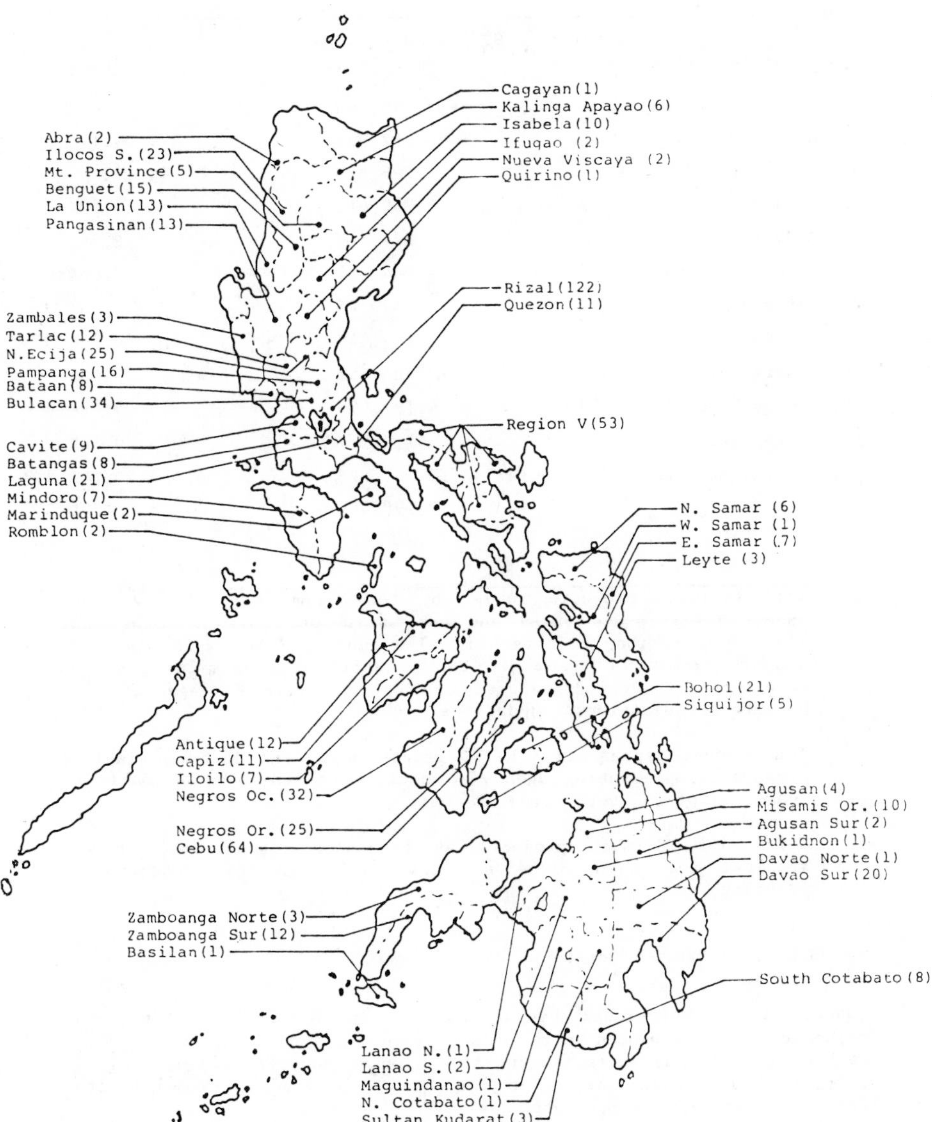

Fig. 1. Philippine Map showing the locations of biogas systems
(Number enclosed in parenthesis indicates number of units)

Table 2 Projects Funded under the National Nonconventional Energy Resources Development Program (NERDP)[a]

(Period: 1977 to Sept. 1984)

	Technology Areas	Project Cost[b]	% Share
1.	Agricultural Wastes	₱17,790,167	23.04%
2.	Direct Solar	28,652,171	37.11%
3.	Integrated Systems	296,850	0.38%
4.	Wind Energy	2,759,615	3.57%
5.	Biogas	2,278,500	2.95%
6.	Alcohol	2,204,229	2.86%
7.	Marsh gas	157,375	0.20%
8.	Hot Springs	745,447	0.97%
9.	Dendrothermal	2,466,556	3.20%
10.	Energy Plants	4,181,596	5.42%
11.	Hydropower	118,077	0.16%
12.	Special assignments/projects[c]	15,547,823	20.14%
		₱77,198,406	100%

a/ This includes projects funded by the Ministry of Energy (MOE) under Grants-in-Aid (GIA) and the USAID-GOP Program. Other R&D and demonstration projects were also funded by MOE with financial support from World Bank and UNDP. Still other similar projects were carried on independently by other government agencies and private organizations/foundations.

b/ This amount represents MOE/BED grants only, and does not include proponent/implementing agency's counterpart, which is normally at least 15% of the total project cost.

c/ This includes such projects as technology promotions, project studies, and other areas not covered by the above technology groupings.

At US $ 1.0 = ₱20.00 (October 1984)

country have operational biogas units. In seven years time, the technology has advanced more than any other nonconventional energy technology. It has attracted the attention of many Phillipino citizens from different walks of life. It has also inspired comment from foreign organizations and individuals.

As of December 1983, it is estimated that there are about 700 biogas system/installations nationwide (see Figure 1). Approximately 600 of these are operational, generating about 26,400 cubic meters per day of biogas, or an energy equivalent of 558 million Btu per day (167

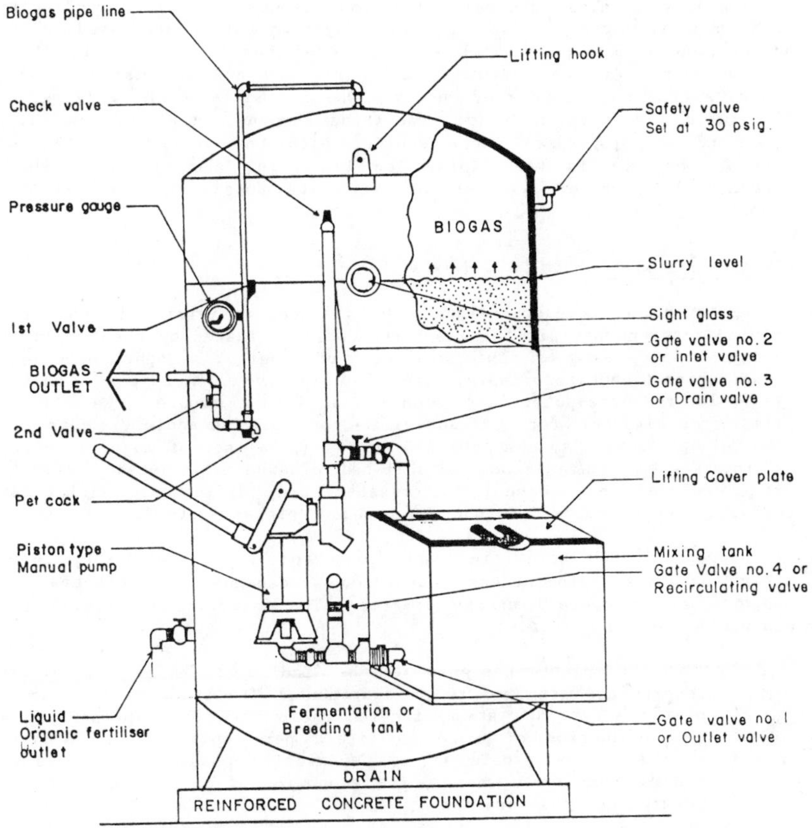

Fig. 2: Cross sectional diagram of a portable biogas system available in the market.

billion Btu/year). About 90% of the biogas energy generated comes from livestock manure (primarily hog manure) while the remaining 10% is from distillery wastes (slops). Another slops-based plant was constructed early this year, and other distilleries are expected to adopt this system. There are approximately 150 large installations (with a gas production capacity of 100 cubic meter/day and above), accounting for about 40-50% of the daily gas output.

With regard to system design and construction, most installations are made of concrete digesters with steel tank floating gasholders. The gasholder/storage is normally independent from the digesters, while the domestic models have gasholders on top of the digester. Portable models for domestic applications are also presently available. These are made of steel plate, with gasholder and storage built-into-one (Figure 2). The system application ranges from direct heating (cooking) to running internal combustion engines (ICE) for electricity or mechanical power generation. Biogas systems have contributed to the energy source mix, either in the form of partial or total (100%) displacement. At least two (2) companies are now in business using biogas as the sole source of energy. One is the famous Maya Farms, an integrated livestock and food processing company, which is now fully dependent on biogas as power source, with the utility as back-up system. Another is the Alpine Ice Plant in Cebu Province, which produces 30 tons of ice per day with its total power requirement supplied by biogas.

RESEARCH AND DEVELOPMENT ACTIVITIES

Research and development and other related activities in the area of biogas technology are being spearheaded/coordinated by the Bureau of Energy Development of the Ministry of Energy. Other agencies/organizations (public/private) are also involved in tapping locally available resource-potentials such as livestock manure. We act as primary coordinator for R&D activities on this technology. For the manure-based biogas system, the R&D component was not of major concern, as this aspect was already established elsewhere. Instead, efforts were centered on technology transfer and diffusion, while R&D activities were centered on non-manure-based biogas systems.

Supportive R&D activities were conducted under the NERDP, while other agencies/organizations conducted independent researches in support of the overall energy program. The following represents the essence of such activities:

- a pilot plant designed to be fed with human waste was constructed attached to an apartment. The study yielded about 62-173 liters/day at a dilution ratio of 1:25, representing a gas production of 6 to 15 liters per capita. This project concluded that while it is technically possible to generate biogas from human wastes, the economics of operation would not permit its proliferation.

- a laboratory digester was set up last year to look into the biogas potentials of _Sargassum_ Sp. This is a marine-based brown algae that abounds in the coastline. Partial results revealed that between 1.7 to 6.2 liters biogas can be obtained per kilogram of sargassum, at a water dilution ratio of 1:3 and a retention period of forty-five days. In either situation, starter was required to enhance digestion of the slurry so as to produce combustible gas.

- a private firm (Central Azucarera de Tarlac) conducted in-house research on the utilization of anaerobic fermentation technology to generate energy from distillery wastes (slops). The system was also viewed as part of the waste treatment and pollution control measures of the company. This undertaking reported a gas yield of 30 liters biogas/liter slops under a thermophillic condition and continuous process. As the system is mechanized to maintain a continuous thermophillic process, about 20-25% of the energy produced is plowed back to the system, representing a net energy gain of about 75-80%. The company has filed a patent claim on the process and is presently operating on the slops-based biogas plant producing 127,118 cubic feet per day.

- the National Institute of Biotechnology and Microbiology (BIOTECH) has likewise studied the biomethanation of crop residues like rice straw, coconut water, and coconut coir dust. For the past two-and-a-half years, efforts were made to study the effect of alkali pre-treatment of crop residues on biomethanation, and to maximize conditions that will produce the greatest amount of biogas. Various alkali preparations were tried and the following are capsulized results:

  - In the case of rice straw, the research revealed that gas production increased from an average of 264 liters/kg dry matter (control) to as high as 421 liter/kg dry matter. The digester was operated at an effective solids content of 50% and maintained for 30 days. The pH was maintained at near neutral. Gas composition varied from 50-70% methane and 50-30% carbon dioxide.

  - For coconut water, results showed that the most diluted cocowater (75%) produced the highest gas volume (41 liters/liter) while the undiluted coconut water produced the least amount of gas (4 liters/liter). Retention time for the undiluted and least diluted cocowater was relatively low. The diluted cocowater produced the highest methane gas composition (36%).

  - The biomethanation of coir dust with 4% and 6% total solids, which were either washed or unwashed was likewise conducted. The digester with 4% produced more gas (250 liter/kg dry matter) than that with 6% total solids (180 liter/kg dry matter). Although gas volume was lower than that produced from rice straw, gas quality of digested coir dust was comparable to that of rice straw. After only two days of digestion, methane gas composition was in the range of 69-72%. The quality of gas was maintained throughout the duration of the study (30 days).

- Water hyacinth (Eichhornia crassipes) is another feedstock in which we are interested. This is a prolific aquatic resource which abounds in lakes, fresh water streams, and estuaries. In the case of Laguna Lake (with an area of 92,500 hectares), water hyacinth pose navigational hazards, restrict potential areas for fish production, and affect productivity as well as the income level of fishermen. Gasifying this material therefore provides not only energy production solutions, but also solutions to other relevant socio-economic problems. A project with the Economic Development Foundation (EDF) is being finalized for the conduct of laboratory experimentation and subsequent construction of a pilot plant on the lakeshore to benefit the villagers in the area.

COMMERCIALIZATION AS A MEANS OF DIFFUSION

As mentioned earlier, biogasification is essentially a proven technology such that establishment of pilot/demonstration plants was a logical step when talking of technology transfer and diffusion. We are now concentrating on encouraging private investors/users to be more active in technology utilization. This is embodied in our "commercialization plan." The plan is actually a tool for initiating the shift of technology from the laboratory to the investor. This "plan" is a market-oriented approach, which assumes "free market forces" as a motivating factor toward eventual technology transfer and adoption, rather than a "shotgun approach". In the plan, a series of activities for each technology is drawn upon and projects are identified or made in order to "manipulate" or determine the answers to several queries a "user" or "seller" may pose to himself if he is to adopt or sell a nonconventional technology such as biogas. This is a phased strategy which aims to create an environment conducive for both "buyers" and "sellers" such that commercialization* is realized. In effect, we are promoting the technology with a minimum bias and a maximum participation of the private sector. While this strategy was conceptualized only in 1982, it matches the biogas technology movements implemented years ago. Aware of this reality, the Bureau of Energy Development has set the following areas of nonconventional energy as priorities:

1) Large Scale Biogas System
2) Utilization of Agricultural Wastes as Boiler Fuel
3) Producer Gas System
4) Large Scale Solar Water Heating
5) Biomass-derived Fuel for Internal Combustion Engine
6) Utilization of Nonconventional Energy in Village and Farms to Enhance Productivity.

To support the faster movement of biogas technology beyond the laboratory and urban centers, we tried to find a means to create an environment conducive to the entry of entrepreneurs. At present, there are at least six (6) companies/organizations involved in the application of biogas technology (Table 3). Four (4) of these companies are involved in direct fabrication and marketing of the technology/ system, while the other two (2) are essentially involved in system design/project management on a consulting basis, especially for large scale capacity. Besides the presence of businessmen, the Ministry of Energy, National Science and Technology Authority, and other governmental organizations assist interested individuals/organizations in coming up with a cost-effective or low-cost biogas system. We maintain a basic instruction manual (for distribution to the public) that would allow every individual to construct his own domestic biogas plant without external or specialized inputs of material and labor/maintenance. This process allows the diffusion of biogas technology to the countryside, either by individual

*Commercialization is achieved once there are a significant number of actual buyers, and there is at least one seller making a profit. In addition, while there may be some government subsidies, the social benefit of using the technology must not be lower than the social cost of producing or operating it.

Table 3 Private Companies Involved in Production, Marketing, and Construction of Biogas Systems in the Philippines

	Company Address	Product Make	Capacities Digester (m^3)	Biogas (m^3)/day)
1.	Little Giant Mill Supply, Inc. 159-161 Shaw Boulevard Mandaluyong, Metro Manila	B.I. steel sheets	2.1-8.00 m^3	1.0-8.0
2.	Sanamatic Tanks Mfg. Corp. 2750 Taft Avenue Manila	Concrete digester tank with rubberized sheeting (cover)	8.0-1,500 m^3	-
3.	Biogas Incorporated 11 Galang St., GSIS Heights Matina Davao City	steel drum	0.68	1.0
4.	Rokar Metalcraft 18 Reynaldo St., Tierra Bella Subd., Tandang Sora Avenue, Quezon City	B. I. steel sheets	-	1.00 to 16
5.	Maya Farms, Inc* Angono, Rizal	concrete tanks	Medium to large capacities	
6.	Applied Microbiological Research Laboratory**(AMRL) Central Azucarera de Tarlac (CAT) San Miguel, Tarlac	-	-	-

* Maya Farms maintains a Bionergy Consulting group, primarily involved in the design and construction of large or industrial scale biogas plants in the Philippines.

** AMRL/CAT is a private firm extending consulting services for the construction of a biogas plant facility utilizing distillery wastes.

"constructing" of the system or "buying" of the system in the market. The route followed to effect the technology diffusion is a matter of economic decision where financial (investment) constraints pose as a stumbling block. For socio-economically sound undertakings, the Ministry of Energy through BED/NCRD may endorse the project to appropriate financing institutions or directly finance the venture on a case-to-case basis.

THE ECONOMICS OF BIOGAS SYSTEM

If we are to categorize the criteria used by potential users of biogas technology, economics of operation is the most dominant factor that would lead to the technology adoption process. For domestic application, the economics of operation is basically clear due to the simplicity and homogeneity of application. In terms of investment and cost of operation, there exists an economics of scale on any biogas system but, regardless of the size, the degree of utilization of the system (either in terms of daily input or gas usage) finally dictates return on investment. It is unfortunate to note that certain biogas installations appear to be economically unattractive in actuality, not because of the technology per se, but for reasons of underutilization, of inappropriate sizing, and system design itself.

Looking at biogas technology from an economic standpoint, investment data and economics of operation of selected cases are herein presented. In the case of the manure-based biogas system, the question how much money to allot or invest is inherent to a specific site, primarily considering such factors as: distribution system, mode of gas utilization, energy cost at site, sludge utilization, interest rate, etc. Thus, instead of presenting a case on biogas system economics of production site, an investment based parameter is presented in Table 4. It must be noted that monetary values tabulated represent the cost of the digester and gasholder alone, or a system designed to produce direct heat.

As shown in Table 4, the investment requirement for a 100-sow unit is P0.235 million, as against P0.709 million for a 500-sow unit. It is clear in this information that while gas production capacity increased by five times in favor of the latter, the investment requirement has multiplied by only about three times. As a result, estimated capital cost per thousand Btu for a 100-sow unit is more expensive than for a 500-sow unit by a factor of about 1.6 times. Clearly, this data shows that under Philippine conditions, the issue is not really on operational cost to produce energy from manure, but on availability of funds to invest in the technology and derivation of substantial benefits from the system. The data further shows that while the technology is attractive for the large scale or corporate user, the same is relatively economically unattractive to small or domestic users who are accustomed to cheap wood or other investment-free energy sources (e.g., wood and other agricultural residues).

Table 4 Estimated Cost of Construction of a Concrete-Based Biogas Plant with Manure (Hog) Feedstock[a/]

Cost Item	C A P A C I T I E S			
	500 sow unit[b/]	100 sow unit[b]	20 sow unit[b]	4 sow unit[b]
A. Cost of the Plant				
1. Material Costs	₱422,037	₱138,702	₱37,531	₱12,352
2. Labor	124,613	50,365	11,261	4,940
3. Administrative Costs	42,840	9,450	2,441	-
4. Contractor's profit/ contingencies	118,300	39,708	10,246	-
Total Cost	₱709,790	₱238,225	₱61,479	₱17,290
B. Gas Production (m^3/day)	450	90	18	3.6
C. Investment per volume of gas produced (₱/m^3/day)	₱1,577	₱2,645	₱342	₱4,803
D. Daily Energy Output[c/] ('000 Btu/day)	9,525	1,905	381	73
E. Capital Outlay/Energy Output (₱/'000 Btu/day)[c/]	74.52	125.05	161.36	236.85

a/ Does not include sludge conditioning plant and cost of land for the digester/gasholder and accessories.

b/ A sow unit is composed of the sow and 10 litters.

c/ Computed at a heating value of 600 Btu/cubic foot.

September 1984 @ ₱18 = 1.0 US$

To have an idea of the cost of operating a domestic biogas system, economic data for a commercially available model is shown in Table 5. This particular data considers only the "energy benefit" as an LPG displacement. It does not include the economic benefit from liquid or dried sludge with its organic fertilizer or soil conditioner value. As such, a biogas system for domestic application is still a luxury, specifically for a capacity range of 1.0 to 3.0 cubic meters/day. Considering the energy benefits alone, buying such a system is not really a worthy investment. For a capacity range of 5.0 to 16.0 cubic meters/day, the investment is relatively justified with a payback period of 2 to 3 years. As we go to a higher capacity range, the investment would be, of course, well justified due to the economies of scale of the technology. To verify the household level impact of the commercially available models, NERDP has dispersed 23 units in the Central Visayas region through its domestic biogas promotion project. Under the project, BED/NCRD financed the acquisition of the units and the recipient will amortize the units for a period of three to five years. This project is being monitored and partial results brought praise from the users. It must be noted that the abovementioned models are "commercially" available, and opportunity for cost reduction exists. Actually, the same quantity and quality of gas can be generated by a "home made" biogas plant. For these reasons, we maintain an easy-to-understand construction manual to enable enterprising villagers to take advantage of the merits of a biogas system.

Table 5 Economics of Operation of Commercially Available Domestic Biogas Models (Costings as of October 1984)

	GAS PRODUCTION CAPACITIES (Cubic Meter Biogas Per Day)					
	1.0	2.0	3.0	5.0	8.0	16.0
A. Investment Costs:						
1. System Cost	₱ 5,000	₱ 8,000	₱ 12,000	₱ 16,000	₱ 25,000	₱ 35,000
2. Pipes/Fittings/ accessories/labor	500	800	1,200	1,600	2,500	3,500
Total Cost	₱ 5,500	₱ 8,800	₱ 13,200	₱ 17,600	₱ 27,500	₱ 38,500
B. Annual Operating Expenses						
1. Depreciation	₱707	1,131	1,697	2,263	3,534	4,950
2. Interest	770	1,232	1,848	2,464	3,850	5,390
3. Maintenance	83	132	198	264	413	578
	₱1,560	₱2,495	₱3,743	₱4,991	₱7,797	₱10,918
C. Annual Gross Benefit	₱1,911	₱3,822	₱5,733	₱10,530	₱15,289	₱30,578
D. Annual Net Benefit	₱351	₱1,327	₱1,990	₱5,539	₱7,492	P₱9,660
E. Payback Period (Years)	15.67	6.63	6.63	3.17	3.67	1.95

Notes: a) System cost is FOB factory. This cost does not include freight/handling charges on the delivery of the unit from the factory to the user/buyer.
b) Liquid/dried sludge which can be used as organic fertilizers is not costed.
c) System benefit is based on the equivalent LPG cost of biogas produced (at ₱11.54/kg)

Assumptions:
1. Depreciation charges - straight line method
2. Economic life - 7 years
3. Salvage value - 10%
4. Interest rate - 14%
5. LPG Cost - ₱11.54/kg.
6. Annual maintenance charges - 1.5% of system cost

US$1 = ₱20.00 (October 1984)

For a large size biogas system using distillery wastes (slops) as feedstock, the case of Central Azucarera de Tarlac in San Miguel, Tarlac, is herein presented. The plant consists of a polyurethane-insulated bioreactor (digester) with a capacity of 1.6 million liters and generating 3.59 million liters/day of biogas. The plant (which utilizes only one-fifth of the daily slops production) is operated on a thermophillic condition and continuous anaerobic digestion process. As the plant is mechanized, net energy gain is about 80%, with the other 20% utilized internally to maintain the thermophillic and continuous operation. The economics of the operation for this system are shown in Table 6. BED through the NERDP, is now

Table 6 Cost Analysis of the Anaerobic Digestion System of the Central Azucarera de Tarlac

1. Investment Cost: - ₱1,400,000.00

2. Gross Income:

a) Projected Biogas to be generated:

Slops discharged from distillery	-	600,000 lts./day
Volume of slops processed	-	120,000 lts./day
Volume of slops available for future expansion	-	480,000 lts./day
Biogas recovery rate	-	30 lts./liter of slops
Heating value	-	590 Btu/ft^3
Biogas generated	-	127,118 ft^3/day
Heating value of fuel oil	-	38,320 Btu/liter

b) Theoretical Fuel Equivalent of Biogas Produced:

$$\frac{127,118 \times 590}{38,320} \times .95 = \underline{1,859} \text{ liters fuel oil/day}$$

Actual average volume of industrial diesel oil (IDO) displaced by biogas at the Refinery kilns = 850 liters/day.
Actual average volume of bunker oil displaced at the Boiler = 900 liters/day.
Price of IDO = ₱4.20 per liter
Price of bunker oil = ₱2.60 per liter
∴ The toal savings due IDO and bunker oil displaced is -
850 x ₱4.20 = ₱3,570.00
900 x ₱2.60 = ₱2,340.00

TOTAL = ₱5,910.00

Assuming 270 days net operation per year -
∴ Total annual fuel savings = 5,910.00 x 270 = ₱1,595,700

3. Annual Operating Costs:

Depreciation.	₱126,000.00
Repairs and Maintenance	28,000.00
Salaries and Wages.	50,566.00
Utility	103,900.00
Insurance	7,000.00
Interest.	210,000.00
	₱525,466.00

4. Economic Feasibility:

a) Net Income

Gross Income	₱1,595,700.00
Annual Operating Cost.	525,466.00
Net Income before Tax.	1,070,234.00

$$\% \text{ Return on Investment} = \frac{₱1,070,234.00}{1,400,000} \times 100 = 76\%$$

$$\text{Payout Time} = \frac{₱1,400,000.00}{1,070,234.00} = 1.3 \text{ years}$$

Adapted from P.F. Ventura (1984)

negotiating a research grant with CAT to conduct a study that would maximize gas yield from the slops, at the same time pinpointing the optimum retention period. These factors are vital toward reduced capital outlay on the digester component.

AVAILABLE INCENTIVES

In order to encourage private entrepreneurs to go into energy-generation through biogas technology, tax and related incentives are integral to the program. As mandated by Presidential Decree 1068, parties involved in the R&D and utilization of nonconventional energy resources shall have the following privileges:

- Costs incurred in the establishment and construction of nonconventional energy conversion facilities, or equipment duly certified by the Bureau of Energy Development may, at the option of the taxpayer, be directly chargeable to expense, and shall be fully deductible as such from gross income in the year wherein such expenses were incurred.

- Exemption from payment of tariff duties and compensating tax on the importation of machinery, equipment, spare parts, and all materials, required in the establishment and construction of nonconventional energy facilities or equipment.

The same Presidential Decree enjoined government financing institutions to give high priority to applications for financial assistance by individuals/enterprises/industries participating in the program to accelerate research, development, and utilization of nonconventional energy resources duly recommended by the Ministry of Energy.

Several other attractive incentives are also authorized under the Energy Priorities Plan (EPP) of the Investments Priority Program (IPP) of the Board of Investments (BOI). More privileges can be derived in this program once the project or venture is accredited by BOI/IPP. This particular incentive plan is applicable to "would be" entrepreneurs, as well as to those existing establishments planning to install nonconventional energy devices and/or systems as part of their business operations.

There are a number of industries where the energy resources exist and technology for energy conversion is available but the application is either limited or not economically attractive. The majority of the livestock (mainly piggery and poultry) companies, for instance, are merely "producers" whose production or process requirements are not energy-intensive compared to integrated firms (e.g., Maya Farms). To attract these potential nonconventional energy producers, a Cabinet Bill on interconnection is being drafted. The proposed bill is essentially similar to the Public Utility Regulation Policies Act (PURPA) of the United States whereby the local utilities are mandated to buy the energy produced by a company within its franchise area. Hopefully, this bill would not only enhance attainment of in-plant energy self-sufficiency among selected firms but would also enable the firm to sell excess energy which, in effect, would advance the payback period of the system. Apart from producing the much needed energy, the scheme would also alleviate the problem of waste disposal of polluting effluent, as in piggeries, alcohol distilleries, timber companies, rice mills, etc.

CONSTRAINTS AND POSSIBILITIES

While it is true that the technology has demonstrated its techno-economic feasibility, especially for large-scale installations, a number of factors posed constraints toward full scale conversion of livestock and other feedstock into useful domestic or commercial energy. Some of the constraints encountered are as follows:

1. Like any other nonconventional energy system, high initial investment on a biogas system scares away enthusiastic villagers and large scale user alike. There is a tendency to buy or adopt the technology due to attractive economics of operation, but investment requirements present some deterrent.
2. Since most of the potential biogas producers/users are plain livestock producers (they sell unprocessed product), there exists a limited biogas fuel, and attractive economies of scale can only be imagined. This connotes availability of energy "in situ," but the plants' energy demand could not justify the immediate recovery of investment. As such, convincing livestock companies to adopt the biogas system is more of an entrepreneurial (mainly economic) decision than one technical in nature.
3. The cheap (if not free) cost of energy from wood and other agricultural waste is a deterrent to the introduction of the biogas system in the rural areas in spite of the availability of feedstock for biogasification.
4. Due to low investment cost and land (lagoon area availability among livestock establishments), acceptability of biogasification as a pollution control system has not really gained wide popularity.
5. The problem of manure collection, especially in the case of carabao and cattle, as well as the demand for some other application, (e.g., chicken dung is used in fish pond) inhibits potential users from adopting the technology in spite of its economic attractiveness.

For rural applications, experiences have shown that diffusion efforts, or style, may not be the same as in the urban areas. Even assuming availability of feedstock for gasification, prevalence of pollution, due to the presence of manure everywhere, as well as the proof that the biogas is much better than wood or charcoal fuel, assurance that the technology would be adopted as part of their (rural folks) lifestyle would not be a full reality. In the context of a rural setting, biogas technology is a "product" that should not be "sold" or "diffused" on the basis of its IRR value, monetary savings, or simple payback period. Such is the case in many circumstances: monetary asset to "buy" or "invest" in the technology is either absent or is not the priority, considering limited financial resources. To a certain extent, there are few domestic scale installations in the countryside who are motivated by status symbol and availability of extra funds. In either case, the technology diffusion issue could not be rated. Given substantial feedstock for gasification in a rural setting, biogas systems are justified under conditions of cost effectiveness (in terms of investment and operation), ease of operation, availability of a soft financing scheme, pollution problems, and high fuel cost (LPG, wood, or charcoal).

IMMEDIATE GOALS AND TARGETS

With the continuous and spiralling cost of fossil fuels, the need to develop the potentials of biogas technology is a goal of the Ministry of Energy. The animal manure-based biogas system will continue to be the primary one used. Large scale capacities or installation with industrial impacts shall be a major concern. Biogas generation from marine resources (Sargassum Sp.) shall be advanced to pilot or semi-commercial scale within a year. Activities relating to the gasification of water hyacinth will also be started this year, with an end view to a pilot plant next year. Other agricultural wastes will also continue to be possibilities for research studies, either on a laboratory or pilot scale.

As a contributor to the overall energy mix, biogas technology is expected to contribute about 22,684 barrels of fuel oil equivalent (FOE) for the year 1984, progressing to 31,284 FOE, and 45,684 in 1988 and 1992, respectively (see Table 7). Involvement of the Ministry of Energy through direct financing, is expected to be 1985 only. Promotional indirect and related activities shall be carried out to enable the private sector to be solely responsible for biogas projects by 1986 and onward.

SUMMARY AND CONCLUSIONS

The past seven years was a fruitful period in terms of local developments relating to biogas technology. Starting with anaerobic fermentors as pollution control measures, the technology has prospered. Livestock manure, as feedstock for biogas generation, has now been commercially adopted in the Philippines. Small and large-scale installations can be seen in practically all parts of the country. There are about 600 operational installations throughout the country, generating about 26,400 cubic meters per day of direct heat energy. We have not prospered very much in terms of number of installations in the early 80s. In terms of investment recovery, payback period is in the range of two to four years for large systems and three to five years for domestic models. At least six (6) companies are now in active business, catering to the biogas energy systems market.

For its part, the Ministry of Energy (MOE) will continue to play a major role in tapping the potentials of biogas technology, especially among commercial piggeries/poultries. Efforts will be concentrated on technology diffusion, in the case of manure and slops-based systems, while R&D will look into other non-manure feedstock. Biogasification of distillery wastes will be a major concern among the non-manure feedstocks, with a view of establishing more large-sized plants. In R&D, aquatic materials such as Sargassum Sp. and water hyacinth will be studied. Any breakthrough on this type of feedstock will surely be of great help to our coastal and island villagers where wood fuel is scarce. Microbiological studies relating to biogasification of other

Table 7 Physical Targets on Biogas Energy Systems

	1983	1984	1985	1986	1987	1988	1989	1990	1991	1992
1. Number of MOE Projects (All Large Units)		1	2	-	-	-	-	-	-	-
2. Private Installations:										
Large		2	2	4	6	10	10	10	10	8
Small		10	10	15	20	25	30	35	40	45
Aggregate Capacity (BOE)*		1,100	1,400	1,500	2,200	3,500	3,600	3,700	3,800	3,300
Cumulative BOE* Contribution	21,584	22,684	24,084	25,584	27,784	31,284	34,884	38,584	42,384	45,684
3. Project Cost:										
Private (₱'000)		1,650	1,650	3,175	4,700	7,625	7,750	7,875	8,000	6,725
Government (₱'000)		700	1,400	-	-	-	-	-	-	-
4. Manpower Requirements:										
Pre-operation		35	40	50	70	100	110	120	130	130
Operation (cumulative)		613	635	662	700	755	815	880	950	1,019

*Capacity is expressed in terms of biogas generation in barrels of fuel oil equivalent.

agricultural waste will also be encouraged. To support the industry's growth, assistance to biogas systems manufacturers/businessmen will be intensified, through product promotion and technical assistance, to effect product development/improvement. A commercial model will be tested for susceptibility to non-manure feedstock to widen the product's market, and hopefully improve corporate profitability. Seminars/workshops and training courses for technical personnel and extension workers will form part of the overall effort to promote the widespread use of biogas systems. With these efforts, we expect to have biogas technology as a vital contributor to the total energy supply and a profit factor in the private sector.

REFERENCES

1. Kilayko, G.U. 1984. The National Nonconventional Energy R&D and Commercialization Thrusts. Ministry of Energy.

2. National Institute of Biotechnology and Microbiology (BIOTECH). 1984. Biomethanation of Agricultural Residue. Los Banos, Laguna.

3. Bureau of Energy Development (BED/NCRD). Various memoranda/Reports. 1979. Fort Bonifacio, Makati, Metro Manila.

4. Ministry of Energy. 1979. National Nonconventional Energy Resources Development Program. (Annual Report). Manila.

5. Little Giant Mill Supply Co., Inc. 1984. Bukhay Biogas Model. Mandaluyong, Metro Manila.

6. Terrado, E.N. Biogas Production: Technologies and Environmental Impacts. Ministry of Energy. Manila.

7. Tata Energy Documentation and Information Center. 1984. Biogas News. Bombay, India.

8. Meneghini dos Santos, D.U., et. al. 1983. Productivity of Water Hyacinth. Compania Energetica de Sao Paolo. Sao Paolo, Brazil.

9. National Census and Statistics Office (NCSO). 1978. Philippine Statistical Yearbook. Manila.

10. Bureau of Agricultural Economics. 1984. Livestock Inventory Report. Quezon City.

11. Orcullo, N.A. 1984. Opportunities for Fuelizing Distillery Wastes Through Anaerobic Fermentation. Bureau of Energy Development. Makati, Metro Manila.

12. UP Institute for Small Scale Industries/Bureau of Energy Development. 1983. Research on the Barriers to the Popular Use of Selected Nonconventional Energy Technologies. (Terminal Report. SA-067-84). Quezon City.

13. Economic Development Foundation (EDF). 1974. Distillery Slops Treatment. (Victorias Milling Company).

14. Asian Development Foundation (EDF). 1974. Distillery Slops Treatment. (Victorias Milling Company).

15. Laxamana, S.F. 1982. Current Distillation and Waste Management Practices in Philippine Distilleries. Central Azucarera de Tarlac.

16. National Academy of Sciences (NAS). 1982. The Potential for Alcohol Fuels in Developing Countries. National Academy Press. Washington, D.C.

17. Ventura, P.F. 1984. Distillery Slops Treatment Methods Used at Central Azucarera de Tarlac. San Miguel, Tarlac.

18. International Energy Projects. 1984. Bioenergy Systems Report. P.O. Box 391, Front Royal, Virginia, USA.

BIOGAS TECHNOLOGY: RESEARCH, DEVELOPMENT, AND DIFFUSION PROJECTS IN SRI LANKA

A. Amaratunga, University of Peradeniya, Sri Lanka

ABSTRACT

This paper summarizes the historical development of biogas technology in Sri Lanka and the efforts that have been made to develop and propagate the technology. Research, development and extension work carried out by the main agencies which have participated in these activities are outlined.

HISTORICAL ASPECTS

The earliest available evidence of an organized attempt to produce biogas in Sri Lanka surfaced in the 1950s at a farm school in Kundasale. It appears that the attempt had not been pursued beyond the initial stages of experimentation, due to an explosion. Since then, there does not seem to have been any activity in this field until 1972.

During early 1973, a small fixed-dome type of unit was put into operation in the laboratories of the Department of Civil Engineering at the University of Peradeniya. The raw material used was cattle dung and the gas was used for routine undergraduate experiments involving the determination of the calorific value of the gas. The unusual technique of gas production aroused much interest and a program of research was carried out to determine the gas-generating capacities of various raw material and the composition of biogas. Hardly any literature was available on this subject in the country at that time.

Following the drastic increase in the price of oil during 1973, a great deal of enthusiasm was generated in using biogas as an alternative source of fuel. The Industrial Development Board of Sri Lanka has obtained some sketches of a type of unit used in India and made an attempt at popularizing this concept by setting demonstration units in schools. Much publicity was given through newspaper articles, and a number of interested individuals, too, attempted to build various adaptations of this type of unit for domestic use. This program, although it gave initial impetus to create an awareness among a wide cross-section of people, was not very successful because there were certain deficiencies in the design of the biogas units, and also because the demonstration units were put up in urban schools where the required raw materials were not readily available.

The digesters soon fell into disuse. With the gradual acceptance of the increase in the cost of oil, there was very little extension work carried out during the next two or three years.

On the international scene, however, a tremendous effort was being put into bringing together people and knowledge from various countries. Workshops and conferences at local and international levels were being organized in various parts of the world and information was becoming more readily available in Sri Lanka. With the wealth ofinformation coming from Asian countries, particularly regarding experiences in India and the People's Republic of China, it became possible to adapt the available knowledge to suit conditions prevailing in Sri Lanka. Numerous institutions were devoting attention to the possibilities of encouraging the use of biogas as a fuel, with varying degrees of success.

The work at the University of Peradeniya, by around 1975, revolved around the concept of Integrated Biogas Systems. The biogas generator was treated as the core of the system and animal houses, fish ponds, duck ponds and algae ponds were introduced as valuable appendages. In such a system, the biogas fuel component constitutes only a fraction of the total potential of the system. This concept was adopted on a practical scale in 1976 at the In-service Training Centre of the Department of Agriculture at Peradeniya, with technical assistance from the University. The project proved to be more successful than expected, considering the financial and other resources which were available at the time, and still continues to serve as a valuable demonstration project. The gas is used for cooking for about one hundred people, and the enriched fertilizer has made possible the cultivation of a variety of crops.

Around this same time, the United Nations Environment Program was establishing a Rural Energy Centre at Pattiyapola where an approximately $90m^3$ biogas unit of the movable dome type was constructed with the intention of using the gas to generate electricity. (Solar and wind energy too were intended to be harnessed to produce electricity to be supplied to this remote village; the Rural Energy Centre, was expected to be a model demonstration centre for the South East Asian region, illustrating the feasibility of utilizing renewable resources of energy).

Since 1978, with the widespread availability of information on biogas technology, as practiced in the People's Republic of China, several fixed dome digesters have been constructed using common burnt clay bricks. Not all these have operated successfully, often due to a lack of understanding of the structural behaviour of the composite unit. During the past two years, several newly built model villages have been provided with domestic-size units; the main use of the gas is for fuel for cooking.

RESEARCH AND DEVELOPMENT WORK

The research work done at the various institutions has dealt mainly with the following aspects:

1. Gas yields from various raw materials

Beside cattle dung, which forms a readily available raw material in rural areas, and is perhaps the most acceptable source of abundant methane bacteria, the raw materials which have been investigated include rice straw and water weeds such as salvinia and water hyacinth.

2. Development of gas utilizing equipment for rural use

Work in this field has included the investigation of the use of burnt clay gas burners and lamp components. Some work has been carried out on the use of oxy-biogas welding, which has potential for application in rural workshops where oxy-acetylene welding is sometimes carried out by producing acetylene in makeshift units, using calcium carbide in discarded steel barrels.

3. Investigation of construction aspects and structural behaviour of fixed dome digesters

These studies have included investigations into the use of various types of materials for digester construction, various techniques of construction, and analysis of stress conditions in fixed- dome digesters.

4. Development of Integrated Systems

Integrated Biogas Systems have been under evaluation for about seven years, and seem to offer the most rational approach to biogas propagation and utilization.

TRAINING PROGRAMS AND EXTENSION WORK

The training programs carried out in Sri Lanka may be divided into the two following distinctive types:

1. Training of persons engaged in agriculture and animal husbandry

These programs are sponsored by the Department of Agriculture as part of an in-service staff training scheme and consist of about a one week residential course (lectures and demonstrations). These courses are intended to provide information and adequate technical know-how to persons who are in a position to engage in propagational work. Some of these courses have been conducted at the In-service Training Centre at Peradeniya, where the integrated system has always provided a source of stimulation to the participants.

2. Training of technical persons

The University of Peradeniya conducts a regular course for final-year students in the Faculty of Engineering on various aspects of biogas technology. This forms part of a course in renewable energy resources, and has now been in operation for about five years.

CONCLUDING REMARKS

A number of institutions in Sri Lanka are engaged in activities related to biogas technology. Although in most instances biogas units are built for the purpose of obtaining a cooking fuel only, training efforts are being directed at showing the multifold advantages of integrated systems. Indeed, in certain situations, even if the gas is allowed to go to waste, there could still be a net benefit.

The initial cost of setting up a digester unit is beyond the financial means of the vast majority of those to whom an integrated system could eventually bring many improvements in their quality of living. Even when units have been given entirely free of charge, some have been allowed to go into disuse through lack of maintenance effort. The greatest single impediment to the rapid and widespread propagation of the technology in Sri Lanka would appear to be sociological rather than technological.

BIOGAS PROGRAM OF THAILAND

Sompongse Chantavorapap
Energy Research and Development Division, National
Energy Administration, Ministry of Science, Technology and Energy

ABSTRACT

The biogas program is one of the most active in the national Alternative Energy Study and Development Plan. At present, there are about 5,000 family-size units and around 10 community size-plants. In collaboration with Belgium, the feasibility of generating biogas from food and agro-industries wastewaters is being conducted. Another 'biogas resource management' demonstration project will be implemented in cooperation with the Italian government. A third activity is being supported by USAID. Plans and priorities relating to research, development, demonstration, promotion and popularization have been clearly set and are being implemented.

INTRODUCTION

Thailand is an agricultural country and, as such, produces several hundred million tons of agricultural products annually. Of these products, used directly as food and raw material for agricultural industry, more than one-hundred million tons of residue are left, and are dumped into the river as a solid or liquid waste causing pollution hazards and increasing environmental problems.

In Thailand, the annual production of rice straw is 30-40 million ton (mton), rice husk 3-4 mton, maize stover 5-10 mton, baggasse 10-15 mton, molasses 0.6 mton, kenaf stalk 0.5 mton, animal manure 57 mton, and cassava residue 10-15 mton (cellulitic, starch, and/or sugary residue).

Dwindling fuel, food, and feed reserves mandate a serious effort toward converting the tremendous resources of organic residue into useful products. The anaerobic-digestion process conversion of organic residues, such as animal manure, wastewater into fuel (biogas), fertilizer, and feed is the most active program in Thailand's Alternative Energy Study and Development Plan.

THE BIOGAS PROGRAM

For the purpose of discussion, the biogas program of Thailand can be divided into five main tasks. Under each task, information about ongoing and planned activities is detailed.

Resource Investigation and Assessment

This task includes a research study for the purpose of ranking potential resources for biogas production. Field investigations would also be performed to determine present and future use, as well as availability, accessibility, necessity, and viability of these biomass resources.

Preliminary evaluation reveals that the total availability of manure in Thailand is approximately 57 mton out of which the collectable portion is estimated at 30 mton (with percentage collectable of cattle manure 50, pigsty and fowl droplet 75). It further shows that 300,000 households in 6,700 villages own more than six head of cattle, and the rest own only 1-2 buffalo, for draught power use. There are tens of milk cow raising farms and several hundred pig and fowl raising farms. Presently, there are about 5,000 family-size biogas digesters, each generating 1-2 m^3/day of biogas for household cooking, and 10 community size biogas digesters, each generating 20-30 m^3/day for running autogenerators for village households, electric lighting, and for running autopump for water pumping of community crop plot irrigation. Such magnitude is deemed insignificant when compared with the overall potential of biogas technology in Thailand.

A large number of food and agricultural industries are spread over Thailand, producing significant amounts of heavily polluted wastewater. These wastewaters would be purified by means of continuous anaerobic digestion processes, producing simultaneously a great quantity of biogas. In this regard, five factories have been investigated (alcohol, paper mill, fruit canning, sugar mill, and tapioca starch). This was done in collaboration with Belgium specialists, under the "RTG - GKB Biogas from Industrial Wastewaters Project" effluent sampling to analyze needed information for the preparation of the feasibility study which will be submitted in June 1985.

In Thailand, a plant using the Biotim, a process of N.V. Biotim Company, Belgium is under construction for a molasses-based alcohol factory, named Thai Alcohol. This factory, which is producing 200,000 l/day alcohol, generates 2,500 m^3/day wastewater. The Biotim A plant will recover about 60,000 Nm^3/day biogas from the waste, representing an equivalent amount of about 35 tons/day fuel oil. This will cover about half of the factory's energy need. This plant will be the largest of its kind in the world. The total investment cost for the plant is estimated at about US$ 2,000,000 (from which US$ 500,000 in foreign exchange), whereas the value of the recovered biogas amounts to about US$ 2,500,000 per year. Other Biotim A plants are under negotiation, one for a starch factory and another for a brewery factory.

Need Identification

This task does not only include determination of need for biogas technology for producing fuel, but also the coproduction of a fertilizer and feed.

In Thailand, energy consumption centers around large urban areas. The Bangkok Metropolitan Area (BMA) alone uses 10 and 6 times as much electricity and petroleum as the area outside BMA. The use of energy in the rural area where most of the people live is very low. Woodfuel in the form of firewood and charcoal is used for cooking and kerosene for lighting. As mentioned, most of the families own 1-2 buffalo for draught power. A micro biogas digester producing biogas from a manure of 1-2 buffalo for lighting, reducing the burden to buy kerosene would be most appropriate for the majority of the rural families.

Increasing the income of a village participant during the off-plantation season is a present government policy through its job creation project, starting from January and ending in June each year. The priority job in the project is a water-related development activity, i.e., pond-deepening for keeping rain water, and channel improvement for conveying of water. The author believes that an adoption of biogas resource management to improve the agricultural production in an integrated manner, with partial support from government, and village participation to help raise and sustain the income of each village participant who becomes a shareholder, with only management provision from the government to be a viable solution. When an increase in income is an incentive, more development activities in the project can be created. Implementation of the biogas resource management scheme, under "the RTG-Italian Government Demonstration of Biomass Pilot Plants Project" would be started in early 1985. It is a two-and-a-half year project, with US$1.5 million grant fund from the Italian government and US$ 0.375 million counterpart fund from the Royal Thai government.

Research and Development (R & D)

This task would lean as much as possible toward adaptive R & D. It would be carried out according to priorities relevant to the use of available indigenous biogas resource and the need of the country. Improvement of biogas production and the development of low cost and mass production design is of primary interest and prerequisite for widespread adoption of biogas installations for household, farm and factory activities. Activities under the task would be as follows:

- Sampling and analysis of potential substrates to determine their pertinent characteristics (pH, COD, BOD, TS, and other parameters) for biogas, fertilizer, and feed production.

- Biochemical and microbiological study of prospective substrates. Findings would be bench tested in a biogas digester simulating field conditions.

- Development of a low-cost biogas production and utilization system, within the reach of an average farmer, with emphasis on the use of durable materials and mass production.

- Prototype development of an integrated system for a small farmer owning 1/2 - 1 hectare of land to optimize land use and diversify cropping and livestock-raising pattern, with self sufficiency in fuel (biogas), fertilizer, and food.

- Popularization of biogas technology, especially in rural area. Methodology, ways and means of popularization, such as a mobile biogas popularization units, would be tested. Attitude and response of potential farmers to adopt biogas technology would be determined.

Demonstration and Promotion (D & P)

The prime focus of the task would be to establish and demonstrate a proven biogas prototype (family size) or model (community size) in a selected site to determine its performance as well as to persuade livestock owners to look upon manure not so much as a waste, but as a resource. Several incentives such as training by video, biogas construction and operation manuals, financial support and/or technical services, would be provided to qualified individuals free of a charge. At present, there are 184 demonstration units with full financial support from the National Energy Administration (NEA), and 3,330 promotion units with one-third financial support from NEA in 54 provinces. It is planned that 30 demonstration units and 600 promotion units would be established in 1985. Evaluation survey of about 20% of the digesters supported has been undertaken since March 1984. It is expected that a report with a set of conclusions and a set of practical recommendations to direct the promotion in Central Thailand is due for release in December, 1984; and in Northeastern, Northern and Southern Thailand in September, 1985.

Promotion and Popularization (P & P)

The promotion of the NEA-supported prototypes and models would be amplified to a large scale if the outcome of the evaluation survey is favorable. The accomplishment of the task, as outlined, would provide an output for an investment project formulation, an extension program preparation, and a support study for widespread application of biogas technology in Thailand.

CONCLUSION

The biogas program of Thailand is being supported and coordinated by the NEA through its Subcommittee on Biogas Development Planning and Coordination, designated by the Board of National Energy Commission. The program comprises various tasks and each task encompasses various activities. At least 5,000 digesters have been built since 1980. Subsidy for promoted household digesters is made available by the NEA to cover 30% of the capital cost. It is also supporting the community biogas system. Incentives in the form of free technical support and supervision of installation of a livestock-raising farmer and a slaughterhouse owner is also provided. To implement the program, the NEA allocates an annual budget of about US$ 500,000, under its Alternative Energy Study and Development Project. More allocations would be required when it is implementing the projects in cooperation with the designated agencies of the Belgian and Italian Governments. An accomplishment report, covering the technical and economical feasibility of treating Thailand's industrial wastewaters by utilizing biogas technology (under the RTG - USAID Renewable Nonconventional Energy Project No. 493-0304), is going to be issued in November 1984.

PROBLEMS CONCERNING BIOGAS PRODUCTION AT FARM LEVEL IN ITALY

P. Balsari, P. Bonfanti, E. Bozza, and P. Sangiorgi
Istituto di Ingegneria Agraria
Milano, Italy

ABSTRACT

This paper shows the current problems tied to the introduction of anaerobic digestion in farming. It should be emphasized, however, that the agricultural sector is relatively poor and has many difficulties compared to the industrial sector.

The major hurdles to be overcome are due to the nonuniform quantity and quality of the excreta, to the lack of constant heat or electricity demand in most breeding farms, low availability of skilled help, and the low added value typical of this sector. The right approach seems to be low-cost simplified plants having low efficiency but short payback periods.

FOREWORD

About 100 anaerobic digestion plants for animal rearing effluents exist in Italy (93 according to the ENEA report of February 1983).

The earliest rural applications date to 1979. From that time, basic and applied research has been carried out to define the performances and technical and economic feasibility of such plants.

For some years, the Institute of Agricultural Engineering of the University of Milan has been conducting a research program, funded by the Lombardy Region, to monitor the performances of some digestion plants installed in the Region.

The problems found by this research can be subdivided into three groups: (1) those occurring upstream of the process, (2) those involving the plant itself, and (3) those concerning output utilization.

PROBLEMS UPSTREAM OF THE PROCESS

These problems are due to:

- Input characteristics: Any variation in livestock diet affects excreta production and composition. For cattle, at the present state of research, the lignin content of the fodder and excreta seems to be the main factor that determines the suitability of such excreta for biogas production.

- Actually, lignin is not degraded by the anaerobic bacteria and because of its structure, it holds part of the digestible organic matter and therefore prevents its breakdown.

  A variation of lignin content in the fodder can affect biogas output by as much as ±15% (Figure 1). This occurs in biogas plants containing the excreta of cattle fed with fresh fodder in the spring and summer but with hay in the winter;

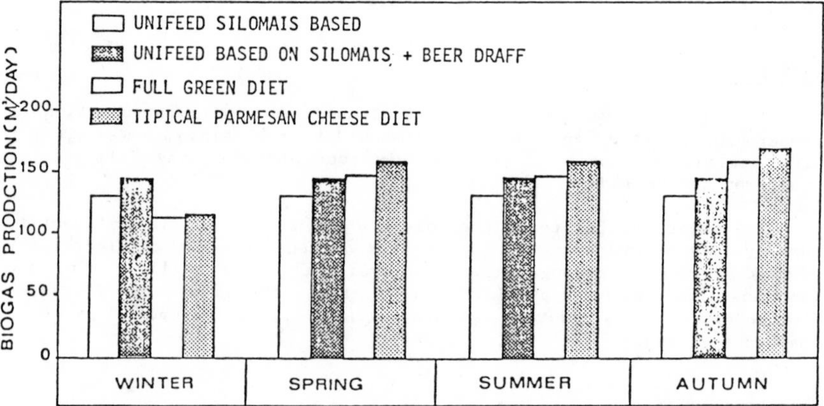

Figure 1 Biogas Yield Seasonal Variations In A Dairy Cattle (115 Heads) Fed By Different Diets

- Excreta dilution: The excreta removal system critically affects the dry matter content. The manure can be greatly diluted by the water used to hose down the stables. In swine breeding farms, the excreta are especially watered down by washing, rain, and other infiltration.

  For an equal volume of slurry, since only the slurry's volatile solids (VS) produce biogas, the higher the dilution the lower the VS and consequently the lower the biogas output. For this reason, larger digesters require more heat and higher power to produce a given amount of biogas from diluted excreta. This translates into costlier equipment with a longer pay-back period.

  Furthermore, its use of slurries containing less than 3% total solids has proved uneconomical under northern Italian weather conditions (average winter temperature: 0 to +5°C), as the biogas output is not enough to ensure plant self-sufficiency during the winter (Table 1).

Table 1 Energy Balance of An Anaerobic Digestion Plant Related to the Slurry Concentration

Slurry concentration	Cleaning water	Total biogas yield	Temp. Diff.	Direct energy requirement		Indirect energy requirement	Energy available	
				Heating	Others			
(TS %)	(1/Head)	(MJ/m^3)	(°C)	(MJ/m^3)	(MJ/m^3)	(MJ/m^3)	(MJ/m^3)	(kJ/kg TS)
8.5	0	512	25	147	26	20	319	3,760
		512	20	120	26	20	346	4,080
		512	15	90	26	20	376	4,430
		512	10	60	26	20	406	4,780
5	5	301	25	147	26	18	111	2,210
		301	20	120	26	18	138	2,750
		301	15	90	26	18	168	3,350
		301	10	60	26	18	197	3,980
3.4	10	205	25	147	26	17	16	460
		205	20	120	26	17	43	1,250
		205	15	90	26	17	72	2,130
		205	10	60	26	17	102	3,000
2.1	15	127	25	147	26	*	-46	-2000
		127	20	120	26	*	-19	- 920
		127	15	90	26	*	11	510
		127	10	60	26	*	41	1,930
1	20	60	25	147	26	*	-112	-11,240
		60	20	120	26	*	- 85	- 8,550
		60	15	90	26	*	- 56	- 5,560
		60	10	60	26	*	- 26	- 2,570

*Indirect energy requirements have not been calculated since balance is negative when considering direct energy requirements only.

NOTE: Specific biogas production = 0.3 Nm^3/kg VS; retention time = 15 days; boiler efficiency = 0.7.

Slurry dilution also varies with the seasons. Readings taken by the Institute of Agricultural Engineering in a swine farm equipped with mechanical slurry removal show that the total solids vary from 7.5% in winter to 3.8% in summer.

This variation is mainly due to the average ambient temperature, which affects the animals' water intake and consequently slurry dilution (Figure 2).

- Slurry storage before loading: The time interval from excreta production to its loading in the digester can vary widely, from a few hours to several weeks, depending on the slurry removal and storage system. Because of the slurry's transformation during this interval, the fraction of organic matter that is more easily attacked by the bacterial flora becomes smaller, and the organic matter available for digester feeding becomes lower, the longer the storage time.

  How these phenomena affect biogas output is hard to establish. They are, however, believed to be more important for swine than for cattle slurry, since the latter contains a large amount of complex organic molecules (e.g., lignin and hemicellulose) that break down more slowly.

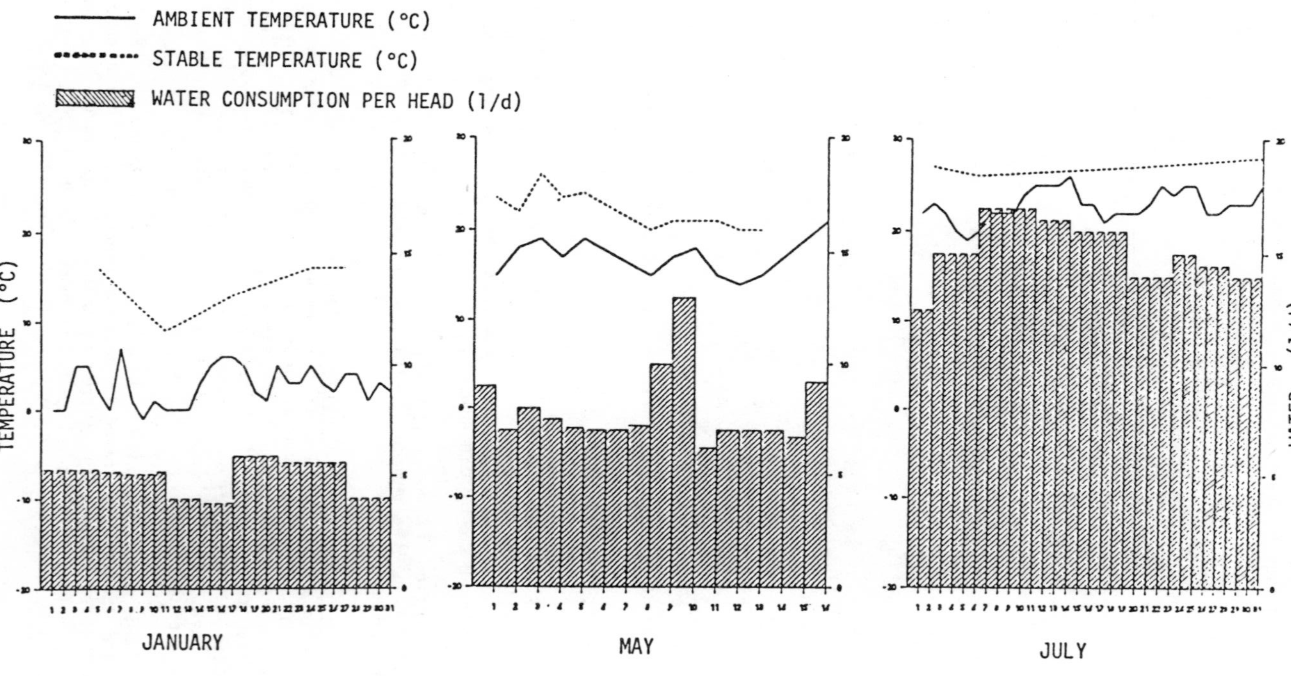

Figure 2 Correlation between temperature and water consumption in three different months

As an example, Figure 3 shows the VS/TS ratio of some samples of slurry taken from different spots in the same farm. The data show a steady decrease of VS content.

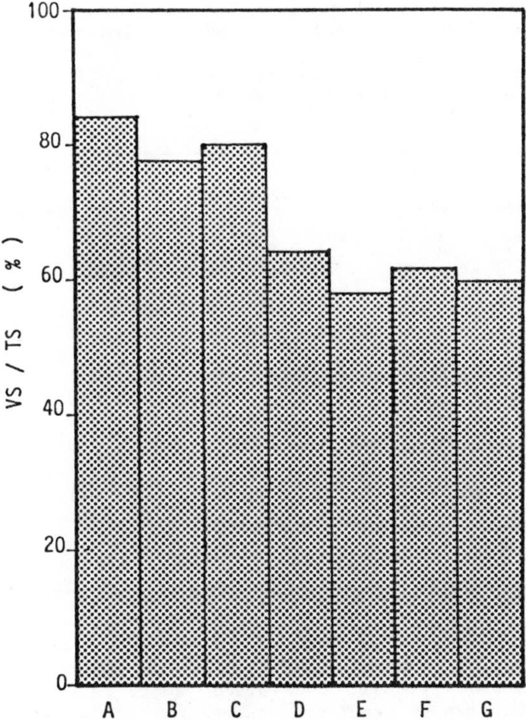

Figure 3 Varation of the VS/TS ratio of slurry (of cattle and swine) drawn in different point of the anaerobic digestion process line:

A) cattle faeces
B) cattle manure drawn during feeding area cleaning by slurry scraper
C) swine faeces
D) swine manure drawn from the pit under the slatted floor
E) swine manure before beeing sent to the digester
F) cattle and swine slurry mixture drawn at the digester inlet
G) effluent from the anaerobic digestion plant

Another problem connected to both slurry removal and storage systems is the cooling of the slurry before loading the digester. Table 2 shows the temperature recorded at the same hour on a September day in different spots of a closed-cycle swine breeding farm. The differences due to the different removal techniques are evident, while the temperature drop due to storage time is less evident because of the relatively high ambient temperature.

Table 2 Temperatures Registered Simultaneously One Day in September in a Closed-Cycle Swine Breeding System

Point of temperature measurement	°C
Ambient temperature	22
Manure just after defecation	37
Slurry coming from pit under slatted floor	26
Slurry coming from external litter alley cleaned by water	20
Storage pit of slurry coming both from pit under slatted floor and litter alley	23
Slurry input in digester	22
Slurry input in digester in February	8

- <u>Slurry sampling</u>: Proper design leading to correct operation cannot be done without a thorough analysis of the slurry. However, bibliographic data are often used since sample collection is not always feasible (proper sampling requires one storage pit for all stable and piggery waste and calls for proper mixing and grinding equipment) and quite costly, as it should be carried out at different times and places. The importance of a proper sampling procedure is shown by the results of a week's sampling by the Institute of Agricultural Engineering at a swine breeder's. The chemical analyses show total solids (TS), volatile solids (VS), biological oxygen demand (BOD), and chemical oxygen demand (COD) to vary during the week as well as during the day, and proves the necessity of multiple sampling by time and location for obtaining reliable results.

PROBLEMS IN THE DIGESTION PLANT

It should be pointed out that most plants are based on city and industrial waste treatment techniques, and not on the specific requirements of agriculture, therefore, they are exceedingly complex and expensive. This was confirmed by the Institute's monitoring campaign: It was shown that most problems are from operating difficulties rather than the process itself (Figure 4). In fact,

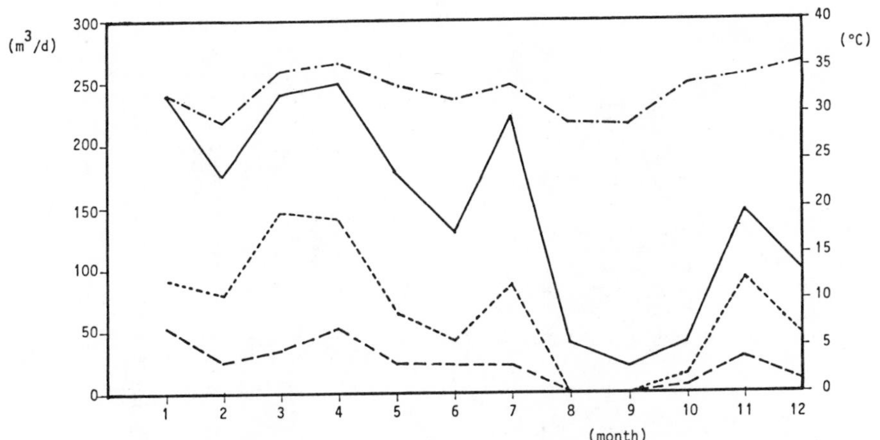

Figure 4 Average daily values of a swine slurry plant in 1983 (the two periods with a lower biogas yield are due to the scanty plant maintenance)

——— TOTAL BIOGAS YIELD (m^3/d)   – – – – HOUSEHOLD BIOGAS CONSUMPTION (m^3/d)

.... COGENERATOR BIOGAS CONSUMPTION (m^3/d)   —·—·— DIGESTER TEMPERATURE (°C)

proper operation is often hampered by lack of training and poor services by the manufacturing firms.

Further difficulties occur due to the components. A wrongly selected pump or pipe can easily fail or get clogged, thereby interrupting the process. Therefore, thorough knowledge of the effluents is required for treatment, and their time-dependent variations of both amount and quality.

Last but not least comes the high cost of these plants, which mostly generate a constant output of biogas while a farm's energy demand, besides being quite low, is also extremely variable with time. Economic and operative considerations indicate that simplified low-cost facilities, retrofitted to existing structures, constitute a sound alternative.

OUTPUT UTILIZATION PROBLEMS

These problems concern the effluent availability, and use of the effluents and the energy output. The latter poses the greatest problems since a farm, especially a cattle or swine fattening farm, will hardly make full use of the output (kW or kcal) of the anaerobic digestion process.

Poor utilization of electrical energy is due to the farm's demand pattern, which exhibits two or three daily peaks (Figure 5). It appears from this demand pattern that the soundest alternative would be a discontinuously running co-generation system to meet the peak demand in full. This, however, runs counter to the fact that co-generation systems need to run continuously to prevent corrosion and wear.

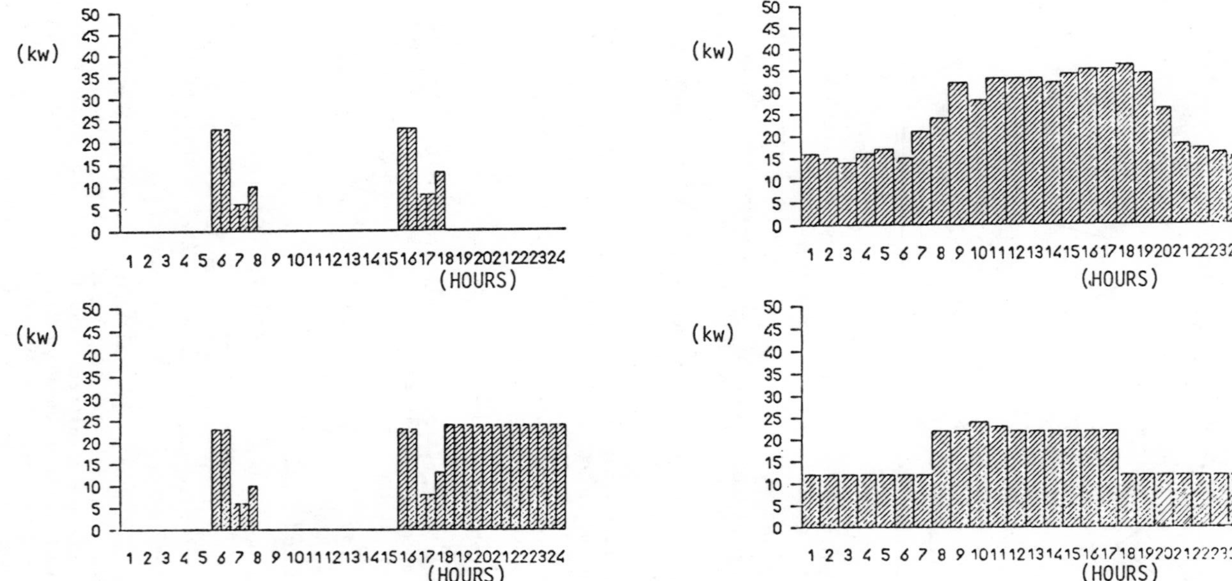

Figure 5 Daily energy requirements in two sample days: on the left for a fattening swine breeding, on the right for a closed cycle swine breeding

Boilers too, when repeatedly started and stopped, unavoidably build up corrosive deposits. Gas purification would be an answer, but would also make the plant more complex and costlier to install and service.

The fertilizer value of the slurry is also difficult to evaluate. Digested slurry has a high agronomical value, and the question is how to make best use of it. Mention should be made here of the experiments made by some Italian researchers. They found that to obtain the same maize grain yield, the amount of fertilizer to be applied in October is three times that required in the spring. While the slurry can be fully and optimally used in certain periods, its effectiveness is insufficient or nil at other times. This further compounds the difficulty of determining the fertilizing value of the slurry, which represent (and we emphasize) the most important output of an anaerobic digestion plant from an energy and economic angle.

REFERENCES

Sangiorgi, F. and P. Bonfanti. 1980. Problemi connessi con la costruzione di impianti di biogas nell'esperienza italiana. Convegno "Produzione di energia alternativa in agricoltura" S.I.M.A.. Milano, Italia.

Balsari, P., P. Bonfanti, A. Ferrari, C. Sorlini, and P. Vigano. 1983. Influenza delle tecniche di allevamento sulla possibilite di controllo e di ottimizzazione del processo di digestione anaerobica. Convegno "Energia e Agricoltura." Milano, Italia.

Balsari, P., P. Bonfanti, E. Bozza, and F. Sangiorgi. 1983. Problems Arising from Commercial Biogas Plants Built in the Lombardy Region. Poster session, Third International Symposium on Anaerobic Digestion. Boston, Massachusetts.

Balsari, P., P. Bonfanti, E. Bozza, and F. Sangiorgi. 1983. Evaluation of the Influence of Animal Feeding on the Performances of a Biogas Installation (Mathematical Model). Poster session, Third International Symposium on Anaerobic Digestion. Boston, Massachusetts.

Balsari, P., P. Bonfanti, E. Bozza, F. Sangiorgi and P. Vigano. 1983. Slurry Management: Results of Swine Slurry Analysis. Poster session, European Symposium on Anaerobic Waste Water Treatment. Noordwijkerhout.

Balsari, P., P. Bonfanti, E. Bozza, and F. Sangiorgi. 1983. Slurry Management: Influence of Cattle Diet (Mathematical Model). Poster session, European Symposium on Anaerobic Waste Water Treatment, Noordwijkerhout.

Sangiorgi, F., P. Balsari, P. Bonfanti, E. Bozza, and M. Lazzari. 1983. Digestions anaerobica in agricoltura: inserimento delle tecnologie nelle aziende. Convegno "Biogas domani." Bologna, Italia.

ANAEROBIC DIGESTION IN PORTUGAL

M.J.T. Carrondo, Universidade Nova de Lisboa, Portugal

ABSTRACT

In this short presentation, the present situation relating to existing or under construction plants as well as research and development (R&D) endeavors are briefly outlined. Several anaerobic digestion plants are used in conjunction with domestic wastewater treatment and industrial wastes, but agricultural wastes have not received any similar attention. Recently, however, a number of demonstration plants have been installed. R&D endeavors are proceeding concurrently on advanced types of biogas producing systems.

INTRODUCTION

Although very well suited for biogas utilization, Portugal does not have a well-developed methane producing system. Portugal depends on external suppliers for more than 70% of its energy. As a not too industrially developed country, environmental concerns have not forced treatment of agricultural and agroindustrial wastes by anaerobic pathways, leading to methane. Agriculture consists generally of small farms; this again has not proved a driving force for biogas development. Finally, since the majority (about two-thirds) of its population live within 20 miles of the Atlantic Ocean, much of its domestic wastewater is not treated in a system incorporating biogas production.

Thus, it appears appropriate to subdivide this short report on existing producing/utilizing units, and research and development efforts now taking place.

EXISTING OR UNDER CONSTRUCTION PLANTS

On the domestic wastewater treatment front, there are plenty of small-scale plants for less than 5,000 inhabitants where dual purpose Imhoff type, sedimentation anaerobic digestion tanks are utilized. Under no circumstances is biogas from these systems used for energy production. For larger scale population, there are three plants, one close to Lisbon, one in the south and one in the north, each serving 50,000 to 70,000 inhabitants. Classical mesophilic digesters are installed, and energy is used only for warming incoming sewage sludge and keeping the heat balance.

For industrial wastes, only three are known to the author -- in the packaging industry, at major hen and egg producing operation, (a less classical plug-flow type) -- and at a large canner where biogas is produced and used for energy. More recently, as we near joining the European Economic Community and thus adopt its environmental standards, more industries are becoming interested in the production of biogas as a byproduct, and some engineering projects are being designed to implement anaerobic digestion mainly for environmental "solutions" with energy only as a second consideration.

For agriculture-related wastes, there is no such thing as an Indian or Chinese biogas plant. Although one small plant has been in operation in the northern part of the country for 30 years, this is almost the only exception to the rule of anything being done in this direction.

RESEARCH AND DEVELOPMENT

Apart from environmental incentives becoming more acute, the recently published Energetic National Plan (Ministry of Industry) emphasized the need to use biogas to supply approximately 2% of the national energy needs by the end of the century. Thus, the Ministry of Industry, through its National Laboratory for Industrial Technology (LNETI), has begun a major effort for the development and demonstration of biogas digestion for energy production. This includes training and case-by-case industrial support, and assembling a few demonstration plants.

On the demonstration side, the Ministry for the Environment has recently opened one plant where manures, straws, and other wastes from the Zoological Garden of Lisbon are digested to produce biogas for water warming. Similarly, with the help of the French Embassy, the Central Region Coordination Commission has built a biogas demonstration plant not too far from Coimbra. In the same region, the Merec program, sponsored by U.S. AID, also intends to include a project for biogas utilization.

At a more fundamental level, our program on the engineering-microbiology-biochemical control of digesters uses double-phase fixed-film reactors to increase the methane production and decrease reactor size and digester failure. This program is heavily supported by the U.S. AID and the wastes used include cane molasses slops and agricultural (namely corn stover) residues. Another Lisbon university, the Instituto Superior Tecnico, uses completely mixed reactors for biogas production from algae and the University of Evora, in the south, deals with piggery wastes in a similar manner. The Department of the Environment of the University of Aveiro is also testing the use of anaerobic filters for agroindustrial (namely cheese whey and wine distillation) wastes. For both the University of Aveiro and ourselves, there are already industries willing to assemble pilot plants prior to full-scale utilization of our research results.

EMBRAPA'S FOOD-FEED-BIOENERGY PRODUCTION SYSTEMS

Levon Yeganiantz, Interamerican Organization for Corporation in Agriculture (IIACA)
and
Adhemar Brandini, Brazilian Agricultural Research Corporation (EMBRAPA)

ABSTRACT

Embrapa's concept of integrated and decentralized food-feed-bioenergy production on a farm is presented. The various systems that have been implemented and operated for demonstration purposes are briefly described. The bioenergy component in these systems is based on a mix of agricultural residues gasification to Gasogene (poor gas producer), fermentation to alcohol, and biodigestion to biogas.

INTRODUCTION

Brazil has a population of 125 million, a gross national product of US \$275.4 x 10^9 and is considered the tenth largest economy in the world.

Due to an abrupt slowdown in international lending to all developing countries in the second half of 1983, Brazil faced serious financial problems similar to those of other Latin American countries. At the end of 1984, Brazil's foreign debt had reached US \$ 92 billion with the sharpest increase in short-term debt. The total of debt service was estimated to be: US \$18 x 10^9. Out of this US \$10 x 10^9 is the interest payments and the US \$8 x 10^9 in amortization. More than half of Brazilian exports come from agricultural raw and processed products. About 40% of imports of more than US \$19 x 10^9 consist of petroleum and its derivatives.

At present, alcohol is used by a fleet of 12 million cars (10% of the total) that consume 54,000 barrels of alcohol/day (43,000 barrel petroleum equivalent (bep/day)). In addition, the consumption of anhydrous alcohol as an extender of regular gasoline (in a proportion of 23.24% in 1984) is estimated at 50,000 barrels of alcohol/day (39,000 bep/day). At a rate of US \$30/barrel of petroleum, nearly US \$900 million of total imports will be substituted. This constitutes 17% of total petroleum imports.

At present, alcohol only directly substitutes for gasoline. Since Brazil has shifted to the import of heavy oil, the refinery production level of diesel oil has been increased from 23 to 34% of imported oil, resulting in a 30% increase in diesel availability due to alcohol substitution of gasoline. Still, a certain gasoline surplus and diesel oil deficit is expected. The agricultural sector directly uses about 18% of total diesel oil, and the transportation of agricultural inputs and outputs is based on diesel trucks.

Increasing dependence on petroleum imports from foreign countries exposes all sectors of the national economy to unpredictable, extremely damaging interruptions in the normal availability of these fuels. Due to the increasing foreign debt of many oil-importing countries, these interruptions can result from the problems of obtaining new forms of loan repayment for importing needed commodities such as petroleum. Such interruptions can be especially disastrous for agriculture and food processing where fuel needs are seasonal and cannot be postponed without the risk of losing entire crops.

INTEGRATED FOOD-FEED-BIOENERGY PRODUCTION SYSTEMS

Brazil regards the application of science and technology to agriculture as a major means of achieving economic and social progress. It is believed that the future of Brazilian agriculture depends on becoming predominantly knowledge intensive, and that agricultural and related research has to be given special emphasis to achieve substantial growth in production. In the 11 years since the reorganization of the national agricultural research system, marked by the founding of EMBRAPA (Brazilian Agricultural Research Corporation), agricultural science has become a powerful, productive force in Brazilian society. According to Eliseu Alves (1982): EMBRAPA proposes to change the focus of the relationship between advanced and developing countries from "technology transfer to science transfer."

When possible, EMBRAPA uses the systems approach in its research, attempting to bring together various components of farm production systems. Figure 1 represents a simplified self-explanatory flow chart of "Integrated Food-Feed-Bioenergy Production Systems." (For details see A.G. Netto and L. Yeganiantz, 1982.)

To avoid competition between traditional food crops and sugar cane, the predominant energy crop, the use of alternative raw materials (such as sweet sorghum, cassava, sugarbeet, vegetable oil crops, and agricultural plant and animal residues in integrated farm bioenergy pilot demonstration projects) is being encouraged.

It is not expected that all the components presented in the flow chart will be included in the same production unit. EMBRAPA has several systems implanted in various research centers designed to provide energy self-sufficiency for these centers. These systems, in addition to generating part of the fuel needed by the experimental stations, provide detailed data and are used for demonstration purposes. They will also eventually become training groups for extension personnel and selected farmers. The systems now in operation are:

1. National Center for Corn and Sorghum Research, at Sete Lagoas, Minas Gerais: small still (sugarcane, sweet sorghum), electricity generator, biogas plant, pelletizer, gasogene (poor gas producer), dryers, alcohol-powered farm machinery. Fish production using biofertilizer.

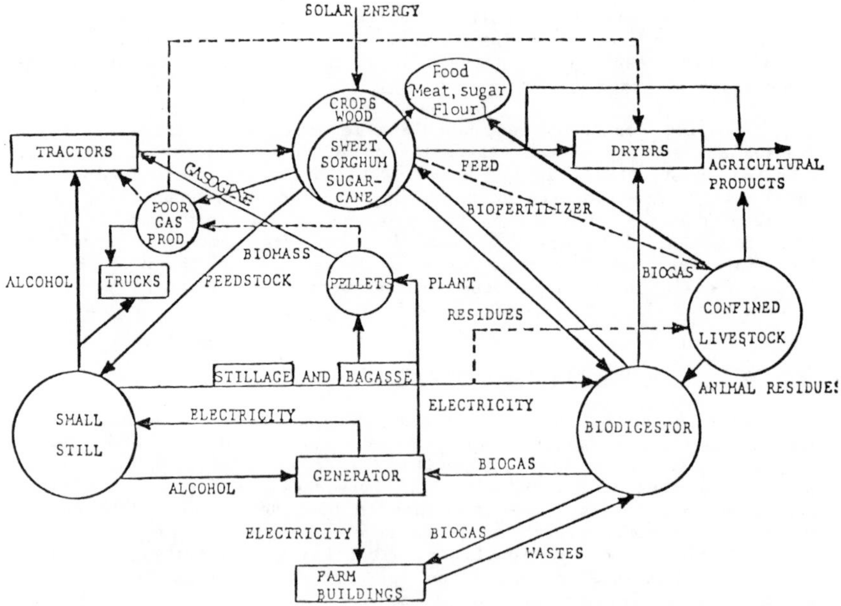

Figure 1 EMBRAPA'S Integrated Food-Bioenergy Production Systems

2. Pelotas Experiment Station, at Pelotas, Rio Grande do Sul: small still (sugarbeet, sweet sorghum), confined livestock feeding operation, biogas plant, dryers.

3. National Center for Beef Cattle Research at Campo Grande, Mato Grosso do Sul: small still (sugarcane, sweet sorghum), confined livestock operation, biogas plant, electricity generator, alcohol-powered farm machinery.

4. National Research Center for Cassava and Fruits at Cruz das Almas: Bahia: small still (cassava), biogas plant, confined animal operation.

5. National Center for Dairy Cattle Research at Coronel Pacheco, Minas Gerais: biogas plant, electricity generator, dairy farm using equipment powered by bioenergy.

6. National Research Center for Rice and Beans at Goiania, Goias, small still, gasogene-powered farm machinery and irrigation system, alcohol-powered tractors, seed dryers using wood gasifiers, biogas plant using crop residues.

7. National Center for Horticultural Research at Brasilia: small still, gasogene-powered farm machinery.

8. State Agricultural Research Unit at Aracaju, Sergipe: small still, confined livestock, biogas plant, electricity generator using biogas, alcohol-powered farm machinery.

In addition, the National Food Technology Center (CTAA) at Rio de Janeiro is studying, at the laboratory level, the fuel potential of oil esters of various native and traditional oil crops. The field tests using standard diesel tractors have been initiated at four research centers based on ethyl esters of soybean oil. Preliminary results indicated generally satisfactory performance, although efficiencies are expected to be slightly less than with conventional diesel fuel.

Brazilian scientists are working hard on the unsolved problems of producing and conserving energy in agriculture and on ways of supplying part of the liquid fuel needs of other sectors to keep the country from remaining overly dependent on foreign oil. These scientists are studying ways of producing sweet sorghum, cassava, sweet potato, and sugarbeet more efficiently for alcohol production. They are optimistic about sweet sorghum as a source of commercially-produced ethanol because of its favorable characteristics.

Since sweet sorghum will ratoon (sprout again) from its own roots, agronomists hope to find a short-season variety that will mature quickly so that it can be harvested early enough to allow a second crop to grow before the cool season. Usually no additional tillage is needed for this second crop, and neither herbicides nor fertilizers need be applied.

The ultimate aim of this research is to give farmers the know-how to produce two crops of sweet sorghum per year. Since an acre of sweet sorghum can be turned into 200-300 gallons of ethanol by a commercial plant, the results should be quite beneficial.

The desirable characteristics of sweet sorghum varieties for ethanol production according to Schaffert and Gourley (1982) are: (1) high biomass yield; (2) high percentage of fermentable sugars along with combustible organics; (3) comparatively short growth period; (4) drought tolerance; (5) low fertilizer requirements; (6) production of grain for food or feed use; and (7) the possibility of complete mechanization.

Microdistillery technology for ethanol production from sugar cane can be adapted and utilized to produce from sweet sorghum.

Scientists are also studying the possibilities of converting sweet potatoes into alcohol. Sweet potatoes and cassava have a potentially high yield in gallons per acre. Because they can be stored longer than sweet sorghum, they can provide a commercial ethanol plant with a year-round source of fuel alcohol.

An alcohol fuel plant can be considered a separate profit center installed on a farm, or it can be integrated with the farming operation in the same fashion as a grain drying bin, a forage chopper, or a small feed mixing plant. Thus farms that begin to manufacture alcohol become more diverse and integrated in their own operations while also becoming more tightly integrated with other sectors of the economy. This implies that any farm operation having the potential for alcohol production needs to make a careful assessment of its entire operation including both physical or operational and fiscal or economic analysis. The most viable operations will be those with maximum flexibility coupled with good marketing and economic planning. The profit may not be maximized, but economic opportunities will be optimized in much the same way that futures hedging can be used to reduce risk to the farmer.

The introduction of an anaerobic digester on a farm points out the opportunities for optimizing energy utilization. It is imperative to look carefully at the total farm picture before deciding on the capacity of an alcohol fuel plant. It becomes necessary to look at the entire farm operation in order to develop an integrated technical and financial approach to the use of renewable energy systems. The most important need is to make people think in energy terms, then alcohol fuels and other renewable sources of energy will be effectively integrated into farms and communities.

Energy in agriculture is not only what makes a tractor move; it includes what makes a cow or hen produce or what makes an irrigation system function. It also turns on the light in the barn, dries the grain, cools the milk, enriches the soil, and kills the corn earworm. In other words, petroleum and its derivatives are major ingredients of chemicals and fertilizers as well as of farm fuels.

FINAL OBSERVATIONS

The farmer can produce fuel and other energy-based inputs from his own land. As a result, the energy crisis is an opportunity for farmers and agricultural research workers to turn their creative minds to shaping energy alternatives to suit the needs of various types of farms making use of on-farm resources. EMBRAPA's Farm Energy Self-Reliance Program was structured to develop community-based approach to agricultural research.

A small-scale "On Farm Energy System" aimed at energy-self-sufficiency farm production units may involve some sacrificing of per hectare productivity associated with specialization, and result in decentralizing and diversifying production processes where the potential of increasing food and agricultural raw material production per unit of energy is high.

The results indicate that the social costs of unexpected oil price increases for agricultural production units using an integrated food-feed bioenergy system compared to traditional food-feed systems will be significantly lower, justifying the efficiency-flexibility trade-off.

The example of Brazil has to be a powerful stimulus for all non-oil-producing countries to consider their own options for developing energy. It is beyond question that developing countries must develop their national energy sources at least to satisfy the energy needs of the food production sector.

The issue of "Integrated and Decentralized Food-Feed Bioenergy Production Systems" developed jointly by public and private sectors is symbolic of a greater issue: the preservation of liberty and social equity by maintaining some independence from the "Biogas System." According to Lovins:

"Small energy systems suited to particular niches can mimic the strategy of ecosystem development, adapting, and hybridizing in constant coevolution with a broad front of technical and social change. Large systems tend to envolve more linearly like single specialized species (dinosaurs?) with less genotypic diversity and greater phenotypic fragility" (A. Lovins, 1976).

REFERENCES

Alves, E.R.A. 1982. Brazil's Program for Development of Agricultural Researchers, International Agricultural Development Service Report. Washington, D.C.

Dias, M.C.S., H.V. Richter, L. Yeganiantz, A. Gorgatti Netto. 1982. Implementation of Energy Self-Reliance in Agriculture: Brazilian On-Farm Systems. Paper presented at the Fifth International Alcohol Conference, New Zealand.

Netto, A. G. and E.R. Da Cruz. February 1982. Brazilian Strategies for Increasing Energy Resources. The Case of Biomass Substitutes for Liquid Fuels. Paper presented at Biomass Substitutes for Liquid Fuels--An Interamerican Symposium Workshop. Campinas.

Netto, A. G. and L. Yeganiantz. 1982. EMBRAPA'S Food-Feed Bioenergy Production Systems. A Joint Government-Industry Research Venture. Brasilia, Department of Planning Methodology, Brazilian Agricultural Research Corporation.

Lovins, A. 1976. Scale Centralization and Electrification in Energy System. In Future Strategies for Energy Development: A Question of Scale: Oak Ridge: Associated Universities.

Schaffert, R.E. and L.M. Gourley. 1982. Sorghum as Energy Source. Sete Lagoas. (mimeographed).

BIOGAS TECHNOLOGY IN COSTA RICA: ITS DEVELOPMENT

G. Chacon, University of Costa Rica

ABSTRACT

This short presentation outlines the biogas technology (BGT) developments in Costa Rica. Particular emphasis is given to the University of Costa Rica's relevant activities.

BGT IN COSTA RICA

In 1978, the Department of Chemical Engineering at the University of Costa Rica (UCR) started a project centered on the development and diffusion of BGT throughout the nation. The project's purpose was to determine if the process fulfilled the necessities and the social and economical requirements of the region under actual operating conditions.

From 1979 to 1981, UCR worked jointly with the Central American Institute of Technology and Industrial Research (ICAITI) of Guatemala. Two prototypes with a capacity of 8 m^3 were built from concrete blocks with concrete covers for experimentation and demonstration.

Such construction proved to be cheaper and easier to build. One prototype was built in a warm zone and fed with pig manure. The other was installed in a cold zone and utilized cow dung. The behavior of these biodigesters was acceptable, but presented difficulties in gas storage[1,2]. The project provided an opportunity for the training of chemical engineering and agronomy students during their graduation works.

Later on, UCR decided to find a new material of construction that would reduce the investment costs. It was found that PVC sheets have the desired characteristics of: insulation, gas permeability, and mechanical and wear resistance. This material is also easy to acquire in the local market. Construction of this type of digester is easy, quick, and requires little operating labor. The material's flexibility allows digester volume changes, maintains anaerobiosis, and presents no inconveniences if it is used for a liquid deposit.

Several tests were carried out with digesters using sheets of 12, 30, and 40 mils thickness, with the collaboration of Nicolas Scheerer (Chomes, Puntarenas, Costa Rica). At the request of UCR , the local PVC producers manufactured sheets of 30 mils thick for digester construction. The digesters were built within a wooden structure covered with galvanized iron sheets or with the same PVC material but thinner (12 mils), in order to protect them.

During the last few years, the UCR has built 30 digesters, especially in technical agricultural high schools, and has trained technicians in this process.

Other institutions in the country have established programs on the building and spreading of anaerobic digesters. They include: Instituto Tecnologico de Costa Rica, the Ministry of Agriculture, and Central American Institute of Technology and Industrial Research. In 1984, the Ministry of Energy and Mines, under the Energy and Energetic Planning Office established guidelines on biogas technology development.

The Government of the Republic of Costa Rica, has decided to advocate dissemination of the technology, assist private industries in order to permit building and maintenance of digesters, establish a consultation entity with representatives from different interested institutions, and aid in establishing a publication program on biogas digesters, their construction, and uses.

Since 1982, UCR has allocated part of its research program to some raw materials for methanogenic anaerobic fermentation of national interest, e.g., coffee pulp, ripe banana peel, and green banana. Studies have been made on: the biogas potential, the behavior to the digestion, and the requirements for such substrates[3,4,5]. Research was done at the experimental levels of: 250 cm^3, 0.02, 0.2, and 2 m^3. Experimentation at larger sizes could not have been realized because of difficulties in forming the reactive mass (on inoculum) which can take several months.

REFERENCES

1. Chacon, G., and L.E. Pacheco. 1979. Aprovechamiento de los desechos organicos en la produccion de energia. Ingenieria y Ciencia Quimica. Costa Rica 2(3):63.

2. Chacon, G., and J.F. Calzada. 1980. Un diseno de digestores de biogas a nivel rural. Ingenieria y Ciencia Quimica. Costa Rica. 4(20):68.

3. Chacon, G., and A.I. Moreles. 1981. Produccion de biogas a partir de pulpa de cafe. Oficina del Cafe (OFICAFE). San Jose.

4. Messeguer, C.M., F. Silesky, and G. Chacon. 1983. Capacidad de la cascara de Banano para la produccion de biogas. Igenieria y Ciencia Quimica. Costa Rica. 7:(3-4):44.

5. Chacon, G. y J.L. Fernandez. Capacidad de la pulpa de cafe en la produccion de biogas. Turrialba. (In press).

JAMAICA COUNTRY PAPER

W.L.A. Boyne
Ministry of Mining and Energy, Kingston, Jamaica

ABSTRACT

The facts that Jamaica's per capita energy consumption is one of the highest in the developing countries and that there is almost complete reliance on fuel importation mandates development of prospective indigenous energy resources. Bioenergy, with biogas as a component, constitutes one of the priorities in the national energy plan. The biogas program, its achievements since 1980, and major findings as well as future plans are outlined.

INTRODUCTION

In Jamaica, the Conservation Program for public and private sectors (including auditing and retrofitting) is well advanced.

INDIGENOUS ENERGY SOURCES

This component is prioritorized in the National Energy Plan. It is based on the following least cost alternatives:

- Hydro-electric;
- Peat mining;
- Substitution of coal -- alternatives being considered are coal water slurry, co-generation and stand alone plants;
- Bioenergy; and
- Solar technologies.

Bioenergy

The preliminary bio-mass survey completed in 1982 indicated that approximately 30 percent of Jamaica's non-bauxite energy utilization can be supplied from bioenergy, including agricultural waste, urban waste, fuelwood for electricity, and biogas. New developments (1984) involving energy cane, which promises a major contribution, were not considered in the present context.

Biogas Program

The Energy Sector Assessment report of 1980 indicated a demand of 20,000 households and approximately 200 medium-sized plants. The report also proposed that the program be developed within a five-year plan. This period should be used in popularization, commercialization, and demonstration of the technology along with training experts.

OVERVIEW

The Jamaica energy situation is characterized by several factors. Jamaica is a net-importer of fossil fuel; 99 percent of the total energy demand is being supplied from imported fuel. Approximately one-half of the imported oil is utilized by the bauxite and alumina industry, with half being utilized in industrial, commercial, and residential endeavors. The per capita use of energy in Jamaica is one of the highest in the developing world.

The development of biogas as well as any other energy source in Jamaica must, therefore, form part of a "systems approach," i.e., in consideration of the least cost alternative, within a national energy policy and plan.

Energy Policy

The general components of the overall energy policy are:

- Institutional Development: to plan coordinate and manage energy programs;
- Energy Conservation: as the short term alternative; and
- The identification and development of indigenous energy resources.

The first two components are well developed in that a ministry is established to promote institutional development, and the energy technicians and farmers are involved in the technology, research, and development; the evaluation of alternative designs; the evaluation of wastes and economic evaluations, to promote the most adaptable and optimized systems for local conditions.

ACHIEVEMENTS

Achievements since 1980 are in two sub-programs, namely:

1) A Ministry of Mining and Energy and the Latin American Energy Organization (OLADE) program which constructed nine household units on farms in different sections of the island ranging in volume from 6-12m^3. The objectives are to demonstrate to and gain acceptance by the rural farmers (completed in 1982).

2) The Ministry of Mining and Energy and the Scientific Research Council program constructed eight medium sized plants ranging from

45-67 m^3 in volume, with the objective of commercializing the technology.

These institutions continue to conduct research and development work on evaluations, economics, materials, and adaptations. The plants are used for lighting, cooking, water pumping, and fertilizer production for farms and the production of algae for fish ponds.

FINDINGS

1) Evaluations indicate that the most suitable plant type for Jamaica is the modified Chinese design, but in general, the cost is too high for the rural poor.

2) A comprehensive attitudinal survey shows that most farmers are acceptable of biogas systems.

3) The possible energy contribution from available waste is approximately 200 GWh/yr or 10 percent of the island electricity in the year 2000.

4) As a follow-up to the 1982 biomass survey, a 1984 survey identifies the areas of mass concentration of wastes for use in biogas technology, and the selection of a pilot area for development.

PLANS

The following plans are now initiated as a consequence of the findings;

- A joint program between the government of Jamaica and OLADE will construct 18 plants on farms, develop a training program for farmers, and establish a US $250,000 revolving loan program at low interest to small farmers. The systems will be sized, designed, and supervised by the Ministry of Mining and Energy and the Scientific Research Council with construction by the farmers island wide.

- The selection of a pilot area will include a 240 kW electrical generating plant connected to the National grid, and it will produce seven tons/day of fertilizer for farmers. The installation has the added benefit of utilizing the waste of a coffee plant which presently pollutes a major river which is a future source of domestic water for the Kingston Metropolitan area;

- The construction of two 80 m^3 plants for pig and poultry farms; and a number of small household units of 6 m^3 for farmers.

- In an ongoing energy sector assistance program between the USAID and the government of Jamaica, funds are available to private sector industries for the development of alternative energy activities, including biogas, of approximately US $10 million.

CONCLUSIONS

The physical achievements in Jamaica in biogas technology have been small. However, it must be appreciated that the development is within an approach that concentrates on least-cost alternatives within an energy policy having as its objectives:

- Improving the quality of life of the rural poor; and

- Implementing an energy plan to utilize alternative energy sources as a way to replace 60 percent of the imported fossil fuel by the year 2000; in which biogas only forms a subcomponent.

An overall energy plan is being updated by Argonne National Laboratories. It will indicate specific energy contributions and unit costs of all the Jamaica energy options including biogas. The biogas program will quite likely be developed in two categories (depending on the results of the Energy Plan update), namely: commercialized unit-sized plants which are economically feasible and can substitute for imported fuel, and secondly small-scale plants for rural farms through low interest loans. The contribution to the national energy budget will be continuously updated.

AUTHOR INDEX